Hands-On Digital
Signal Processing

Hands-On Digital Signal Processing

Fred Taylor
Jon Mellott

University of Florida
Gainesville, Florida

and

The Athena Group, Inc.
Gainesville, Florida

McGRAW-HILL

New York   San Francisco   Washington, D.C.   Auckland   Bogotá
Caracas   Lisbon   London   Madrid   Mexico City   Milan
Montreal   New Delhi   San Juan   Singapore
Sydney   Tokyo   Toronto

McGraw-Hill

A Division of The McGraw·Hill Companies

1   2   3   4   5   6   7   8   9   0   DOC/DOC   9   0   3   2   1   0   9   8

P/N   063319-3
PART OF
ISBN 0-07-912965-X

The sponsoring editor for this book was Stephen S. Chapman and the production supervisor was Clare Stanley. It was set in Century Schoolbook by J. K. Eckert & Company, Inc.

Printed and bound by R. R. Donnelley & Sons Company.

This book is printed on recycled, acid-free paper containing a minimum of 50% recycled, de-inked fiber.

Contents

Preface

Digital signal processing, or DSP, is defined to be that body of knowledge associated with the study of systems that create, modify, and manipulate signals using digital technology. The signal domain embraced by DSP currently includes everything from audio to radar, and simple signals to complex images. The technology used to exercise control over this signal environment can range from a small desktop computer, executing an elementary algorithm, to very complicated dedicated digital processors. Today, virtually every facet of audio and video information processing, controls, communications, speech processing, multimedia, biomedicine, transportation, acoustics, and instrumentation is being influenced by the science of DSP and its attendant technology. DSP has, in fact, become the enabling technology for many of the products and devices that define our modern life.

The theoretical foundations of DSP were laid down in the mid-1960s. Early attempts to forge an identity for this new field of study came painfully slow. The defense community became an early adopter of DSP for use in radar and sonar applications, while commercial interests were, at best, subdued. The two major obstacles retarding the DSP movement were a lack of DSP-aware engineers and the absence of a viable implementation technology. While the popular general-purpose minicomputers of the era were sufficient to develop and test DSP algorithms, converting them into hardware solutions was problematic. Systems based on dedicated logic or general-purpose microprocessors were too expensive, complex, or slow to garner much interest. The enabling technology that finally forced the DSP revolution was the DSP microprocessor. It is the DSP microprocessor, which made its first appearance in the late 1970s, that made DSP a pervasive technology. This remarkable piece of technology, then and now, has provided opportunities to develop powerful solutions ranging from sophisticated medical imaging systems to the lowly modem.

DSP, as a discipline, can be traced back only a few decades, which makes it a relatively young science. In fact, it can be claimed that DSP is one of the first and most important spin-offs of the digital era.

Today, the worldwide market for DSP technology is measured in the billions of dollars and is exhibiting double-digit annual growth. The rapid growth and importance of DSP has created a high demand for a new class of practitioner who has far more breadth than that expected of many technical disciplinarians. A DSP engineer is required to exhibit competency in DSP theory, hardware, algorithms, and applications. Because of the rarity of these skills and the high demand for DSP products, DSP engineers and scientists are in great demand.

Academia has been slow to respond to this new force. Higher education has approached the problem in a top-down manner beginning at the graduate level. The last two decades have seen most institutions of higher education offering a collection of courses in the DSP area. They are generally oriented toward the study of algorithms and are generally are weak in exposing students to problems relating to implementation and DSP microprocessors in particular. This, in many respects, reflects the theoretical research interests of graduate faculty who too often equate DSP with DSP algorithms. More recently, we have witnessed DSP penetrating the undergraduate curriculum and appearing as industrial short courses. The emphasis here is admittedly foundational and builds upon material generally introduced in a traditional signals and systems course. Such courses seem to have attracted the interest of undergraduate students when the studies involve the use of DSP microprocessors. This cadre of students knows, quite correctly, that hands-on experience with technology opens career opportunities.

These observations and experiences motivate this book. This, by itself, is probably insufficient motivation to justify adding yet another text to the long list of DSP titles unless it brings something new a tangible to the reader. Over past decade, the authors have also become increasingly aware that graduating engineers where having considerable difficulty translating academic theory into practice. Unfortunately, the path from neophyte to expert is often a difficult transition due to the fact that academic models, fashioned under ideal assumptions, often do not translate well to the real world. The book has made a commitment, therefore, to forge a strong bridge that links theory with practice in a manner that the engineer can use and the student can understand. It is in this vein that both authors bring their collective academic and industrial experiences to bear on this problem. While the authors take a rather holistic view of DSP, they are always mindful that the objective of most engineering efforts is the design and development of successful solutions. While attempting to provide the reader with a balanced view of the essentials of DSP, a stronger effort has been made to demystify the design process. We hope that we have achieved this goal.

It was in this spirit that a project was proposed which resulted in this book, *Hands-On Digital Signal Processing*. It was important to the authors to respond faithfully to challenges that motivated this project, namely providing the reader with a comprehensive exposure of both DSP analysis and synthesis methods and techniques. It was decided, from prior experience, that the only way to achieve this goal was to integrate the text closely with a comprehensive synthesis and analysis software package. We were very fortunate to have the latest version of MONARCH for Windows (from The Athena Group, Inc.) available for this project. The package supplied with the text is a fully functional version of MONARCH for Windows, except it cannot save files. Nevertheless, in this form, it remains an ideal learning tool for use in a hands-on study of DSP. MONARCH has soundboard support and thus allows the reader to include audio demonstrations along with simulations to clarify fundamental concepts that are difficult to comprehend using only mathematics. Linking the text with the software also facilitates the study of nontrivial DSP problems.

The text is organized broadly into discrete-time and digital system studies. Chapter 1 provides the reader with a review of basic concepts normally found in an introductory signals and systems course. Chapters 2 and 3 cover the study of discrete-time signals using z-transforms and frequency transforms. This material is necessary reading for anyone having only a limited exposure to discrete-time system analysis. Chapter 4 covers the more popular methods in implementing discrete-Fourier transforms and similar transforms. Chapters 5 and 6 cover the study of discrete-time systems using z-transforms and frequency transforms. This material is highly recommended to anyone not having a strong formal exposure to these topics. In Chap. 7, details on implementing the fundamental DSP convolution operation are presented. This material will reappear throughout the text in the form of digital filters. To prepare the way to understanding how digital filters are implemented, linear discrete-time systems are studied at a macro level in Chap. 8. Chapter 9 interprets linear discrete-time systems from the important state variable standpoint. State variables provide a unifying mathematical framework in which systems can be studied in detail. DSP technically assumes a digital implementation of a discrete-time system model. The importance and impact of the statement is explored in Chap. 10. The first important class of digital filter is presented in Chap. 11 and is called the *finite impulse response* (FIR) filter. The implementation of the FIR filter is developed in Chap. 12. The same pattern is presented in Chaps. 14 and 15 for the second important class of digital filter, called the *infinite impulse response* (IIR) filter. The study of digital filters takes a serious turn in Chaps. 15, 16, and 17, which address problems that arise when finite precision digital hardware is used to implement a digital filter. Lastly, Chap. 18 introduces the concept of the multirate filter, an area of growing importance.

Collectively, the material covers a wide range of DSP topics and implementation questions. We present this material in a hands-on forum that is intended to actively engage the reader in the study of DSP. We do this because we firmly believe in the ancient Chinese proverb, *"I hear, I forget; I see, I remember; I do, I understand."*

Fred Taylor
Jon Mellott

About the Authors

Dr. Fred J. Taylor is Professor of Electrical and Computer Engineering at the University of Florida and president of The Athena Group, Inc. He has authored 9 textbooks and 2 monographs, contributed encyclopedia and handbook chapters, and written over 100 scholarly papers on engineering topics.

Dr. Jon Mellott is Director of ASIC Technologies for The Athena Group, Inc. The Athena Group, Inc. is a supplier of digital signal processing solutions to industry, government, and education. Dr. Mellott has extensive experience in algorithm and system design for high-performance digital signal processing systems.

1

Signals and Systems

1.1 Introduction

Signal processing refers to the art and science of creating, modifying, manipulating, and displaying signal information and attributes. Since the dawn of time, man has been the quintessential signal processor whose signals consisted of primitive sounds and physical gestures communicated across line-of-sight channels. Signal processing was performed by one of the most powerful signal processors ever developed, the human brain. As humans have evolved, other elements were added to man's signal processing environment, such as information coding in terms of intelligent speech* and the written word. In time, communication links expanded from local to global. It was the introduction of electronics to the signal processing scene, however, that enabled the modern information revolution. Consider, as the defining example, the plain old telephone system (POTS), which consists of a signal source (human speaker), a sensor (microphone), a communication channel (wired and wireless channels), signal processor (filters), and response system (the listener) and how it has revolutionized man's life. These remain the basic elements in an end-to-end signal processing system.

1.2 DSP Technology

Regardless of the information source, or the machines used to process it, engineers and scientists have habitually attempted to reduce sig-

---

* *They received use of the five operations of the Lord, and in the sixth place He assigned them understanding, and in the seventh speech so they could express the meaning of their cognitations.* Ecclesiasticus 17:5.

nals to a set of parametric values that can be mathematically manipulated, combined, dissected, analyzed, or archived. This obsession has been fully actualized with the advent of the digital computer. One of the consequences of this fusion of man and machine has been the development of a field called *digital signal processing,* or DSP. While many trace its beginnings to the first iterative computing algorithms, others will claim that the honor belongs to Laplace, Fourier, or Gauss. Admittedly, they were laying the foundation for signal processing, but these early pioneers were all missing the important "D-word," namely *digital.* During the 1950s and into the 1960s, digital computers were considered to be far too costly to be used as a signal analysis or laboratory tool. In 1965, however, Cooley and Tukey introduced an algorithm that has come to be known as the *fast Fourier transform* (FFT). The FFT algorithm was indeed a breakthrough in that it recognized both the strengths and weaknesses of a classic von Neuman general-purpose digital computer and used this knowledge to craft an efficient computer algorithm for computing Fourier transforms. What should be equally appreciated is that, simultaneously, the cost of computing had rapidly plummeted due to the invention of the minicomputer and integrated circuits (ICs). As a result, a general-purpose computer, for the first time, could be economically used in signal processing applications.

It is interesting to note that digital filters also made their first appearance in the mid-1960s using discrete and IC logic. Their initial expense and limited programmability restricted their use to narrowly defined applications. These were, however, simply precursors of what was about to take place.

In the late 1970s, a true revolution began with the introduction of the first-generation DSP microprocessor represented by Intel's 2920. This rapidly led to a second, third, and fourth generation of devices from Texas Instruments, Motorola, AT&T, Analog Devices, NEC, and others, which commercially defined a viable technology we now call DSP. DSP is therefore a relatively young science which, in just two decades, has become a major economic and technological force. The DSP hardware market alone is increasing at an annual rate of 40 percent. DSP solutions are now routinely developed using commercial off-the-shelf (COTS) DSP microprocessors and FASICs,[*] along with a growing trend for designers to develop their own ASIC solutions. There is now an abundance of DSP software design and development tools that

---

[*] Application specific integrated circuits (ASICs) are defined by the solution provider, not the manufacturer. FASICs are factory ASICs having a specific application that is defined by the manufactureer for embedding into systems being developed by the solution provider.

serve this industry. DSP has become an enabling technology for high-speed, low-cost data communications (modems), digital controllers, wireless (cellular telephony and other personal communications services), video compression, multimedia and entertainment (audio and video), plus a host of other applications. At the core of this revolution are the tens of thousands of scholars and technologists who now refer to themselves as DSP engineers and scientists. They are hybrids in that they must be competent in the traditional systems engineering studies, possess strong skills in computer hardware and software engineering, and be formally exposed to the mechanics of DSP. They, like DSP technology itself, are new and still in the formative stage. All that can be accurately predicted at this time is that DSP will be one of the principal economic driving forces of the 21st century

1.3 Signal Taxonomy

Many signals possess a mathematical representation called a *model* defined in terms of a collection of variables and coefficients. Simple signals can often be expressed as a function of a single variable, such as time in $x(t) = \cos(\omega_0 t)$. The signal $x(t)$ can also be modeled as the solution to a second order ordinary differential equation (ODE) $d^2x(t)/dt^2 + \omega_0^2 x(t) = 0$, with appropriate initial conditions. Another version would model a sinusoid as a function of three variables—e.g., amplitude, time, and phase—as in $x(A,t,\phi) = A\cos(\omega t + \phi)$. Images are often classified as two-dimensional signals that can be expressed as a function of two spatial parameters such as $f(x_1,x_2) = \cos(\omega x_1 + \omega x_2)$. A simple black-and-white image can be represented as a two-dimensional signal, where at the coordinates (x,y), the image intensity is $i(x,y)$. Signals of higher dimension also exist. For example, a black-and-white video or motion picture signal has three dimensions: two spatial and one temporal. The Dow Jones industrial average, for example, is a function of 30 economic variables and is a classic example of a multidimensional, or M-D signal. The list continues indefinitely.

There are instances when the mathematical representation of a signal can be assumed to be exact, and the signal is said to be *deterministic*. For example, $x(t) = \sin(\omega_0 t)$ exactly models a specific sinusoid. There are other times, however, when an exact representation cannot be developed or justified. A *random* signal changes in such an unpredictable manner that it defies being accurately modeled by an equation. In these cases, parameters called *statistics* are used to quantify one's ignorance about the exactness of a random signal.

Signals produced by physical devices, or systems, are called *causal signals*. Physical devices are assumed to have a finite history; that is, they did not preexist time. One can assume that a causal signal generator was activated, or turned on, at some finite time, $t = 0$. Thereafter, the generator will produce a signal $y(t)$ that satisfies

$$y(t) = \begin{cases} x(t) & \text{for } t \geq 0 \\ 0 & \text{for } t < 0 \end{cases} \tag{1.1}$$

Signals that are not causal are called *noncausal* or *anticausal* signals. If, for example, you observe that the ceiling light (a signal) goes on before you turn the switch, you may have empirical evidence of the existence of a noncausal signal. However, if human speech is considered to be a signal, and your friend starts giving you advice before you ask for it, you are simply observing human nature.[*] While noncausal signals are not products of a physically realizable signal generation mechanisms, they will play a significant role in the mathematical study of real-world signals and systems. For example, the signal $x(t) = \cos(\omega_0 t)$ persists for all time [i.e., $t \in (-\infty, \infty)$] and is obviously an important noncausal mathematical test signal.

An important property possessed by some systems is *linearity*. Strictly speaking, a system is said to be *linear* if it exhibits properties of *homogeneity* and *additivity* (sometimes called *superposition*). If a system is not linear, then it is said to be nonlinear. In general, the study of nonlinear systems is far more difficult than the study of linear systems. Fortunately, many practical problems are linear or may be treated as linear. Consider, for example, a system that for each input, $x(t)$, produces a corresponding output, $y(t)$. That system is said to be *homogenous* if, given a scalar α, the input $\alpha x(t)$ produces an output $\alpha y(t)$. Now, suppose a system takes as inputs $x_1(t)$ and $x_2(t)$, and produces outputs $y_1(t)$ and $y_2(t)$, respectively. That system is said to be *additive* if the input $x_1(t) + x_2(t)$ produces an output of $y_1(t) + y_2(t)$. Frequently, these two properties will be lumped together. Given a system, any two scalars α_1 and α_2, and any two input functions, $x_1(t)$ and $x_2(t)$, that produce outputs $y_1(t)$ and $y_2(t)$, respectively, the system is linear if the input $\alpha_1 x_1(t) + \alpha_2 x_2(t)$ produces the output $\alpha_1 y_1(t) + \alpha_2 y_2(t)$. In summary,

$$y(t) = H[x(t)] \xrightarrow{\text{homogeneity}} \alpha y(t) = H[\alpha x(t)] \tag{1.2}$$

and

---

[*] *He who answers before listening—that is his folly and his shame.* Proverbs 18:13

$$y_1(t) = H[x_1(t)]$$
$$y_2(t) = H[x_2(t)]$$
$$\xrightarrow{\text{additivity}} y_1(t) + y_2(t) = H[x_1(t) + x_2(t)] \quad (1.3)$$

where H is called the *system function*. If a system is both homogenous and additive, then the system is linear; this is illustrated in Fig. 1.1.

Time invariance is another property that effects the study of systems. Again, consider the response of a system to an arbitrary input $x(t)$, to be the output $y(t)$. If the input signal is translated in time by t_0, and the resulting output signal is also translated by t_0 (i.e., input $x(t - t_0)$ causes an output $y(t - t_0)$), then the system is said to be *time invariant*. If a system is not *time invariant*, it is said to be *time variant* or *time varying*. Many real-world systems may be treated as if they were time invariant, even if they are not. A system that is both linear and time invariant is frequently said to be an LTI (linear, time-invariant) system.

1.4 Continuous-Time Signals

Continuous-time or *analog* signals are defined on a continuum of points in both the independent and dependent variables. A signal is said to be *continuous* if its derivative is defined everywhere, and it is said to be *discontinuous* if it is not. A signal is called *piecewise-continuous* if it is discontinuous only at isolated (discrete) points (e.g., a rectangular pulse).

Continuous-time signals can also be qualified in terms of statistics. The *average* or *mean* value of an arbitrary continuous-time signal $x(t)$ is given by

$$\langle x(t) \rangle = \lim_{T \to \infty} \frac{1}{2T} \int_{-T}^{T} x(t)\,dt \quad (1.4)$$

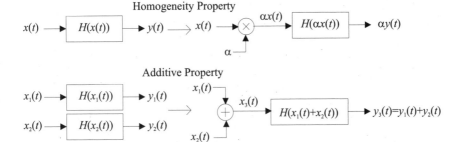

Figure 1.1   Example of a generic linear system.

Another common signal statistic is called the *mean-squared value,* which is given by

$$\langle x^2(t) \rangle = \lim_{T \to \infty} \frac{1}{2T} \int_{-T}^{T} |x(t)|^2 dt \tag{1.5}$$

The square root of the mean squared value is called the *root mean squared* or *rms* value. The mean squared value also leads to another second-order statistic that has significant importance in signal analysis, called *variance,* which is given by

$$\sigma_{x(t)}^2 = \lim_{T \to \infty} \frac{1}{2T} \int_{-T}^{T} |x(t) - \langle x(t) \rangle|^2 dt \tag{1.6}$$

or equivalently,

$$\sigma_{x(t)}^2 = \langle x^2(t) \rangle - \langle x(t) \rangle^2 \tag{1.7}$$

Variance is a measure of how "spread out" the values of the signal are with respect to the mean value over time. These formulas apply regardless of whether the signal under analysis is deterministic or random.

The energy contained in a deterministic $x(t)$ is given by E_x where

$$E_x = \lim_{T \to \infty} \int_{-T}^{T} |x(t)|^2 dt \tag{1.8}$$

From energy, the power in signal $x(t)$, denoted P_x, can be computed as

$$P_x = \lim_{T \to \infty} \frac{1}{2T} \int_{-T}^{T} |x(t)|^2 dt \tag{1.9}$$

A signal having a finite energy $E_x < M$ can immediately be seen to have a power value $P_x = 0$ from Eq. (1.9). Observe also that the power in a deterministic signal is also equal to the mean squared value given by Eq. (1.5).

1.5 Discrete-Time Signals

The set of discrete-time events $t = kT_s$ are called the *sample instances*. The value of $x(t)$ at sample instance $t = kT_s$ is called the *sample value*, has a value $x(t = kT_s) = x[k]$, and is denoted $x[k]$. A collection of sample values is called a *time-series*. The value of T_s is called the *sample period*, and $f_s = 1/T_s$ is called the *sample rate* or *sample frequency* and is measured in sample-per-second units, usually written "Sa/s," although "Hz" is commonly used interchangeably.[*] A discrete-time series is continuously resolved along the dependent axis (amplitude) and is discretely resolved along the independent axis (time). That is, the values of the time-series are discretely defined along the independent (time) axis at the time instances $t = kT_s$, but the value of $x[k]$ can be anything. Discrete-time signals can be created by passing a continuous signal through an electronic device called a *sample-and-hold* circuit. An ideal sample-and-hold device, or *sampler*, instantaneously captures and saves the value of a signal at each sample instant $t = kT_s$. Discrete-time signals can also be produced by computing algorithms such as those studied in discrete mathematics. An early discrete data generator was provided in 1202 by the Italian mathematician Leonardo da Pisa (a.k.a. Fibonacci, or literally, *Blockhead*) who posed a formula to model the number of newly born rabbits, assuming that an adult pair, after mating, would produce another pair. The *Fibonacci sequence*, given by $F_n = F_{n-1} + F_{n-2}$, for the initial conditions $F_0 = 1$, $F_{-1} = 0$ produces a discrete-time sequence $\{1,1,2,3,5,8,13,21,34,55,...\}$ which is Fibonacci's prediction of the rabbit population.

Numerous other types of discrete-time sequences arise in the fields of economics, biology, calculus, statistics, physics, and many others. The engineering study of physically meaningful discrete-time signals can be traced back to post-World War II telephone and sampled data control systems. During the early days of the cold war, strategic bombers needed to be prepared to fly missions lasting tens of hours while navigating with a high degree of precision. What was required was a precise autopilot that would sense slowly varying analog signal from the airframes ailerons and rudder and convert them into corrective control signals. Designing an analog circuit that could operate drift-free for 24 hours was virtually impossible. It was discovered that, if the signal was modulated, simple drift-free AC amplifiers could be used to implement the control channel. The simplest modulation scheme is an alternating ± 1 sequence at a known frequency of f_s. The signal could be demodulated at the receiver and in the process over-

---

[*] Technically, the unit *hertz* is equivalent to 1 /second, which is *not* strictly equivalent to samples/second.

come the drift problem. This gave rise to a technology called *sampled-data control*.

The most enduring technology emerging from this era is found in telephony. It was discovered that a number of distinct discrete time-series could be interlaced (i.e., *time-division multiplexed*) onto a common channel, thereby increasing the channel's capacity in terms of subscribers per line (and thereby increasing the billing potential). Combining multiple time-series sampled at a relatively slow sample rate into a more dense time-series sampled at a higher sample rate has been a mainstay of telephony and communications in general ever since. From this important discovery, we gain Shannon's sampling theorem (see Sec. 1.7), which is fundamentally important to all phases of DSP.

Discrete-time signals can also be quantified in terms of their statistics. The sample *average*, or *mean*, value of a discrete-time signal $x[k]$ is given by

$$\langle x[k] \rangle = \lim_{N \to \infty} \frac{1}{2N+1} \sum_{k=-N}^{N} x[k] \tag{1.10}$$

Another important signal parameter is called the sample *mean squared* statistic which is given by

$$\langle x^2[k] \rangle = \lim_{N \to \infty} \frac{1}{2N+1} \sum_{k=-N}^{N} (x[k])^2 \tag{1.11}$$

The square root of the mean squared value is called the root mean squared or rms value. The mean and mean-squared values are used to compute the sample variance, which is given by

$$\sigma_{x[k]}^2 = \lim_{N \to \infty} \frac{1}{2N+1} \sum_{k=-N}^{N} (x[k] - \langle x[k] \rangle)^2 \tag{1.12}$$

or equivalently,

$$\sigma_{x[k]}^2 = \langle x[k]^2 \rangle - \langle x[k] \rangle^2 \tag{1.13}$$

If $x[k]$ is the kth element of a finite array, or string, of measurable sample values, then the time-series statistics can be estimated using a digital computer.

Example 1.1 Statistics

Problem Statement. A signal is given by $x[k] = u[k] + \cos(\omega_0 k)u[k]$, where $\omega_0 \neq 0$, and $u[k]$ is a unit step function defined by:

$$u[k] = \begin{cases} 1 & \text{if } k \geq 0 \\ 0 & \text{otherwise} \end{cases}$$

What is the mean, mean-squared, and variance of $x[k]$ for $k \geq 0$?

Analysis and Conclusions. Direct application of Eqs. (1.10) and (1.11) yields:

$$\text{mean } \langle x[k] \rangle = \lim_{N \to \infty} \frac{1}{N+1} \sum_{k=0}^{N} [1 + \cos(\omega_0 k)]$$

$$= \lim_{N \to \infty} \frac{1}{N+1} \left[\sum_{k=0}^{N} 1 + \sum_{k=0}^{N} \cos(\omega_0 k) \right]$$

$$= 1 + 0 = 1$$

$$\text{mean-squared } \langle x^2[k] \rangle = \lim_{N \to \infty} \frac{1}{N+1} \sum_{k=0}^{N} [1 + \cos(\omega_0 k)]^2$$

$$= \lim_{N \to \infty} \frac{1}{N+1} \sum_{k=0}^{N} [1 + 2\cos(\omega_0 k)]^2$$

$$= \lim_{N \to \infty} \frac{1}{N+1} \sum_{k=0}^{N} \left[1 + 2\cos(\omega_0 k) + \frac{1}{2} + \frac{\cos(2\omega_0 k)}{2} \right]$$

$$= 1 + 0 + \frac{1}{2} + 0 = \frac{3}{2}$$

It immediately follows from Eq. (1.13) that the variance of $x[k]$ is given by $\sigma^2 = (3/2) - (1)^2 = 1/2$.

Computer Study. The analytical results shown above can be approximated by computing the statistics for a 1024-sample biased cosine having a frequency $f = f_s/16$ (i.e., $x[k] \approx (u[k] + \cos(2\pi k/16)u[k])$. The following sequence of Siglab commands implements the statistical test:

```
>x=1+mkcos(1,1/16,0,2^10)
>avg(x)
1
>avg(x^2)
1.5
>var(x)
0.5
```

The same test can be applied to a normally (Gaussian) distributed zero mean unit variance random variable (i.e., $N(0,1)$) with the following sequence of Siglab commands (data will vary due to the seeding of the random number generator):

```
>x=gn(2^10)
>avg(x)
-0.0696979        # ~0
>avg(x^2)
0.97257           # ~1
>var(x)
0.967712          # ~1                                     ❖
```

1.6 Digital Signals

Digital signals are discrete-time signals which are also quantized along the dependent axis (amplitude). Digital signals are naturally produced by a digital computer. Here, by definition, all calculations are performed using arithmetic units which round or truncate signal variables to words of finite (discrete) precision. Another way of producing digital signals is by passing a discrete-time sampled signal $x[k]$ through an *analog-to-digital* converter (ADC or A/D). An ADC consists of two parts: the sample-and-hold circuit, which physically converts an analog signal to a discrete-time signal, and a comparator, which maps the discrete-time value to an equivalent digital word. An n-bit ADC will quantize a sample value into one of 2^n finite values. The mechanism by which a digital signal is converted back into a continuous signal is called *interpolation* which generally involves the use of a *digital-to-analog* converter (DAC).

The ADC process is implemented on a single chip or, in some cases, embedded into a more capable chipset. An ADC is classified by its precision measured in bits (typically 6 to 20 bits) and by its conversion speed, which is technology dependent and can range from gigahertz values down to kilohertz. Errors and the quality of the conversion are specified in terms of local (least significant bit or LSB) to global measures of linearity, aperture uncertainty, and so forth. It is generally considered to be unrealistic to design ADCs that operate at slow conversion rates, since implementing the requisite low-frequency analog sample-and-hold circuits would be prohibitively expensive. If sampling at low frequencies is required, then signals can be oversampled at a rate consistent with that provided by a commercially available ADC, and the unwanted samples can simply be discarded (this will be called *decimation* in Chap. 18). ADCs are also grouped by type (e.g., flash, sequential approximation, Σ-Δ). In some cases, special nonlinear analog amplifiers are placed in front to the ADC to logarithmically compress the analog dynamic range of the signal presented to the ADC. They typically take the form of μ-*law companders* in North America and *A-law companders* in Europe. More specifically, the μ-law compressor is given by

$$|y[k]| = \frac{\log(1 + \mu|x[k]|)}{\log(1 + \mu)} \qquad (1.14)$$

where μ is typically $\mu = 225$. The A-law rule is given by

$$|y[k]| = \frac{\log(1 + A|x[k]|)}{\log(A)} \qquad (1.15)$$

where A is typically $A = 87.56$. Other analog systems may also need to be placed in front to the ADC to limit the highest frequency presented to the converter. This device will be called an *anti-aliasing* filter.

DACs are also graded by their speed, linearity, and precision. In addition, some DACs come supplied with integrated analog multipliers to perform multiplicative waveform shaping or modulation operations. In both ADC and DAC cases, many commercial products are available, spanning a wide range of cost, precision, and bandwidth choices.

Analyzing the errors introduced by quantization is important to the study of many practical DSP systems. Suppose the amplitude domain over which a discrete-time signal $x[k]$ is defined is double-ended and is bounded by $\pm M$ (i.e., $-M \leq x[k] < M$), then the *quantization step-size Q* is given by

$$Q = \frac{2M}{2^n} \qquad (1.16)$$

which is often measured in volts per bit. The real sample value of $x[k]$ would be quantized to a discrete or digital value $x_Q[k] = dQ$ if $dQ \leq x[k] < (d + 1)Q$, where d is an integer such that $d \in [-2^{n-1}, 2^{n-1} - 1]$. If the input is single ended (unsigned), then $Q = M/2^n$ and $d \in [0, 2^n - 1]$. ADCs produce a binary n-bit approximation of $x[k]$ which, in this case, would correspond to the value d. A digital designer would interpret the quantized value d as an unsigned integer or other digitally formatted word. The DSP engineer, however, would interpret the ADC output as the signal value $x_Q[k] = dQ$.

The difference between quantized version of $x[k]$, namely $x_Q[k] = dQ$, and $x[k]$ is called the *quantization error* and is mathematically given by

$$e[k] = x_Q[k] - x[k] \tag{1.17}$$

If n is sufficiently large ($n \geq 4$ bits), the error $e[k]$ can be approximated to be a uniformly distributed random process as shown in Fig. 1.2. There are two cases considered in Fig. 1.2. The first corresponds to the situation where $x[k]$ is always quantized downward to the nearest value of $x_Q[k] = dQ \leq x[k]$ (i.e., truncation). Here d is an integer such that $d \in [-2^{n-1}, 2^{n-1} - 1]$. The second case corresponds to always quantizing $x[k]$ to dQ or $(d + 1)Q$, depending on which is nearer (i.e., rounding). The statistics of the two error models summarized in Table 1.1 are seen to differ in mean and maximal error. These values, of course, can be made very small by increasing the size of data wordlength n, with an attendant economic and possible computational bandwidth penalty.

Table 1.1 Quantization Error Statistics

Policy	Figure	Max. |error|	Error range	Error mean	Error variance
Truncation	1.2a	$<Q$	$e[k] \in [0, Q)$	$Q/2$	$Q^2/12$
Rounding	1.2b	$Q/2$	$e[k] \in [-Q/2, Q/2)$	0	$Q^2/12$

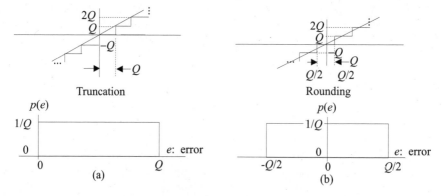

Figure 1.2 Uniform truncation error models assuming (a) rounding downward and (b) rounding.

The error variance, given by $\sigma^2 = Q^2/12$, is sometimes interpreted in terms of a standard deviation σ, or in bits, using the decimal-to-binary conversion formula,

$$\text{error variance (in bits)} = \log_2(\sigma) = \log_2(Q) - 1.79 \qquad (1.18)$$

In decibels, the error variance is

$$\text{error variance (dB)} = 20 \log_{10}(\sigma) = [20 \log_{10}(Q) - 10.8] \qquad (1.19)$$

Thus if $Q \sim 2^{-m}$, the standard deviation of the quantization error leaving an ADC would be on the order of $\log_2(\sigma) = -m - 1.79 \sim -(m + 2)$ bits, which is about two bits less than the least significant bit. In decibels, the error would be on the order of $20 \log_{10}(\sigma) \sim (-6m - 10.8)$ dB.

Example 1.2 Quantization

Problem Statement. A double-ended signal, $x(t)$, having a known range $|x(t)| \le \pm 15$ V is sent to a ± 15 V input range four-bit ADC that rounds to the nearest quantization level. Compute the error statistics and verify experimentally.

Analysis and Conclusions. The quantization step size Q for a signal having a ± 15 V swing (i.e., $M = 15$) digitized by a four-bit ADC, is given by $Q = 2 \times 15/2^4 = 1.875$ V/bit. Ideally, the maximal error is bounded by $Q/2 = 0.9375$, the mean error is zero, and the variance is $\sigma^2 = Q^2/12 = 0.3$, which corresponds to $\log_2(\sigma) = -0.89$ bits of uncertainty. Interpreting this in the context of the example problem, the effective *statistical precision* of the ADC would not be four bits but rather 4.89 bits.

Computer Study. The s-file "quant.s" simulates the quantization of an ADC using rounding. The simulation is accomplished using the function quant(x,m,n) where x is a time-series $x[k]$, and m defines the input range (i.e., the input of the converter is expected to be in the range of $\pm m$). A computer generated sinusoidal signal $x[k]$ having a ± 15 dynamic range ($m = 15$), is presented to an $n = 4$-bit ADC with the output shown in Fig. 1.3. The effect of quantization is readily apparent. Notice that the computed statistics are in close agreement with their predicted values.

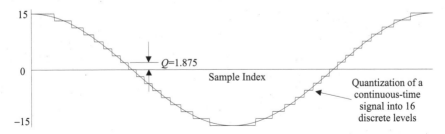

Figure 1.3 Simulation of the quantization of a sinusoid using a four-bit rounding ADC.

```
>include "chap1\quant.s"
>x=mkcos(15,1/1001,0,1001)
>quant(x,15,4)
Quantization level (units/bit) 1.875
Calculated maximum error 0.936366
Calculated average error 0.00749251
Calculated error variance 0.268385
Predicted error variance 0.292969
```

The soundboard program "quant.sb" simulates a sinusoid sampled using a two-, four-, and eight-bit ADC using the function quant(f,n,b), where f is the normalized frequency of the sinusoid, n is the length of the sinusoid in samples, and b declares the precision of the soundboard in bits ($b = 8$ or 16). The signal length n is chosen to be sufficiently long so as to provide audible playing time at the soundboard sample rate (e.g., $n = 10,000$). The signal is quantized using a simulated eight-, four-, and two-bit ADC, displayed, and then played through the soundboard in descending order of precision. The effects of quantization can be distinctly heard during playback. The following sequence of commands will create a simulation with a normalized frequency of $f = 1/50$, n = 10,000, and a 16-bit soundboard.

```
>include "chap1\quant.sb"
>quant(1/50,10000,16)
```

The nonlinear distortion due to quantization is audibly apparent.   ❖

The three classes of signals considered to this point are continuous-time (analog), discrete-time (sampled-data), and digital. These signals are compared in Fig. 1.4. Contemporary electronic systems typically contain a mix of analog, discrete, and digital signals and systems. Digital signals, however, are increasingly dominant in this mix. Applications that were once considered to be exclusively analog, such as sound recording and reproduction, have become digital. The wireless communications industry is replacing analog IF and audio sections with DSP elements wherever possible. Images and video are now routinely coded as digital signals. Discrete-time systems, as defined, are rarely found in use today except as elements of an ADC or DAC. Discrete-time control systems have also been largely replaced by DSP systems which are now the principal technology in many servo motor control applications. Analog systems remain viable in optical and RF applications that operate at very high frequencies. Nevertheless, the objective is to convert to digital as soon as possible.

The struggle between analog and DSP will continue into the future. It is clear to many that DSP will become the dominant signal processing technology in the twenty-first century. It is commonly assumed that the attributes of analog and digital signal processing systems compare as follows:

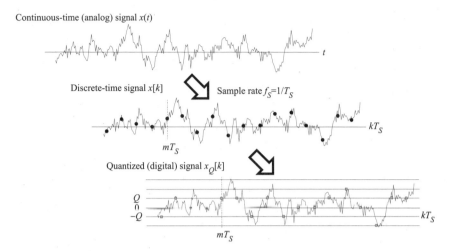

Continuous-time (analog) signal $x(t)$

Discrete-time signal $x[k]$ Sample rate $f_S{=}1/T_S$

kT_S

mT_S

Quantized (digital) signal $x_Q[k]$

kT_S

mT_S

Figure 1.4 Signal hierarchy consisting of analog, discrete-time or sampled, and digital or quantized processes.

- Both analog and digital systems can generally be fabricated as highly integrated semiconductor systems. Compared to analog, digital devices can take full advantage of submicron technologies and are generally more electronically dense, providing both economic and performance advantages.

- As semiconductor technologies shrink (deep submicron) and signal voltages continue to decline (1.25 V), the intrinsic signal-to-noise ratio found at the transistor level will render these devices essentially useless as analog systems (e.g., equivalent three-bit precision per transistor) but would not compromise their value as digital processors.

- Digital systems can operate at extremely low frequencies, which would require unrealistically large capacitor and resistor values if implemented as analog systems.

- Digital systems can be designed with increasing precision with an incremental increase in cost, whereas the precision of an analog system is physically limited.

- Analog systems can operate at extremely high frequencies (e.g., microwave and optical frequencies) which exceed the maximum clock rate of a digital devices.

- Digital systems can be easily programmed to change their function whereas reprogramming analog systems is extremely difficult.

- Digital signals can be easily delayed and/or compressed, an effect that is difficult to achieve with analog signals.

- Digital systems can easily work with dynamic ranges far in excess of 60–70 dB (10–12 bits), which is the maximum limit for a typical analog system.

- Analog systems need periodic adjustment (due to temperature-drift, aging, and so on), whereas digital systems do not require alignment.

- Digital systems do not have impedance-matching requirements, whereas analog systems do.

- Digital systems are less sensitive to additive noise, as a general rule.

1.7 Sampling Theorem (Shannon)

One of the most important scientific advancements in the first half of the twentieth century was due to Claude Shannon. Many of Shannon's inventions affect our lives today. One of his more amusing creations was a black box which, when activated with a switch, would extend a green hand outward and turn the switch off. Of greater value is his celebrated sampling theorem. Shannon worked for the telephone company and as such was interested in maximizing the number of billable subscribers that could simultaneously use a copper telephone line. One approach would be to sample the individual subscriber's conversations, interlace them together, place them on a common telephone wire, and reconstruct the original message at the receiver. Today, we refer to this as time-division multiplexing (TDM). Shannon established the rules that govern sampling as well as define the reconstruction procedure. Shannon's theory of sampling is both elegant and critically important to all discrete-time and DSP studies.

Sampling theorem

Suppose that the highest frequency contained in a continuous-time signal $x(t)$ is $f_{max} = B$ Hz. Then, if $x(t)$ is sampled periodically at a rate $f_s > 2B$, the original signal, $x(t)$, can be *exactly* recovered (reconstructed) from the sample values $x[k]$ using the *interpolation* rule

$$x(t) = \sum_{k = -\infty}^{\infty} x[k]h(t - kT_s) \qquad (1.20)$$

where $h(t)$ has a $\sin(x)/x$-shaped envelope and is given by

$$h(t) = \frac{\sin(\pi t/T_s)}{\pi t/T_s} \qquad (1.21)$$

Here, $h(t - kT_s)$ is a delayed version of $h(t)$, delayed by kT_s seconds. Observe also that $h(t)$ persists for all time (i.e., $h(t)$ is noncausal). The lower bound on the sampling frequency, f_s, is $f_r = 2B$, and is called the *Nyquist sample rate*. It is important to remember that the Sampling Theorem requires that the sampling frequency be strictly greater than the Nyquist sample rate, or $f_s > f_r$. The frequency $f_N = f_s/2 > B$ is likewise important and is called the *Nyquist frequency* or, as it is sometimes referred to, the *folding frequency*. It should be noted that when discussing discrete-time systems, frequencies are specified using *normalized frequency* units. The normalized frequency is defined to be ratio of the actual frequency versus the sampling frequency, as given by

$$\theta = \frac{f}{f_s} \tag{1.22}$$

where $|f| < f_s/2$ so that $-0.5 < \theta < 0.5$ where 0.5 corresponds to the Nyquist frequency. The normalized frequency is also referred to as the *digital frequency* and is often used to simplify the analysis of a discrete-time of digital system.

Equation (1.20) will later be interpreted as a convolution (filtering) operation which is demonstrated in Fig. 1.5. Here, an analog signal $x(t)$ is sampled at a rate of f_s samples per second to produce the time series of sample values $x[k]$. Individual interpolation filters, $h(t - kT_s)$, are weighted (scaled) by the sample values $x[k]$ and then summed with the other responses according to Shannon's sampling theorem, resulting in a perfectly reconstructed copy of $x(t)$.

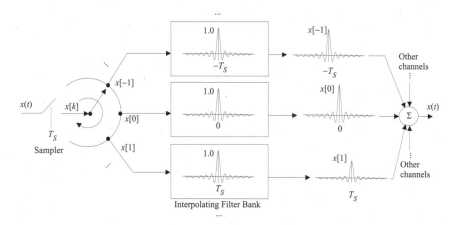

Figure 1.5 Shannon interpolator interpreted as a sampler producing a time-series, interpolation filters, and reconstruction.

The relationships between two or more Shannon interpolation filters $h(t - kT_s)$ are shown in detail in Fig. 1.6. Observe that each delayed version of an interpolating filter contains a null (i.e., zero value) at sample instant $t = kT_s$, except for the kth interpolating filter, which has a value of unity at that sample instant. Therefore, at every sampling instant, the value of $x(t)$ is given by $x(kT_s) = x[k]$. Elsewhere, the interpolation functions $x[k]h(t - kT_s)$ collectively reconstruct the original analog signal $x(t)$.

Example 1.3 Shannon Interpolation

Problem Statement. A simple cosine wave is sampled at a rate f_s to form a time-series $x[k]$. Use computer simulation to reconstruct a facsimile of the original input using Shannon interpolation formula with suitable approximations.

Analysis and Conclusions. Shannon's sampling theorem states that if the signal is band limited to B Hz and is sampled at a rate at least $f_s > 2B$ Hz, the original $x(t)$ can be perfectly reconstructed from its sample values. Unfortunately, a computer study can only approximate this reconstruction, since the Shannon interpolating filters are noncausal (infinitely long response). Nevertheless, remarkably good results can be obtained.

Computer Studies. An approximate Shannon interpolator has been created using s-file "shannon.s" which:

1. Accepts a highly oversampled input, an n-sample time-series $x[k]$, assumed to be sampled at a high rate, f_0. That is, the time-series $x[k]$ is assumed to be a densely sampled version of $x(t)$.

2. Resamples the oversampled signal to form a new time-series $s[k]$ at a lower (practical) sample rate f_s, where $f_0 = pf_s$, and $f_s \ll f_0$.

3. Processes the time-series $s[k]$ with approximate Shannon's interpolating filters by invoking the function interp(x,p,m), where p is an integer

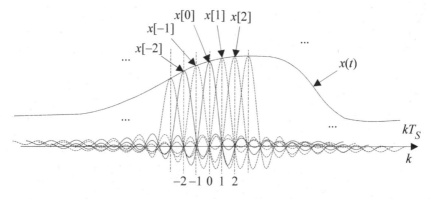

Figure 1.6 Shannon interpolation filters shown over a finite interval of time showing their interrelations.

that is greater than one, and m is the number of samples to be displayed along the discrete-time axis where $m < n - (p/2)$.

4. Displays the outcome.

Using a simple cosine wave and $p = 6$, the simulated results are shown in Fig. 1.7. The interpolated signal is seen to be a good facsimile of the original signal. The degradation at the outside edges is due to using data records of finite length to replace those of infinite length. The implementation code is given by:

```
>include "chap1\shannon.s"
>x=mkcos(1,1/128,0,1024)
>interp(x,6,1000)                                    ❖
```

The s-file shannon.sb is designed to work with soundboards. The function in the s-file may be executed using the following format: interp(x,p,n,b,c) where x is a time-series of length m to be interpolated to a signal of length $n < m$, p is previously defined, b is the precision of the soundboard in bits ($b = 8$ or 16), and c is a time-series extension parameter used to create a concatenated signal of length $N = 2^c n$, which is sent to the soundboard. The values of c and n should be defined to ensure that the audible signal is intelligible when played. Furthermore, to promote a smooth interface between concatenated data section, the signal $x[k]$ should have a period that is a divisor of n. The following code will produce a graph, shown in Fig. 1.7, and send the original, resampled ($p = 6$), and interpolated signals to an eight-bit sound board. The length of the soundboard records is $N = 2^4 \times 1000 = 16{,}000$ samples.

```
>include "chap1\shannon.sb"
>x=mkcos(1,1/100,0,1024)
>interp(x,6,1000,8,4)
```

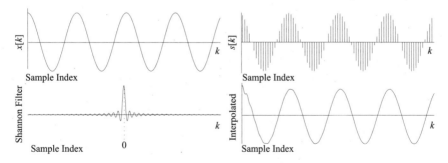

Figure 1.7 Shannon interpolator experiment consisting of an analog signal facsimile of $x(t)$ (i.e., $x[k]$), a resampled signal $s[k]$ for $p = 6$, and the reconstructed signal using a Shannon interpolator.

It should be noted that the resampled signal is of poor quality, whereas the Shannon interpolated signal is close to the original.

The previous example demonstrates a systemic weakness of the Shannon interpolator. Specifically, reconstruction requires an infinite number of Shannon interpolation filters required, each of which is infinitely long. This is obviously impractical from a digital implementation viewpoint, since it would require an infinite amount of memory to store the acquired data and an infinite amount of time to accumulate each point in the interpolated sequence. As such, alternative interpolation schemes must be found where a successful model is one that can be easily implemented in digital hardware. One of the simplest approximations to an ideal Shannon interpolator is called the *zero-order hold,* which is shown in Fig. 1.8. A zero-order hold simply "holds" the value of $x[k]$ for an interval T_s beginning at $t = kT_s$. Therefore, the interpolated signal is a piecewise constant approximation of the true $x(t)$ and is given by the formula

$$x(t) \approx x[k] = x(kT_s) \text{ for } t[(k)T_s, (k+1)T_s) \tag{1.23}$$

It should be apparent that the quality of the zero-order hold approximation of $x(t)$ is directly correlated to the value of T_s. If $x(t)$ is rapidly changing in time, T_s must be extremely small for a piecewise constant approximation of $x(t)$ to be a reasonably good facsimile of $x(t)$.

Another important practical interpolation procedure is the first-order hold. The first-order hold linearly interpolates the values of $x(t)$ between two adjacent samples $x[k]$ and $x[k+1]$ using the slope-intercept approximation formula

$$x(t) \approx x(kT_s) + \Delta_x(t - kT_s) \tag{1.24}$$

where

$$\Delta_x = \frac{x[(k+1)T_s] - x(kT_s)}{T_s} \tag{1.25}$$

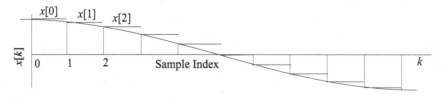

Figure 1.8 Zero-order hold window.

The interpolation formula defined in Eq. (1.24) is interpreted in Fig. 1.9. Again the quality of the interpolation is seen to be again correlated to the value of T_s, but to a lesser degree than found in the zero-order hold case.

A third interpolation method uses a lowpass filter, called a *smoothing filter*. A smoothing filter is a lowpass filter that permits only small incremental changes of Δ_x to appear at the filter's output. As a result, the interpolated signal is said to be smoothed over an interval T_s. In practice, the passband of the interpolating filter needs to be "tuned" to the expected rate of change of the signal $x(t)$. The more varying the nature of $x(t)$, the greater must be the bandwidth of the interpolating filter. In practice, interpolation is frequently performed using analog smoothing (lowpass) filters attached to the output of a digital to analog (DAC) converter. The DAC output is a piecewise constant or zero-order hold process. Passing this signal through a low-order analog lowpass filter can produce a quality reproduction of $x(t)$. The design of such filters is presented in later chapters.

Example 1.4 Interpolation

Problem Statement. Compare, using simulation, the zero-order, first-order, and lowpass interpolation techniques.

Analysis and Conclusions. Assume that a discrete-time signal $x[k]$ is an oversampled version of an analog signal $x(t)$. Specifically, let $x[k]$ be sampled at a rate $f_0 = df_s$, where $d \gg 1$, which is assumed to be well above the Nyquist frequency for $x(t)$. Let $s[k]$ be a resampled version of $x[k]$ given by $s[k] = x[kd]$. The interpolated results are displayed in Fig. 1.10, which shows that there is a hierarchy of interpolation quality.

Computer Study. The s-file "s&hold.s" performs the interpolation of a sinusoid for $d > 1$ using all three interpolation algorithms using the function "hold(x,d)". The data shown in Fig. 1.10 corresponds to the case where $d = 32$ (low sample rate) and $d = 2$ (highly oversampled). It can be immediately seen that, for high sample rates relative to the Nyquist frequency, all schemes are essentially equal. However, at slow rates, the zero-order hold only coarsely approximates $x(t)$, the first-order hold is better, and the lowpass interpolating

Figure 1.9 First-order hold interpolation of a discrete-time function.

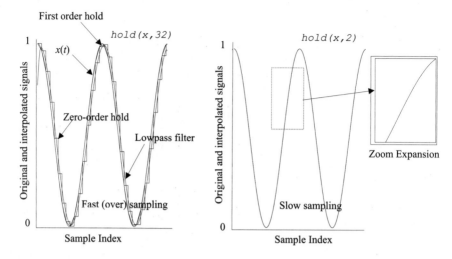

Figure 1.10 Interpolation demonstration for (a) slow sample rate ($d = 32$) and (b) high sample rate ($d = 2$) case. Shown are a sinewave $x(t)$ and a reconstructed signal from the sample values of $x(t)$.

filter is significantly better. Also, notice that the smoothing filter creates a slightly delayed version of $x(t)$. This effect will be explained when filters are studied.

```
>include "chap1\s&hold.s"
>x=mkcos(1,2/1001,0,1001)
>hold(x,32)
>hold(x,2)
```

The s-file "s&hold.sb" is a soundboard-aware version of s&hold.s. The format of the function contained in the s-file is hold(x,d,b,c), where x is again an n-sample time-series, d is as before, b is the precision of the sound-board in bits ($b = 8$ or 16) and c defines the length of the concatenated signal sent to the soundboard ($N = 2^c n$). The period of $x[k]$ should be chosen to be a divisor of n to promote a smooth transition between concatenated n-sample data records. Upon execution, a graph similar to that shown in Fig. 1.10 is produced, and the original, sampled, the N-sample original, zeroth- and first-order hold, and lowpass interpolated signals are sent to the soundboard. The zeroth-order held signal is inferior to the others, with the lowpass filter generally performing best.

```
>include "chap1\s&hold.sb"
>x=mkcos(1,10/1000,0,1000)
>hold(x,64,8,4)    # d=64 slow sampling
>hold(x,32,8,4)    # d=32 medium sampling
>hold(x,2,8,4)     # d=2 fast sampling
```
❖

1.8 Aliasing

It has been established that perfect reconstruction can take place if a band-limited signal is sampled at a rate faster than the Nyquist sample rate (i.e., $f_s > 2B$). A logical question to ask is what would happen if the Nyquist sampling theorem is violated? If a signal is sampled at or below the Nyquist sample rate, a phenomenon called *aliasing* can occur which will introduce *aliasing errors* into the reconstruction process. Aliasing, as the name implies, means that a sampled signal can be impersonated by another sampled signal. This condition can be easily motivated by considering two signals, $x(t) = \cos(k\omega_0 t)$, where k is an integer, and $y(t) = 1$. Assume that both signals are sampled at a rate $f_s = 2\pi\omega_0$. The resulting time-series are given by $x[k] = \{1,1,1,...\}$ and $y[k] = \{1,1,1,...\}$, which are identical. That is, the sampled versions of the two functions cannot be distinguished from each other. The DC process $y[k]$ is obviously sampled well above its Nyquist rate, while the sampling of $x(t)$ at f_s violates the Sampling Theorem. Convention states that $x[k]$ is an aliased version of $x(t)$. In Sec. 2.4, aliasing will be reinvestigated in a quantitative manner.

Example 1.5 Aliasing

Problem Statement. Consider the sinusoidal signal $x(t) = \cos(\omega_0 t)$ sampled at a rate $f_s = 1$ Hz. Determine if aliasing can occur over a range of frequencies ω_0.

Analysis and Conclusions. A number of $x[k] = \cos(\omega_0 k)$ cases are summarized in Table 1.2 for $f_s = 1$ Hz. In Sec. 2.4, it will be shown that the aliased signal is given by $y[k] = \cos(\omega_1 k)$ where $\omega_1 \equiv \omega_0 \bmod(2\pi)$.

Consider the instance $x_0[k] = \cos(1.8\pi k)$, or $\omega_0 = 1.8\pi$. Then $\omega_1 \equiv \omega_0 \bmod(2\pi)$ $= 1.8\pi \bmod(2\pi) = -0.2\pi$. Therefore, $x_0[k]$ impersonates $x_1[k] = \cos(-0.2\pi k) = \cos(0.2\pi k)$. Notice also that if $x_0[k] = \sin(1.8\pi k)$, or $\omega_0 = 1.8\pi$, then $\omega_1 \equiv \omega_0 \bmod(2\pi) = 1.8\pi \bmod(2\pi) = -0.2\pi$. Here, $x_0[k]$ impersonates $x_1[k] = \sin(-0.2\pi k) = -\sin(0.2\pi k)$.

Computer Study. The effect of aliasing, as presented in Table 1.2, can be experimentally studied. The following sequence of Siglab commands demonstrates the two special cases previously analyzed.

Table 1.2 Aliasing Cases

Signal	Min. sample freq.	Aliased?	Aliased freq., ω_1
$\cos(0.2\pi k)$	>0.2 samples/second	No	n/a
$\cos(0.4\pi k)$	>0.4 samples/second	No	n/a
$\cos(\pi k)$	>1.0 samples/second	Yes	π
$\cos(1.6\pi k)$	>1.6 samples/second	Yes	-0.4π
$\cos(1.8\pi k)$	>1.8 samples/second	Yes	-0.2π

```
>x1=mkcos(1,0.1,0,51)
>x9=mkcos(1,0.9,0,51)
>ograph(x1,x1,x9,x9)          # cosine test
>x1=mksin(1,0.1,0,51)
>x9=mksin(1,0.9,0,51)
>ograph(x1,x1,x9,x9)          # sine test
```

The results are reported in Fig. 1.11 and are in agreement with the calculated results.   ❖

Example 1.6 Audio Example of Aliasing

Problem Statement. Investigate the effect of aliasing using a soundboard and a linearly swept FM signal called a *chirp*.

Analysis and Conclusion. A chirp is a linearly swept FM signal having a time-varying frequency given by $\omega = \omega_0 + \Delta_w k$. If $\omega_0 = 0$ and $\Delta_w = 2\pi/N$, then over N-samples the chirp will sweep out a normalized instantaneous frequency range $\omega \in [0,2\pi]$. This is called an *up-chirp* because its frequency linear increases in time. If the sampling frequency is 1 Hz, the sampled signal will appear as an up-chirp sweeping out a linear frequency range over $\omega_a \in [0,\pi]$ for $k \in [0,N/2]$. Over $k \in [N/2,N-1]$, however, the aliased signal would be appear as a down-chip covering the range $\omega_a \in [-\pi,0]$ from the relationship $\omega_a \equiv \omega \bmod(2\pi)$.

Computer Study. The s-file "`chirp.s`" will create a chirp using the function `chirp(n,f0,s0,s1)` where n is the number of samples, f_0 is the called the *offset frequency*, $s_0 + f_0$ is the beginning of the swept range, and $s_1 + f_0$ is its end. The sequence of Siglab commands shown below will play a chirp through a B-bit soundboard. The length of the chirp should be chosen to be as large as practical, and the amplitude of the chirp should be set to $A = 0.9 \times 2^{B-1}$. Assuming a 16-bit soundboard, a chirp ranging over $f \in [0,f_s]$ is created and played as follows.

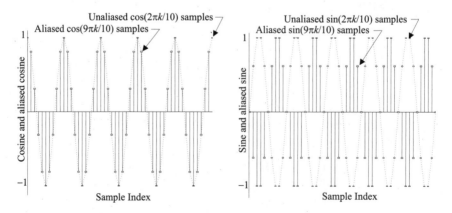

Figure 1.11 Example of aliased versions of $x[k] = \cos(1.8\pi k)$ and $x[k] = \sin(1.8\pi k)$.

```
>include "chap1\chirp.s"
>B=16          # Soundboard precision in bits
>x=0.9*(2^(B-1))*chirp(10^5,0,0,1)
>wavout(x)
```

Playing the chirp will produce an up chirp for the first half of the record followed by a down chirp for the second half.   ❖

1.9 Elementary Signal Properties

Signals can be simple or complex, deterministic or random, symmetric or asymmetric. Regardless, it is often desirable to study these signals in modified form. The important manipulation rules and signal and system properties are summarized in Table 1.3. An example of a time-scaled sinusoid is shown in Fig. 1.12.

Figure 1.13 displays an example of even and odd symmetry. The *even part* of an arbitrary continuous-time signal is given by $x_e[k] = 1/2(x[k] + x[-k])$. The *odd part* is given by $x_o[k] = 1/2(x[k] - x[-k])$. In fact, any arbitrary signal $x[k]$ can be represented as $x[k] = x_e[k] + x_o[k]$.

Of the signal properties listed in Table 1.3, periodicity is of major importance. A signal $x[k]$ is said to be *periodic* with period $T = NT_s$,

Table 1.3 Signal and System Properties

Transformation or property	Effect
Reflection	$y[k] = x[-k]$
Time shift	$y[k] = x[k - k_0]$
Time scaling	$y[k] = x[rk]$, r an integer
Even symmetry	$x[-k] = x[k]$
Odd symmetry	$x[-k] = -x[k]$
Periodicity	$x[k] = x[k + N]$, period N
Linearity	$\alpha y_1[k] = H(\alpha x_1[k])$ and $\beta y_2[k] = H(\beta x_2[k]) \leftrightarrow$ $\alpha y_1[k] + \beta y_2[k] = H(\alpha x_1[k] + \beta x_2[k])$
Time invariance	$y[k] = H(x[k]) \leftrightarrow y[k - k_0] = H(x[k - k_0])$ for all k_0

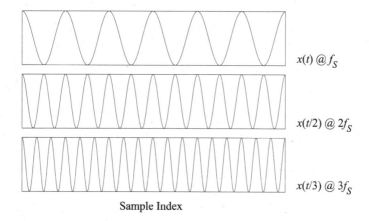

$x(t) @ f_S$

$x(t/2) @ 2f_S$

$x(t/3) @ 3f_S$

Sample Index

Figure 1.12 Scaled sinusoid showing relationship between temporal scaling and frequency.

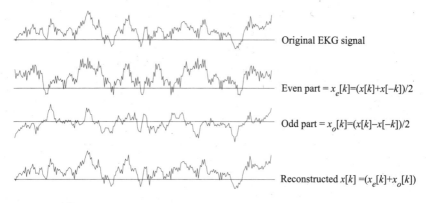

Original EKG signal

Even part = $x_e[k]=(x[k]+x[-k])/2$

Odd part = $x_o[k]=(x[k]-x[-k])/2$

Reconstructed $x[k] =(x_e[k]+x_o[k])$

Figure 1.13 From top to bottom, a signal $x[k]$, even component $x_e[k]$, odd component $x_o[k]$, and reconstructed $x[k]$ signal from the sum of the even component and the odd components.

where $T_s = 1/f_s$, if N is the smallest non-zero integer such that $x[k] = x[k + N]$ for all k. If a periodic signal has period T, then $1/T$ is called its *natural* or *fundamental frequency*. If a signal is not periodic, then it is said to be *aperiodic*. A *pseudo-periodic* signal has verifiable periodic behavior over a finite interval of time. A *quasi-periodic* signal is almost periodic over a time record. A periodic signal, it should be remembered, is technically of infinite duration and is defined over the interval $k \in (-\infty, \infty)$. Therefore, non-zero causal signals cannot be periodic. Signals of finite length, say N-samples, can be self-replicated to create a pseudo-periodic signal of period N-samples and of length kN-samples. Replication, in the case of discrete-time signals, can be accomplished by concatenating a finite time-series $x[k]$ with itself k-times.

1.10 Elementary Signals

The mathematical study of signals is often premised on representing complicated signals in terms of a set of *elementary functions* or *signal primitives.* They serve as both a basis for building high-order signals as well as providing a commonly agreed set of definitions. Elementary signals can be combined linearly or nonlinearly, scaled, shifted or translated, delayed, reversed, and extended to synthesize new signals. Basic elementary one-dimensional (1-D) signals are developed in this section.

1.10.1 Unit impulse

The *Kronecker impulse,* or *impulse,* or *delta* function, is defined to be

$$\delta[k] = \begin{cases} 1 & \text{if } k = 0 \\ 0 & \text{otherwise} \end{cases} \tag{1.26}$$

and is shown in Fig. 1.14. Notice that, unlike the *Dirac impulse* response $\delta(t)$ used in the study of continuous-time systems, the Kronecker delta function is finite everywhere.

1.10.2 Unit step

A discrete-time unit step function is defined to be

$$u[k] = \begin{cases} 1 & \text{if } k \geq 0 \\ 0 & \text{if } k < 0 \end{cases} \tag{1.27}$$

and is shown in Fig. 1.15.

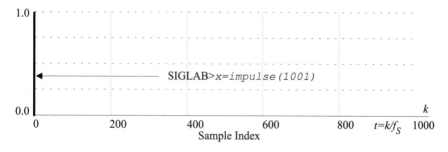

Figure 1.14 Kronecker impulse function along with simulation code.

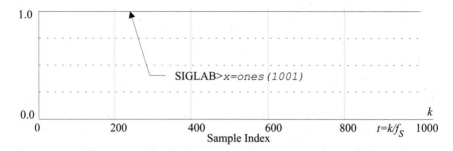

Figure 1.15 Discrete-time unit step function along with simulation code.

1.10.3 Rectangular pulse

A discrete-time *rectangular pulse* is given by

$$\text{rec}[k{:}K] = \begin{cases} 1 & \text{if } k \le k < K \\ 0 & \text{otherwise} \end{cases} \tag{1.28}$$

and is shown in Fig. 1.16 along with simulation code.

1.10.4 Triangular pulse

A discrete-time triangular function is given by

$$\text{tri}[k{:}K] = \begin{cases} 1 - \left|\dfrac{k}{K}\right| & \text{if } |k| \le K \\ 0 & \text{otherwise} \end{cases} \tag{1.29}$$

and is shown in Fig. 1.17.

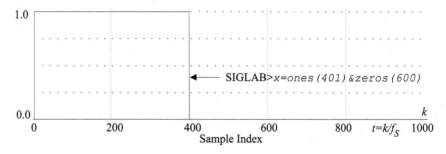

Figure 1.16 Discrete-time rectangular function.

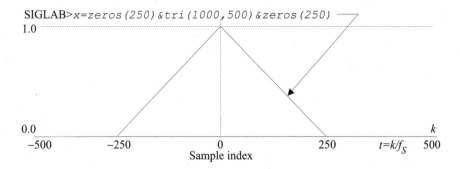

Figure 1.17 Discrete-time triangular function along with simulation code.

1.10.5 Harmonic oscillation

A fundamentally important signal is the familiar *sinusoidal (harmonic) oscillator*. A real causal discrete-time harmonic oscillation of frequency ω_0 radians per second is given by

$$x[k] = \sin(\omega_0 k T_s)u[k] \quad \text{discrete-time sine wave} \qquad (1.30)$$

$$x[k] = \cos(\omega_0 k T_x)u[k] \quad \text{discrete-time cosine wave} \qquad (1.31)$$

and is shown in Fig. 1.18.

1.10.6 Real exponential functions

Real exponential function can represent the energy decay of a natural system or the energy expansion in the case of an instability. A causal real discrete-time exponential is given by

$$x[k] = e^{\sigma k}u[k] \qquad (1.32)$$

and is shown in Fig. 1.19.

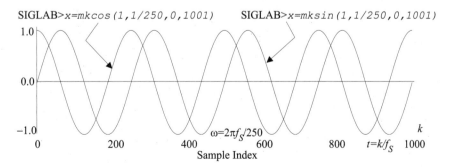

Figure 1.18 Discrete-time triangular function along with simulation code.

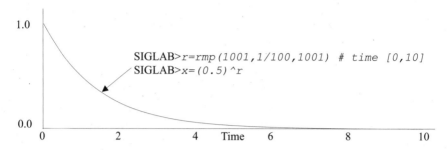

Figure 1.19 Real exponential function along with simulation code.

1.10.7 Complex exponential functions

Complex exponential function can represent oscillatory decaying or expanding processes. A causal complex discrete-time exponential is given by

$$x[k] = e^{(\sigma + j\omega)k}u[k] \tag{1.33}$$

and is shown in Fig. 1.20.

1.11 Two-Dimensional Signals

Two-dimensional signals often reside in a spatial domain. Black-and-white images, for example, would have an intensity Q at coordinates (m,n) (denoted $Q[m,n]$). A 2-D signal $Q[m,n]$ is said to be separable if

$$Q[m, n] = Q_1[m]Q_2[n] \tag{1.34}$$

Elementary 2-D signals can also be used to construct higher-order signals. Examples of 2-D elementary signals are images, complex I-Q channel communication system, and the like.

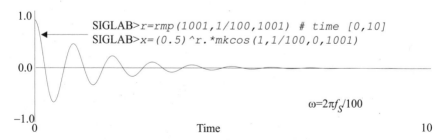

Figure 1.20 Complex exponential function along with simulation code.

1.11.1 Unit Impulse

A 2-D discrete-time *Kronecker impulse function* is defined to be

$$\delta[m, n] = \begin{cases} 1 & \text{if } m = n = 0 \\ 0 & \text{otherwise} \end{cases} \tag{1.35}$$

1.11.2 Unit step

A 2-D discrete-time *unit step function* is defined to be

$$u[m, n] = \begin{cases} 1 & \text{if } m \geq 0 \text{ and } n \geq 0 \\ 0 & \text{otherwise} \end{cases} \tag{1.36}$$

1.11.3 Causal exponential

A 2-D causal discrete-time *exponential function* is defined to be

$$Q[m, n] = a^m b^n u[m, n] \tag{1.37}$$

which is also separable, since $Q[m,n] = (a^m u[m])(b^n u[n])$.

Example 1.7: 2-D Signal

Problem Statement. Compute and display a 2-D causal exponential for $|a| < 1$ and $|b| < 1$.

Analysis and Conclusions. The data in Fig. 1.21 corresponds to the 2-D signal displayed in the positive first quadrant.

Computer Study. The following sequence of Siglab commands computes and displays a 2-D causal exponential for $|a|<1$ and $|b|<1$.

```
>a = 0.95; b = 0.9
>twod = zeros(32,32)       # create dummy storage array
>for (i = 0:31); for (j = 0:31)
>twod[i,j] = (a^i)*(b^j)
>end; end
>graph3d(twod)             # 2-D graph                    ❖
```

1.12 Summary

In this chapter, signals were classified as being continuous, discrete, or digital. Digital signals were interpreted as quantized discrete-time signals carrying with them an attendant quantization error. Important system properties such as linearity and time-invariance were de-

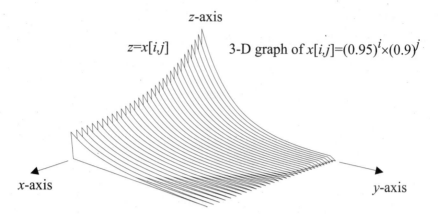

Figure 1.21   3-D display of a 2-D causal exponential for $|a| < 1$ and $|b| < 1$.

fined. The fundamentally important Shannon Sampling Theorem was developed and shown to have two parts. One relates the establishment of minimal sampling rate, and the second defines the reconstruction process. Finally, elementary discrete-time signals and signal properties were introduced.

1.13 Self-Study Problems

1. Nyquist Sampling. Compute the Nyquist sampling rate and Nyquist frequency for a signal given by

$$x(t) = \sum_{k=0}^{9} \cos(2\pi k t)$$

2. Periodicity. Determine the period of $x(t) = \cos^2(\omega_0 t)$.

3. Signal Model. A discrete-time model for a savings account is given by $b[k] = d[k] + 1.01\, b[k-1]$ where $b[k]$ is the balance and $d[k]$ is the deposit for month k. The factor 1.01 represents the monthly interest. Thus, $b[k]$ is equal to the previous month's balance $b[k-1]$ multiplied with the interest plus the deposit $d[k]$. What is the monetary growth of the account started at $k = 0$ with a $10,000 initial deposit and an additional $100 deposited every month? What is the value of the account at 12 and 24 months?

4. Quantization Error. A digital signal is an approximation of a discrete-time signal. The quality of the approximation is determined by the number of bits used by the digital word $x_D[k]$ to represent a discrete-time sample $x[k]$. The difference between the digital and discrete-time value at sample instant k will be defined to be $e[k] = x_D[k] - x[k]$. If the digital word contains a sufficient number of fractional bits of precision ($\geq$6-bits, typically), and if arithmetic rounding is used, then the error process is assumed to be zero mean with a variance of $\sigma^2 = Q^2/12$ where Q is called the quantization step-size. If a ± 5 V range is digitized with an eight-bit analog-to-digital converter, then $Q = 10/28 = 0.39$ V per bit. If a unit amplitude sinewave is digitized using the ± 5 V input range ADC, what is the quantization error variance? Conduct a numerical experiment to test your hypothesis. Repeat using 12- and 16-bit ADCs.

2

Mathematical Representation of Discrete-Time Signals

2.1 Introduction

Digital signals can be produced by presenting a continuous-time (i.e., analog) signal to an ADC. Here it is assumed that a causal analog signal $x(t)$ is sampled at a rate f_s which is above the Nyquist rate to produce an unaliased causal discrete-time series $x[k]$, namely

$$x(t) \xleftrightarrow{f_s} x[k] \tag{2.1}$$

Other digital signal production methods are based on discrete mathematics and digital arithmetic.

2.2 Difference Equations

The study of DSP systems is often introduced using a discrete-time signal motif. One of the common means of modeling a discrete-time signal is as a *difference equation* (DE). Creating signals using DEs is exemplified in Sec. 2.13. Specifically, an Nth order DE signal generator has the general form:

$$x[k] - \sum_{n=1}^{N} a[n]x[k-n] = 0 \tag{2.2}$$

where

$$x[0] = x_0 \quad x[-1] = x_{-1} \quad \dots \quad x[-(N-1)] = x_{-(N-1)} \tag{2.3}$$

and $x[-j]$ is the jth initial condition. The shape, or envelope, of the produced time-series is defined by the characteristics of the DE, while uniqueness is established by the initial conditions.

Example 2.1: Difference Equation

Problem Statement. Verify that the first-order DE given by

$$x[k+1] - ax[k] = 0$$

with $x[0] = x_0$, generates the exponential discrete time-series

$$x[k] = x_0 a^k u[k]$$

where $u[k]$ is a unit step.

Analysis and Conclusions. Observe that $x[0] = x_0 a^0 u[0] = x_0$, as required. Inductively, note that $x[k+1] = x_0 a^{(k+1)} u[k+1] = a(x_0 a^k) = ax[k]$. Therefore, the DE generates the indicated time-series. ❖

The solution of a linear difference equation can be approached in a manner similar to that historically used to solve an Nth order homogeneous ordinary differential equation (ODE),

$$\sum_{i=0}^{N} a_i \frac{d^i x(t)}{dt^i} = 0 \tag{2.4}$$

based on the use of eigenfunctions. Recall that the eigenfunctions of an ODE are assumed to have the form $\phi_n(t) = e^{s_n t}$, $n \in [1, N]$, where s_n is an *eigenvalue* of the ODE's characteristic equation. The general solution is given by

$$x(t) = \sum_{i=1}^{N} \alpha_i \phi_i(t) = \sum_{i=1}^{N} \alpha_i e^{s_i t} \tag{2.5}$$

where the coefficients α_i, and therefore the solution, can be obtained using the method of undetermined coefficients.

A similar approach can be applied to the problem of solving homogeneous difference equations of the form

$$\sum_{i=0}^{N} a_i x[k-i] = 0 \tag{2.6}$$

The *eigenfunctions* of an Nth order difference equation take the form $\Phi_n[k] = \lambda_n^k$, $n \in [0,N]$, where λ_n is a root, or eigenvalue, of the DE's characteristic equation

$$\sum_{i=0}^{N} a_i \lambda_n^i = 0 \qquad (2.7)$$

The DE's homogeneous solution can be expressed in terms of a linear combination of eigenfunctions $\Phi_n[k] = \lambda_n^k$, namely

$$x[k] = \sum_{i=1}^{N} \alpha_i \Phi_i[k] = \sum_{i=1}^{N} \alpha_i \lambda_i^k \qquad (2.8)$$

where the coefficients α_i are computed using the method of undetermined coefficients.

Example 2.2: Eigenfunctions

Problem Statement. Consider the second order DE given by $x[k] = (3/4)x[k-1] - (1/8)x[k-2]$, with $x[0] = 1$, $x[-1] = 0$. What is the homogeneous solution $x[k]$?

Analysis and Conclusions. The characteristic equation is given by $\lambda^2 - (3/4)\lambda + 1/8 = (\lambda - 1/2)(\lambda - 1/4)$. The eigenvalues of the DE are therefore $\lambda_1 = 1/2$, and $\lambda_2 = 1/4$. From Eq. (2.8), it follows that

$$x[k] = (\alpha_1 \lambda_1^k + \alpha_2 \lambda_2^k) = \left[\alpha_1 \left(\frac{1}{2} \right)^k + \alpha_2 \left(\frac{1}{4} \right)^k \right]$$

Substituting the known initial conditions into the above results leads to

$$x[0] = (\alpha_1 + \alpha_2) = 1$$

and

$$x[-1] = (\alpha_1 (1/2)^{-1} + \alpha_2 (1/4)^{-1}) = (2\alpha_1 + 4\alpha_2) = 0$$

which can be expressed in matrix-vector form as

$$\begin{bmatrix} 1 & 1 \\ 2 & 4 \end{bmatrix} \begin{bmatrix} \alpha_1 \\ \alpha_2 \end{bmatrix} = \begin{bmatrix} 1 \\ 0 \end{bmatrix}$$

Solving the matrix equation for α_1 and α_2 results in $\alpha_1 = 2$ and $\alpha_2 = -1$. Therefore, $x[k] = 2(1/2)^k - 1(1/4)^k$. As a test of the solution, one notes that $x[0] = 1$ and $x[-1] = 0$ as required. Furthermore, as an additional check, upon substituting $x[k]$ into the given DE, one obtains:

$$x[k] = \frac{3}{4}x[k-1] - \frac{1}{8}x[k-2]$$

$$= \frac{3}{4}\left[2\left(\frac{1}{2}\right)^{k-1} - 1\left(\frac{1}{4}\right)^{k-1}\right] - \frac{1}{8}\left[2\left(\frac{1}{2}\right)^{k-2} - 1\left(\frac{1}{4}\right)^{k-2}\right]$$

$$= 2\left(\frac{1}{2}\right)^{k} - 1\left(\frac{1}{4}\right)^{k}$$

as required. ❖

Whether dealing with ODEs or DEs, eigenfunctions are generally considered to be an inefficient solution methodology. *Laplace transforms* are generally considered to be the preferred method for solving an ODE. Once an ODE has been Laplace transformed, the solution $x(t)$ can be obtained using algebra and the assistance of a standard table Laplace transforms. The same paradigm applies to the study of discrete-time signals and systems.

2.3 z-Transform

The venerable Laplace transform has, for decades, been a primary analysis tool in the study of continuous-time signals. When discrete-time signals first made their appearance in sampled-data control and time-multiplexed communication systems, they were studied using Laplace transforms. The connection between continuous- and discrete-time signals was, by then, well established using the delay theorem of Laplace transforms, which states that if

$$x(t) \xleftrightarrow{L} X[s] \tag{2.9}$$

then

$$x(t - kT_s) \xleftrightarrow{L} e^{-skT_s}X(s) \tag{2.10}$$

Since a time-series can be represented in *sample value* form as

$$x[k] \leftrightarrow \{x[0], x[1], x[2], \ldots\} \tag{2.11}$$

it therefore follows that

$$x[k] = x[0]\delta[k] + x[1]\delta[k-1] + x[2]\delta[k-2] + \ldots = \sum_{n=0}^{\infty} x[n]\delta[k-n] \tag{2.12}$$

The Laplace transform of Eq. (2.12) is seen to be given by

$$X(s) = x[0] + x[1]e^{-sT_s} + x[2]e^{-2sT_s} + \dots = \sum_{n-0}^{\infty} x[n]e^{-nsT_s} \qquad (2.13)$$

The fundamental problem with this representation is that each delay requires the insertion of a delay operator of the form e^{-snT_s}. This can be very tedious. This notational problem gave rise to the introduction of a shorthand representation of the ubiquitous e^{-snT_s}, namely

$$z = e^{sT_s} \qquad (2.14)$$

which is better known by its popular name, the *z-operator*. Equation (2.14) also gives rise to the *delay operator* which is given by

$$z^{-1} = e^{-sT_s} \qquad (2.15)$$

The z-operator provides the foundation for the z-transform, which is fundamental to the study of discrete-time signals and systems. To the student of complex variables, the z-transform is called a *conformal* (angle-preserving) mapping. To a DSP engineer, the z-transform is a design and analysis tool.

Equation (2.14) defines a relationship between points in the s-plane and z-domain, which is graphically interpreted in Fig. 2.1. Some of the more salient observations are:

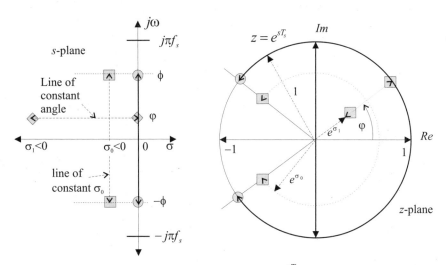

Figure 2.1 Mapping of s-plane into the z-plane under $z = e^{sT_s}$.

- The $s = j\omega$ imaginary axis in the s-plane maps, under $z = e^{j\omega}$, onto the periphery of the unit circle in the z-plane.
- The mapping to the periphery of the unit circle is unique for $s = j2\pi f$, $f \in [-f_s/2, f_s/2)$ (i.e., plus or minus the Nyquist frequency).
- The point $s = 0$ (DC) maps to $z = e^{j0} = 1$.
- The Nyquist frequency $f_N = \pm f_s/2$, corresponds to the points $s = \pm j\omega = \pm j2\pi f_N$ in the s-plane and

$$z = e^{\pm j2\pi(f_s/2)T_s} = e^{\pm j\pi} = -1$$

in the z-plane.

- Trajectories from the stable left-hand portion of the s-plane are mapped to trajectories interior to the unit circle in the z-plane.

These results can be further explored by considering the complex exponent $sT_s = \sigma + j\varphi$. Then, $z = e^s = re^{j\varphi}$ where $r = e^\sigma$. For $\varphi = k2\pi + \varphi_0$, it follows that

$$z = re^{j(k2\pi + \varphi_0)} = re^{j\varphi_0}$$

It can be seen that, if uniqueness is required, the imaginary part of s should be restricted to a range $\varphi_0 \in [-\pi, \pi)$, which corresponds to bounding the frequency range to plus or minus the Nyquist frequency in the s-plane. For values of s outside this range, the mapping $z = e^{sT_s}$ will "wrap" around the unit circle modulo $(2\pi f_s)$ or modulo (2π) for the normalized frequency case.

2.4 Aliasing

The z-transform provides an opportunity to reinvestigate aliasing which was introduced in Sec. 1.8. It is noted that the z-transform interprets the frequencies along the imaginary axis in the s-plane as $\omega_b \equiv \omega \bmod (2\pi f_s)$ in the z-plane. ω_b is a baseband frequency restricted to the range $\omega_b \in [-\pi f_s, \pi f_s)$ radians per second. In hertz, $f_b \equiv f \bmod (f_s)$, where f_b is again a baseband signal process. What needs to be appreciated is that frequencies located along the periphery of the unit circle correspond to a positive frequency range $f_b^+ = [0, f_{s/2})$ and a negative *baseband* frequency range $f_b^- = [-f_{s/2}, 0)$ as shown in Fig. 2.2. For example, $f = 0.8f_s$ is congruent to the negative baseband frequency $f_b = (-0.2f_s) \equiv f \bmod (f_s)$, where $f \in f_b^+ \cup f_b^-$. Notice also that $f = 1.2f_s$ is congruent to the positive baseband frequency $f_b = 0.2f_s \equiv f \bmod (f_s)$.

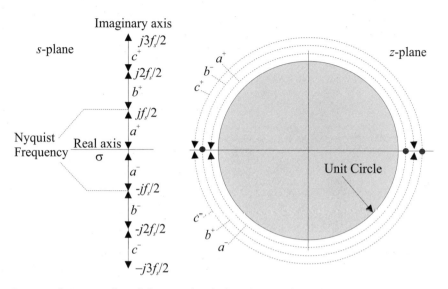

Figure 2.2 Interpretation of the mapping $j\omega$ imaginary axis in the s-plane to baseband frequencies in the z-plane.

Example 2.3: Aliasing

Problem Statement. Interpret the set of frequencies $f \in \{0, f_s/3, 2f_s/3, 3f_s/3, 4f_s/3, 5f_s/3\}$ in the context of their location on the unit circle in the z-plane.

Analysis and Conclusions. The locations of the baseband frequencies f_b are shown in Fig. 2.3. They are determined as follows:

real frequency f $\qquad\qquad\qquad$ $f \in \{0, f_s/3, 2f_s/3, 3f_s/3, 4f_s/3, 5f_s/3\}$

baseband frequency $f_b = f \bmod(f_s)$ $\qquad$ $f_b \in \{0, f_s/3, -f_s/3, 0, f_s/3, -f_s/3\}$ $\quad$ ❖

Example 2.4: Undersampling

Problem Statement. A cosine wave having a period of 32 seconds (i.e., $B = 1/32$ Hz) is used as a test signal. The Nyquist sampling rate is $f_0 = 2B = (1/16)$ Hz; the sampling theorem states that the alias-free sampling frequency f_s must exceed f_0. The signal is sampled below the Nyquist sample rate at a frequency of $(1/31)$ Hz resulting in aliasing. What is the frequency of the aliased signal?

Analysis and Conclusions. The sampling rate was stated to be $f_s = 1/31$, and the signal being sampled has a frequency of $f = 1/32$ Hz, so the signal is grossly undersampled. The aliasing frequency is given by $f_b = f \bmod (f_s)$, $f_b([-1/62, 1/62)$, so

$$\frac{1}{32} \div \frac{1}{31} = 0 \text{ remainder } \frac{1}{32} \left\{ \notin \left[\frac{-1}{62}, \frac{1}{62} \right) \right\}$$

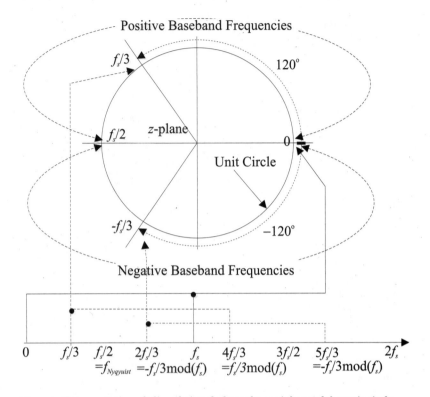

Figure 2.3 Interpretation of aliased signal along the periphery of the unit circle.

$$= 1 \text{ remainder } \frac{1}{31} - \frac{1}{32}$$

$$= 1 \text{ remainder } -\frac{1}{992} \left\{ \notin \left[\frac{-1}{62}, \frac{1}{62} \right) \right\}$$

This is the value experimentally verified in Fig. 2.4. The undersampled signal that results is an aliased signal which can be modeled as a cosine wave, having a period of 31 × 32 = 992 seconds (i.e., 1/992 Hz). Note that the negative frequency cosine at −1/992 Hz would have the same envelope as one at the positive frequency 1/992 Hz. Specifically, the resulting envelope of the new time-series takes on the appearance of a signal having a frequency well below the Nyquist sampling frequency, as shown by the dotted curve in Fig. 2.4.

Computer Study. The following program performs the required tests, which are reported in Fig. 2.4.

```
>include "chap2\alias.s"
```

❖

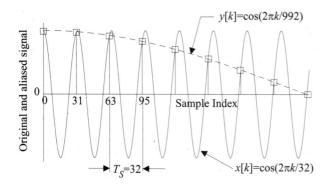

Figure 2.4 Original sine wave with period of 32 seconds being sampled once every 31 samples ($<f_0$), resulting in an aliased signal $y[k]$ that impersonates a signal having a period of 992 samples.

2.5 Existence

The mapping rule $z = e^{sT_s}$ gives rise to the z-transform. The *two-sided z-transform* of a time-series $x[k]$ is formally given by

$$X(z) = \sum_{k = -\infty}^{\infty} x[k]z^{-k} \tag{2.16}$$

The infinite sum described in Eq. (2.16) is a Laurent series and exists for only those values of z for which $X(z)$ is bounded. For a causal time-series, sometimes called a *right-sided sequence,* the *one-sided z-transform* applies and is given by

$$X(z) = \sum_{k = 0}^{\infty} x[k]z^{-k} \tag{2.17}$$

which again exists only if the sum converges.

A time-series may also be of finite duration. In such cases, the z-transform is a finite sum given by

$$X(z) = \sum_{k = m}^{n} x[k]z^{-k} \tag{2.18}$$

Observe that as long as $x[k]$ is bounded, then $X(z)$ is likewise bounded for all finite z. Therefore, $X(z)$ converges over the entire z-plane. Formally, the values of z for which the sums found in Eqs. (2.16) and

(2.17) converges is called the *region of convergence* or ROC of the z-transform. The ROC for Eq. (2.18) is $|z| < \infty$.

A one- and two-sided z-transform sum, on some occasions, can be reduced to a closed-form solution. In such cases, $X(z)$ is said to be a rational function and can be expressed as

$$X(z) = \frac{N(z)}{D(z)} \tag{2.19}$$

where $N(z)$ and $D(z)$ are polynomials in z. The values of z that satisfy $N(z) = 0$ are called the *zeros* of $X(z)$. The values of z that satisfy $D(z) = 0$ are called the *poles* or *characteristic roots* of $X(z)$.

Example 2.5: z-Transform

Problem Statement. Compute the z-transform of a causal unit step and decaying exponential functions given by

$$x_1[k] = u[k] = 1 \text{ for all } k \geq 0$$

$$x_2[k] = a^k u[k]$$

$$x_3[k] = a^k (u[k] - u[k-N])$$

Analysis and Conclusions. From Eq. (2.18), it should also be apparent that the z-transform of the step function is given by

$$X_1(z) = \sum_{k=0}^{\infty} x_1[k] z^{-k} = \sum_{k=0}^{\infty} 1 z^{-k} = 1 + z^{-1} + z^{-2} + \ldots$$

which is finite if $|z^{-1}| < 1$ or, equivalently, $|z| > 1$. Therefore, the region of convergence of $X_1(z)$ is formally given by $|z| > 1$. The ROC is seen to be an annulus (of infinite outside radius) centered about the origin in the z-plane. The infinite sum $X_1(z)$ can also be rewritten in closed form as

$$X_1(z) = 1 + z^{-1} + z^{-2} + \ldots = \frac{1}{1 - z^{-1}}$$

where the pole of $X_1(z)$ is located at $z = 1$. Notice also that the ROC does not include the pole of $X_1(z)$.

For the decaying exponential, $x_2[k]$,

$$X_2(z) = \sum_{k=0}^{\infty} x_2[k] z^{-k} = \sum_{k=0}^{\infty} a^k z^{-k} = \sum_{k=0}^{\infty} (az^{-1})^k = 1 + az^{-1} + a^2 z^{-2} + \ldots = \frac{1}{1 - az^{-1}}$$

converges if $|az^{-1}| < 1$. Therefore, the region of convergence in this case is given by $|z| > |a|$, which again is an annular ring.

For the finite length decaying exponential, $x_3[k]$,

$$
\begin{aligned}
X_3(z) &= \sum_{k=0}^{\infty} x_3[k]z^{-k} = \sum_{k=0}^{N-1} a^k z^{-k} \\
&= \sum_{k=0}^{\infty} a^k z^{-k} - \sum_{k=N}^{\infty} a^k z^{-k} \\
&= \sum_{k=0}^{\infty} a^k z^{-k} - a^N z^{-N} \sum_{k=0}^{\infty} a^k z^{-k} \\
&= \frac{1}{1-az^{-1}} - \frac{a^N z^{-N}}{1-az^{-1}} = \frac{1-a^N z^{-N}}{1-az^{-1}}
\end{aligned}
$$

where, for finite N and a, the ROC is which is the entire z-plane.   ❖

The computation of the z-transform of an arbitrary signal $x[k]$, and the computation of its ROC can be a challenging problem. Fortunately, the z-transform and ROC for many important time-series (called *elementary functions*) are known. The z-transforms of these elementary functions are cataloged, along with their ROCs, in Table 2.1.

The importance of the ROC to the study of discrete-time signal is subject to debate. Some place a great emphasis on quantifying the ROC, while others totally ignore this study. The approach presented admittedly deemphasizes the importance of the ROC and instead simply showcases the significance of the ROC.

Consider a one-sided z-transform of a causal time-series $x_c[k]$ given by Eq. (2.17):

$$
X_c(z) = \sum_{k=0}^{\infty} x_c[k]z^{-k} \tag{2.20}
$$

where the sum is assumed to converge for all $|z| > R$. Now consider using the time-series $x_c[k]$ to synthesize a noncausal time-series $x_n[k]$ given by

$$
x_n[k] = \begin{cases} x_c[-k] & \text{for } k \le 0 \\ 0 & \text{for } k > 0 \end{cases} \tag{2.21}
$$

Table 2.1   *z*-transforms of Elementary Functions

Time-domain	z-transform	Region of convergence
$\delta[k]$	1	Everywhere
$\delta[k-m]$	z^{-m}	Everywhere
$u[k]$	$z/(z-1)$	$|z|>1$
$ku[k]$	$z/(z-1)^2$	$|z|>1$
$k^2 u[k]$	$(z(z+1))/(z-1)^3$	$|z|>1$
$k^3 u[k]$	$z(z^2+4z+1)/(z-1)^4$	$|z|>1$
$a^k u[k]$	$z/(z-a)$	$|z|>|a|$
$ka^k u[k]$	$az/(z-a)^2$	$|z|>|a|$
$k^2 a^k u[k]$	$az(z+a)/(z-a)^3$	$|z|>|a|$
$\sin[bk]u[k]$	$\dfrac{z\sin(b)}{z^2-2z\cos(b)+1}$	$|z|>1$
$\cos[bk]u[k]$	$\dfrac{z[z-\cos(b)]}{z^2-2z\cos(b)+1}$	$|z|>1$
$\exp[akT_s]\sin[bkT_s]u[kT_s]$	$\dfrac{ze^{aT_s}\sin(bT_s)}{z^2-2ze^{aT_s}\cos(bT_s)+e^{2aT_s}}$	$|z|>|\exp(aT_s)|$
$\exp[akT_s]\cos[bkT_s]u[kT_s]$	$\dfrac{z[z-e^{aT_s}\cos(bT_s)]}{z^2-2ze^{aT_s}\cos(bT_s)+e^{2aT_s}}$	$|z|>|\exp(aT_s)|$
$a^k\sin(bkT_s)u[kT_s]$	$\dfrac{az\sin(bT_s)}{z^2-2az\cos(bT_s)+a^2}$	$|z|>|a|$
$a^k\cos(bkT_s)u[kT_s]$	$\dfrac{z[z-a\cos(bT_s)]}{z^2-2az\cos(bT_s)+a^2}$	$|z|>|a|$
$a^k(u[k]-u[k-N])$	$\dfrac{1-a^N z^{-N}}{1-az^{-1}}$	Everywhere

The z-transform of $x_n[k]$ is formally given by

$$X_n(z) = \sum_{k=-\infty}^{0} x_n[k]z^{-k} = \sum_{k=0}^{\infty} x_c[k]z^{k} = \sum_{k=0}^{\infty} x_c[k](z^{-1})^{-k} = X_c(z^{-1})$$

(2.22)

Comparing Eqs. (2.20) and (2.22) leads to the conclusion that if Eq. (2.20) converges for all $|z| > R$, then Eq. (2.22) will converge if $|z| < R$. Note, however, that it may now be possible that $X_n(z)$ can be identical to $X_c(z-1)$, differing only in their ROCs.

Example 2.6: Region of Convergence

Problem Statement. Three related time-series are defined below. The first is a causal decaying exponential time-series $x_c[k]$,

$$x_c[k] = \begin{cases} a^k & \text{for } k \geq 0 \\ 0 & \text{otherwise} \end{cases}$$

where $|a| < 1$. The second is a noncausal, or left-sided sequence, $x_n[k]$ which is defined to be

$$x_n[k] = \begin{cases} b^k & \text{for } k < 0 \\ 0 & \text{otherwise} \end{cases}$$

where $|b| > 1$. The third is a two-sided sequence $x_t[k]$ formed by combining $x_c[k]$ and $x_n[k]$ to form

$$x_t[k] = \begin{cases} x_c[k] & \text{for } k \geq 0 \\ x_n[k] & \text{otherwise} \end{cases}$$

$$= x_c[k] + x_n[k]$$

Compute the ROCs of the three signals.

Analysis and Conclusions. The z-transform of a decaying exponential was computed in Example 2.5 where it was found to be

$$X_c(z) = \frac{1}{1 - az^{-1}}$$

if $|az^{-1}| < 1$. Equivalently the region of convergence is given by $|z| > |a|$, which is an annular ring exterior to the pole location $z = a$. The z-transform of $x_n[k]$ is

$$X_n(z) = \frac{1}{1 - zb^{-1}}$$

but the ROC is now $|z| < |b|$, which is the disk centered at the origin and interior to the pole location $z = b$. Using the linearity property of z-transforms, $X_t(z)$ satisfies

$$X_t(z) = \frac{1}{1 - az^{-1}} + \frac{1}{1 - zb^{-1}} = \frac{2 - az^{-1} - zb^{-1}}{(1 - az^{-1})(1 - zb^{-1})}$$

So, if $|a| < |b|$ the ROC of $X_t(z)$ is the annular ring defined by $|a| < |z| < |b|$, however, if $|a| \geq |b|$, then the ROC of $X_t(z)$ is $\varnothing$ (the empty set) (see Fig. 2.5).   ❖

2.6 Properties of z-Transforms

It is generally assumed that most of important signals can be represented as a collection of one or more elementary functions found in Table 2.1. However, these primitive time-series may have to be altered

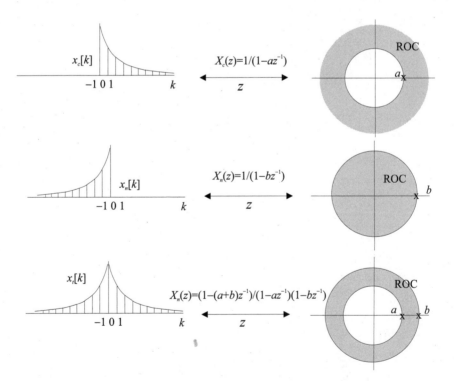

Figure 2.5 Causal and noncausal exponential time-series and their ROCs.

in some form using one or more of the operators, or signal modifiers, listed in Table 2.2.

In addition to the properties listed in Table 2.2, there are several others not listed that can be invaluable to the study of time-series. The first is called the *initial value theorem,* which states

$$x[0] = \lim_{z \to \infty} X(z) \qquad (2.23)$$

if $x[k]$ is causal. The second property is called the *final value theorem* which is given by

$$x[\infty] = \lim_{z \to \infty} (z-1)X(z) \qquad (2.24)$$

Table 2.2   Properties of z-Transforms

Property	Time-series	z-transform
Homogeneity (scaling)	$\alpha x[k]$	$\alpha X(z)$
Additivity	$x_1[k] + x_2[k]$	$X_1(z) + X_2(z)$
Linearity	$\alpha x_1[k] + \beta x_2[k]$	$\alpha X_1(z) + \beta X_2(z)$
Shift (delay)	$x[k-N]$	$z^{-N}X(z) - \displaystyle\sum_{i=1}^{N-1} \zeta^{-N+i}x[i]$
Reversal	$x[-k]$	$X(1/z)$
Complex modulation	$e^{j\theta k}x[k]$	$X(e^{-j\theta}z)$
Multiplication by a complex power series	$w^k x[k]$	$X(z/w)$
Ramping	$kx[k]$	$-z\dfrac{dX(z)}{dz}$
Reciprocal decay	$\dfrac{1}{k}x[k]$	$-\displaystyle\int^z \dfrac{X(\zeta)}{\zeta}d\zeta$
Summation (accumulation)	$\displaystyle\sum_{n=-\infty}^{k} x[n]$	$\dfrac{zX(z)}{(z-1)}$
Periodic extension of signal of length N	$\displaystyle\sum_{n=0}^{\infty} x(k+nN)$	$\dfrac{z^N X(z)}{z^N - 1}$

provided $X(z)$ has no more than one pole on the unit circle, and all other poles are interior to the unit circle.

Example 2.7: Properties of z-Transforms

Problem Statement. Compute the z-transform of $x[k] = k^2 u[k]$.

Analysis and Conclusions. The z-transform of $x[k] = k^2 u[k]$ can be defined in terms of $x_1[k] = ku[k]$ and $x[k] = kx_1[k]$. From Table 2.1, it follows that $X_1(z) = z/(z-1)^2$. From Table 2.2,

$$X(z) = -z\frac{dX_1(z)}{dz} = -z\left[\frac{1}{(z-1)^2} - 2\frac{z}{(z-1)^3}\right] = z\frac{(z+1)}{(z-1)^3}$$

To test the validity of this result, the value of $x[0]$ can be computed using the initial value theorem. Here, $x[0] = \lim_{z \to \infty} X(z) = 0$ as required. The final value is given by

$$x[\infty] = \lim_{z \to 1} (z-1)X(z) = \infty$$

which is expected of a divergent time-series.   ❖

Signal symmetry is another important signal property which gives rise to special properties in the z-domain. A signal is said to have *even symmetry* if

$$x_e[k] = x_e[-k] \tag{2.25}$$

An example of an even symmetry time-series is a noncausal cosine waveform of the form $x[k] = \cos(\omega_0 k)$. If $x[k]$ is an even symmetry, finite duration time-series over $k \in [-M, M]$, and zero elsewhere, then

$$X_e(z) = \sum_{k=-M}^{M} x_e[k]z^{-k} = \sum_{k=-M}^{-1} x_e[k]z^{-k} + x_e[0] + \sum_{k=1}^{M} x_e[k]z^{-k} \tag{2.26}$$

Now, by substituting $n = -k$ into Eq. (2.26), and from the even symmetry of $x[k]$ (i.e., $x[k] = x[-k]$), it follows that

$$X_e(z) = \sum_{k=-M}^{-1} x_e[-k]z^{-k} + x_e[0] + \sum_{k=1}^{M} x_e[-k]z^{-k}$$

$$= \sum_{k=1}^{M} x_e[k]z^{k} + x_e[0] + \sum_{k=1}^{M} x_e[k]z^{-k} \tag{2.27}$$

Now, it follows that

$$X_e(z^{-1}) = \sum_{k=1}^{M} x_e[k](z^{-1})^k + x_e[0] + \sum_{k=1}^{M} x_e[k](z^{-1})^{-k}$$

$$= \sum_{k=1}^{M} x_e[k]z^k + x_e[0] + \sum_{k=-M}^{-1} x_e[k]z^{-k} = X_e(z) \tag{2.28}$$

Thus, if $x[k]$ is even symmetric then $X(z) = X(z^{-1})$. Therefore, if z_0 is a root of $X_e(z) = 0$, then so is $1/z_0$.

An *odd symmetry,* or *antisymmetric,* time-series satisfies the condition

$$x_o[k] = -x_o[-k] \tag{2.29}$$

An example of an antisymmetric time-series is a sampled finite duration sine wave defined in the range $[-M,M]$, and zero elsewhere. Following the procedure used to compute the z-transform for an even symmetry time-series, it can be shown that

$$X_o(z) = -X_o(z^{-1}) \tag{2.30}$$

Thus, if z_0 is a root of $X_o(z) = 0$, then so is $1/z_0$.

Example 2.8: Symmetry

Problem Statement. Verify that the roots of the z-transforms of an even and odd time-series have the relationship shown in Eqs. (2.28) and (2.30). Specifically, consider

(even) $x_e[k] = \delta[k-1] + 2\delta[k] + \delta[k+1]$
(odd) $x_o[k] = \delta[k-1] - \delta[k+1]$

Analysis and Conclusions. The z-transform of $x_e[k]$ is

$$X_e(z) = z^1 + 2 + z^{-1}$$

which has a single root, $z = -1$, of multiplicity two. Observe that

$$X_e(z^{-1}) = z^{-1} + 2 - z^1 = X_e(z),$$

which also has a single root, $z = -1$, of multiplicity two. Repeating the analysis for $x_o[k]$ yields

$$X_o(z^{-1}) = z^{-1} - z^1$$

which has two roots, $z = 1$, and $z = -1$, both with multiplicity one. Observe that

$$X_o(z^{-1}) = -z^{-1} + z^1 = -X_o(z)$$

has the same roots as $X_o(z)$.   ❖

2.7 Two-Dimensional z-Transform

In Chap. 1, 2-D signals were introduced and represented as $x[m,n]$. The 2-D z-transform of $x[m,n]$ is defined to be

$$X(z_1, z_2) = \sum_{m=-\infty}^{\infty} \sum_{n=-\infty}^{\infty} x[m, n] z_1^{-m} z_2^{-n} \tag{2.31}$$

and exists if

$$\sum_{m=-\infty}^{\infty} \sum_{n=-\infty}^{\infty} \left| x[m, n] z_1^{-m} z_2^{-n} \right| < \infty \tag{2.32}$$

A 2-D z-transform is said to be separable if there exists $X_1(z_1)$ and $X_2(z_2)$ such that

$$X(z_1, z_2) = X_1(z_1) X_2(z_2) \tag{2.33}$$

Clearly, from Eq. (2.31), $X(z_1, z_2)$ is separable if $x[m,n]$ is separable.

2.8 Inverse z-Transform

It was previously established that most time-series of interest can be expressed as a collection of modified primitive signals using the information found in Tables 2.1 and 2.2. If the time-series under study does not conform to this model, then the production of its z-transform can be a truly challenging problem. Regardless, by one method or another it shall be assumed that the z-transform of a time-series $x[k]$ is known and is given by $X(z)$. The problem now becomes one of inverting the z-transform $X(z)$ to produce the parent time-series $x[k]$. The tool by which a z-transform $X(z)$ is returned to the time-domain is called the *inverse z-transform*. The inverse z-transform is defined by

$$x[k] = Z^{-1}[X(z)] = \frac{1}{2\pi j} \oint_C X(z) z^{k-1} dz \tag{2.34}$$

where C is a restricted closed path which resides within the ROC of $X(z)$. Solving the integral equation can be a very tedious process. Fortunately, there exists a nonintegral alternative, namely to use Tables 2.1 and 2.2 to produce an inverse transform. In most circumstances, it is not possible to simply lookup the inverse transform of some z-domain function $X(z)$. Instead, it is necessary to decompose the z-domain function into its constituent elementary functions so that its inverse may be determined by table lookup. The most popular means of breaking complicated z-domain functions into components that may be inverted by table lookup is called a *partial fraction expansion*. Another method, which is direct but of limited value is called the *long-division method*. Both are developed below.

2.9 Long Division

Long division can be used to explicitly compute the first few sample values of the inverse z-transform of a given $X(z)$. This can be important in those cases where only the initial samples of a time-series need to be quantified. Since the z-transform of $\delta[k]$ is unity, and the time delay property of z-transforms applied to $\delta[k - N]$ results in a z-transform of z^{-N}, the sample values of $x[k]$ can be read directly from $X(z)$ if it can be expressed polynomial form

$$X(z) = a_0 + a_1 z^{-1} + a_0 + a_1 z^{-2} + \dots \tag{2.35}$$

or more generally,

$$X(z) = \sum_{n = M}^{N} a_n z^{-n} \tag{2.36}$$

where $M \le N$. For example, in Eq. (2.35), the term $a_1 z^{-1}$ has an inverse z-transform of $a_1 \delta[k - 1]$. Therefore, the value of the time series at sample one is simply $x[1] = a_1$.

Now, consider $X(z)$ to be a rational polynomial function defined as the ratio of two polynomials, $X(z) = N(z)/D(z)$, where

$$N(z) = \sum_{k = 0}^{M} b_k z^{-k} \tag{2.37}$$

and

$$D(z) = \sum_{k=0}^{N} a_k z^{-k} \tag{2.38}$$

The rational polynomial $X(z)$ is said to be *proper* if $N \geq M$ and be *strictly proper* if $N > M$. Using long division, the initial sample values of a proper $X(z) = N(z)/D(z)$ can be determined by formally computing

$$a_0 + a_1 z^{-1} + a_2 z^{-2} + \ldots \overline{)\ b_0 + b_1 z^{-1} + b_2 z^{-2} + \ldots}$$

with quotient $\dfrac{b_0}{a_0} + \dfrac{a_0 b_1 - a_1 b_0}{a_0^2} z^{-1} + \ldots$
$$\tag{2.39}$$

The first few values of the quotient are seen to correspond to the sample values of the inverse z-transform of $X(z)$. In the context of Eq. (2.39), the sample values are $x[k] = \{b_0/a_0, (a_0 b_1 - a_1 b_0)/a_0^2, \ldots\}$.

Example 2.9: Long Division

Problem Statement. A discrete, causal time-series has a z-transform

$$X(z) = \frac{1}{1 - az^{-1}}$$

Using long division, determine the values of $x[0]$ and $x[1]$.

Analysis and Conclusions. Using long division, it follows that

$$X(z) = \frac{1}{1 - az^{-1}} = 1 + az^{-1} + a^2 z^{-2} + a^3 z^{-3} + \ldots$$

It can therefore be seen that $x[0] = 1$ and $x[1] = a$.   ❖

2.10 Partial Fraction Expansion

Partial fraction expansion is by far the most popular method of reducing z-transforms to elementary functions for inversion. Assume that $X(z)$ has a general form

$$X(z) = \frac{N(z)}{D(z)} \tag{2.40}$$

where $N(z)$ is an Mth order polynomial, and $D(z)$ is an Nth order polynomial with $M \leq N$ (i.e., $X(z)$ is proper). If $X(z)$ is not proper, then long division can be used to re-express $X(z)$ as the sum of a polynomial in z

and a proper rational remainder function. Assume that $D(z)$ is factorable as

$$D(z) = \prod_{i=1}^{L} (z - \lambda_i)^{n_i} \tag{2.41}$$

where λ_i is the root of the characteristic equation $D(z) = 0$, n_i is the multiplicity of the root λ_i, and

$$N = \sum_{i=1}^{L} n_i \tag{2.42}$$

Since $X(\lambda_i) \to \infty$, λ_i cannot be in the ROC of $X(z)$. Furthermore, if $n_i > 1$, then λ_i is said to be a *repeated* root and if $n_i = 1$, λ_i is said to be *distinct*. If $X(z)$ is proper (i.e., $M \le N$), the *partial fraction*, or *Heaviside* expansion of $X(z)$ is given by

$$X(z) = \alpha_0 + \sum_{i=1}^{L} \sum_{j=1}^{n_i} \frac{\alpha_{i,j} z}{(z - \lambda_i)^j} \tag{2.43}$$

where the coefficients α_0 and $\alpha_{i,j}$ are called *Heaviside coefficients*. Each term in the summation defined by Eq. (2.43) has the assumed form

$$V_{i,j}(z) = \frac{\alpha_{i,j} z}{(z - \lambda_i)^j} = \alpha_{i,j} \Phi_{i,j}(z) \tag{2.44}$$

where each $V_{i,j}(z)$ corresponds to an entry in Table 2.1 scaled by $\alpha_{i,j}$. Suppose the inverse z-transform of $\Phi_{i,j}(z)$ is given in Table 2.1 and is $\phi_{i,j}[k]$, then $v_{i,j}[k] = \alpha_{i,j}\phi_{i,j}[k]$, and $x[k]$ can then be expressed as

$$x[k] = \alpha_0 \delta[k] + \sum_{i=1}^{L} \sum_{j=1}^{n_i} \alpha_{i,j} \phi_{i,j}[k] \tag{2.45}$$

Once the Heaviside coefficients $\alpha_{i,j}$ are computed, $x[k]$ can be determined. The problem, of course, remains one of efficiently computing the Heaviside coefficients $\alpha_{i,j}$. The preferred method of computing the Heaviside coefficients is called *Heaviside's method*.

2.11 Heaviside's Method

Heaviside's method is based on an expansion *(partial fraction expansion)* of $X(z)$ of the form

$$X(z) = \alpha_0 + \frac{z\alpha_{i, n_i}}{(z-\lambda_i)^{n_i}} + \dots + \frac{z\alpha_{j, n_j}}{(z-\lambda_j)^{n_j}} + \dots + \frac{z\alpha_{j, k}}{(z-\lambda_j)^{k}} + \dots \qquad (2.46)$$

It can be observed that all the numerator terms are multiplied by z in order to maintain a one-to-one correspondence with the elements of Table 2.1. To simplify the computation of the Heaviside coefficients, the numerator's z dependence is often suppressed by computing the partial fraction expansion of $X'(z) = X(z)/z$ instead of $X(z)$. The partial fraction expansion of $X'(z) = X(z)/z$ is seen to be

$$X'(z) = \frac{X(z)}{z} = \frac{\alpha_0}{z} + \frac{\alpha_{i, n_i}}{(z-\lambda_i)^{n_i}} + \dots + \frac{\alpha_{j, n_j}}{(z-\lambda_j)^{n_j}} + \dots + \frac{\alpha_{j, k}}{(z-\lambda_j)^{k}} + \dots \qquad (2.47)$$

Observe that α_{j, n_j} can be isolated by forming

$$X'(z)(z-\lambda_j)^{n_j} = \frac{X(z)(z-\lambda_j)^{n_j}}{z}$$

which results in

$$X'(z)(z-\lambda_j)^{n_j} = \frac{X(z)(z-\lambda_j)^{n_j}}{z}$$

$$= \frac{(z-\lambda_j)^{n_j}\alpha_0}{z} + \frac{(z-\lambda_j)^{n_j}\alpha_{i, n_i}}{(z-\lambda_i)^{n_i}} + \dots$$

$$= \alpha_{j, n_j} + \dots + (z-\lambda_j)^{n_j-k}\alpha_{j, k} + \dots \qquad (2.48)$$

If Eq. (2.48) is evaluated at $z = \lambda_j$, then all the terms except α_{j, n_j} contain a zero multiplier. That is,

$$X'(z)(z-\lambda_j)^{n_j}\Big|_{z \to \lambda_j} = \frac{X(z)(z-\lambda_j)^{n_j}}{z}\Big|_{z \to \lambda_j}$$

$$= \frac{(z-\lambda_j)^{n_j}\alpha_0}{z}\Bigg|_{z\to\lambda_j} + \frac{(z-\lambda_j)^{n_j}\alpha_{i,n_i}}{(z-\lambda_i)^{n_i}}\Bigg|_{z\to\lambda_j}$$

$$+ \alpha_{j,n_j} + \ldots + (z-\lambda_j)^{n_j-k}\alpha_{j,k}\Big|_{z\to\lambda_j} + \ldots$$

$$= \alpha_{j,n_j}$$

$$(2.49)$$

Or, more concisely,

$$\alpha_{j,n_j} = \lim_{z\to\lambda_j}\left[\frac{(z-\lambda_j)^{n_j}X(z)}{z}\right] \qquad (2.50)$$

Suppose that $n_j > 1$, then the other terms involving the root $z = \lambda_j$ also need to be evaluated. To accomplish this, consider again Eq. (2.49) expressed as

$$X'(z)(z-\lambda_j)^{n_j} = \frac{X(z)(z-\lambda_j)^{n_j}}{z}$$

$$= \sum_{\forall i \neq j} \text{terms in } (z-\lambda_j) + \ldots + \alpha_{j,k}$$

$$+ \alpha_{j,n_j-1}(z-\lambda_j) + \ldots + (z-\lambda_j)^{n_j-k}\alpha_{j,k} \qquad (2.51)$$

It can be observed that α_{j,n_j-1} is linear in z. Therefore, α_{j,n_j-1} can be isolated using a simple derivative:

$$\alpha_{j,n_j-1} = \lim_{z\to\lambda_j}\left\{\frac{d}{dz}\left[\frac{(z-\lambda_j)^{n_j}X(z)}{z}\right]\right\} \qquad (2.52)$$

Continuing this line of reasoning, it follows that the production rule for α_{j,n_j-2} is given by

$$\alpha_{j,n_j-2} = \lim_{z\to\lambda_j}\left\{\frac{1}{2}\frac{d^2}{dz^2}\left[\frac{(z-\lambda_j)^{n_j}X(z)}{z}\right]\right\} \qquad (2.53)$$

and, in general, $\alpha_{j,k}$ is given by

$$\alpha_{j,k} = \lim_{z \to \lambda_j} \left\{ \frac{1}{(n_j-k)!} \frac{d^{(n_j-k)}}{dz^{(n_j-k)}} \left[\frac{(z-\lambda_j)^{n_j} X(z)}{z} \right] \right\} \tag{2.54}$$

Therefore, if the root λ_j is repeated with multiplicity n_j, Eq. (2.50) plus the $n_j - 1$ derivative equations given by Eq. (2.54) are required to determine the n_j Heaviside coefficients $\alpha_{j,k}$, for $k \in [1, n_j]$.

Example 2.10: Nonrepeated Roots

Problem Statement. Suppose $X(z) = z^2/[(z-1)(z-e^{-T})]$, where $x[k]$ is causal and $T > 0$. What is $x[k]$?

Analysis and Conclusions. It is apparent that all poles of $X(z)$ are not repeated and are $z = 1$ and $z = e^{-T}$. Therefore

$$X(z) = \alpha_0 + \alpha_1 \left(\frac{z}{z-1} \right) + \alpha_2 \left(\frac{z}{z-e^{-T}} \right)$$

and

$$x[k] = \alpha_0 \delta[k] + (\alpha_1 + \alpha_2 e^{-kT}) u[k]$$

where α_0 corresponds to the multiplier of a Kronecker delta function in the time domain (i.e., $\delta[k] \leftrightarrow 1$), α_1 is the weight applied to a step-function (i.e., $u[k] \leftrightarrow z/(z-1)$), and α_2 scales an exponential time-series (i.e., $a^k u[k] \leftrightarrow z/(z-a)$). The Heaviside coefficient production rules, based on Eq. (2.50), are specified in terms of factors of $X'(z) = X(z)/z$ where

$$\alpha_0 = \lim_{z \to 0} \left[\frac{z X(z)}{z} \right] = \lim_{z \to 0} \left[\frac{z^2}{(z-1)(z-e^{-T})} \right] = 0$$

$$\alpha_1 = \lim_{z \to 1} \left[\frac{(z-1)X(z)}{z} \right] = \lim_{z \to 1} \left[\frac{z}{(z-e^{-T})} \right] = \left[\frac{1}{(1-e^{-T})} \right]$$

$$\alpha_2 = \lim_{z \to e^{-T}} \left[\frac{(z-e^{-T})X(z)}{z} \right] = \lim_{z \to e^{-T}} \left[\frac{z}{(z-1)} \right] = \left[\frac{e^{-T}}{e^{-T}-1} \right] = \left[\frac{1}{(1-e^{T})} \right]$$

For the purpose of discussion, assume that $e^{-T} = 0.5$. Then $\alpha_1 = 1/(1-0.5) = 2$, and $\alpha_2 = 1/(1-2) = -1$. Finally,

$$X(z) = \alpha_1 \left(\frac{z}{z-1} \right) + \alpha_2 \left(\frac{z}{z-e^{-T}} \right)$$

and

$$x[k] = (\alpha_1 + \alpha_2 e^{-kT})u[k]$$

Computer Study. The s-file "pf.s" performs a partial fraction expansion of $X(z) = N(z)/D(z)$. The syntax is given as "pf(n,d)" where n is an array of coefficients of $N(z)$, and d is an array of coefficients of $D(z)$. The program pf.s can be used to evaluate $X'(z) = X(z)/z$ in partial fraction form once all the parameters are specified. For the purpose of numerical analysis, let $e^{-T} = 0.5$. Then the roots of the denominator of $X'(z) = X(z)/z$ are distinct and are 0, 1, and 0.5. The partial fraction expansion of $X(z)$ is given by

$$X(z) = \alpha_0 + \alpha_1\left(\frac{z}{z-1}\right) + \alpha_2\left(\frac{z}{z-0.5}\right)$$

and the partial fraction expansion of $X'(z) = X(z)/z$ satisfies:

$$X'(z) = \frac{X(z)}{z} = \alpha_0\frac{1}{z} + \alpha_1\left(\frac{1}{z-1}\right) + a_2\left(\frac{1}{z-0.5}\right)$$

The following sequence of Siglab commands is used to compute the Heaviside coefficients:

```
>include "chap2\pf.s"
>a=[1,0,0]              # numerator z^2
>b=[1,-1.5,0.5,0]       # denominator (z)(z-1)(z-0.5)=z^3-1.5z^2+0.5z
>pfexp(a,b)
pole @ 0+0j with multiplicity 1 and coefficients: 0+0j => a_0
pole @ 0.5+0j with multiplicity 1 and coefficients: -1-0j => a_2
pole @ 1+0j with multiplicity 1 and coefficients: 2+0j => a_1
```

The Heaviside coefficients are seen to be $\alpha_0 = 0$, $\alpha_1 = 2$, and $\alpha_2 = -1$. Therefore

$$X(z) = 2\left(\frac{z}{z-1}\right) - \left(\frac{z}{z-0.5}\right)$$

or $x[k] = [2 - (0.5)^k]u[k]$. ❖

For the purpose of completeness, consider the following example involving repeated roots.

Example 2.11: Repeated Root Polynomial

Problem Statement. Suppose that the previous example is slightly modified to read $X(z) = z^2/(z-1)^2$, which has a repeated root (i.e., $n_1 = 2$). The partial fraction expansion of $X(z)$ is given by

$$X(z) = \alpha_0 + \alpha_{1,1}\left(\frac{z}{z-1}\right) + \alpha_{1,2}\left[\frac{z}{(z-1)^2}\right]$$

What are α_0, $\alpha_{1,1}$, $\alpha_{1,2}$, and $x[k]$?

Analysis and Conclusions. The partial fraction expansion for $X'(z) = X(z)/z$ is given by

$$X'(z) = \frac{X(z)}{z} = \alpha_0 \frac{1}{z} + \alpha_{1,1}\left(\frac{1}{z-1}\right) + \alpha_{1,2}\left[\frac{1}{(z-1)^2}\right]$$

and

$$x[k] = \alpha_0 \delta[k] + (\alpha_{1,1} + \alpha_{1,2}k)u[k]$$

The Heaviside coefficients are given by

$$\alpha_0 = \lim_{z \to 0} \frac{zX(z)}{z} = 0$$

$$\alpha_{1,2} = \lim_{z \to 1} \frac{(z-1)^2 X(z)}{z} = \lim_{z \to 1}(z) = 1$$

$$\alpha_{1,1} = \lim_{z \to 1} \frac{d}{dz}\left[\frac{(z-1)^2 X(z)}{z}\right] = \lim_{z \to 1}\left[\frac{d(z)}{dz}\right] = 1$$

which results in a partial fraction expansion given by

$$X(z) = \frac{z}{z-1} + \frac{z}{(z-1)^2}$$

By table lookup, time-series $x[k]$ is

$$x[k] = (1+k)u[k]$$

Computer Study. The s-file "pf.s" can again be used to provide a partial fraction expansion of $X'(z) = X(z)/z$, as given above. Since the denominator polynomial of $X'(z)$ has roots at zero and one (multiplicity two), the Siglab "poly" function is used to generate the canonical form of the denominator polynomial. The following sequence of Siglab commands is used to generate the Heaviside coefficients for the partial fraction expansion of $X'(z)$.

```
>include "chap2\pf.s"
>n=[1,0,0]          # numerator z^2
>d=poly([0,1,1])    # denominator z(z-1)(z-1) in polynomial form
>d # D(z)=z^3-2z^2+z
1   -2   1   0
>pfexp(n,d)         # evaluate partial fraction expansion
pole @ 0+0j with multiplicity 1 and coefficients: 0+0j=> a_0
pole @ 1+0j with multiplicity 2 and coefficients: 1-0j 1+j0=> a_12
and a_11
```

The results here corresponds to the analytically derived values.   ❖

When discrete-time signals are represented with higher-order polynomials, analytical manipulations can become very tedious. In such cases, computer-generated analysis is often preferred. However, due to computational imprecision, the results should be checked before they are accepted.

Example 2.12: Higher-Order Polynomial

Problem Statement. The z-transform of a causal time-series is assumed to be given by:

$$X(z) = \frac{3z^3 - 5z^2 + 3z}{(z-1)^2(z-0.5)}$$

What is the partial fraction expansion of $X(z)$ and the time-series $x[k]$?

Analysis and Conclusions. The partial fraction expansion of $X(z)$ is given by

$$X(z) = \alpha_0 + \alpha_1 \frac{z}{(z-0.5)} + a_{2,1} \frac{z}{(z-1)} + \alpha_{2,2} \frac{z}{(z-1)^2}$$

and

$$x[k] = \alpha_0 \delta[k] + [\alpha_1(0.5)^k + \alpha_{2,1} + \alpha_{2,2}k]u[k]$$

where α_0 corresponds to the weight applied to a Kronecker delta function (i.e., $\delta[k]$), α_1 weights $z/(z - 0.5)$, which corresponds to an exponential $(0.5)^k u[k]$, $\alpha_{2,1}$ scales $z/(z - 1)$, corresponding to a step function $u[k]$, and $\alpha_{2,2}$ is applied to $z/(z-1)^2$, corresponding to a ramp, $ku[k]$. The partial fraction expansion of $X'(z)$ $= X(z)/z$ is given by

$$X'(z) = \frac{X(z)}{z} = \alpha_0 \frac{1}{z} + \alpha_1 \frac{1}{(z-0.5)} + \alpha_{2,1} \frac{1}{(z-1)} + \alpha_{2,2} \frac{1}{(z-1)^2}$$

The roots of the denominator of $X'(z)$ are located at zero with multiplicity one, 0.5 with multiplicity one, and one with multiplicity two. The Heaviside coefficient production rules for $X(z)$ are given by Eqs. (2.50) and (2.54). The coefficients are computed to be

$$\alpha_0 = \lim_{z \to 0} \frac{zX(z)}{z} = 0$$

$$\alpha_1 = \lim_{z \to 0.5} \frac{(z-0.5)X(z)}{z} = \lim_{z \to 0.5} \left[\frac{3z^3 - 5z^2 + 3z}{z(z-1)} \right] = 5$$

$$\alpha_{2,2} = \lim_{z \to 1} \left[\frac{(z-1)^2 X(z)}{z} \right] = \lim_{z \to 1} \left[\frac{3z^3 - 5z^2 + 3z}{z(z-0.5)} \right] = 2$$

$$\alpha_{2,\,1} = \lim_{z \to 1} \frac{d}{dz} \left[\frac{(z-1)^2 X(z)}{z} \right]$$

$$= \lim_{z \to 1} \left\{ \frac{9z^2 - 10z + 3}{z(z - 05)} - \frac{(3z^3 - 5z^2 + 3z)(2z - 0.5)}{[z(z - 0.5)]^2} \right\} = -2$$

which results in $x[k] = (5(0.5)^k - 2 + 2k)u[k]$. The candidate solution can be checked using the initial value theorem which gives

$$x[0] = \lim_{z \to \infty} X(z) = 3$$

which agrees with the computed value of $x[0]$.

Computer Study. The s-file "pf.s" can again be used to provide a partial fraction expansion of $X'(z) = X(z)/z$. The denominator of $X(z)/z$ has roots at zero, 0.5, and one (multiplicity two).

```
>include "chap2\pf.s"
>n=[3,-5,3,0] # numerator 3z^3-5z^2+3z
>d=poly([0,1,1,0.5]) # denominator (z)(z-1)(z-1)(z-0.5)
>d
1    -2.5    2    -0.5    0
>pfexp(n,d) # evaluate partial fraction expansion
pole @ 0+0j with multiplicity 1 and coefficients: 0+0j => a_0
pole @ 0.5+0j with multiplicity 1 and coefficients: 5+0j => a_1
pole @ 1+0j with multiplicity 2 and coefficients: 2+0j -2+0j
   =>a_22 and a_21
```

This is seen to agree with the manual calculations.   ❖

To this point, the partial fraction example problems have had one thing in common: the coefficient on the Kronecker delta terms (i.e., α_0) has always had a computed value of zero. This is the result of the numerator of $X(z)$, namely $N(z)$, being devoid of a quotient term of the form $(b_0/a_0)z^0$ as given in Eq. (2.39). The next example will include such a term.

Example 2.13: Kronecker Output

Problem Statement. The z-transform of a causal sequence $x[k]$ is given as

$$X(z) = \frac{(z+1)^3}{(z+0.5)(z-0.5)^2} = \frac{z^3 + 3z^2 + 3z + 1}{(z+0.5)(z-0.5)^2} = 1 + \frac{\left(\frac{7}{2}\right)z^2 + \left(\frac{13}{4}\right)z + \left(\frac{7}{8}\right)}{(z+0.5)(z-0.5)^2}$$

which is seen to possess a quotient term of the form $\alpha_0 z^0 = 1$. What is the partial fraction expansion of $X(z)$ and the time-series $x[k]$?

Analysis and Conclusions. The partial fraction expansion of $X(z)$ has the general form

$$X(z) = \alpha_0 + \alpha_1 \frac{z}{(z+0.5)} + \alpha_{2,1}\frac{z}{(z-0.5)} + \alpha_{2,2}\frac{z}{(z-0.5)^2}$$

resulting in a time series of the form

$$x[k] = \alpha_0\delta[k] + [\alpha_1(-0.5)^k + \alpha_{2,1}(0.5)^k + \alpha_{2,2}k(0.5)^k]u[k]$$

where α_0 corresponds to the weight applied to a Kronecker delta function (i.e., $\delta[k]$), α_1 weights $z/(z+0.5)$, which corresponds to an oscillating exponential $(-0.5)^k u[k]$, $\alpha_{2,1}$ scales $z/(z-0.5)$, an exponential $(0.5)^k u[k]$, and $\alpha_{2,2}$ is applied to $z/(z-0.5)^2$, which corresponds to $k(0.5)^k u[k]$ in the time-domain. The partial fraction expansion of $X'(z) = X(z)/z$ has the form

$$X'(z) = \frac{X(z)}{z} = \alpha_0\frac{1}{z} + \alpha_1\frac{1}{(z+0.5)} + \alpha_{2,1}\frac{1}{(z-0.5)} + \alpha_{2,2}\frac{1}{(z-0.5)^2}$$

The Heaviside coefficients are computed to be

$$\alpha_0 = \lim_{z\to 0} \frac{zX(z)}{z} = 8$$

$$\alpha_1 = \lim_{z\to -0.5} \frac{(z+0.5)X(z)}{z} = \lim_{z\to -0.5}\left[\frac{z^3 + 3z^2 + 3z + 1}{z(z-0.5)^2}\right] = -0.25$$

$$\alpha_{2,2} = \lim_{z\to 0.5}\left[\frac{(z-0.5)^2 X(z)}{z}\right] = \lim_{z\to 0.5}\left[\frac{z^3 + 3z^2 + 3z + 1}{z(z-0.5)}\right] = 6.75$$

$$\alpha_{2,1} = \lim_{z\to 0.5}\frac{d}{dz}\left[\frac{(z-0.5)^2 X(z)}{z}\right] = -6.75$$

which results in the candidate solution

$$x[k] = \{8\delta[k] - 0.25(-0.5)^k - 6.75(0.5)^k + 6.75k(0.5)^k\}u[k]$$

Direct evaluation of $x[0]$ produces $x[0] = 8 - 0.25 - 6.75 = 1$. The initial value theorem can be used to verify this data point:

$$x[0] = \lim_{z\to \infty} X(z) = 1$$

Computer Study. The s-file "`pf.s`" is used to provide a partial fraction expansion of $X'(z) = X(z)/z = N'(z)/D'(z) = (z+1)^3/[z(z+0.5)(z-0.5)^2]$. The roots of $D'(z) = z(z+0.5)(z-0.5)^2$ are zero, -0.5, and 0.5 (multiplicity two). The following sequence of Siglab commands can be used to determine the Heaviside coefficients for the partial fraction expansion of $X(z)$.

```
>include "chap2\pf.s"
>n=[1,3,3,1]            # N´(z)=N(z)
>d=poly([0,-0.5,0.5,0.5]) # D´(z)=zD(z)
>d                      # Denominator z^4-0.5z^3-0.25z^2+0.25z
[1,-0.5,-0.25,0.125,0]
>pfexp(n,d)             # evaluate partial fraction expansion
pole @ -0.5+0j with multiplicity 1 and coefficients: -0.2500+0j
  => a_1
pole @ 0+0j with multiplicity 1 and coefficients: 8.000+0j => a_0
pole @ 0.499985+0j with multiplicity 2 and coefficients:
  6.75+0j-6.75+0j =>a_22 and a_21
```

These results agree with the analytically derived values.   ❖

To this point, only real roots have been considered. It is often the case that the roots of the denominator of $X(z) = N(z)/D(z)$ are complex. In such cases, the Heaviside evaluation methodology can be directly applied except the results are now complex numbers.

Example 2.14: Complex Roots

Problem Statement. Consider $X(z)$ to be given as

$$X(z) = \frac{3z^3 - 1.5z}{z^2 - z\cos(\pi/6) + 0.25}$$

What is the partial fraction expansion of $X(z)$ and what is the time-series $x[k]$?

Analysis and Conclusions. The roots of $D(z)$ may be determined directly using

$$z = \frac{\cos(\pi/6) \pm \sqrt{\cos(\pi/6)^2 - 1}}{2}$$

The nonrepeated complex roots are a conjugate pair, located at $0.433 \pm j0.25 = 0.5e^{\pm j\pi/6}$. Let $\lambda = 0.433 - j0.25 = 0.5e^{-j\pi/6}$. The partial fraction expansion of $X(z)$ is formally defined by

$$X(z) = \alpha_0 + \alpha_1 \frac{z}{z-\lambda} + \alpha_2 \frac{z}{z-\lambda^*}$$

where λ^* is the complex conjugate value of λ. The partial fraction expansion of $X'(z) = X(z)/z$ is given by

$$X'(z) = X\frac{(z)}{z} = \alpha_0 \frac{1}{z} + \alpha_1 \frac{1}{z-\lambda} + \alpha_2 \frac{1}{z-\lambda^*}$$

and the Heaviside coefficients are defined by

$$\alpha_0 = \lim_{z \to 0} \frac{zX(z)}{z} = 0$$

$$\alpha_1 = \lim_{z \to \lambda} \frac{(z-\lambda)X(z)}{z} = \lim_{z \to \lambda} \left[\frac{3z^2 - 1.5z}{z(z-\lambda^*)} \right]$$

$$\alpha_2 = \lim_{z \to \lambda^*} \left[\frac{(z-\lambda^*)X(z)}{z} \right] = \lim_{z \to \lambda^*} \left[\frac{3z^2 - 1.5z}{z(z-\lambda)} \right] = \alpha_1^*$$

and α_1 and α_2 are complex conjugate pairs. Specifically, $\alpha_1 = 1.5 - j0.401924 \cong$ $1.55e^{-j\pi/12}$. Evaluating $x[k]$ in a piecemeal fashion, one notes that $x[k] = x_1[k] + x_2[k]$, where

$$x_1[k] = \alpha_1 Z^{-1}\left(\frac{z}{z-\lambda} \right) = \alpha_1(\lambda)^k u[k] \cong 1.55e^{-j(\pi/12)}(0.5)^k e^{jk\pi/6}$$

$$x_2[k] = \alpha_2 Z^{-1}\left(\frac{z}{z-\lambda^*} \right) = \alpha_2(\lambda^*)^k u[k] \cong 1.55e^{j(\pi/12)}(0.5)^k e^{jk\pi/6}$$

Therefore

$$x[k] = x_1[k] + x_2[k]$$

$$= 1.55e^{j(\pi/12)}(0.5)^k e^{-jk(\pi/6)} + 1.55e^{j(\pi/12)}(0.5)^k e^{-jk\pi/6}$$

$$= 1.55(0.5)^k (e^{-j\pi/12 - jk\pi/6} + e^{j\pi/12 + jk\pi/6})$$

$$\cong 3.1(0.5)^k \cos(k\pi/6 + \pi/12)$$

The time-series $x[k]$ is seen to be a phase delayed harmonic oscillation with an exponentially decaying envelope. As a check, the candidate solution provides $x[0] = 3$, and from the initial value theorem,

$$x[0] = \lim_{z \to \infty} X(z) = 3$$

as well.

Computer Study. Using s-file "pf.s" the partial fraction expansion of $X'(z) =$ $X(z)/z$ can be computed. The poles of $X'(z) = X(z)/z$ are $z = 0$ and $z = 0.433 \pm$ $j0.25$. Using the following sequence of Siglab commands, the Heaviside coefficients for the partial fraction expansion of $X(z)$ can be determined.

```
>include "chap2\pf.s"
>n=[3,-1.5,0]              # Numerator: 3z^2-1.5z
>d=[1,-cos(pi/6),1/4,0]    # Denominator: z^3-z cos(pi/6)+0.25z
>pfexp(n,d)                # evaluate partial fraction expansion
pole @ 0+0j with multiplicity 1 and coefficients: 0+0j => a_0
pole @ 0.433013-0.25j with multiplicity 1 and coefficients:
1.5-0.401924j => a_1
```

```
pole @ 0.433013+0.25j with multiplicity 1 and coefficients:
1.5+0.401924j => a_2
```

These results agree with the manually computed results.   ❖

2.12 Series Expansions

In some instances, one may arrive at a function in the z-domain that does not have an obvious expansion into elementary functions from which an inverse may be obtained. In these cases, a series expansion may be helpful.

A function $X(z)$ is said to be *analytic* in an open set if it has a derivative at each point in that set. $X(z)$ is said to be analytic at a point $z = z_0$ if it is analytic over some neighborhood of z_0. If $X(z)$ is analytic at a point z_0 and is analytic over the open disk defined by $|z - z_0| < R$, then it can be expressed as a Taylor series having the form

$$X(z) = \sum_{m=0}^{\infty} a_m(z - z_0)^m \qquad (2.55)$$

for all z such that $|z - z_0| < R$, where

$$a_m = \frac{1}{m!} \frac{d^m X(z)}{dz^m} \bigg|_{z = z_0} \qquad (2.56)$$

for all $m \in \{0,1,2,...\}$. For the special case where $z_0 = 0$, the Taylor series is called a *Maclaurin series*.

Example 2.15: Taylor Series

Problem Statement. Consider the z-domain function $X(z) = e^z$. Using Taylor series determine the time-series $x[k]$?

Analysis and Conclusions. The function $X(z) = e^z$ is continuous and differentiable everywhere. Consequentially, it is analytic at all points $z \in \mathbf{C}$. In many cases, expansion of the function into a useful series is easiest if a Maclaurin series expanded about $z_0 = 0$ is used. Using Eq. (2.56), the coefficients for the series found in Eq. (2.55) can be determined to be

$$a_m = \frac{dme^z}{dzm} \bigg|_{z = 0} = 1$$

so the Maclaurin series for $X(z) = e^z$ is

$$X(z) = e^z = \sum_{m-0}^{\infty} \frac{z^m}{m!} = 1 + z + \frac{z^2}{2} + \dots$$

This sum converges for all $z \in C$. From the Maclaurin series expansion the time series can be determined using the lookup tables. In this case, the time series is

$$x[k] = \frac{1}{(-k)!} u[-k]$$

Notice that this time series is noncausal. Using the time-reversal property, $X(1/z) \overset{z}{\leftrightarrow} x[-k]$, a causal signal can be constructed.   ❖

For the case where $X(z)$ is not analytic at z_0, the Taylor series expansion cannot be used. It is, however, possible to produce a series expansion using a Laurent series. Suppose that the function $X(z)$ is analytic over an open annulus about z_0, $R_1 < |z - z_0| < R_2$. Let C denote a simple, closed contour oriented in the clockwise (positive) direction, contained in the annulus. Then, at each point z in the annulus, $X(z)$ has the series representation

$$X(z) = \sum_{m=0}^{\infty} a_m (z - z_0)^m + \sum_{m=1}^{\infty} \frac{b_m}{(z - z_0)} \qquad (2.57)$$

or

$$X(z) = \sum_{m=-\infty}^{\infty} c_m (z - z_0)^m \qquad (2.58)$$

for $R_1 < |z - z_0| < R_2$, where

$$a_m = \frac{1}{2\pi j} \int_C \frac{X(z) dz}{(z - z_0)^{m+1}}, \text{ for } m \in \{1, 2, 3, \dots\}$$

$$b_m = \frac{1}{2\pi j} \int_C \frac{X(z) dz}{(z - z_0)^{-m+1}}, \text{ for } m \in \{1, 2, 3, \dots\}$$

$$c_m = \frac{1}{2\pi j} \int_C \frac{X(z) dz}{(z - z_0)^{m+1}}, \text{ for } m \in \{\dots, -2, -1, 0, 1, 2, \dots\} \qquad (2.59)$$

Either of the forms in Eq. (2.57) or (2.58) is referred to as the *Laurent series expansion* of $X(z)$. There are several means of evaluating the contour integrals in Eq. (2.59); however, a full treatment of all available methods to evaluate these integrals is beyond the scope of this text.[1]

2.13 Waveform Synthesis

The discrete-time signals analyzed earlier in the chapter can be derived from sampling a continuous-time signal or created using digital hardware. Digital signal synthesizers often appear in the form of a numerically controlled oscillator (NCO)/modulator (NCOM) or digital signal and waveform generator.

There are basically two methods of synthesizing a signal using digital devices. The first is called the *direct* method which, for signals of short duration or of a periodic nature, stores a signal in a memory device (called a *wavetable*) on a sample-by-sample basis. During run-time, all or part of the signal is played from memory as a *memory table lookup* (TLU). A classic example of this signal generator policy is simply a device that saves N-sample signal $x[k]$ in a shift register (equivalently a wavetable) as shown in Fig. 2.6. The attraction of this method is that it will allow any signal that can be stored to be played back. This technique is frequently used in synthesizers and musical instruments.

The N-sample signal can also be easily converted into a period-N signal, $x_p[k]$ and played using a TLU operation by simply repeating an N-sample finite sequence $x[k]$ a shown in Fig. 2.6. The periodic signal, $x_p[k]$, would have a z-transform of the form

$$X_p(z) = \sum_{n=0}^{\infty} X(z)z^{-nN} = \frac{X(z)}{1 - z^{-N}} \qquad (2.60)$$

Equation (2.60) can be seen to correspond to a periodic signal which can be implemented using an N-element circular shift register network shown in Fig. 2.6. A periodic signal generator can also be implemented using a *modulo*(N) address generator, also shown in Fig. 2.6.

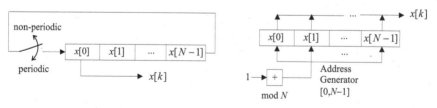

Figure 2.6 Direct implementation of a periodic time-series showing, on the left, an aperiodic (finite length) and periodic direct signal generator and, on the right, a randomly accessed direct periodic or aperiodic signal generator.

Here it is assumed that the data has been sequentially loaded into an N-word wavetable or shift-register. Data is read from the register in a circular *modulo*(n) manner.

One of the most important applications of direct synthesis is NCO/NCOM implementation. NCOs and NCOMs are commonly found in communications systems. The direct TLU approach uses memory to store the mapping

$$x[k] = \cos(\Phi[k]) \qquad (2.61)$$

For a constant frequency cosine, $\Phi[k] = \omega_0 k$, which implies that the phase angle can be generated using a modular adder. Using various combinations of adders and a lookup table, complicated phase modulated signals can be created, up to the precision of the table. An example of a table lookup cosine synthesizer is given in Fig. 2.7. The cosine data is stored in a table (usually a ROM) that has an M bit input with a W bit output (i.e., 2^M words of W bits each). When designing a digital sinusoid synthesizer it should be remembered that the two significant parameters are M and W as shown in Fig. 2.7. The selection of M and W is eventually reflected in the difference between the synthesized sinusoid and the ideal sinusoid and is easily quantified. The lookup table actually needs to store only one quadrant of the cosine data; the other quadrants are determined by symmetry arguments. The synthesizer is driven by a phase accumulator. For a fixed-frequency sinusoidal sequence, the output of the phase accumulator is the sequence $\Phi[k] = \omega_0 k$, or $\Phi[k] = (\Delta\phi)k$, where $\Delta\phi = 2\pi\theta$ can be thought of as the phase change per sample. Since the table has an M bit input and only

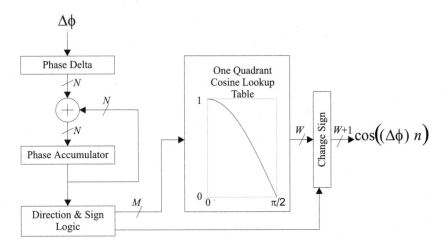

Figure 2.7 Block diagram of a typical table lookup cosine synthesizer.

contains one quadrant of the cosine, the phase accumulator must provide at least $M + 2$ bits of resolution, where the two most significant bits determine the quadrant and the next most significant M bits determine the TLU location (quadrant dependent). Since the precision of the phase accumulator determines the precision to which the frequency of the sinusoid generator can be set, it is often desirable to assign the data width of the phase accumulator, and $\Delta\phi$ (N bits in Fig. 2.7) be greater than $M + 2$ bits. In no case, however, would it be desirable to allow the phase accumulator to be smaller than $M + 2$ bits since this would result in under utilization of the cosine table. The precision to which the sinusoidal generator can be set, in terms of the digital frequency is given as

$$\left|\Delta\phi_{\text{min step}}\right| = \frac{1}{2^{N-1}} \tag{2.62}$$

where, again, $N \geq M + 2$.

The direct cosine synthesizer in Fig. 2.7 may be configured to synthesize a sinusoid at a single frequency by setting the $\Delta\phi$ parameter. In some systems it is desirable to change $\Delta\phi$ with respect to time. For example, in communications, a variety of modulation schemes, both analog and digital, rely on modulation of the phase parameter. A system such as that shown in Fig. 2.7 can be used to directly synthesize an IF (intermediate frequency) signal. Modulation schemes that rely exclusively on phase include analog phase modulation (PM) and frequency modulation (FM) and digital frequency shift keying (FSK) and m-ary phase shift keying (mPSK).

An implementation problem often attendant with the design of a direct digital signal synthesizer is called *beating*. Beating naturally occurs as the synthesized frequency approaches the Nyquist frequency. While the Nyquist limit represents a theoretical bound applicable to reconstructing a signal from its sample values, problems arise when synthesizing a facsimile of an analog signal from a set of predefined sample values. This can be illustrated using simulation to study the beating that occurs in attempting to synthesize a cosine sequence at a frequency that is slightly less than half the Nyquist frequency and another just less than the Nyquist frequency. The results of this simulation are shown in Fig. 2.8. Notice the aggressive beating (i.e., non-constant amplitude envelope) at a normalized frequency of $\theta = 63/128$, and the reduced beating for a lower frequency case at $\theta = 31/128$.

Figure 2.9 illustrates in greater detail how the beating phenomenon occurs. An "analog" sinusoid is plotted for $f = (63/128)f_s$. Each sample of a synthesized digital sinusoid at the same frequency [$f = (63/128)f_s$ as in Fig. 2.7] is shown in the graph as a vertical bar. Notice how the

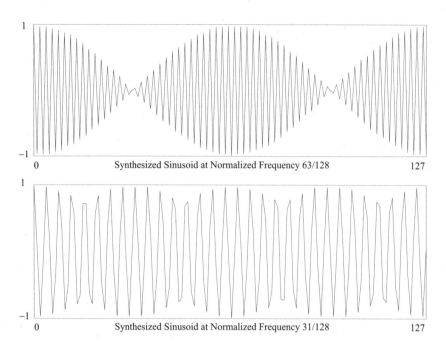

Figure 2.8 Beating effects in a direct digital synthesized sinusoid.

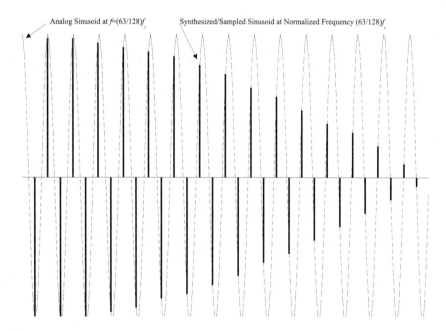

Figure 2.9 Detailed diagram of beating effects in a direct digital synthesized sinusoid.

magnitude of the samples decreases, and each sampled value is the same as that of the analog sinusoid at the sample instance. In Fig. 2.7, the sinusoid synthesized at the normalized frequency θ = 63/128 appears to have a beating frequency of 64 samples. The very same phenomenon can be observed when sampling an analog signal close to the Nyquist frequency. In many applications, beating may not be problematic, while in others it may be very significant. In those applications where the deep beating (see Fig. 2.9) may cause significant problems, the synthesized (or sampled) signal should be assigned a practical upper frequency limit of one-half of the Nyquist frequency (i.e., $|f| \leq f_s/4$ or $|\theta| \leq 1/4$). The effect of this limitation can be seen in Fig. 2.8, where a signal is synthesized at the normalized frequency θ = 31/128, which is just below one-half of the Nyquist frequency, exhibits only minor beating distortion.

Example 2.16: Qualitative Demonstration of Audio Beating Effects

Problem Statement. Demonstrate that the beating phenomenon illustrated in Figs. 2.8 and 2.9 can also be illustrated in the audio domain.

Analysis and Conclusions. The beating frequency of a synthesized sinusoid is related to how close the signal frequency is to the Nyquist frequency. The generated sinusoidal sequence is given as $x[k] = \cos(2\pi\theta k)$, where

$$x[k] = \cos 2\pi\left[\frac{1}{2} - \left(\frac{1}{2} - \theta\right)\right]k$$

$$= \left(\cos 2\pi\frac{1}{2}k\right)\left(\cos 2\pi\left[\theta - \frac{1}{2}\right]k\right) + \left(\sin 2\pi\frac{1}{2}k\right)\left(\sin 2\pi\left[\theta - \frac{1}{2}\right]k\right)$$

$$= (\cos \pi k)\left(\cos 2\pi\left|\theta - \frac{1}{2}\right|k\right)$$

since $\sin(\pi k) = 0$, and the cosine function is even. This expression explains the beating effect seen in Figs. 2.8 and 2.9. From the second term of the above product of cosines, the normalized frequency θ may be directly selected so that a specific beat frequency is produced.

Computer Study. Use the function "cossyn" contained in the s-file "cossyn.s" to synthesize a signal at the Nyquist frequency and two just below the Nyquist frequency. The "cossyn(f,n,m,w,pq)" function is defined in the s-file "cossyn.s". The parameter f is the normalized frequency of the sinusoid to be synthesize, n is the length of the signal to be generated, m is the number of address bits in the quarter quadrant lookup table (see M in Fig. 2.7), w is the width of the cosine lookup table (see W in Fig. 2.7), and pq, the phase quantization, is the width of the phase accumulator (see N in Fig. 2.7). The generated signals should be played through a soundboard using a sampling rate 11,025 Hz and an 8-bit data word.

```
>include "chap2\cossyn.s"
>x=cossyn(0.5,8192,6,6,16)          # Gen sinusoid at Nyquist freq
>y=cossyn(63/128,8192,6,6,16)       # Gen cos with beat @theta=1/64
>z=cossyn(4095/8192,8192,6,6,16)    # Gen cos with beat @theta=1/4096
>wavout(x) # Use lowest sampling rate
>wavout(y) # Use lowest sampling rate
>wavout(z) # Use lowest sampling rate
```

The first signal played is free of beating. The second signal played should have a rapid beating that makes it sound radically different from the first tone. The third signal played should have a relatively slow beating, sounding similar to the first tone except with the volume control being rapidly modulated.   ❖

Example 2.17: Direct Sinusoid Synthesis

Problem Statement. Suppose that a direct sinusoidal synthesizer, such as that shown in Fig. 2.7, operates at a clock rate of 100 MHz, and the size of the single quadrant lookup table is 64 words. Determine the minimum size of the phase accumulator to ensure 1 kHz frequency resolution. Experimentally determine the signal-to-noise ratio for a synthesized cosine at 1 MHz using the above assumptions and lookup table quantization levels of four, five, and six bits.

Analysis and Conclusions. The information given is $f_s = 100$ MHz and $f_s\,\theta_{\text{min step}} = 1$ kHz. Therefore,

$$\theta_{\text{min step}} = \frac{1\text{ kHz}}{100\text{ MHz}} = 10^{-5}$$

Using Eq. (2.62), N can be determined in such a way that $\theta_{\text{min step}}$ is less than or equal to 10^{-5}. It follows that:

$$N = \left(\log_2\left(\frac{2\pi}{\theta_{\text{min step}}}\right)\right) + 1 = 20.26$$

or, $N = 21$ bits.

Computer Study. Use the function "cossyn" contained in the s-file "cossyn.s" to synthesize the required signals.

```
>include "chap2\cossyn.s"
>x=mkcos(1,1/100,0,128)        # Generate the 'ideal' signal
>x4=cossyn(1/100,128,6,4,18)   # Synthesize the signal with w=4
>x5=cossyn(1/100,128,6,5,18)   # Synthesize the signal with w=5
>x6=cossyn(1/100,128,6,6,18)   # Synthesize the signal with w=6
>graph(x,x4,x5,x6)             # Graph results
>-10*log(sum((x-x4)^2)/128)    # Determine SNR in dB for w=4
23.1353
>-10*log(sum((x-x5)^2)/128)    # Determine SNR in dB for w=5
30.3982
```

```
>-10*log(sum((x-x6)^2)/128)   # Determine SNR in dB for w=6
38.0881
```
❖

The second method of signal synthesis, called *indirect*, relies on an algorithm to synthesize a signal. Using a general purpose DSP microprocessor or dedicated hardware, an algorithm is implemented which produces a desired time-series. For example, the z-transform of a causal decaying sine or cosine signal are given by

$$x_{\sin}[k] = a^k \sin(\omega_0 k)\, u[k]; \; x_{\cos}[k] = a^k \cos(\omega_0 k)\, u[k] \qquad (2.63)$$

where $a \in (0,1]$
$\quad \omega_0 = 2\pi f_0 / f_s$
$\quad f_0$ = frequency of oscillation
$\quad f_s$ = sampling frequency

For $a = 1$, a pure causal sinusoid results. The transform representation of the two generators are given by

$$X_{\sin}(z) = \frac{a\,\sin(\omega_0)z^{-1}}{1 - 2a\,\cos(\omega_0)z^{-1} + a^2 z^{-2}}$$

$$X_{\cos}(z) = \frac{1 - a\,\cos(\omega_0)z^{-1}}{1 - 2a\,\cos(\omega_0)z^{-1} + a^2 z^{-2}}$$

$$(2.64)$$

The second-order sinusoidal generators are interpreted in block diagram form in Fig. 2.10 and will be referred to in Chap. 14 as a *filter*. The z-transform of the signal found at locations $x[k]$ of Fig. 2.10 is given by

$$X(z) = \frac{1}{1 - 2a\,\cos(\omega_0)z^{-1} + a^2 z^{-2}} \qquad (2.65)$$

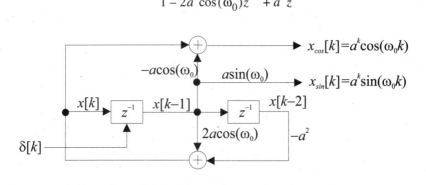

Figure 2.10 Indirect sinusoidal signal generator.

By linearly combining the signals $x[k]$, $x[k-1]$ and $x[k-2]$, sinusoids of various amplitudes and phase shifts can be created. Notice also that the system shown in Fig. 2.10 is being driven by a Kronecker delta function $\delta[k]$. The production mechanism is demonstrated in the next example.

Example 2.18: Signal Generation

Problem Statement. Produce a cosine and sine time-series using the indirect method.

Analysis and Conclusions. Equation (2.65) can be interpreted as the difference equation:

$$x[k] = 2a \, \cos(\omega_0)x[k-1] - a^2 x[k-2]$$

where it is assumed that $x[0] = 1$ (initial condition on the left-most shift register) and $x[-1] = 0$ (initial condition on the right-most shift register). This is equivalent to driving the second order signal generator with an impulse function. For $\omega_0 = 0.25\pi$ and $a = 1$, $2\cos(\omega_0) = 1.414$, the trace of the values of signal $x[k]$, produced by the algorithm, is given by

k	−1	0	1	2	3	4	5	6	7	8	9	
$x[k]$	0	1	1.414	1	0.414	0	−0.414	−1	−1.414	−0.414	0	...

It also follows that

$$x_{\cos}[k] = x[k] - a \, \cos(\omega_0)x[k-1]$$

$$x_{\sin}[k] = a \, \sin(\omega_0)x[k-1]$$

The creation of $x[k]$, $x_{\cos}[k]$, and $x_{\sin}[k]$, for $\omega_0 = \pi/4$ and $\omega_0 = \pi/16$ are shown in Fig. 2.11. Notice that the responses have a sinusoidal shape and that the quality of the synthesized signal improves with oversampling (i.e., $f_0 \ll f_s$).

Computer Study. The s-file "siggen.s" produces an N-sample time-series representation of $x[k]$, $x_{\cos}[k]$ and $x_{\sin}[k]$, for a given ω_0 using the function siggen(omega0,N) where omega0 is the normalized frequency of the sinusoid to be synthesized and N is the signal length.

```
>include "chap2\siggen.s"
>x=siggen(pi/4,16)
>x=siggen(pi/16,301)
```

The results are displayed in Fig. 2.11. The synthesized sinusoid can be evaluated by playing it through a soundboard. The following Siglab commands can

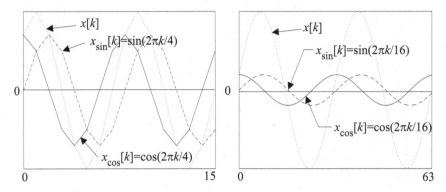

Figure 2.11 Indirect sinewave synthesis for $\omega_0 = \pi/4$ and $\pi/16$.

be used to play a 10,000-sample 345 Hz sinewave through a sixteen-bit sound-board with a sampling rate of 11,025 Hz.

```
>x=siggen(pi/16,10000)    # unit amplitude signal
>wavout(2^14*x)           # amplify by 2^14                      ❖
```

The indirect sinusoidal generator carries with it a caveat. First, it must comply with the requirements imposed by the sampling theorem, namely $2\pi\omega_0 \leq f_{Nyquist}$. For the case where $a = 1$, roundoff errors inherent to a digital system can cause the realized poles to fall outside the unit circle resulting in an unstable filter. As a result, a is generally chosen to be slightly interior to the unit circle and the generator is reset periodically before it has a chance to decay by an appreciable amount.

Example 2.19: Tone Generator

Problem Statement. Telephone touch tone dialers are based on a *dual-tone multifrequency* (DTMF) protocol. When the keys on a 3 × 4 keypad are pressed, the system generates two tones, one according to the column of the key pressed, and one according to the row of the key pressed. The frequencies of the tones used are summarized in Table 2.3. Implement the DTMF digitally.

Table 2.3   DTMF Tone Code Table

Key	1209 Hz	1336 Hz	1477 Hz
696 Hz	1	2	3
770 Hz	4	5	6
852 Hz	7	8	9
941 Hz	*	0	#

Analysis and Results. Since the DTMF tone is the linear combination of two separate tones, the DTMF tone generator can be constructed using the encoding shown Table 2.3. The critical frequencies, namely $f_L \in \{696, 770, 852, 941\}$ Hz and $f_U \in \{1209, 1336, 1477\}$ Hz, can be transmitted over a "plain old telephone system" (POTS) line.

Computer Study. A DTMF "five" is produced by superimposing tones at 770 Hz and 1336 Hz. This DTMF signal at a sampling rate of 5 kSa/s is generated by the sequence of Siglab commands shown below. Notice that the function `dtmf(f1,f2,fs,n)` is entered from the command line. The generated signal is shown in Fig. 2.12.

```
>function dtmf(f1,f2,fₛ,n)
>>dtmf=mkcos(1,f1/fₛ,0,n)+mkcos(1,f2/fₛ,0,n)
>>end
>graph(dtmf(770,1336,5000,1001))
```

The synthesized DTMF tone for the number five can be evaluated by playing it through a soundboard. The following command can be used to play the tone through a sixteen-bit soundboard with a sampling rate of 11,025Hz.

```
>wavout(2^14* dtmf(770,1336,11025,10000)) # amplify by 2^14   ❖
```

2.14 Summary

In Chap. 2, a fundamental discrete-time signal analysis tool, called the z-transform, was developed along with mechanisms to generate discrete-time signals. The z-transform is used in the study of discrete-time signals in the identical manner that the Laplace transform is used in study of continuous-time signals. The z-transform of primitive signals are generally cataloged along with their ROCs in tables. Returning from the transform-domain to the discrete-time-domain can be accomplished using long-division or through a Heaviside partial-fraction expansion. The z-transform will be extensively used throughout the remaining chapters as both a modeling and analysis tool.

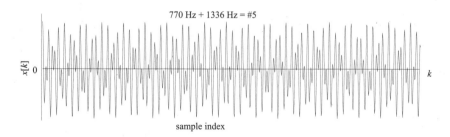

Figure 2.12 DTMF tone complex for five.

2.15 Self-Study Problems

1. z-Transform   Derive the z-transform for the following signals and compute their ROC.

$\delta[k - m]$

$(1/2)^k u[k]$

$(1/2)^k \cos(2\pi k/10)u[k]$

What conditions will need to be satisfied in order to use the initial and final value theorem to numerically check your result?

2. Modulation Theorem   Verify the modulation theorem found in Table 2.2.

3. Inverse z-Transform   Determine $x[k]$ for the following (assume all signals are causal):

$X(z) = 1 + z^{-1} + 6z^{-5} + 8z^{-10} + 4z^{-12}$

$X(z) = 1 - z^{-1} + 6z^{-5} - 8z^{-10} + 4z^{-12}$

$X(z) = 1/(z - 0.5)^3$

$X(z) = z^3/(z - 0.5)^3$

4. Inverse z-Transform   Determine $x[k]$ explicitly for $k = [0,1,2]$ ($x[k]$ causal) and graph $x[k]$ for all k:

$X_1(z) = (1 + z^{-3})/z(1 - 0.5z^{-1})(1 - 0.25z^{-1})$

$X_2(z) = 4(1 - z^{-1} + z^{-2})/(2 + 5z^{-1} + 2z^{-2})$

$X_3(z) = (1 + z^{-1})^3/(1 - z^{-1})^3$

$X_4(z) = (z + 1)^4/(z-1)^2(z^2 + 1)$

2.16 Reference

1. Churchill, R.V., Brown, J.W. *Complex Variables and Applications*, 5/e. New York: McGraw-Hill, 1990.

3

Frequency Domain Representation of Discrete-Time Signals

3.1 Introduction

Laplace and Fourier transforms are intrinsic to the study of continuous-time signals. In Chap. 2, the z-transform was developed as a discrete-time counterpart of the Laplace transform. In Chap. 3, the discrete-time version of the continuous-time Fourier transform (CTFT) will be introduced. Recall that the CTFT pair is given by the analysis equation,

$$X(e^{j\omega}) = \int_{-\infty}^{\infty} x(t)e^{-j\omega t}dt \qquad (3.1)$$

and the *synthesis equation*,

$$x(t) = \frac{1}{2\pi} \int_{-\infty}^{\infty} X(e^{j\omega})e^{j\omega t}d\omega \qquad (3.2)$$

where ω is called the *analog frequency* and is measured in radians per second, and $X(e^{j\omega})$ is called the spectrum of $x(t)$. The Fourier transform exists if and only if $x(t)$ is absolutely integrable, i.e.,

$$\int_{-\infty}^{\infty} |x(t)|dt < \infty \qquad (3.2a)$$

The Fourier transform, while being a fundamentally important scientific tool for centuries, displayed a systemic weakness once we entered the digital computer era. Notice that the production of the spectrum of $x(t)$ requires the knowledge of the signal over all time. This is an unrealistic assumption in the context of digital computing. A modification of the Fourier transform, called the *continuous-time Fourier series* (CTFS), simplified the computational problem by restricting the study to only periodic continuous-time signals $x_p(t)$ where $x_p(t) = x_p(t + T)$ for all time t. The CTFS series *analysis equation* is given as

$$c_n = \frac{1}{T}\int_0^T x(t)e^{-jn\omega_0 t}\,dt \tag{3.3}$$

where ω_0 is called the *fundamental frequency* and is given by $\omega_0 = 1/T$, c_n is the nth harmonic and corresponds to the signal activity located at frequency $f_n = n/T$, or $\omega_0 = 2\pi n/T$. The existence of the Fourier series is governed by the *Dirichlet conditions,* which must be true in order for the Fourier series to exist. The Dirichlet conditions are:

1. The function $x_p(t)$ must be absolutely integrable over any period; in other words,

$$\int_t^{t+T} |x(\tau)|\,d\tau < \infty$$

   for all finite t.

2. The function $x(t)$ must have a finite number of maxima and minima over any period.

3. The function $x(t)$ must have a finite number of discontinuities over any period.

The synthesis equation is given by

$$x_p(t) = \sum_{n=-\infty}^{\infty} c_n e^{jn\omega_0 t} \tag{3.4}$$

While the limits of integration are finite compared to infinite limits of a Fourier transform, the production of c_n would nevertheless require a tremendous effort on the part of a general-purpose digital computer. In addition, an infinite number of harmonics c_n would need to be computed for each $x_p(t)$.

The development of the *discrete Fourier transform*, or *DFT* allowed the Fourier transform to become a useful tool using a digital computer. The impact of the DFT was to replace integration and summation over infinite bounds with summation over finite bounds, thus making the DFT much more practical than the CTFT and CTFS for implementation in digital computing environments. However, despite the algorithmic improvements offered by the DFT, the DFT originally remained too computationally expensive for use in most problems. This changed in 1965 when J.W. Cooley and J.W. Tukey introduced the now celebrated *fast Fourier transform* or *FFT*.[1] The FFT implementation of the DFT radically reduced the computational complexity of the DFT, thus allowing the technique to be broadly applied using the available general-purpose computing hardware. For over a quarter century the FFT has been a critical element in the study of speech, communication, video, biomedical, and a host of other signal classes. In this chapter a collection of spectral analysis tools, including the FFT, will be developed for the study of discrete-time signals.

3.2 Discrete-Time Fourier Transforms

The discrete-time equivalent of a Fourier transform is the *discrete-time Fourier transform* (DTFT). The DTFT defines a method of computing a Fourier representation of a discrete-time signal $x[k]$. The DTFT *analysis equation* satisfies

$$X(e^{j\varpi}) = \sum_{k=-\infty}^{\infty} x[k]e^{-jk\varpi} \qquad (3.5)$$

where $\varpi \in [-\pi, \pi)$ is called the *normalized digital frequency*. Normalization is defined with respect to the sampling frequency, f_s, with the actual analog frequency given by ω, so that

$$\varpi = \frac{\omega}{f_s} = \omega T_s \qquad (3.6)$$

The DTFT, defined by Eq. (3.5), exists if $x[k]$ is absolutely summable or of finite energy. That is, $X(e^{j\varpi})$ exists if

$$\sum_{k=-\infty}^{\infty} |x[k]| < \infty \qquad (3.7)$$

or

$$\sum_{k=-\infty}^{\infty} |x[k]|^2 < \infty \tag{3.8}$$

If the signal $x[k]$ is known analytically, the bounds defined by Eqs. (3.7) or (3.8) can often be computed analytically. If $x[k]$ is an arbitrary time-series, it is unrealistic to assume that either infinite sum bounds can be computed digitally. Nevertheless, if $X(e^{j\varpi})$ exists, the resulting $X(e^{j\varpi})$ is called the *spectrum* of the time-series $x[k]$. As the sample period $T_s \to 0$ ($f_s = 1/T_s$, therefore $f_s \to \infty$), the summation found in Eq. (3.5) will converge to the integral equation

$$\lim_{T_s \to 0} X(e^{j\varpi}) = \lim_{T_s \to 0} \sum_{k=-\infty}^{\infty} x[k]e^{-jk\varpi} \to \int_{-\infty}^{\infty} x(t)e^{-j\omega t}dt = X(e^{j\omega}) \tag{3.9}$$

which is recognized to be the CTFT of $x(t)$.

The DTFT is also closely related to the z-transform. In particular, the two-sided z-transform of a time-series $x[k]$ was given by [see Eq. (2.16)]

$$X(z) = \sum_{k=-\infty}^{\infty} x[k]z^{-k} \tag{3.10}$$

Evaluating Eq. (3.10) along the circular arc lying on the periphery of the unit circle in the z-plane {i.e., $z = e^{j\varpi}$, $\varpi \in [-\pi,\pi)$}, results in

$$X(z)\Big|_{z=e^{j\varpi}} = \sum_{k=-\infty}^{\infty} x[k](e^{-jk\varpi}) = X(e^{j\varpi}) \tag{3.11}$$

Equation (3.11) is recognized to be a DTFT given in Eq. (3.5). Technically, it should be noted, that the arc $z = e^{j\varpi}$ must belong to the region of convergence of $X(z)$ if $X(z)$ is guaranteed to exist. This, for example, would preclude the use of this method to compute the DTFT of a causal sinusoid, when its z-transform is given by

$$\sin(bkT_s)u(kT_s) \overset{z}{\leftrightarrow} \frac{z\sin(bT_s)}{z^2 - 2\cos(bT_s) + 1}$$

with an ROC given by $|z| > 1$. Hence the arc $z = e^{j\varpi}$ would not belong to the ROC of $X(z)$.

Example 3.1: DTFT of an Exponential Time-Series

Problem Statement. What is the DTFT of the causal time-series $x[k] = a^k u[k]$?

Analysis and Conclusions. The z-transform of $x[k] = a^k u[k]$ is known to be given by $X(z) = z/(z - a)$ if $|a| < 1$. The ROC is given by $|z| > |a|$. If $z = e^{j\varpi}$ and $|a| < 1$, then it follows from Eq. (3.11) that

$$X(e^{j\varpi}) = \left. \frac{z}{z - a} \right|_{z = e^{j\varpi}} = \frac{e^{j\varpi}}{e^{j\varpi} - a}$$

which is interpreted in Fig. 3.1. At the normalized frequency given by $\varpi = 2L\pi + \omega_0$, it follows that $X(e^{j\varpi}) = X(e^{j(2\pi L + \omega_0)}) = X(e^{j\omega_0})$ for all integer values of L. The maximum values of $X(e^{j\varpi})$ are seen to occur at frequencies $\varpi \in (-\infty, \ldots, -2\pi, 0, 2\pi, \ldots, \infty)$ as shown in Fig. 3.1. Observe that the DTFT spectrum is periodic with a normalized frequency of 2π. The baseband spectrum is defined over a normalized frequency range $\varpi \in [-\pi, \pi)$, which corresponds to a real frequency range of $f \in [-f_s/2, f_s/2)$ (i.e., $\pm$ Nyquist frequency).

Computing $X(e^{j\varpi})$ using a general-purpose digital computer would require that

$$X(e^{j\varpi}) = \sum_{k = -\infty}^{\infty} a^k e^{-jk\varpi} u[k] = \sum_{k = 0}^{\infty} a^k e^{-jk\varpi}$$

be numerically evaluated. This is obviously a far more challenging problem than simply evaluating the known z-transform $X(z)$, in closed form, along the circular arc $z = e^{j\varpi}$. ❖

Example 3.2: DTFT of a Discrete-Time Rectangular Pulse

A discrete-time rectangular pulse $r[k]$ symmetrically distributed about $k = 0$ is shown in Fig. 3.2 and can be mathematically defined as $r[k] = x[k - K/2]$ where $x[k] = \text{rect}(k/K)$. The finite length time-series $x[k]$ is absolutely summable which ensures that the DTFT exists. The DTFT satisfies

$$X(e^{j\varpi}) = \sum_{k = -\infty}^{\infty} x[k] e^{-jk\varpi} = \sum_{k = -K/2}^{K/2} e^{-jk\varpi} = e^{-j\varpi K/2} \sum_{m = 0}^{K} e^{jm\varpi}$$

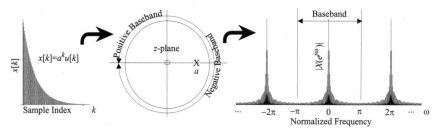

Figure 3.1 Discrete-time Fourier transform (DTFT) of $x[k] = a^k u[k]$ showing periodic behavior.

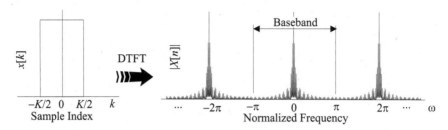

Figure 3.2   DTFT of a symmetric rectangular pulse.

The finite sum can be defined in terms of

$$\sum_{n=0}^{N} x^n = \sum_{n=0}^{\infty} x^n - \sum_{n=N+1}^{\infty} x^n = \sum_{n=0}^{\infty} x^n - x^{N+1} \sum_{n=0}^{\infty} x^n = \frac{1}{1-x} - x^{N+1}\frac{1}{1-x} = \frac{1-x^{N+1}}{1-x}$$

The DTFT equation can therefore be expressed as

$$X(e^{j\omega}) = e^{-j\omega(K/2)}\frac{(1-e^{j\omega(K+1)})}{(1-e^{j\omega})} = \frac{\sin[(\omega(K+1))/2]}{\sin(\omega/2)}$$

which is seen to have a periodically repeated $\sin(x)/x$ envelope with a period of 2π radians per second.

Computer Study. The s-file "ex3-2.s" contains a procedure to compute and graph the magnitude DTFT frequency response of the rectangular pulse $x[k]$ using the format "dtft(K)," where K governs the width of the pulse and is measured in sample instances. For $K = 7$, the magnitude frequency response is computed over the frequency range $f \in [-9f_N, 9f_N]$ where $f_N = f_s/2$. The spectrum is displayed in Fig. 3.2.

```
>include "chap3\ex3-2.s"
>dtft(7)
```
❖

The example emphasizes the fact that the DTFT spectrum is periodic with a normalized period 2π and is assumed to be symmetrically distributed across the baseband range $\omega_b \in [-\pi,\pi)$. The normalized baseband range $\omega_b \in [-\pi,\pi)$ corresponds to a real baseband range $f_b \in [-f_s/2, f_s/2)$. In general, due to periodicity, an arbitrary frequency f is mapped into a baseband frequency under $f_b \equiv f \bmod(f_s)$.

The mapping of the spectrum of a signal back into the discrete-time domain is defined by the *inverse discrete-time Fourier transform*, which is given by

$$x[k] = \frac{1}{2\pi}\int_{-\pi}^{\pi} X(e^{j\omega k})d\omega \qquad (3.12)$$

which is also called the *synthesis equation*. The limits of integration, namely $\omega \in [-\pi, \pi)$, reflect the normalized Nyquist restriction of the DTFT baseband spectrum. That is, from Shannon's sampling theorem, it is assumed that all the signal energy in the frequency domain is restricted to the normalized baseband range $\omega \in [-\pi, \pi)$. The mathematical production of energy outside this range is purely artificial and is called an *artifact*.

Many of the properties of the CTFT and CTFS carry over to the discrete-time Fourier transform. The more important properties are summarized in Table 3.1.

One of the more interesting properties listed in Table 3.1 is *Parseval's theorem* for the DTFT. It may be recalled that the energy found in a continuous-time signal can be equated to the energy distributed in

Table 3.1   Properties of a Discrete-Time Fourier Transform of a Discrete-Time Signal

Discrete-time series	Discrete-time Fourier transform	Remarks
$\displaystyle\sum_{m=0}^{L} a_m x_m[k]$	$\displaystyle\sum_{m=0}^{L} a_m X_m(e^{j\omega})$	Linearity
$x[k-q]$	$X(e^{j\omega})e^{-jq\omega}$	Time-shift
$x^*[-k]$	$X^*(e^{j\omega})$	Time-reversal
$x[k]e^{j\omega_0}$	$X[e^{j(\omega-\omega_0)}]$	Modulation, ω_0 real
$\displaystyle\sum_{k=-\infty}^{\infty} x[k]y^*[k]$	$\displaystyle\frac{1}{2\pi}\int_{-\pi}^{\pi} X(e^{j\omega})Y^*(e^{j\omega})\,d\omega$	Parseval's theorem
$x_{\text{even}}[k]$	$X(e^{j\omega}) = x_{\text{even}}[0] + 2\displaystyle\sum_{k=0}^{\infty} x_{\text{even}}[k]\cos(k\omega)$	Even time-series
$x_{\text{odd}}[k]$	$X(e^{j\omega}) = x_{\text{odd}}[0] - j2\displaystyle\sum_{k=0}^{\infty} x_{\text{odd}}[k]\sin(k\omega)$	Odd time-series

the frequency domain using Parseval's theorem for the CTFT. The DTFT version of Parseval's theorem plays the same role and is given by

$$\sum_{k=-\infty}^{\infty} |x[k]|^2 = \frac{1}{2\pi} \int_{-\pi}^{\pi} |X(e^{j\omega k})|^2 d\omega \qquad (3.13)$$

which states that the energy in a time-series $x[k]$ can be also computed given knowledge of the signal's DTFT. The term $|X(e^{j\omega})|^2$ is referred to as the *energy spectrum* and represents the distribution of energy in the frequency domain on a per harmonic basis.

While the DTFT theory is fundamentally important from a conceptual level, it does possess a major drawback. It is apparent that the DTFT is defined in terms of an infinite summation, and produces a spectrum that is defined at all points on the interval $[-\pi,\pi)$, which has an infinite number of points. Furthermore, computing the inverse DTFT involves integration. What is obviously needed is a more efficient and compact spectral computing formula.

3.3 Discrete-Time Fourier Series (DTFS)

The DTFT can be applied to an arbitrary time-series $x[k]$ which is absolutely summable or of finite energy. This, at first glance, seems to be a realistic constraint. However, many important signals do not satisfy these conditions. A unit step, for example, is neither absolutely summable nor of finite energy. Periodic signals, such as a sinusoid, would also not be absolutely summable or of finite energy. Therefore, the DTFT cannot be directly used to compute the spectral representation of periodic or other important signals without modification. The DTFT of a time-series of period N, denoted $x_p[k]$, can be expressed in terms of a modification of Eq. (3.5) which exploits known properties of a periodic signal.

Recall that a CTFS representation of a signal $x(t)$, having a period T, was based on the use of complex exponential *periodic basis functions* of the form

$$\phi(t) = e^{-j\omega_0 t} \qquad (3.14)$$

where $\omega_0 = 2\pi/T$. For discrete-time signals, a discrete-time exponential basis function is given in the compact *W-notation* by

$$W_N^m = e^{-j2\pi m/N} \qquad (3.15)$$

The complex exponential W_N^{-nk} is periodic in *both* k and n with period N. That is,

$$W_N^{-nk} = e^{j\left(\frac{2\omega nk}{N}\right)} = e^{j\left\{\frac{2\pi[(nk)\bmod(N)]}{N}\right\}} \qquad (3.16)$$

Using the CTFS as motivation, assume that the periodic time-series $x_p[k]$ can be expressed as a linear combination of periodic discrete-time basis functions and is given by

$$x_p[k] = \sum_n c_n W_N^{-nk} \qquad (3.17)$$

where the set of coefficients c_n are called the *spectrum* of the periodic time-series $x_p[k]$, and the bounds of summation are not yet defined. Assume now that $x_p[k]$ is periodic with period N and is sampled above the Nyquist frequency. From Eq. (3.16) it is clear that only N contiguous values of n are needed to define a unique set of basis functions. For $n \in \{0,1,2,...,N-1\}$, Eq. (3.17) can be rewritten as the discrete-time Fourier series (DTFS) synthesis equation

$$x_p[k] = \sum_{n=0}^{N-1} c_n W_N^{-nk} \qquad (3.18)$$

The restriction for $n \in \{0,1,2,...,N-1\}$ can be justified by appealing to the *Nyquist sampling theorem*. Recall that it has been assumed that the highest frequency component in a time-series to be transformed is located at B Hz. This, in turn, establishes the Nyquist frequency bound $-f_s/2 < -B \le f \le B < f_s/2$. A periodic, and therefore noncausal, signal having a period N would repeat every T_0 seconds, where

$$T_0 = \frac{N}{f_s} = N T_s \qquad (3.19)$$

The frequency $f_0 = 1/T_0$, or $\omega_0 = 2\pi f_0$, is called the *fundamental* or *first harmonic* frequency. The frequency $k\omega_0 = 2\pi k f_0$ is called the kth *harmonic*. The highest harmonic would be the $(N-1)$th harmonic with the next harmonic, namely the Nth harmonic, corresponding to the sample frequency $f_s = N/T_0 = 1/T_s$. This states that the highest harmonic which can contain signal energy, based on Shannon's sampling theorem, is the $(N-1)$th. This relationship between harmonic index and corresponding baseband (real) frequency can be further developed. A periodic time-series, $x_p[k]$, namely a cosine oscillating at the mth harmonic frequency, can be expressed as

$$x[k] = \cos\left(\frac{2\pi kmf}{N}\right) = \text{Re}(W_N^{km}) = \text{Re}(e^{-j2\pi km/N}) \qquad (3.20)$$

which is seen to complete m cycles in N samples. A cosine wave at the Nyquist frequency $f_s/2$ Hz corresponds to the time-series

$$x_p[k] = \cos(\pi k) = \text{Re}(W_N^{kN/2}) = \text{Re}(e^{-j\pi k}) \qquad (3.21)$$

The first $N/2$ harmonics, namely those indexed by $n \in \{0,1,2,\dots,(N/2)$ $-1\}$, correspond to mappings of W_N^{nk} to the positive frequency arc in the z-plane shown in Fig. 3.3. The next $N/2$ harmonics, namely those indexed over $n \in \{N/2,\dots,N-1\}$, correspond to the negative frequency arc shown in Fig. 3.3. According to Eq. (3.16), the nth nonbaseband harmonic (i.e., $|n| > N/2$) is equivalent, or congruent to, the mth baseband harmonic where $m \equiv n \mod(N)$ for $m \in \{-N/2,\dots,N/2-1\}$. Therefore, Eq. (3.18) provides a formula for reconstructing a periodic time-series based on knowledge of the signal's spectrum. The problem is producing the spectrum (i.e., computing coefficients c_n).

The DTFS coefficients can be defined in terms of Eq. (3.18) by multiplying both sides of the equation by the complex exponential W_N^{mk} to form

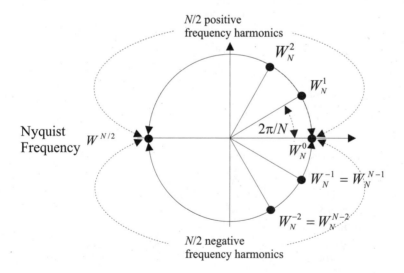

Figure 3.3 Interpretation of harmonic indices relative to the unit circle.

$$x_p[k]W_N^{mk} = \sum_{n=0}^{N-1} c_n W_N^{(m-n)k} \tag{3.22}$$

Summing both sides of Eq. (3.22) over $k \in \{0,1,2,\ldots,N-1\}$, gives

$$\sum_{k=0}^{N-1} x_p[k]W_N^{mk} = \sum_{k=0}^{N-1}\sum_{n=0}^{N-1} c_n W_N^{(m-n)k} = \sum_{n=0}^{N-1} c_n \sum_{k=0}^{N-1} W_N^{(m-n)k} \tag{3.23}$$

The rightmost summation found in Eq. (3.23) satisfies

$$\sum_{n=0}^{N-1} W_N^{(m-n)k} = \begin{cases} N & \text{if } m-n \in \{0, \pm N, \pm 2N, \ldots\} \\ 0 & \text{Otherwise} \end{cases} \tag{3.24}$$

Evaluating Eq. (3.23) for each $m \in \{0,1,2,\ldots,N-1\}$, one obtains the DTFS analysis equation

$$c_n = \frac{1}{N}\sum_{k=0}^{N-1} x[k]W_N^{kn} \tag{3.25}$$

The analysis equation can be used to compute the spectrum of a periodic time-series $x_p[k]$ as the sum of complex products.

Example 3.3: Discrete-Time Fourier Series

Problem Statement. Consider the periodic sinusoidal signal given by

$$x_p[k] = \sin\left(\frac{2\pi Kk}{N}\right) = \frac{1}{2j}\left(e^{j2\pi Kk/N} - e^{-j2\pi Kk/N}\right)$$

where the rational fraction $r = K/N$ is bounded by $0 \le r = K/N < 1/2$. What is the DTFT of $x_p[k]$?

Analysis and Conclusions: The DTFT of $x_p[k]$ is given by

$$c_n = \frac{1}{N}\sum_{k=0}^{N-1} x_p[k]W_N^{nk} = \frac{1}{N}\frac{1}{2j}\sum_{k=0}^{N-1} e^{-j2\pi nk/N}\left(e^{j2\pi Kk/N} - e^{-j2\pi Kk/N}\right)$$

$$= \frac{1}{N}\frac{1}{2j}\sum_{k=0}^{N-1}\left(e^{j2\pi k(K-n)/N} - e^{-j2\pi k(K+n)/N}\right)$$

which has two non-zero cases [see Eq. (3.24)]. First, if $n = K$, then $c_K = 1/2jN$. Next, if $k = -K$ or equivalently, $k = N-K$, then $c_{-K} = c_{N-K} = -1/2jN$. Also, observe that $c_K = -c_{-K}$. The spectrum is graphically interpreted in Fig. 3.4 in the complex frequency domain. ❖

The properties of the DTFS follow from those developed for the DTFT. They are summarized in Table 3.2 for a periodic time-series $x_p[k]$.

The DTFS is seen to have a number of practical advantages over the DTFT. The most significant of these is that the infinite summation in the DTFT [Eq. (3.5)] is replaced by a finite sum in the DTFS [Eq. (3.25)], and that the integration in the inverse DTFT [Eq. (3.12)] is replaced by a finite sum in the DTFS [Eq. (3.17)]. This means that the DTFS and its inverse can be computed using a general-purpose digital

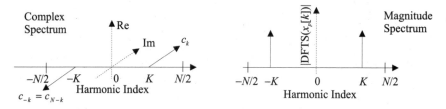

Figure 3.4 Spectrum of a discrete-time periodic sine wave.

Table 3.2   Properties of a Discrete-Time Fourier Series of a Periodic Discrete-Time Signal

Discrete-time series	Discrete-time Fourier transform	Remarks
$x_p[k] = \displaystyle\sum_{m=0}^{L} a_m x_m[k]$	$c_n = \displaystyle\sum_{m=0}^{L} a_m c_{m,n}$	Linearity
$x_p'[k] = x_p[k-q]$	$c_n' = c_n W_N^{kq}$	Time-shift
$x_p'[k] = x_p^*[k]$	$c_n' = c_{-n}^*$	Complex conjugate
$x_p'[k] = x_p^*[-k]$	$c_n' = c_n^*$	Time-reversal
$x_p'[k] = x[k] W_N^{rk}$	$c_n' = c_{n+r}$	Modulation
$X_{\text{even}}[k]$	$c_n = \text{Re}(c_n)$; real part	Even time-series
$x_{\text{odd}}[k]$	$c_n = \text{Im}(c_n)$; imaginary part	Odd time-series

computer. However, the DTFS requires that the signal under study be periodic. Technically this means that the time-series $x[k]$ must be known for all time, and therefore, be noncausal.

The DTFT and DTFS are intimately interrelated if the signal under study is truly periodic. The linkage between the continuous spectrum of a DTFT and discrete spectrum of a DTFS will be shown to be similar to the relationship existing between a continuous-time signal and a discretely sampled time-series. Recall, from Eq. (1.20), that the discrete-time process can be returned to a continuous-time domain using a $\sin(x)/x$-type interpolator. To derive the relationship between the DTFT and DTFS, note that the z-transform of an N sample section of a periodic time-series $x_p[k]$, is given by

$$X(z) = \sum_{k=0}^{N-1} x_p[k] z^{-k} \tag{3.26}$$

Since $x_p[k]$ is periodic, it has a DTFS given by $X[n]$, and

$$x_p[k] = \sum_{n=0}^{N-1} X[n] W_N^{-kn} \tag{3.27}$$

Upon substituting $x_p[k]$ into Eq. (3.26), one obtains

$$X(z) = \sum_{k=0}^{N-1} \left(\sum_{n=0}^{N-1} X[n] W_N^{-kn} \right) z^{-k}$$

$$= \sum_{n=0}^{N-1} X[n] \left(\sum_{k=0}^{N-1} W_N^{-kn} z^{-k} \right)$$

$$= (1 - z^{-N}) \sum_{n=0}^{N-1} \left(\frac{X[n]}{1 - W_N^{-n} z^{-1}} \right) \tag{3.28}$$

Evaluating $X(z)$ along the periphery of the unit circle in the z-plane (i.e., $z = e^{j\omega}$, for $\omega \in [-\pi, \pi)$) results in

$$X(z) \Big|_{z=e^{j\omega}} = X(e^{j\omega}) = \sum_{n=0}^{N-1} X[n] \Phi_N \left(\omega - \frac{2\pi n}{N} \right) \tag{3.29}$$

where

$$\Phi_N(\omega) = \frac{\sin\left(\dfrac{\omega N}{2}\right)}{\sin\left(\dfrac{\omega}{2}\right)} = e^{-j\omega[(N-1)/2]}e^{-j\pi n[1-1/N]} \tag{3.30}$$

Observe that $\Phi_N(\omega)$ has a $\sin(x)/x$-type envelope like that found in the Shannon interpolator [Eq. (1.20)]. It can be seen that $\Phi_N(\omega)$ contains nulls at the discrete frequencies $\omega = 2\pi k/N$ for all $k \neq 0 \bmod(N)$. Notice also that $\Phi_N(\omega)$ serves the same purpose as Shannon's interpolator in that it interpolates a discrete spectrum of a DTFS into a continuous spectrum of a DTFT. The spectral interpolator, $\Phi_N(\omega)$, is shown in Fig. 3.5.

The study of discrete-time aperiodic and periodic signals in the frequency-domain is fundamentally important. Fortunately, many important primitive signals have known precomputed and tabled DTFT or DTFS. However, there is a fundamental limitation to these methods. If only the sample values of signal time-series are known, as would be the case if an arbitrary, possibly noisy signal is sampled using an ADC, then computing the DTFS is technically questionable since the signal may be aperiodic. As a result, producing a spectrum may, in fact, require the use of the infinite sum DTFT. From a practical perspective, the digital computation of a spectrum would have to be based on the analysis of an arbitrary N sample time-series. In the next section, the important *discrete Fourier transform* (DFT) will be introduced as an extension of the DTFS. It will shown that DFT can be computed by a general-purpose digital computer, and as such has become an important spectral analysis tool.

3.4 Discrete Fourier Transform

In general, it is assumed that only signals having a finite number of sample values can be processed by a general-purpose digital computer.

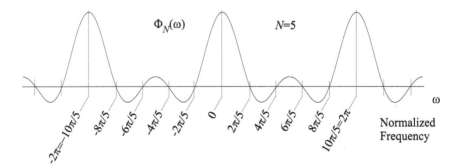

Figure 3.5   Interpolation of a DTFS into a DTFT.

Of the DTFT and DTFS, only the DTFS is defined over a finite number of samples. However a DTFS requires that the signal being transformed be periodic. Unfortunately it is impossible to claim that a signal is periodic without having knowledge of its infinite time history. These realities motivate the important *discrete Fourier transform* or *DFT*.

The DFT is a mapping of an N sample time-series (possibly complex) into the frequency domain containing N harmonics. Since the harmonics are generally complex, the DFT is a mapping of $\mathbf{C}^N \leftrightarrow \mathbf{C}^N$. The DFT of an N sample time-series, $x_N[k]$, is

$$X[n] = \sum_{k=0}^{N-1} x_N[k] W_N^{nk} \tag{3.31}$$

for $n \in \{0,1,2,...,N\text{--}1\}$, where $X[n]$ is called the nth harmonic. The sum in the DFT analysis equation [Eq. (3.31)] exists if all of the sample values of $x_N[k]$ are bounded.

The inverse transform, called the inverse DFT (IDFT) or the DFT synthesis equation, is

$$x_N[k] = \frac{1}{N} \sum_{n=0}^{N-1} X[n] W_N^{-nk} \tag{3.32}$$

for $k \in \{0,1,2,...,N\text{--}1\}$. Like the analysis equation, the sum in the inverse DFT [Eq. (3.32)] exists if all of the sample values $X[n]$ are bounded. Because the DFT and inverse DFT are defined by finite sums of weighted sample values, both may be implemented directly with a digital computer.

What differentiates the DFT from the DTFS are

- the DFT *assumes* signal periodicity while the DTFS requires periodicity, and

- the DFT scales the synthesis equation by $1/N$, whereas the DTFS scales the analysis equation.

The first point may seem unimportant but provides justification for the use of a finite summation transform. This is a subtle point; the assumption of periodicity by the DFT will provide an explanation for some phenomena seen in the transform domain. An illustrative case will be discussed in Example 3.8. The second difference is one of efficiency. While it is apparent that, in both the case of the DFT/IDFT and the case of the DTFS/IDTFS, the $1/N$ scaling could be moved to the opposite transform, convention places the scaling factors as given in this text. In practice, the Fourier transform of a time-series is taken far more often than the inverse transform. Therefore, the digitally inefficient operation of division (i.e., $1/N$) is deferred to the less used mapping [Eq. (3.32)]. However, one must recognize that the harmonics produced by a DFT are N times larger than those produced by a DTFS when a DTFS exists.

Example 3.4: Discrete Fourier Transform

Problem Statement. An arbitrary discrete-time signal $x[k]$ is assumed to contain an N sample section, $x_N[k]$, given by

$$x_N[k] = \sin\left(\frac{2\pi Kk}{N}\right) = \frac{1}{2j}(e^{j2\pi Kk/N} - e^{-j2\pi Kk/N})$$

for $k \in \{0,1,2,...,N-1\}$. What is the DFT of $x_N[k]$?

Analysis and Conclusions. The use of the DFT assumes that $x_N[k]$ extends periodically in time with period N. The DFT of $x_N[k]$ is formally given by

$$X[n] = \sum_{k=0}^{N-1} x_N[k]W_N^{nk} = \frac{1}{2j}\sum_{k=0}^{N-1} e^{-j2\pi nk/N}(e^{j2\pi Kk/N} - e^{-j2\pi Kk/N})$$

$$= \frac{1}{2j}\sum_{k=0}^{N-1}(e^{j2\pi k(K-n)/N} - e^{-j2\pi k(K+n)/N})$$

which has two non-zero cases as discussed in Example 3.3. First, if $n = K$, then $X[K] = N/2j$. Next, if $n = -K$, or equivalently $n = N - K$, then $X[-K] = X[N-K] = -N/2j$. Also observe that $X[-K] = -X[K]$ and that the DFT is N times larger than that produced by a DTFS of the same periodic sine wave studied in Example 3.3.   ❖

The advantage of the DFT is its ability to directly compute a spectrum from N sample values of $x[k]$, say $x_N[k]$, without establishing any mathematical properties of $x_N[k]$ or $X[\omega]$. Thus a DFT can be computed using the complex-multiply intensive algorithm given in Eq. (3.31) (called the *direct method*). To produce the N harmonics of an N sample time-series $x_N[k]$, N^2 complex multiply-accumulations are required. This translates to a finite, but possibly long, computation time. This condition was radically altered with the advent of the *FFT* algorithm, which will be developed in the next chapter.

Example 3.5: DTFT, DTFS, and DFT

Problem Statement. A satellite tumbling with a period $T = 32/f_s$, where $f_s = 1$ Sa/s, emits a bandlimited signal $x[k]$. The signal is processed using an approximate DTFT and DTFS algorithm, and the resulting spectra are saved as $X(e^{j\omega})$ and c_n. The signal is also processed using a DFT, and the resulting spectrum is saved as $X[n]$. Compare $X(e^{j\omega})$, c_n, and $X[n]$.

Analysis and Conclusions. The DTFT produces a continuous spectrum, while the DTFS and DFT spectra are discrete. A discrete spectrum is sometimes referred to as a *line spectrum*. What differentiates the DFT from the DTFT and the DTFS are the properties of the signal space. The signal space is infinite for the DTFT and the DTFS and finite for the DFT. Furthermore, the signal space is periodic with period N for the DFT and the DTFS but is arbitrary for the DTFT. The DFT assumes that the signal to be transformed is of finite duration and period N. If this signal is loaded into an N sample linear and circular shift-register network, as shown in Fig. 3.6, two infinitely long time-series result. One is aperiodic, $x[k]$, and the other is periodic, $x_p[k]$. One has a DTFT and the other a DTFS.

The processes illustrated in Fig. 3.7 can also be described as zero *padding*. Since all that is known are the sample values of an N sample data record $x_N[k]$, it can be used as a seed to produce $x[k]$ using a technique called *zero-padding*. Zero padding refers to appending to a signal of finite length an array of zeros. If enough zeros are padded to a finite data record, a signal that approximates an infinitely long time-series can result. Analyzing this signal using a DFT will approximate a DTFT.

If the signal is periodically extended, rather than zero-padded, a long periodic time-series will result. The DFT of this signal approximates a DTFS and agrees with the DTFS in the limit. The approximate DTFT and DTFT, plus DFT spectra are compared in Fig. 3.7. Since the DFT and DTFS differ by a scale factor, the resulting spectra may be normalized to unity to assist in their comparison. The three normalized spectra are displayed in Fig. 3.7 and are seen to agree with their predicted behavior at the normalized positive baseband discrete frequency locations $\varpi = k\pi/32$ for $k \in \{0,1,2,\dots,15\}$.

Computer Study. The s-file "ex3-5.s" produces the DFT/DTFS and approximates the DTFT using a high-resolution DFT produced by zero padding of the original thirty-two sample record, where the sampling rate is one sample per second. The spectra are displayed in Fig. 3.7. The approximate DTFT spectrum appears to be continuous, while the DFT/DTFS spectra are obviously discrete.

```
>include "chap3\ex3-5.s"
```
❖

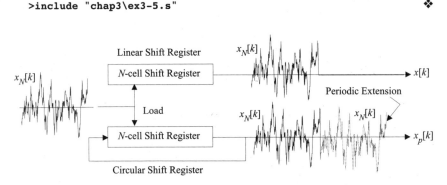

Figure 3.6   Creation of infinitely long time-series from an N sample data record.

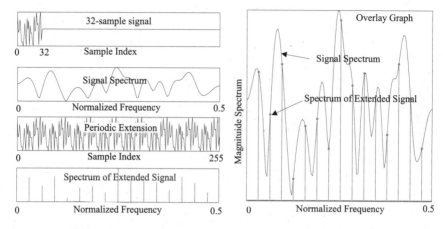

Figure 3.7 DTFT, DTFS, and DFT spectra showing (left) time and frequency behavior, (right) spectra comparison (results may vary due to random test signal).

3.5 Properties of the DFT

The list of DFT properties are found in Table 3.3 and can be seen to bear a strong similarity to those associated to the DTFT and DTFS.

Table 3.3 **Properties of the Discrete Fourier Transform**

Discrete-time series	Discrete Fourier transform	Remarks
$x_N[k] = \displaystyle\sum_{m=0}^{L} a_m x_m[k]$	$X[n] = \displaystyle\sum_{m=0}^{L} a_m X_m[n]$	Linearity
$x_N[(k-q)\mathrm{mod}N]$	$X[N]W_N^{qn}$	Circular time-shift
$x_N^*[-k \bmod N]$	$X^*[n]$	Time-reversal
$x_N[k]W_N^{-qk}$	$X[(n-q)\mathrm{mod}\,N]$	Modulation
$\displaystyle\sum_{k=0}^{N-1} x[k]y^*[k]$	$\dfrac{1}{N}\displaystyle\sum_{n=0}^{N-1} X[n]Y^*[n]$	Parseval's theorem
$x_N^*[k]$	$X^*[-n\mathrm{mod}\,N]$	Complex conjugation

The DFT possesses the symmetry properties of the DTFT and DTFS. *Parseval's theorem,* in the context of a DFT, can be stated as

$$\sum_{k=0}^{N-1} x_N[k]^2 = \frac{1}{N}\sum_{n=0}^{N-1} |X[n]|^2 \qquad (3.33)$$

The term $|X[k]|^2$ is referred to as the *power spectrum* and is a measure of the power in a signal on a per harmonic basis. Notice that the unit of measure is now power instead of energy as found in Eq. (3.13), since the energy in a signal is now averaged over an N sample time interval.

Example 3.6: Parseval's Equation

Problem Statement. An EKG signal sampled at 135 Hz is suspected also to contain 60 Hz contamination (from AC power systems) due to poor signal isolation. It is assumed, due to the biophysics of a human body, that the EKG should contain no appreciable signal power above a few Hertz. Any unusual amount of power detected above these frequencies, say at 60, 120, ...,$k60$ Hz, will be attributed to 60 Hz contamination. Measure power in 60Hz line.

Analysis and Conclusions. The power spectrum of a 1.90 s (256 samples at 135 Sa/s) discrete-time EKG signal record containing 60 Hz contamination is shown in Fig. 3.8. The power spectrum can be seen to have large spectral components located near DC. This is the area of valid EKG activity. However, it can be seen that there are others areas of strong local spectral activity. One is located at 60 Hz.

Computer Study. The s-file "ex3-6.s" contains a procedure to compute and plot the magnitude spectrum (in dB) of a discrete-time signal $x[k]$ over the frequency range $f_b \in [0, 67.5]$ Hz. The procedure "look(f,x)" computes and plots the magnitude spectrum, where $x = x[k]$ is a signal of length 256 samples assumed to be sampled at 135 Sa/s, and f is the frequency where a marker will be placed. The following sequence of commands plots the magnitude spectrum

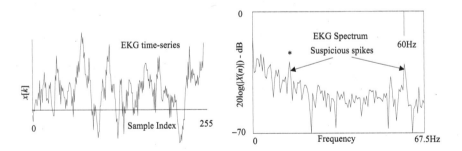

Figure 3.8   Power spectrum of a noisy EKG signal.

of an EKG signal with a marker placed at 60 Hz, followed by another marker
at 15 Hz.

```
>x=rf("signals\ekg1.imp")
>include "chap3\ex3-6.s"
>look(60,x) # Plot spectrum with marker at 60 Hz
>look(15,x) # Plot spectrum with marker at 15 Hz
```

The data shown in Fig. 3.8 clearly show that an abnormally high 60 Hz signal
component is present in the power spectrum of $x[k]$. The first harmonic of
60 Hz, namely 120 Hz, is aliased to 15 Hz. This will be explored in detail in
Example 3.7. In Fig. 3.8, there is clearly a spike at 15 Hz.   ❖

The DFT can parameterized in a straightforward manner. The fundamental parameters that define the precision of a DFT are the sample
rate f_s and the number of samples N transformed. These parameters
are combine to yield the data shown in Table 3.4.

Table 3.4   Discrete Fourier Transform Parameters

Parameter	Notation or units
Sample period	T_s seconds
Record length	N samples or $T = NT_s$ seconds
Number of harmonics	N harmonics
Number of positive (negative) harmonics	$N/2$ harmonics
Frequency spacing between harmonics	$\Delta f = \dfrac{1}{T} = \dfrac{1}{NT_s} = \dfrac{f_s}{N}$ Hz $\Delta\theta = \dfrac{1}{N} = \dfrac{\Delta f}{f_s}$ (normalized, $\Delta\theta \in [0,0.5)$)
DFT frequency (one-sided baseband range)	$f \in \left[0, \dfrac{f_s}{2}\right)$ Hz
DFT frequency (two-sided baseband range)	$f \in \left[-\dfrac{f_s}{2}, \dfrac{f_s}{2}\right)$ Hz
Frequency of the kth harmonic	$f_k = k\Delta f = \dfrac{kf_s}{N}$ Hz

Example 3.7: DFT Properties

Problem Statement. The EKG signal studied in Example 3.6 is sampled at a rate of 135 samples per second. What is the frequency resolution of the 256 sample DFT? What line is closest to 60 Hz? What would the resolution be if the EKG sequence were expanded to 2048 samples using zero padding?

Analysis and Conclusions. A 256 sample DFT would resolve the frequency axis into 128 positive and 128 negative harmonics with a resolution of $\Delta f =$ 135/256 = 0.527 Hz per harmonic. If a 60 Hz component is present, then it should appear in the (60 Hz)/Δf = 113.778 ≈ 114th harmonic location. The peak located at the 114th harmonic is therefore attributable to 60 Hz power frequency noise. The peak at the 27th harmonic translates to an unaliased frequency of $27\Delta f$ = 14.3 Hz, which has no direct physical significance. It is therefore suspected, however, that this peak may be due to aliasing since the harmonic after 60 Hz, namely 120 Hz, is aliased according to the congruence (120 Hz) ≡ (–15 Hz) mod (135 Hz).

Computer Study. The s-file "ex3-7.s" computes and displays an EKG spectrum and superimposes markers representing 60, 120, 180, and 240 Hz. The results are displayed in Fig. 3.9.

```
>ekg=rf("signals\ekg1.imp") # Initialize ekg variable 1st
>include "chap3\ex3-7.s"                                    ❖
```

Another property germane to the study of the DFT is called *leakage*. Recall that a DFT assumes that the signal to be transformed, namely $x_N[k]$, is an N sample time-series which is also *assumed* to be *periodic* with period N. Suppose the actual signal $x[k]$ is not periodic. Then the DFT of $x_N[k]$ will differ from the DTFT of $x[k]$. If $x_p[k]$ is periodic with a period $M \neq N$, then the N point DFT of $x_N[k]$ may also differ from the DTFS of $x_p[k]$. The difference between the spectra is due to discontinuity found at the boundary of N sample intervals leaking into the DFT

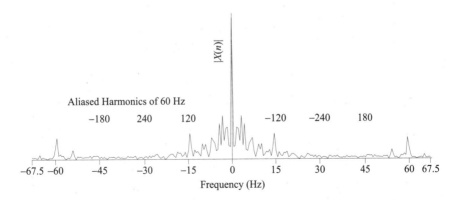

Figure 3.9 EKG spectrum showing the effects of aliasing, showing markers at 60 Hz, and aliases for 120, 180, and 240Hz.

spectrum creating artificial spectral components called *artifacts*. This phenomenon can be illustrated by analyzing the data shown in Fig. 3.10. Shown are two sinusoidal time-series of length N, along with their periodic extension. One time-series completes an integer number of cycles in N samples, and the other does not. The time-series that does not complete an integer number of cycles in N samples, when periodically extended, will demonstrate a discontinuity at the record boundary. The spectra of the two signals is also shown in Fig. 3.10. The DFT of a sinusoidal signal completing an integer number of oscillations in N samples is seen to possess a well defined and localized line-spectrum. The other spectrum exhibits "spreading" of spectral energy about the sinusoid's frequency. The leaked energy from the jump discontinuity found at the N sample boundary can be reduced by increasing the length of the time-series (i.e., N).

Example 3.8: Leakage

Problem Statement. Leakage can result when the signal being transformed does not complete an integer number of cycles within an N sample interval. Consider the two $N = 256$ sample cosine waves shown in Fig. 3.11. Let $x_1[k]$ be a harmonic oscillation having a normalized frequency of $\theta = 12/256$, which means the signal completes twelve full oscillations in 256 samples. Let $x_2[k]$ have a normalized frequency of $\theta = 12.5/256$, which means it completes 12.5 complete oscillations in the same interval. Experimentally display the effects of leakage.

Analysis and Conclusions. The 256-sample DFT assumes that both $x_1[k]$ and $x_2[k]$ are periodically extended in all directions to form $x_N[k] = x[k \bmod (256)]$. The periodic extension of the $\theta = 12/256$ signal is smooth across the N sample boundary. The $\theta = 12.5/256$ signal, however, exhibits an abrupt discontinuity wherever $k \equiv 0 \bmod (256)$. Discontinuities will introduce new frequencies into

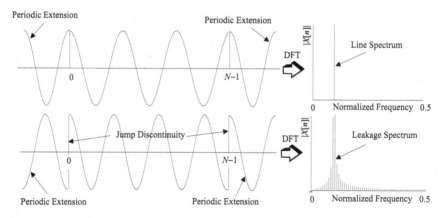

Figure 3.10 Example of leakage and its cause.

the signal's spectrum as shown in Fig. 3.11. Increasing the record length to 512 points, however, allows both signals to complete an integer number of cycles in a 512 sample window, producing a leakage-free DFT as shown in Fig. 3.11.

Computer Study. The following sequence of Siglab commands creates, transforms, and displays the effect of leakage.

```
>x1=mkcos(1,12/256,0,256)    # Non-leakage time series
>x2=mkcos(1,12.5/256,0,256)  # Leakage time series
>x1f=mag(pfft(x1))           # Mag spectrum of non-leakage series
>x2f=mag(pfft(x2))           # Mag spectrum of leakage time series
>graph(x1,x2,x1f,x2f)        # Graph results
>x1=mkcos(1,12/256,0,512)    # Non-leakage time series
>x2=mkcos(1,12.5/256,0,512)  # Non-leakage time series
>x1f=mag(pfft(x1))           # Mag spectrum of non-leakage serioo
>x2f = mag(pftt(x2))         # Mag spectrum of non-leakage series
>graph(x1,x2,x1f,x2f)        # Graph results                    ❖
```

3.6 Windowing

In Sec. 3.3, a connection between the DTFT and DTFS was established in terms of $\sin(x)/x$ interpolation rule. An equivalent question can be posed in the context of gaining agreement between the DFT and the DTFT. Complete agreement can never be gained if $x_N[k]$ is not periodically present in $x[k]$ (i.e., $x[k]$ is periodic with period N). What can be hoped, however, is to minimize the variations between the DFT and DTFT spectra in some acceptable manner.

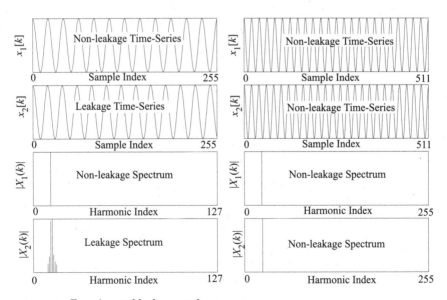

Figure 3.11   Experimental leakage study.

One source of error, or distortion, in a DFT spectrum are artifacts introduced by a signal not having a smooth periodic extension along the time-axis. The basic problem is graphically interpreted in Fig. 3.12 using an arbitrary signal of infinite length $x[k]$ along with its DTFT spectrum. Also shown is a *window* or *gating* function of length N, denoted $w[k]$. The object of the window function is to reduce artifact contamination and to create a finite duration signal $x_N[k]$ which is given by

$$x_N[k] = \{x[k]w[0], x[k+1]w[1], ..., x[k+(N-1)]w[N-1]\} \quad (3.34)$$

whose DFT agrees (in some sense) with the DFT of $x[k]$. Observe that N samples are stripped from $x[n]$, where each retained sample is weighted by a window coefficient. In Chap. 7, the concept of convolution will be introduced and be used to mathematically explain the windowing process in the frequency domain. At this point, it shall be assumed that leakage artifacts can be suppressed by reducing the influence of jump discontinuities at the window boundary. This can be achieved by having the leading and trailing tails of the windowing sequence, $w[k]$, have values near zero. A rectangular window, or gating function, obviously does not satisfy this criteria. Popular windows that

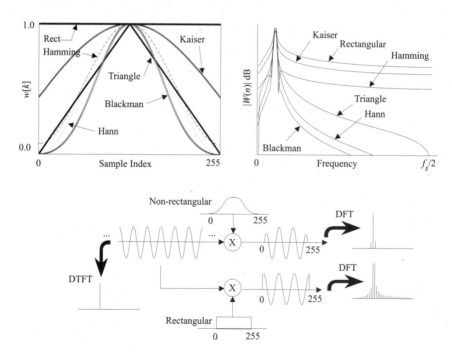

Figure 3.12 Experimental study of windowing on a DFT.

do meet this criteria are listed in Table 3.5 (rectangular included for completeness). All of the listed windows, except the rectangular window, have values near zero locally about $k = 0$ and $k = N - 1$. Window functions will be reinvestigated in Sec. 11.11.

Example 3.9: Window Function

Problem Statement. Compare the shape of the envelope of the windows listed in Table 3.5 and compare their effect of the DFT of the leakage-prone signal studied in Example 3.8.

Analysis and Conclusions. The windows are compared in Fig. 3.12. Except for the rectangular window, all have reduced values near the end-points of the N sample data interval. The DFT of windowed signals is seen to have reduced leakage, as shown in Fig. 3.12.

Computer Study. The following Siglab commands are used to examine the effects of windows.

Table 3.5   Summary of Commonly Used Window Functions

Window name	Window function ($k \in \{0,1,\ldots,N-1\}$
Rectangular	$w[k] = 1$ for $k \in [0, N-1]$
Bartlett (triangular)	$w[k] = \begin{cases} \dfrac{2k}{N-1} & \text{for } k \in \left[0, \dfrac{N-1}{2}\right] \\ 2 - \dfrac{2k}{N-1} & \text{for } k \in \left[\dfrac{N-1}{2}, N-1\right] \end{cases}$
Hann*	$w[k] = \dfrac{1}{2}\left[1 - \cos\left(\dfrac{2\pi k}{N-1}\right)\right]$
Hamming	$w[k] = 0.54 - 0.46\cos\left(\dfrac{2\pi k}{N-1}\right)$
Blackman	$w[k] = 0.42 - 0.5\cos\left(\dfrac{2\pi k}{N-1}\right) + 0.08\cos\left(\dfrac{4\pi k}{N-1}\right)$
Kaiser	$w[k] = \dfrac{I_0\left[\beta\sqrt{\left(\dfrac{N-1}{2}\right)^2 - \left(k - \dfrac{N-1}{2}\right)^2}\right]}{I_0\left(\beta\dfrac{N-1}{2}\right)}$
	where $I_0(\beta)$ is a 0th-order Bessel function

*The Hann window is named after Austrian meteorologist Julius van Hann. It is commonly, and mistakenly, referred to as a "Hanning" window.

```
>x=mkcos(1,12.5/256,0,256) # Leakage time series
>rect=ones(256) # Create windowing sequences...
>hannw=han(256)
>hammw=ham(256)
>triw=tri(512,256)
>blackw=black(256)
>kaiw=kai(2,256) #beta=2
># Graph windows
>ograph(rect,triw,hannw,hammw,blackw,kaiw)
>x1=mag(pfft(x.*rect)) # Create mag spectrum of windows
>x2=mag(pfft(x.*triw))
>x3=mag(pfft(x.*hannw))
>x4=mag(pfft(x.*hammw))
>x5=mag(pfft(x.*blackw))
>x6=mag(pfft(x.*kaiw))
># Graph (log) magnitude spectrum of windows
>ograph(log(x1),log(x2),log(x3),log(x4),log(x5),log(x6))        ❖
```

The window functions described in this section are applied to time-domain data. The duality principle states that the Fourier transforms of time- and frequency-domain signals can be exchanged. That is, refer to Example 3.2 and note that the DTFT of a rectangular pulse has a $\sin(x)/x$ shape. The duality principle states that a $\sin(x)/x$ function in the time domain would have a rectangular pulse shape (called an *ideal filter*) in the frequency domain. Therefore, applying a multiplicative filter to a signal in the frequency domain would have the effect of convolving the time domain signal with inverse Fourier transform of the filter function in the time domain. One of the principal uses of frequency-domain windowing is to mask unwanted frequency components. In particular, suppose two signals $x[k]$ and $y[k]$ have nonoverlapping spectra $X[n]$ and $Y[n]$, respectively. The contribution of $y[k]$ can be annihilated by masking out $Y[n]$ using a window function that has a value of unity over the spectrum of $X[n]$ and zero over $Y[n]$. The reconstructed time-domain image derived from the masked spectrum would contain only samples of $x[k]$. The implementation of a frequency domain mask can also take the form of a band-selective filter (see Chaps. 11 and 13) that can suppress signal energy over specified frequency ranges and accentuate it elsewhere. The spectra of two signal processes may, however, partially overlap. Then a mask can always be used to eliminate the undesired signal energy of one of the signals over the range when the two spectra do not overlap.

Example 3.10: Frequency-Domain Window Function

Problem statement. Suppose a desired 500 Hz tone is heavily contaminated with broadband noise. Show that a frequency domain mask can significantly suppress the effects of the broadband noise process.

Analysis and Conclusions. The spectrum of a 2^{14} sample sinusoidal signal embedded in Gaussian white noise with a −10 dB signal-to-noise ratio (SNR) is reported in Fig. 3.13. Notice that the sinusoid exhibits a line spectrum, whereas the Gaussian noise is broadband. Therefore, it should be possible to null all of the harmonics of the input spectrum, except those located at (or near) the ±500 Hz. For f_s = 11025 Hz, this corresponds to the region around the ±743rd harmonic. Upon building such a mask, or window, and applying it in a multiplicative manner to the original spectrum, the new spectrum shown in Fig. 3.13 results. The masked spectrum is seen to be essentially that of an ideal 500 Hz sinusoid.

Computer Study. The following sequence of Siglab commands is used to produce the data shown in Fig. 3.13. The s-file "mask.s" assumes the presence of a sixteen-bit soundboard having a sample frequency of 11,025 Hz. The procedure used is "mask(snr)" where the *snr* parameter is the signal-to-noise ratio (decimal). Upon execution, the noisy signal is first played followed by the denoised (filtered) frequency masked signal.

```
>include "chap3\mask.s"
>mask(0.1) # -10dB SNR                                    ❖
```

3.7 Two-Dimensional DFT

One of the many uses of the DFT and FFT is to analyze two-dimensional signals and images. A 2-D periodic signal has two sample indices $x_p[k_1,k_2]$ where $k_1 \in \{0,1,2,...,N–1\}$ and $k_2 \in \{0,1,2,...,M–1\}$. A typical example of a 2-D signal is a graphic image where k_1 and k_2 are x-axis and y-axis coordinates. The *two dimensional* DFT (2-D DFT) maps an $N \times M$ matrix of sample values into an $N \times M$ matrix of harmonics where the $[n_1,n_2]$ element is denoted $X[n_1,n_2]$ and $n_1 \in \{0,1,2,...,N–1\}$ and $n_2 \in \{0,1,2,...,M–1\}$. The spatial resolution in the x-direction is $f_x N$ and $f_y M$ in the y-direction, where f_x and f_y are the sample rates along each axis. The 2-D DFT, like its 1-D counterpart, assumes that the signal $x[k_1,k_2]$ is periodic with period N in the x-direction and M in the y-

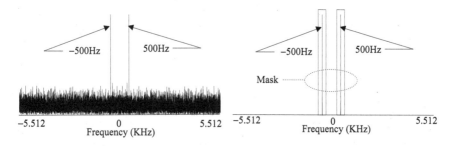

Figure 3.13   Noisy sinusoidal spectrum and frequency masked spectrum.

direction. That is, $x_p[k_1,k_2] = x_p[N + k_1, M + k_2]$. The synthesis equation for a 2-D DFT (i.e., the inverse 2-D DFT) of $x_p[k_1,k_2]$ is formally defined to be

$$x_p[k_1, k_2] = \frac{1}{MN} \sum_{n_1 = 0}^{N - 1} \sum_{n_2 = 0}^{M - 1} X[n_1, n_2] W_N^{-k_1 n_1} W_M^{-k_2 n_2} \qquad (3.35)$$

and the analysis equation is given by

$$X[n_1, n_2] = \sum_{k_1 = 0}^{N - 1} \sum_{k_2 = 0}^{M - 1} x_p[k_1, k_2] W_N^{-k_1 n_1} W_M^{-k_2 n_2} \qquad (3.36)$$

where $X[n_1,n_2]$ is the n_1 harmonic in the x-direction and the n_2 harmonic in the y-direction.

The computation of a 2-D DFT can be accomplished in a piecemeal manner. This is true because the DFT exhibits a property called *separability*. Consider a generalized, L dimensional transform with kernel $g(k_1,k_2,k_3,...,k_L)$,

$$X[n_1, n_2, ..., n_L]$$

$$= \sum_{k_1 = 0}^{N_1 - 1} \sum_{k_2 = 0}^{N_2 - 1} \cdots \sum_{k_L = 0}^{N_L - 1} x(k_1, k_2, ..., k_L) g(k_1 n_1, k_2 n_2, ..., k_L n_L) \qquad (3.37)$$

The transform is said to be *separable* if the kernel, $g(k_1,k_2,k_3,...,k_L)$, can be expressed as a product of L functions as follows,

$$g(k_1, k_2, ..., k_L) = g_1(k_1) g_2(k_2)...g_L(k_L) \qquad (3.38)$$

If the transform shown in Eq. (3.37) is separable, then the summations may be nested such that

$$X[n_1, n_2, ..., n_L] = \left(\sum_{k_1 = 0}^{N_1 - 1} \left(\sum_{k_2 = 0}^{N_2 - 1} \left(\cdots \left(\sum_{k_L = 0}^{N_L - 1} x(k_1, k_2, ..., k_L) g_L(k_L n_L) \right) \cdots \right) g_2(k_2 n_2) \right) \right) g_1(k_1 n_1) \qquad (3.39)$$

While the nesting demonstrated in Eq. (3.39) may seem trivial, it can have a significant impact on the implementation of a multidimensional transform. Also, it is worth noting that in many applications, such as the DFT, the separated kernel functions, g_i, for $i \in \{1,2,3,...,L\}$, are all the same.

By using the separability of the DFT, the 2-D DFT may be computed piecemeal. The rows (columns) of the $N \times M$ data matrix, $x_p[k_1,k_2]$, can first be transformed and the results returned to the row (column) of the $N \times M$ matrix of origin. Once computed, the columns (rows) of the $N \times M$ matrix are then transformed and the spectrum returned to the column (row) of the $N \times M$ matrix of origin. The complex elements of the matrix $X[n_1,n_2]$ are the 2-D DFT of $x_p[k_1,k_2]$.

Using the separability of a transform can actually reduce the computational expense of an algorithm. To illustrate, consider the DFT of an $N \times N$ matrix. Computing the DFT directly, as suggested by Eq. (3.39), requires N^4 multiply-accumulate operations. In comparison, if the separability of the DFT is exploited, then a 1-D DFT of length N, requiring N^2 multiply-accumulate operations, will be computed N times for the row-wise transform and N times for the column-wise transform for a total of $2N^3$ multiply-accumulate operations. For all $N > 2$, $2N^3 < N^4$, so that, for example, a 256×256 transform computed directly requires 128 times as many operations as the same transform computed exploiting the separability property.

The physical meaning of the 2-D spectrum may be better understood using a simple example. Suppose a satellite image of a agriculturally rich region is seen as a collection of interlaced rectangles. Suppose further that the camera resolution is about 0.7 yards. The digitized image is stored in a $10{,}000 \times 10{,}000$ array that corresponds to an area on the order of 16 square miles. The fundamental spatial period along either the x or y axes is on the order of 7,000 yards or about 4 miles. The first harmonic of the 2-D spectrum is at the fundamental spatial frequency of about 4 miles per period (repetition). The fourth harmonic has a period of 1 mile, and so on. All rectangular fields having a geometry of at least $1/4 \times 1/4$ mile in size are found in the harmonics $X[n_1,n_2]$, for $n_1 \leq 16$ and $n_2 \leq 16$, since they have the required spatial frequency.

Example 3.11: Two-Dimensional DFTs

Problem Statement. Compute the 2-D DFT of a square 2-D series $x_p[k_1,k_2]$.

Analysis and Conclusions. The time-domain data base is shown in Fig. 3.14. It is created by "spinning" a symmetric 1-D signal about the origin in a 2-D plane. Once the 2-D database is created, it can be transformed into the frequency domain using a DFT using 1-D operations on the rows and then the columns.

Computer Study. The s-file "spin.s" contains a function to convert a symmetric 1-D signal $x[k]$ into a symmetric 2-D signal matrix, $x[i,j]$, using the function "spin(x)." The 2-D DFT of a 2-D symmetric signal generated by a 1-D symmetric pulse is computed using the following Siglab commands:

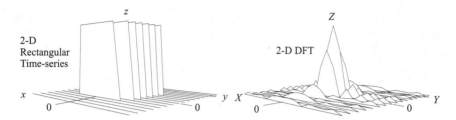

Figure 3.14 Symmetric 2-D symmetric signal and 2-D DFT.

```
>include "chap3\spin.s"
>sig=spin(zeros(28)&ones(8)&zeros(28)) # centered pulse
>graph3d(sig)
>graph3d(mag(fft(fft(sig)'))) # Use separability property          ❖
```

The results are displayed in Fig. 3.14. In the left panel, the 2-D series is displayed. On the right, the 2-D magnitude DFT is shown. The spectrum is seen to be low-frequency dominated and has as $\sin(x)/x$ envelope emanating for the DC line found at the center of the 2-D spectrum.

3.8 Uncertainty Principle

The Fourier transform represents a classic example of the uncertainty principle from physics. Briefly, there is a constant struggle between maintaining high precision in the time and frequency domains. Based on the DFT techniques developed in this chapter, it can be noted that high spectral precision, namely $\Delta f \to 0$, requires that the number of samples to be transformed $N \to \infty$. If the time domain behavior is to be known with high precision, then $T_s \to 0$, or equivalently $f_s \to \infty$. Thus, both the observation time and the sample rate must become unbounded if a signal is to be quantified precisely in the time-frequency plane. However, from a practical standpoint, both the observation interval (i.e., N-samples) and sample rate (i.e., f_s) must be finite. Therefore, there will always be uncertainty in the time-frequency domain.

3.9 Direct Cosine Transform

The *discrete cosine transform (DCT)* is not a DFT, although it can be defined in terms of a DFT, nor is it a time-frequency transform. However, due to its close linkage to the DFT, it will be presented in this section. The DCT is widely used today in image compression (e.g., JPEG) and video compression (e.g., MPEG, H.261, and H.263). The rationale for its use is found in its ability to achieve moderate compression levels with limited complexity. Virtually all of the current image and video compression methods use an 8×8 DCT that is repetitively applied to a larger picture or video frame. Like the DFT, the DCT is a separable transform. As a consequence, most 2-D DCT implementations use the 1-D DCT to form the 2-D DCT. A 1-D, L point DCT is given by

$$y[k] = \frac{C_k}{2} \sum_{i=0}^{L-1} x[i] \cos\left(\frac{(2i+1)k\pi}{2L}\right) \tag{3.40}$$

where $C_0 = 1/(\sqrt{2})$ and $C_k = 1$ for $k \neq 0$. The 2-D $L \times M$ transform is given by

$$y[k_1, k_2] = \frac{C_{k_1} C_{k_2}}{2} \sum_{i=0}^{L-1} \sum_{j=0}^{M-1} x[i, j] \cos\left(\frac{(2i+1)k_1\pi}{2L}\right) \cos\left(\frac{(2j+1)k_2\pi}{2M}\right)$$

$$\tag{3.41}$$

Since the DCT is separable and most multidimensional implementations exploit this property, the computation of a 1-D DCT will be more carefully examined. The DCT may be expressed as a matrix-vector product. For $L = 8$, this product is

$$\begin{bmatrix} y_0\sqrt{2} \\ y_1 \\ y_2 \\ y_3 \\ y_4 \\ y_5 \\ y_6 \\ y_7 \end{bmatrix} = \frac{1}{2} \begin{bmatrix} c_0 & c_0 & c_0 & c_0 & c_0 & c_0 & c_0 & c_0 \\ c_1 & c_3 & c_5 & c_7 & c_9 & c_{11} & c_{13} & c_{15} \\ c_2 & c_6 & c_{10} & c_{14} & c_{14} & c_{10} & c_6 & c_2 \\ c_3 & c_9 & c_{15} & c_{11} & c_5 & c_1 & c_7 & c_{13} \\ c_4 & c_{12} & c_{12} & c_4 & c_4 & c_{12} & c_{12} & c_4 \\ c_5 & c_{15} & c_7 & c_3 & c_{13} & c_9 & c_1 & c_{11} \\ c_6 & c_{14} & c_2 & c_{10} & c_{10} & c_2 & c_{14} & c_6 \\ c_7 & c_{11} & c_3 & c_{15} & c_1 & c_{13} & c_5 & c_9 \end{bmatrix} \begin{bmatrix} x_0 \\ x_1 \\ x_2 \\ x_3 \\ x_4 \\ x_5 \\ x_6 \\ x_7 \end{bmatrix} \tag{3.42}$$

where

$$c_n = \cos\left(\frac{n\pi}{16}\right) \tag{3.43}$$

Since the cosine function has the property that $\cos(\pi - x) = -\cos(x)$, Eq. (3.42) may be changed to

$$
\begin{bmatrix} y_0\sqrt{2} \\ y_1 \\ y_2 \\ y_3 \\ y_4 \\ y_5 \\ y_6 \\ y_7 \end{bmatrix} = \frac{1}{2}
\begin{bmatrix}
c_0 & c_0 & c_0 & c_0 & c_0 & c_0 & c_0 & c_0 \\
c_1 & c_3 & c_5 & c_7 & -c_7 & -c_5 & -c_3 & -c_1 \\
c_2 & c_6 & c_{10} & c_{14} & c_{14} & c_{10} & c_6 & c_2 \\
c_3 & c_9 & c_{15} & c_{11} & -c_{11} & -c_{15} & -c_9 & -c_3 \\
c_4 & c_{12} & c_{12} & c_4 & c_4 & c_{12} & c_{12} & c_4 \\
c_5 & c_{15} & c_7 & c_3 & -c_3 & -c_7 & -c_{15} & -c_5 \\
c_6 & c_{14} & c_2 & c_{10} & c_{10} & c_2 & c_{14} & c_6 \\
c_7 & c_{11} & c_3 & c_{15} & -c_{15} & -c_3 & -c_{11} & -c_7
\end{bmatrix}
\begin{bmatrix} x_0 \\ x_1 \\ x_2 \\ x_3 \\ x_4 \\ x_5 \\ x_6 \\ x_7 \end{bmatrix}
$$

$$(3.44)$$

By permuting the rows of the coefficient matrix in Eq. (3.44), the matrix formulation for the DCT may be reexpressed as

$$
\begin{bmatrix} y_0\sqrt{2} \\ y_2 \\ y_4 \\ y_6 \\ y_1 \\ y_3 \\ y_5 \\ y_7 \end{bmatrix} = \frac{1}{2}
\left[\begin{array}{cccc|cccc}
c_0 & c_0 & c_0 & c_0 & c_0 & c_0 & c_0 & c_0 \\
c_2 & c_6 & c_{10} & c_{14} & c_{14} & c_{10} & c_6 & c_2 \\
c_4 & c_{12} & c_{12} & c_4 & c_4 & c_{12} & c_{12} & c_4 \\
c_6 & c_{14} & c_2 & c_{10} & c_{10} & c_2 & c_{14} & c_6 \\
c_1 & c_3 & c_5 & c_7 & -c_7 & -c_5 & -c_3 & -c_1 \\
c_3 & c_9 & c_{15} & c_{11} & -c_{11} & -c_{15} & -c_9 & -c_3 \\
c_5 & c_{15} & c_7 & c_3 & -c_3 & -c_7 & -c_{15} & -c_5 \\
c_7 & c_{11} & c_3 & c_{15} & -c_{15} & -c_3 & -c_{11} & -c_7
\end{array}\right]
\begin{bmatrix} x_0 \\ x_1 \\ x_2 \\ x_3 \\ x_4 \\ x_5 \\ x_6 \\ x_7 \end{bmatrix}
$$

$$(3.45)$$

The upper four rows of the coefficient matrix in Eq. (3.45) have mirror symmetry about the vertical partition. The lower four rows of the same coefficient matrix also have symmetry about the vertical partition, except for a change in sign. Also notice that the $\sqrt{2}$ factor can be factored into the first row of the coefficient matrix. Given the symmetry of Eq. (3.45), it is possible to reduce the computation to two smaller matrix-vector products,

$$
\begin{bmatrix} y_0 \\ y_2 \\ y_4 \\ y_6 \end{bmatrix} = \frac{1}{2}
\begin{bmatrix}
c_0/(\sqrt{2}) & c_0/(\sqrt{2}) & c_0/(\sqrt{2}) & c_0/(\sqrt{2}) \\
c_2 & c_6 & c_{10} & c_{14} \\
c_4 & c_{12} & c_{12} & c_4 \\
c_6 & c_{14} & c_2 & c_{10}
\end{bmatrix}
\begin{bmatrix} x_0 + x_7 \\ x_1 + x_6 \\ x_2 + x_5 \\ x_3 + x_4 \end{bmatrix}
$$

$$(3.46)$$

and

$$
\begin{bmatrix} y_1 \\ y_3 \\ y_5 \\ y_7 \end{bmatrix} = \frac{1}{2} \begin{bmatrix} c_1 & c_3 & c_5 & c_7 \\ c_3 & c_9 & c_{15} & c_{11} \\ c_5 & c_{15} & c_7 & c_3 \\ c_7 & c_{11} & c_3 & c_{15} \end{bmatrix} \begin{bmatrix} x_0 - x_7 \\ x_1 - x_6 \\ x_2 - x_5 \\ c_3 - x_4 \end{bmatrix}
$$

(3.47)

This implementation of the DCT is seen to require half of the multiplication operations required by Eq. (3.45) (32 versus 64). In trade, 8 extra addition/subtraction operations are required. Given this simplified implementation of the DCT, the following properties are apparent:

- The DCT is real multiply-accumulate intensive.

- It requires limited data preprocessing ($x_i \pm x_{L-i-1}$).

- It has a coefficient matrix constructed from a small, permuted set of coefficients.

3.10 Short-Time Fourier Transform

A basic problem faced in the field of spectral estimation is the analysis of signals having a time-varying spectrum. Naturally occurring signal processes, such as speech and vision, fall into this class, along with man-made signals such as a radar chirp. In a dynamic signal environment, a time-series may have to be observed over all time to capture the full character of the signal. This would suggest the need for an infinitely long DTFT given by

$$
X(e^{j\omega}) = \sum_{k=-\infty}^{\infty} x[k]e^{-jk\omega}
$$

(3.48)

The fact that the DTFT requires that the signal under study be infinitely long is impossible from a digital implementation viewpoint. Practicality suggests that only signals of finite duration can be digitally processed in realistic time using realistic resources. Unfortunately, the uncertainty principal states that the shorter the signal observation interval, the poorer is the resulting resolution of the Fourier transform. A compromise is the called the *short time Fourier transform* or STFT. The STFT uses a sliding window, $w[k]$, to provide a finite cover of a time-series. The window, for example, may take the form of a N sample uniform (rectangular) or Hamming window. The purpose of the window is to make the computation practical (i.e., finite length) and shape the spectrum.

The STFT of a signal $x[k]$ is given by

$$X_{STFT}(e^{j\omega}, s) = \sum_{k=-\infty}^{\infty} x[k]w[k-s]e^{-jk\omega} \tag{3.49}$$

That is, $X_{STFT}(e^{j\omega},s)$ is defined over an N sample window sliding beginning at sample s. It should be clear that the STFT spectrum will not generally agree with $X(e^{j\omega})$, the DTFT of $x[k]$. However, with the proper choice of window and window length, the STFT can produce useful results. The STFT, however, must be interpreted in the context of the approximation and assumptions used to produce an STFT which can be interpreted in terms of a uniform time-frequency grid or tiling diagram, such as that shown in Fig. 3.15. Here it is assumed that the window being used is of length N, and that over each N sample interval, the signal's spectrum is assumed to be stationary. If the spectrum is rapidly changing, a shorter observation interval N must be used along with an attendant loss in spectral precision—a consequence of the uncertainty principle.

The recovery of the windowed time-series is given by

$$x[k]w[k-s] = \frac{1}{2\pi} \int_{-\pi}^{\pi} X_{STFT}(e^{j\omega}, s)e^{jk\omega} d\omega \tag{3.50}$$

and, if $w[0] \neq 0$, it follows that

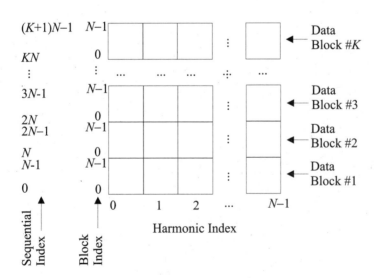

Figure 3.15   Time-frequency space representation of an STFT.

$$x[s] = \frac{1}{2\pi w[0]} \int\limits_{-\pi}^{\pi} X_{STFT}(e^{j\omega}, s) e^{jk\omega} d\omega \tag{3.51}$$

Example 3.12: STFT

Problem Statement. Investigate the STFT and the uncertainty principle using a chirp signal of the form

$$x[k] = \cos(2\pi f_0 k + 2\pi \Delta_0 k^2)$$

where f_0 is the normalized starting frequency and Δ_0 is the normalized range of the chirp.

Analysis and Conclusions. The instantaneous frequency of the chirp is

$$f[k] = f_0 + 2\Delta_0 k$$

For small Δ_0, the instantaneous change in frequency is likewise small and, as a result, a long window can be used resulting in a STFT with high precision in the frequency domain. If, however, Δ_0 is large, then a small-duration window must be used to preserve the assumption that the signal is stationary within the window. The resulting spectral precision will be low; however, the localization of spectral features in time will be high. An example of these trade-offs is shown in Fig. 3.16 for a 4096 sample chirp with Δ_0 small ($\Delta_0 = 0.05$) and large ($\Delta_0 = 0.25$). The STFT window is a 512 sample sliding uniform (rectangular) window. The displayed STFT spectra was computed for windowed data sets $x[k{:}k + 511]$ for $k = 512r$, $r\in\{0,1,2,...,7\}$. The STFT, for Δ_0 small, exhibits within each 512 sample uniform window strong tonal activity which is nearly monotone. For Δ_0 large, the spectrum associated with each 512 sample uniform window is now smeared and contains many frequencies. The STFT spectra in this case would be of limited value in understanding the structure of the original signal.

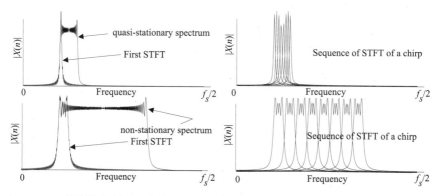

Figure 3.16 DTFT (left) and STFT (right) of a chirp showing the spectrum of a nearly stationary and a wideband process plus the individual STFTs for each case.

Computer Study. The s-file "stft.s" contains two functions, "stft(x,w)" and "pstft(x,w)," which compute the two-sided STFT and positive (one-sided) STFT of the sequence x using the window w. The s-file "chirp.s" contains the function "chirp(n,f0,s0,s1)" which can be used to create a chirp of length n, with center frequency f_0, starting frequency relative to f_0 of s_0, and ending frequency of s_1 relative to f_0, where all frequencies are normalized with respect to the sampling frequency. The following sequence of Siglab commands will illustrate a STFT with a slowly swept chirp and a rapidly swept chirp.

```
>include "chap3\stft.s"
>include "chap3\chirp.s"
>x1=chirp(4096,0.1,0,0.05) # Generate slowly swept chirp
>x2=chirp(4096,0.1,0,0.25) # Generate rapidly swept chirp
>w=ones(4096/8)      # Generate uniform window
>y1=mag(pstft(x1,w)) # Generate 1-sided magnitude STFT of x1
>y2=mag(pstft(x2,w)) # Generate 1-sided magnitude STFT of x2
># Show approx DTFT and STFT of an individual windowed data
  record
>ograph(mag(pfft(x1)),mag(pfft((w.*x1[0:511])&zeros(7*512))))
># 8 STFT records
>ograph(y1[7,],y1[6,],y1[5,],y1[4,],y1[3,],y1[2,],y1[1,],y1[0,])
># Show approx DTFT and STFT of an individual windowed data
  record
>ograph(mag(pfft(x2)),mag(pfft((w.*x2[0:511])&zeros(7*512))))
># 8 STFT records
>ograph(y2[7,],y2[6,],y2[5,],y2[4,],y2[3,],y2[2,],y2[1,],y2[0,]❖
```

3.11 Nonparametric Spectral Estimation

The DFT is an example of a nonparametric spectral estimator, since it operates on time-series sample values rather than a mathematical signal model. This provides a practical spectral analysis tool. Nonparametric spectral estimators provide an estimate of the spectral content of a signal based purely on the signal's sample values. It does not make any assumptions regarding the statistical properties of the signal under study or exploit any *a priori* signal information. As a result, it is generally considered to be a "crude" estimate.

Before discussing nonparametric spectral estimation techniques in detail, it is necessary to introduce some important definitions. Practical implementations of nonparametric spectral estimators rely on a variety of statistical assumptions that must be taken into account in order for the estimator's output to be meaningful.

Let X denote the outcome of a random experiment such as the flip of a (fair) coin. It is obvious that the probability $X =$ heads is equal to the probability that $X =$ tails and that the probability of each is one-half. Formally, that is

$$\Pr_X \text{ (heads)} = \Pr_X \text{ (tails)} = \frac{1}{2} \tag{3.52}$$

where the notation $\Pr_X(x)$ denotes the probability that $X = x$ for some x, or $\Pr(X = x)$. In general, *probability* is a value defined on the closed interval $[0,1]$, where an event with probability zero will never occur, assuming the sample space is finite, while an event with probability one will always occur. The probability of each possible outcome of the coin toss experiment is obvious by virtue of our intuitive ability to take the limit of an infinite number of experiments. In practice, if one were to measure the probability of each outcome for a finite number of coin toss experiments, the measured probabilities would be unlikely to be exactly one-half. Computing a probability by taking the limit as the number of experiments goes to infinity is referred to as an *ensemble average*. This is *not* the same as taking a time average. Since the ensemble average depends upon an infinite number of experiments it is inherently anti-causal. However, most classical statistical theory is based upon ensemble averages.

The *probability distribution function* is defined by

$$F_X(x) = \Pr(x \le X) \tag{3.53}$$

and is a monotonically increasing function from zero to one. The *probability density function* is defined as the derivative of $F_X(x)$,

$$f_X(x) = \frac{\partial F_X(x)}{\partial x}$$

The *mean* or *expected value* of a random variable X is denoted $E\{X\}$ and is given by

$$E\{X\} = \int_{-\infty}^{\infty} x f_X(x) dx = \bar{X} \tag{3.54}$$

The mean is sometimes called the *first moment* of X. Another useful statistical parameter is a measure of the deviation of the signal about the mean of the signal. This measure is usually called the *variance* or *second central moment* of X. It is defined as

$$\sigma_X^2 = \int_{-\infty}^{\infty} |x - \bar{X}| f_X(x) dx \tag{3.55}$$

and may alternately be expressed as

$$\sigma_X^2 = E\{(X - \bar{X})^2\} = E\{X^2\} - \bar{X}^2 \tag{3.56}$$

The expected value operation is based upon the knowledge of $F_X(x)$ or $f_X(x)$, which in turn is based upon an ensemble average. To estimate the statistical properties of a measured random variable, it is necessary to use time averages rather than ensemble averages. In particular, to estimate the mean of X on the time interval $[t - T,t]$, the following expression is used

$$\mathcal{E}\{X\} = \frac{1}{T} \int_{t-T}^{t} X(\tau) d\tau \tag{3.57}$$

The "$\mathcal{E}$" operator is used to denote a time average. Just as the mean is estimated using a time average in Eq. (3.57), the variance may be estimated by replacing the expectation operator in Eq. (3.56) with the time average operator "$\mathcal{E}$."

A discrete time signal may be thought of as a *random process* where the value of a sequence, $x[k]$, for any particular k may be thought of as a random variable. When estimating statistics with signals from real systems, it is impractical to use the expected value operation, since it requires information from an infinite number of experiments when only one is available.

The *autocorrelation* of a random process at two different indices, m, and n, is

$$r_{xx}[m, n] = E\{x[m]x^*[n]\} \tag{3.58}$$

A random process is said to be *wide-sense stationary* (WSS) if its mean is constant over all time and the autocorrelation is dependent only upon the difference $m - n$. If a random process is WSS then its autocorrelation is frequently written as

$$r_{xx}[k] = E\{x[n + k]x^*[n]\} \tag{3.59}$$

where, by definition, n may be any value. To estimate the autocorrelation of a finite time-series, the biased[*] autocorrelation function is used as follows:

$$\hat{r}_{xx}[k] = \begin{cases} \dfrac{1}{N-k} \displaystyle\sum_{n=0}^{N-k-1} x[n+k]x^*[n] & \text{for } 0 \le k \le N-1 \\[4mm] \dfrac{1}{N-|k|} \displaystyle\sum_{n=0}^{N-|k|-1} x[n]x^*[n+|k|] & \text{for } -(N-1) \le k < 0 \end{cases} \tag{3.60}$$

To substitute time averages for ensemble averages, the time averages must be good estimates of ensemble averages. In particular, if for any single sequence of the process ensemble the process statistics can be determined from the time averages, then the process is said to be *ergodic*. The biased autocorrelation estimate in Eq. (3.60) depends on an assumption of both wide-sense stationarity and ergodicity. In general, ergodicity and wide-sense stationarity are important process properties which allow the application of statistical methods developed assuming ensemble averages to real systems where only a single time-series is available.

3.12 Periodogram

The power spectrum or power spectral density (PSD) of a discrete-time signal, $x[k]$, is given by

$$P_{xx}(\varpi) = \lim_{n \to \infty} E\left\{ \frac{1}{2N+1} \left| \sum_{k=-N}^{N} x[k]e^{-jk\varpi} \right|^2 \right\} \tag{3.61}$$

where "E" denotes the expected value, and ϖ is the normalized frequency. It can be seen that the power spectrum is based on the expected value of the magnitude squared value of a DTFT, normalized by the record length. From the definition of PSD given by Eq. (3.61), a number of techniques have been developed to estimate the PSD of a real signal.

The *periodogram* estimate of the power spectrum associated with a time-series $x[k]$, is based only on an N sample data record. The periodogram is defined to be

---

* Bias refers to the average deviation of an estimate from its true value. Since the autocorrelation is defined by an infinitely long time-series, and Eq. (3.60) is based on a finite number of samples, a bias results. Bias is often traded off against error variance by slight modifications of the estimator (e.g., applied scale factor). Only increasing N (size of the sample space) will simultaneously reduce the error bias and variance.

$$\tilde{P}(\varpi) = \frac{1}{N} \left| \sum_{k=0}^{N-1} x[k] e^{-jk\varpi} \right|^2 \qquad (3.62)$$

The summation in Eq. (3.62) is recognized to be the DTFT of an infinite sequence, $x[k]$, windowed by a rectangular window. The implicit windowing of the periodogram expression in Eq. (3.62) results in a smearing of the power spectral density estimate. This occurs because the point-wise multiplication of the data series, $x[k]$, by the rectangular windowing sequence in the time domain corresponds to *convolution* of the DTFT of $x[k]$ with the DTFT of the rectangular window in the frequency domain. Convolution will be developed in more detail in Chap. 5. The smearing function in the frequency domain, namely the DTFT of a rectangular pulse, was discussed in Example 3.2. If the rectangular pulse is of length N samples, then its DTFT has an envelope of $|\sin(\varpi N/2)/\sin(\varpi/2)|$, which is periodic with period 2π, and has a repeated $\sin(x)/x$ shape (see Fig. 3.2). The null-to-null width of the lobe centered about $\varpi = 0$, also known as the *main lobe*, is $4\pi/N$. While one would expect the power spectrum of a sinusoid at frequency ϖ_0 to be a line at ϖ_0, the periodogram will produce a function with an envelope $|\sin(\varpi N/2)/\sin(\varpi/2)|$. Since the main lobe has a width of $4\pi/N$, it is impossible for a periodogram to resolve two frequencies that are closer than $2\pi/N$. The need to resolve signals at two closely spaced frequencies must be considered when selecting a record length for a periodogram.

Example 3.13: Rectangular Window Effect

Problem Statement. Verify that a periodogram is not be able to resolve peaks in the power spectrum separated by a normalized frequency of $\varpi < 2\pi/N$.

Analysis and Results. An experimental examination of the blurring effect introduced by a finite rectangular window, using two pure harmonic oscillations, is reported in Fig. 3.17. Notice that when the two harmonic oscillations are separated by a width of 10Δ, where $\Delta = 2\pi/N$, the two peaks are easily discernible in the computed power spectrum. When the width is 2Δ, the peaks continue to be distinctly visible. However, when the width is Δ, the periodogram is seen to lose its ability to resolve the discrete sinusoids.

Computer Study. The s-file "psd.s" uses a procedure with format "psd(d,n)," where d is the frequency separation between two sinusoids, given by $(2\pi d/n)$, and n is the length of a rectangular window in samples, to produce the results shown in Fig. 3.17. Execute the following Siglab commands to generate the plots shown in Fig. 3.17.

```
>include "chap3\psd.s"
>psd(10,2^10) # tones separated by 10 harmonics
>psd(2,2^10)  # tones separated by 2 harmonics
>psd(1,2^10)  # tones separated by 1 harmonics
```
❖

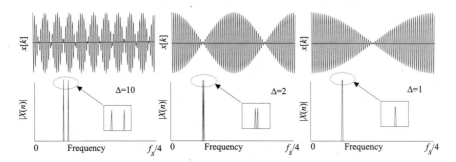

Figure 3.17 Two-tone process and spectra (with zoom insert) for $\Delta = 10$, 2, and 1.

The power spectral density defined in Eq. (3.61) may also be expressed as

$$P_{xx}(\omega) = \sum_{k=-\infty}^{\infty} r_{xx}[k]e^{-jk\omega} \tag{3.63}$$

which is the DTFT of the autocorrelation sequence. Interestingly, the autocorrelation may be recovered from the power spectral density using

$$r_{xx}[k] = \int_{-\pi}^{\pi} P_{xx}(\omega)e^{j\omega k} \tag{3.64}$$

The transform pair given by Eqs. (3.63) and (3.64) are referred to as the *Weiner-Khintchine theorem*. The periodogram given in Eq. (3.62) may also be expressed using the biased autocorrelation estimate, $\hat{r}_{xx}[k]$, as

$$\tilde{P}(\omega) = \sum_{k=-(N-1)}^{N-1} \hat{r}_{xx}[k]e^{-jk\omega} \tag{3.65}$$

The expected value of $\tilde{P}(\omega)$ is given by

$$E(\tilde{P}(\omega)) = \sum_{k=-(N-1)}^{N-1} E(\hat{r}_{xx}[k])e^{-jk\omega} \tag{3.66}$$

where $E(\hat{r}_{xx}[k])$ is related to the actual unbiased correlation function $r_{xx}[k]$ through

$$E(\hat{r}_{xx}[k]) = \left(1 - \frac{|k|}{N}\right)r_{xx}[k] = w_{tri}[k]r_{xx}[k] \qquad (3.67)$$

That is, the true correlation function is weighted by a triangular window, w_{tri} (also called a *Bartlett window*), to become the biased autocorrelation function implying the that data found at high lags (i.e., $|k| \approx N$) is overly weighted.

It should be recalled that the power spectrum of the signal $x[k]$ can be defined as the discrete-time Fourier transform of the autocorrelation function $r_{xx}[k]$ (*Wiener-Khintchine* relationship). Upon substituting Eq. (3.67) into Eq. (3.65), it can be concluded that the periodogram is a somewhat distorted version of the true power spectrum. The distortion, or as it is sometimes called "blurring," is due to the triangular window $w_{tri}[k]$, which reflects the bias in the result. The blurring can be mitigated somewhat by increasing the size of the window (i.e., N). N should be chosen so that the main lobe of the Fourier transform of the window function is considerably smaller that the width of the spectral peaks in $\tilde{P}(\varpi)$. The main lobe width of a triangular window is approximately $4\pi/N$ wide, which means that a periodogram will not be able to resolve peaks in the power spectra separated by less than $2\pi/N$. This result is consistent with the rectangular window result developed in the context of the periodogram expression in Eq. (3.62).

A special case that lends itself to analysis assumes that the input is a white Gaussian noise process. If $N \rightarrow \infty$, then

$$\lim_{N \to \infty} E(\tilde{P}(\varpi)) = P_{xx}(\varpi) \qquad (3.68)$$

which defines the mean to be asymptotically unbiased. The asymptotic variance of periodogram is given by

$$\lim_{N \to \infty} \mathrm{var}(\tilde{P}(\varpi)) = P_{xx}^2(\varpi) \qquad (3.69)$$

which states that, even in the limit, the variance of the periodogram is equal to the square of the power spectrum. The periodogram is, therefore, an inconsistent estimator of the power spectrum, which explains why the periodogram often is seen to exhibit high fluctuations.

Example 3.13: Periodogram

Problem Statement. A "white" Gaussian random process has an ideal power spectrum which is flat, that is, $P_{xx}[\varpi] = c$. Compute and compare the periodogram of a 64, 256, and 1024 sample white Gaussian time-series. Compare the results with the theoretical results.

Analysis and Results. The experimental results are reported in Fig. 3.18 (results will vary due to use of a random time-series). The power spectrum detail is seen to be directly related to the number of samples used in its construction. The power spectrum of a white random process is theoretically flat or constant at all frequencies. This is seen not to be the case for small records, although the $N = 1024$ sample power spectrum exhibits power at each harmonic. Even extremely long records do not produce a completely flat spectrum, but rather a spectrum that contains many spectral lines.

Computer Study. The data displayed in Fig. 3.18 is produced by executing the sequence of Siglab commands shown below.

```
>x=gn(1024)                     # Gaussian noise record
>p64=mag(fft(x[0:63]))^2/64     # Periodogram of 64 pt record
>p256=mag(fft(x[0:255]))^2/256  # Periodogram of 256 pt record
>p1024=mag(fft(x))^2/256        # Periodogram of 1024 pt record
>graph(p64,p256,p1024)          # Graph results              ❖
```

3.13 Bartlett Periodogram

Example 3.13 demonstrated that the PSD of a single record where white noise is present can contain significant power spectrum fluctuations. To reduce the fluctuations in the power spectrum estimate, ensemble averaging of periodograms can be used. Since real ensemble averaging is not available for a single real time-series, ensemble averaging must be simulated using a time average. To perform time averaging of periodograms, it must be assumed that the time-series is ergodic over the total record. Consider the partitioning of a time-series $x[k]$ of length MN into M sets of N samples each. That is, the sample $x[k] = x[iN + j]$ is the jth sample ($j \equiv k \bmod (N)$) of the ith data record for $i \in \{0,1,2,\ldots,M{-}1\}$. Then the *Bartlett periodogram* is constructed by averaging M periodograms with nonoverlapping data records of length N and is given by

$$\tilde{P}_B(\omega) = \frac{1}{MN} \sum_{m=0}^{M-1} \left| \sum_{n=0}^{N-1} x[mM + n]e^{-j\omega n} \right|^2 \tag{3.70}$$

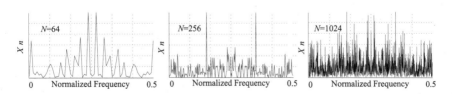

Figure 3.18 Power spectrum of a Gaussian time-series of length $N = 64$, 256, and 1024.

Example 3.14: Bartlett Periodogram

Problem Statement. Suppose $x[k]$ is the sum of a Gaussian white noise process with a variance of four and a symmetric squarewave (i.e., ±1) with a frequency of $\theta = 1/16$ (i.e., a period of 16 samples) having a duty cycle of one-half. Compute the Bartlett periodogram of a time-series where $N = 512$ is the record length for each individual periodogram, and the number of periodograms to average is given by $M \in \{1,8,16,32\}$. Compare the results.

Analysis and Results. The power spectrum of the random time-series is ideally a constant for all harmonics. The power spectrum of the squarewave has a line at the frequency of the squarewave and lines of declining value at odd multiples (harmonics) of the fundamental frequency of the squarewave. The experimental results are shown in Fig. 3.19. Observe that for the $M = 1$ periodogram, only the fundamental harmonic is clearly visible and that the noise floor is not very flat. For the $M = 8$ Bartlett periodogram, the harmonic at three times the fundamental frequency of the square wave is now visible and the noise floor is flatter than in the $M = 1$ case. For the $M = 16$ and $M = 32$ Bartlett periodograms, an improving picture of the spectrum is developed with a flatter noise floor.

Computer Study. The data shown in Fig. 3.19 were produced by the sequence of Siglab commands shown below. The s-file "`bartlett.s`" contains a function to compute the Bartlett periodogram using the function format "`bartlett(x,n,m)`" where x is at least an nm sample time-series, n is the number of data to use for each periodogram, and m is the number of periodograms to ensemble average. Each constituent periodogram is computed using a DFT.

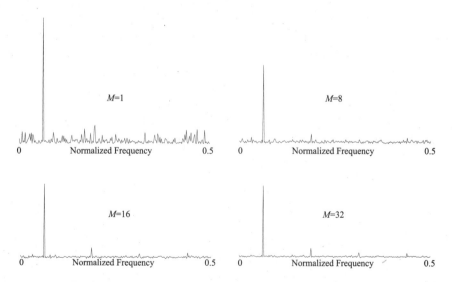

Figure 3.19 Bartlett periodogram power spectrum estimate.

```
>include "chap3\bartlett.s"
>x=sq(16,1/2,16384)+2*gn(16384) # Square wave plus noise
>graph(bartlett(x,512,1))  # Periodogram (M=1, N=512)
>graph(bartlett(x,512,8))  # Bartlett Periodogram (M=8, N=512)
>graph(bartlett(x,512,16)) # Bartlett Periodogram (M=16, N=512)
>graph(bartlett(x,512,32)) # Bartlett Periodogram (M=32, N=512) ❖
```

3.14 Welch Periodogram

The Bartlett periodogram method partitioned a signal into the set of M distinct (nonoverlapping) N sample data records. Each constituent periodogram in the Bartlett periodogram had an implied square window. A logical variation on this theme is to allow other windowing functions besides the square window. The obvious approach is to modify the Bartlett periodogram in Eq. (3.70) to directly allow arbitrary windowing functions, $w[k]$, to be applied:

$$\tilde{P}_B(\omega) = \frac{1}{MN} \sum_{m=0}^{M-1} \left| \sum_{n=0}^{N-1} x[mM+n]w[n]e^{-j\omega n} \right|^2 \qquad (3.71)$$

Any likely windowing function (see Table 3.5) selected is intended, in part, to eliminate the jump discontinuity that usually results from the assumption of periodic extension. Elimination of the jump discontinuity is accomplished by tapering the window to zero at its ends. This has important consequences in the context of the modified Bartlett periodogram in Eq. (3.71), namely that the application of such windows can lead to significant destruction of signal (and, thus, possible spectral) content. To illustrate, consider the case of an AM signal as shown in Fig. 3.20. Here, a carrier sequence of 1024 samples is modulated by a raised sinusoid producing the modulated carrier shown. Next, a plot of four nonoverlapping Hamming windows, each 256 samples long, is shown. These represent a concatenated Hamming window function, $w[k]$, where $N = 256$ and $M = 4$. The product of the sequence of Hamming windows and the modulated carrier sequence is shown next. Notice that approximately half of the signal content is annihilated. Had the modulating waveform varied in frequency, then the modified Bartlett periodogram given by Eq. (3.71) might have missed significant signal content.

To combat the signal annihilation problem illustrated in Fig. 3.20, the modified Bartlett periodogram given in Eq. (3.71) may be further modified to allow overlapping data segments. By overlapping segments for the periodogram the signal energy may be conserved over almost the entire record length. The *Welch periodogram* is given by

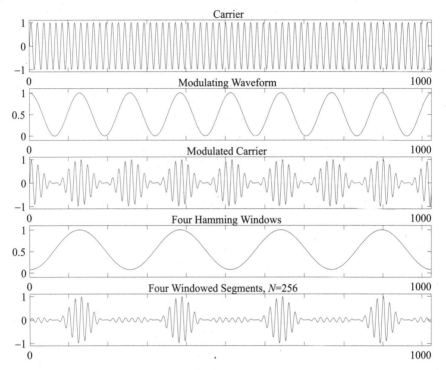

Figure 3.20 Construction of an amplitude modulated signal and the application of four nonoverlapping Hamming windows to the signal.

$$\tilde{P}_w(\omega) = \frac{1}{MNR} \sum_{m=0}^{M-1} \left| \sum_{n=0}^{N-1} x[mL+n]w[n]e^{-j\omega n} \right|^2 \qquad (3.72)$$

where N is the number of samples in each data segment, M is the number of periodograms to ensemble average, L is the number of data to skip from segment to segment, where $0 < L \le N$ and R is a window scaling factor derived from the window sequence, $w[k]$, and is given by

$$R = \sum_{k=0}^{N-1} w^2[k] \qquad (3.73)$$

If $M = 1$ and the windowing function is a rectangular window then the Welch periodogram of Eq. (3.72) is the same as the periodogram, Eq. (3.62). If $M > 1$, $L = N$, and the windowing function is rectangular then the Welch periodogram of Eq. (3.70) is the same as the Bartlett periodogram of Eq. (3.72). Frequently, the segment overlap will be discussed in terms of the percentage of overlap from segment to segment.

In general, if the overlap is said to be p percent then $L = (1 - p/100\%)N$.

Example 3.16: Welch Periodogram

Problem Statement. Consider once again $x[k]$ to be the sum of a Gaussian white noise process with variance four and a symmetric squarewave with a frequency of $\theta = 1/16$ (i.e., a period of 16 samples) having a duty cycle of one-half. Compute the Welch estimate using a Hamming window with $N = 512$, and 25 percent overlap for $M \in \{1,8,16,32\}$. Compare with the Bartlett periodogram.

Analysis and Results. The experimental outcomes are shown in Fig. 3.21. All of the periodograms plotted are at the same scale. It can be seen that the power spectrum produced using the Welch periodogram has a somewhat flatter noise power spectrum due greater smearing effect of the Hamming window compared to the rectangular window. Another aspect of the increased smearing by the Hamming window is that the higher harmonics (at five and seven times the fundamental frequency) do not rise as far out of the noise floor as that of the comparable Bartlett periodogram. In fact, all of the peaks of the Welch periodograms are lower than their Bartlett periodogram counterparts. This is easily explained: since the peaks in the Welch periodogram are wider, they must be shorter for there to be the same area (energy) under the curve. Otherwise, the effect of the Hamming window versus the rectangular window is seen to be marginal in this case.

Computer Study. The s-file "welch.s" was used to create the graphs shown in Fig. 3.21. The function "welch(x,w,l,m)" computes a Welch periodogram where x is the input time-series, w is an N sample window function, and l is an integer defining the segment-to-segment skip, and m is the number of segments to ensemble average. The following sequence of Siglab commands creates the graphs shown in Fig. 3.21.

```
>include "chap3\welch.s"
>x=2*gn(16384)+sq(16,1/2,16384) # Create signal plus noise
>wr=ones(512)
>wh=ham(512)
>graph(welch(x,wh,384,1)   # Welch per. with 25% ovlp, M=1
>graph(welch(x,wh,384,8)   # Welch per. with 25% ovlp, M=8
>graph(welch(x,wh,384,16)  # Welch per. with 25% ovlp, M=16
>graph(welch(x,wh,384,32)  # Welch per. with 25% ovlp, M=32
>graph(welch(x,wr,512,1)   # Bartlett periodogram, M=1
>graph(welch(x,wr,512,8)   # Bartlett periodogram, M=8
>graph(welch(x,wr,512,16)  # Bartlett periodogram, M = 16
>graph(welch(x,wr,512,32)  # Bartlett periodogram, M = 32          ❖
```

3.15 Blackman-Tukey Estimator

The Blackman-Tukey method computes a power spectrum estimate based on a manipulation of the autocorrelation function. Referring to the periodogram expression in Eq. (3.62), recall that

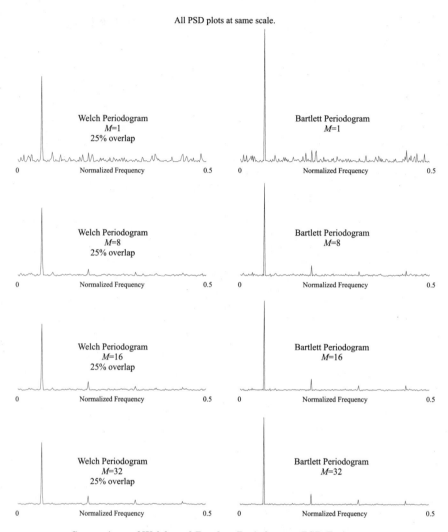

Figure 3.21 Comparison of Welch and Bartlett Periodogram PSD Estimators.

$$\tilde{P}(\varpi) = \sum_{k=-(N-1)}^{N-1} \hat{r}_{xx}[k]e^{-jk\varpi} \qquad (3.74)$$

which is based on the biased estimate of the autocorrelation function $\hat{r}_{xx}[k]$. One of the problems associated with this autocorrelation function is that is based on only a few samples for large lags. The Black-man-Tukey algorithm attempts to compensate for this problem by weighting the autocorrelation terms at large lags less heavily. Specifically, the Blackman-Tukey estimator is given by

$$\tilde{P}_{BT}(\omega) = \sum_{k=-(N-1)}^{N-1} \hat{r}_{xx}[k]w[k]e^{-jk\omega} \qquad (3.75)$$

where $w[k]$ is a window function of width $2M + 1$ where $M \le N - 1$, the length of the correlation interval. The choice of window is arbitrary but only a triangular window will guarantee that the resulting power spectrum estimate is positive (i.e., $\tilde{P}_{BT}(\omega) \ge 0$). Other windows can lead to a power spectrum estimate which is negative (physically impossible). The window function has the effect of reducing the variance of the power spectrum estimate but at the price of creating a larger bias. The triangular window applied to the Blackman-Tukey estimator results in

$$\tilde{P}_{BT}(\omega) = \sum_{k=-M}^{M} \hat{r}_{xx}[k]\left(1 - \frac{|k|}{N}\right)^{-jk\omega} \qquad (3.76)$$

The variance of the estimator is

$$\text{var}(\tilde{P}_{BT}(\omega)) \approx \frac{2M}{3N}P_{xx}^2(\omega) \qquad (3.77)$$

Blackman and Tukey have suggested that M be limited to one-tenth of N (or less).

In summary, the fundamental problem with nonparametric estimator is that it can blur the power spectral signature of a signal. These classical nonparametric estimators remain popular due to their simplicity and ease of computation. Proakis and Manolakis[2] compared these basic nonparametric methods using a quality factor Q given by

$$Q = \frac{\left\{E[P_{xx}(f)]^2\right\}}{\text{var}[P_{xx}(f)]} \qquad (3.78)$$

The basic methods are compared in Table 3.6 for M representing the total length of a data record, N the number of samples per segment, and $K = M/N$ being the number of segments.

Table 3.6   Comparison of Spectral Estimators

Method	Qualifiers	Δf	Q
Bartlett periodogram	Based on 3 db main lobe approximation	$\Delta f = 0.9/M$	$Q = 1.11N\Delta$
Welch	Triangular window, 50% overlap	$\Delta f = 1.28/M$	$Q = 1.39N\Delta f$
Welch	Triangular window, 0% overlap	$\Delta f = 0.78/M$	$Q = 1.39N\Delta f$
Blackman-Tukey	Triangular window, length $2M - 1$	$\Delta f = 0.64/M$	$Q = 2.34N\Delta f$

3.16 Capon

For completeness, the Capon method is often included in a list of basic nonparametric estimators. The Capon power spectrum estimate is given by

$$P_c(\phi) = \frac{1}{e^H(\phi)\hat{R}_{xx}^L e(\phi)}$$ (3.79)

where

$$e^H(\phi) = (1, e^{-j\phi_0}, e^{-j2\phi_0}, ..., e^{-j(N-1)\phi_0})$$

$$\hat{R}_{xx}^L = \text{cov}(x[k])$$ (3.80)

and $\text{cov}(x[k])$ is an $L \times L$ covariance matrix. L is generally chosen to be small compared to N, the length of $x[k]$, to avoid computing covariance terms with long lags. The Capon method is generally considered to be of higher precision that the others but suffers from a possible high error variance due to the uncertainty associated with the covariance matrix calculation.

3.17 Summary

In Chap. 3, the concept of a discrete-time frequency-domain representation of a signal was introduced. The discrete-time Fourier transform (DTFT) and discrete-time Fourier series (DTFS) were extensions of the continuous-time Fourier transform and series (CTFT, CTFS). The study of the DTFT, DTFS, and DFT was introduced. All produced periodic spectra. The DTFT was shown to be applicable for arbitrary discrete-time signals. If the discrete-time signal under study is periodic, then a simpler DTFS can be used. When only a finite duration discrete-time signal is known, the DFT is used with an implicit assumption of periodicity. The Fourier transform choices are summarized in Fig. 3.22 and Table 3.7. Of these transforms, the DFT is the most popular due to the fact that is easily computed using a general-purpose digital computer. The preferred embodiment of the DFT will be developed in the next chapter.

3.18 Brief Biography of Jean Baptiste Joseph Fourier (1768–1830)

Fourier was the son of a tailor and orphaned at the age of eight. Educated at a nonsectarian military college where he excelled in math, he considered entering the priesthood when the French Revolution broke out. Fourier joined the revolution, which, in its aftermath, left France without a viable intelligentsia. Fourier himself had several close encounters with the guillotine. The persecution of scientists was finally abated through the intervention of Napoleon. Napoleon, in fact, en-

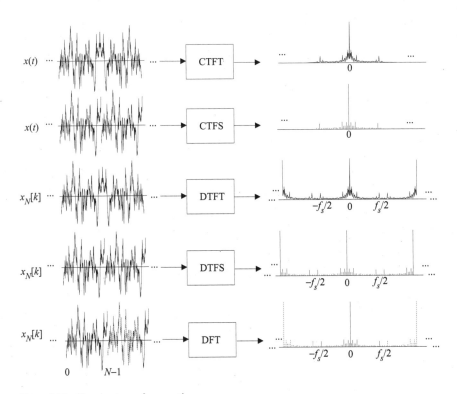

Figure 3.22 Fourier transform options.

Table 3.7 Summary of Fourier Transform Cases

Transform	Signal	Type	Duration	Resolution	Spectrum	Type	Domain	Resolution
CTFT	$x(t)$	arbitrary	∞	continuous	$X(f)$	arbitrary	∞	continuous
CTFS	$x_p(t)$	periodic	∞	continuous	c_n	arbitrary	∞	discrete
DTFT	$x[k]$	arbitrary	∞	discrete	$X(e^{j\omega})$	periodic	$[-f_s/2, f_s/2]$	continuous
DTFS	$x_p[k]$	periodic	∞	discrete	c_n	periodic	$[-f_s/2, f_s/2]$	discrete
DFT	$x_N[k]$	periodic*	N	discrete	$X[n]$	periodic	$[-f_s/2, f_s/2]$	discrete

*= assumed periodic

couraged the academic production scientists, which benefited Fourier who, at the age of 26, was appointed to the faculty of the new Ecole Normale.

Napoleon, in 1789, visited Egypt with a group of scientists including Fourier. Fourier was appointed to the new Institut d'Egypte, which became famous for its work with the Rosetta stone. Returning to France in 1801, Fourier's career advanced and retreated along with Napoleon's exiles and along with his several changes of political loyalties. In 1807, while serving as the Prefect of Grenoble, Fourier claimed the work that would make him famous. His studies in heat transfer were collected in a paper that introduced what we now would consider to be the foundations of orthogonal functions and the Fourier series. As was the custom of the time, papers were judged by the eminent scholars of the day. They were, in this case, Lagrange, Laplace, Legendre, and a fourth scholar. They found the paper innovative but lacking mathematical elegance and rigor. Lagrange, however, did not believe the results were correct and vigorously opposed its publication over the support of the other three judges. Fourier, unfortunately, could not provide a rigorous defense for his work (Dirichlet later completed the proof). The paper would wait 15 more years to see print in the text *Théorie analytique de la chaleur.*

3.19 Self-Study Problems

1. DTFT properties Given that $x_p[k]$ is a periodic signal of period N (samples) having a DTFS given by c_n, what is the DTFS of $y[k]$, where:

$y[k] = x_p[k - m]$

$y[k] = x_p[k] + x_p[-k]^*$

$y[k] = (-1)^k x_p[k]$

2. EKG signal A 256 sample EKG data record, as found in Example 3.6, was sampled at a rate of 135 Hz. A 256-sample DFT is computed. Specify what the frequency resolution is and which harmonics are closest to 33.2 Hz and −45.7 Hz. If a frequency resolution greater than or equal to 2.2 Hz per harmonic is required, what would be the minimum time-series length N?

3. Leakage Leakage was observed when a signal did not complete an integer number of oscillations in an N sample interval. Create three cosine waves of period 3.25/64 with lengths of 64, 128, and 256. Measure the power ratio of the peak spectral line to the leakage power as a function of N (hint: use Parseval's equation). Make conclusions about how leakage is related to signal length.

4. Compression Examine the power spectrum of an EKG and locate the contiguous harmonic indices over which are contained 90 percent of the total signal power. How are harmonics are tagged as disposable, and what would the compression ratio be if only the power-important harmonics are kept?

3.20 References

1. Cooley, J.W. and Tukey, J.W. "An Algorithm for Machine Calculation of Complex Fourier Series." *Mathematics of Computation* (American Mathematical Society), vol. 19, no. 90, 1965.
2. Proakis, J.G. and Manolakis, D.G. *Introduction to Digital Signal Processing.* New York: Macmillan, 1988.

4

Computing Discrete Fourier Transforms

4.1 Introduction

In the previous chapter, the DFT was introduced. The benefit of an N-point DFT, in comparison to the DTFT and DTFS, was one of practicality. The direct implementation of a DFT simply requires that N^2 complex multiply-accumulates (MACs) be performed per transform, whereas the DTFT and DTFS require infinite summations or integration of continuously defined functions. Even with reasonably fast computers, a long DFT can be a computationally expensive task. This condition dramatically changed with the advent of the *fast Fourier transform*, or FFT. It should be appreciated, however, that "fast" has a relative meaning in this case. The FFT, while being faster than a direct calculation, remains computationally expensive. In addition, the FFT also incurs some overhead penalty. As a general rule, the advantage of a software-based FFT over a direct DFT cannot be realized unless $N \geq 32$. Finally, high-bandwidth real-time FFTs remain elusive even with today's digital computers and application specific integrated circuits (ASICs) that have been developed expressly to perform FFTs.

4.2 The Fast Fourier Transform

The DFT assumed a periodic time-series $x_N[k]$ with period N, and defines a mapping of the complex N sample series in the time-domain into a complex N sample series in the frequency-domain (i.e., $\mathbf{C}^N \leftrightarrow \mathbf{C}^N$). The DFT is formally given by

$$X[n] = \sum_{k=0}^{N-1} x_N[k] W_N^{nk} \qquad (4.1)$$

for $0 \leq n < N$, $X[n]$ is called the nth harmonic, and the complex exponential $W_N = e^{-j2\pi/N}$ corresponds to a phase shift. The inverse DFT is given by

$$X_N[k] = \frac{1}{N} \sum_{n=0}^{N-1} X[n]W_N^{-nk} \tag{4.2}$$

for $0 \leq k < N$. It can be seen that the same linear algebraic computational framework can be used to implement both Eq. (4.1) and Eq. (4.2). Let $x_N[k]$ be expressed as a vector,

$$\mathbf{x}_N = \begin{bmatrix} x_N[0] \\ x_N[1] \\ \cdots \\ x[N-1] \end{bmatrix} \in \mathbf{C}^N \tag{4.3}$$

and let $X[n]$ be expressed as a vector,

$$\mathbf{X} = \begin{bmatrix} X[0] \\ X[1] \\ \cdots \\ X[N-1] \end{bmatrix} \in \mathbf{C}^N \tag{4.4}$$

Define an $N \times N$ matrix of complex coefficients,

$$\mathbf{T} \in \mathbf{C}^{N \times N} \tag{4.5}$$

where each element of $\mathbf{T}$, T_{ij}, is defined to by

$$T_{ij} = W_N^{(i-1)(j-1)} \tag{4.6}$$

for $i,j \in \{1,2,3,\ldots,N\}$. $\mathbf{T}$ is clearly symmetric. It then follows that

$$\mathbf{X} = \mathbf{Tx} \tag{4.7}$$

and

$$\mathbf{x} = \mathbf{T}^{-1}\mathbf{X} \tag{4.8}$$

However, from Eq. (4.2), it is known that

$$\mathbf{x} = \frac{1}{N}\mathbf{VX} \tag{4.9}$$

where $\mathbf{V} \in \mathbf{C}^{N \times N}$, and V_{ij}, is defined by

$$V_{ij} = W_N^{-(i-1)(j-1)} \tag{4.10}$$

From Eqs. (4.6), (4.9), and (4.10), it follows that

$$\mathbf{T}^{-1} = \frac{1}{N}\mathbf{V} = \frac{1}{N}\mathbf{T}^* = \frac{1}{N}\mathbf{T}^H \tag{4.11}$$

where $\mathbf{T}^H$ is the Hermitian transpose (conjugate-transpose) of $\mathbf{T}$ and $\mathbf{T}^*$ is the conjugate of $\mathbf{T}$. Therefore,

$$\mathbf{x} = \mathbf{T}^{-1}\mathbf{X} = \frac{1}{N}\mathbf{T}^*\mathbf{X} = \frac{1}{N}\mathbf{T}^H\mathbf{X} \tag{4.12}$$

Example 4.1: The Discrete Fourier Transform

Problem Statement. Compute the DFT of a four-point time-series, $x_4[k] = \{1,1,0,0\}$, and the inverse DFT using the matrix formulation.

Analysis and Conclusions. The four-point matrix-determined DFT is given by

$$\mathbf{X} = \begin{bmatrix} 1 & 1 & 1 & 1 \\ 1 & W_4^1 & W_4^2 & W_4^3 \\ 1 & W_4^2 & W_4^4 & W_4^6 \\ 1 & W_4^3 & W_4^6 & W_4^9 \end{bmatrix} \mathbf{x} = \begin{bmatrix} 1 & 1 & 1 & 1 \\ 1 & -j & -1 & j \\ 1 & -1 & 1 & -1 \\ 1 & j & -1 & -j \end{bmatrix} \begin{bmatrix} 1 \\ 1 \\ 0 \\ 0 \end{bmatrix} = \begin{bmatrix} 2 \\ 1-j \\ 0 \\ 1+j \end{bmatrix}$$

The inverse mapping is given by

$$\mathbf{x} = \frac{1}{4}\begin{bmatrix} 1 & 1 & 1 & 1 \\ 1 & W_4^{-1} & W_4^{-2} & W_4^{-3} \\ 1 & W_4^{-2} & W_4^{-4} & W_4^{-6} \\ 1 & W_4^{-3} & W_4^{-6} & W_4^{-9} \end{bmatrix} \mathbf{X} = \frac{1}{4}\begin{bmatrix} 1 & 1 & 1 & 1 \\ 1 & j & -1 & -j \\ 1 & -1 & 1 & -1 \\ 1 & -j & -1 & j \end{bmatrix} \begin{bmatrix} 2 \\ 1-j \\ 0 \\ 1+j \end{bmatrix} = \begin{bmatrix} 1 \\ 1 \\ 0 \\ 0 \end{bmatrix}$$

as expected. ❖

It was argued in Chap. 3 that the direct computation of all N harmonics of $X[n]$ required $M_1 = N^2$ complex multiply-accumulates per transform. Suppose that N is a radix-two number given by $N = 2^n$. Then the time-series $x_N[k]$ can be logically partitioned into even and odd indexed data blocks of length $N/2$ samples each. Consider, then, the following partitioning:

$$x_{N/2}^e[k] = x_N[2k] \quad \text{even}$$

$$x_{N/2}^o[k] = x_N[2k+1] \quad \text{odd}$$

(4.13)

for $0 \le k < N/2$. Substituting the even and odd signal components into Eq. (4.1), the DFT, results in

$$X[n] = \sum_{k=0}^{N-1} x_N[k] W_N^{nk}$$

$$= \sum_{k=0}^{N/2-1} x_N[2k] W_N^{n2k} + \sum_{k=0}^{N/2-1} x_N[2k+1] W_N^{n(2k+1)}$$

$$= \sum_{k=0}^{N/2-1} x_N[2k] W_N^{n2k} + \sum_{k=0}^{N/2-1} x_N[2k+1] W_N^{n2k} W_N^n$$

(4.14)

The complex exponential found in Eq. (4.14) may be simplified by noticing that the identity

$$W_N^{2m} = e^{-2\pi(2m)/N} = e^{2\pi m/(N/2)} = W_{N/2}^m$$

(4.15)

may be substituted into Eq. (4.14) to produce

$$X[n] = \sum_{k=0}^{N/2-1} x_{N/2}^e[k] W_{N/2}^{nk} + W_N^n \sum_{k=0}^{N/2-1} x_{N/2}^o[k] W_{N/2}^{nk}$$

$$= X_{N/2}^e[n] + W_N^n X_{N/2}^o[n]$$

(4.16)

where $X_{N/2}^e[n]$ is an $N/2$-point DFT of the even samples of $x_N[k]$ and $X_{N/2}^o[n]$ is an $N/2$-point DFT of a time series formed by the odd samples of $x_N[k]$. Recall that an N-point DFT is also periodic with period N. Therefore, $X_{N/2}^e[n]$ and $X_{N/2}^o[n]$ are also periodic in the frequency domain with period $N/2$. That is,

$$X_{N/2}^e[n] = X_{N/2}^e[N/2+n] \text{ and } X_{N/2}^o[n] = X_{N/2}^o[N/2+n]$$

To produce $X_{N/2}^e[n]$ and $X_{n/2}^o[n]$ using a direct DFT would require $(N/2)^2$ complex MACs for each $N/2$-point DFT. Excluding the multiplication by W_N^n found in Eq. (4.16), a direct DFT would require $M_2 = 2(N/2)^2 = N^2/2$ complex MACs per transform. This is half the previous MAC count (i.e., M_1) required to implement a direct N-point DFT. If this trick halved the MAC count, consider using it again. If the $N = 2^n$ time-series $x_N[k]$ is subdivided into four $N/4$ sample data-blocks, the DFT complexity could be reduced to $M_4 = (N/4)^2 + (N/4)^2 + (N/4)^2 + (N/4)^2 = N^2/4$, or a quarter of the direct N-point DFT complexity. Continuing this line of reasoning, dividing the $N = 2^n$ sample time-series into $N/2$ two-point time-series, would produce a DFT formula having a complexity on the order of $M_{N/2} = N\log_2(N)$.

Example 4.2: A Four-Point DFT

Problem Statement. Compute the DFT of a four-point time-series $x_4[k] = \{1,1,0,0\}$, using a radix-four ($N = 4$) and a radix-two ($N = 2^2$) DFT.

Analysis and Conclusions. The four-point DFT algorithm is given by

$$X[n] = \sum_{k=0}^{3} x_4[k] W_4^{nk}$$

where

$X[0] = x_4[0] + x_4[1] + x_4[2] + x_4[3]$

$\quad = 2$

$X[1] = x_4[0] + x_4[1]W_4^1 + x_4[2]W_4^2 + x_4[3]W_4^3 = x_4[0] - jx_4[1] - x_4[2] + jx_4[3]$

$\quad = 1 - j$

$X[2] = x_4[0] + x_4[1]W_4^2 + x_4[2]W_4^4 + x_4[3]W_4^6 = x_4[0] - x_4[1] + x_4[2] - x_4[3]$

$\quad = 0$

$X[3] = x_4[0] + x_4[1]W_4^3 + x_4[2]W_4^6 + x_4[3]W_4^9 = x_4[0] + jx_4[1] - x_4[2] - jx_4[3]$

$\quad = 1 + j$

which is the result obtained in Example 4.1. The radix-two approach partitions the four-sample signal into even and odd data blocks, as found in Eq. (4.16). This results in

$$X[n] = \sum_{k=0}^{1} x_4[2k] W_2^{kn} + W_4^n \sum_{k=0}^{1} x_4[2k+1] W_2^{kn}$$

$$= X_2^e[n] + W_4^n X_2^o[n]$$

where

$$X_2^e[0] = x_4[0] + x_4[2]W_2^0 = x_4[0] + x_4[2] = 1$$

$$X_2^e[1] = x_4[0] + x_4[2]W_2^1 = x_4[0] - x_4[2] = 1$$

$$X_2^o[0] = x_4[1] + x_4[3]W_2^0 = x_4[1] + x_4[3] = 1$$

$$X_2^o[1] = x_4[1] + x_4[3]W_2^1 = x_4[1] - x_4[3] = 1$$

Then forming

$$X[n] = X_2^e[n] + W_4^n X_2^o[n]$$

results in

$$X[0] = X_2^e[0] + W_4^0 X_2^o[0] = X_2^e[0] + X_2^o[0] = 2$$

$$X[1] = X_2^e[1] + W_4^1 X_2^o[1] = X_2^e[1] - jX_2^o[1] = 1 - j$$

$$X[2] = X_2^e[2] + W_4^2 X_2^o[2] = X_2^e[2] - X_2^o[2] = 0$$

$$X[3] = X_2^e[3] + W_4^3 X_2^o[3] = X_2^e[3] + jX_2^o[3] = 1 + j$$

as expected.   ❖

4.3 Decimation-in-Time FFT

A graphical interpretation of an eight-point DFT is shown in Fig. 4.1.
The first step, shown in Fig. 4.1a, corresponds to dividing $x_N[k] = x_8[k]$
into two time-series of length $N/2 = 4$ each, which are denoted by
$x_{N/2}^e[k]$ and $x_{N/2}^o[k]$ in Eq. (4.13), or $x_4^e[k]$ and $x_4^o[k]$ for the case
where $N = 8$. This act is called *decimation* and gives rise to the nomenclature, *decimation-in-time-DFT* or *DIT DFT*. The two four-point DFTs
are labeled $X_{N/2}^e[n]$ and $X_{N/2}^o[n]$, respectively, in Eq. (4.16) which, in
this case, are denoted $X_4^e[n]$ and $X_4^o[n]$. The two groups of four harmonics found at the output of $X_4^e[n]$ and $X_4^o[n]$ are then combined according to Eq. (4.16). Observe that the outputs of $X_4^o[n]$ are weighted
by W_8^n for $n \in \{0,1,2,...,7\}$.

The next step, shown in Fig. 4.1b, is a conversion of each of the
four-point DFTs, namely $X_4^e[n]$ and $X_4^o[n]$, into two two-point DFTs.
Consider, for demonstrative purposes, the reduction of $X_4^e[n]$ into two
two-point DFTs. The four-point time-series presented to $X_4^e[n]$ is
$x_4^e[k] = \{x_4^e[0], x_4^e[1], x_4^e[2], x_4^e[3]\} = \{x_8[0], x_8[2], x_8[4], x_8[6]\}$, which
can be partitioned, after Eq. (4.13), into the two two-point time-series

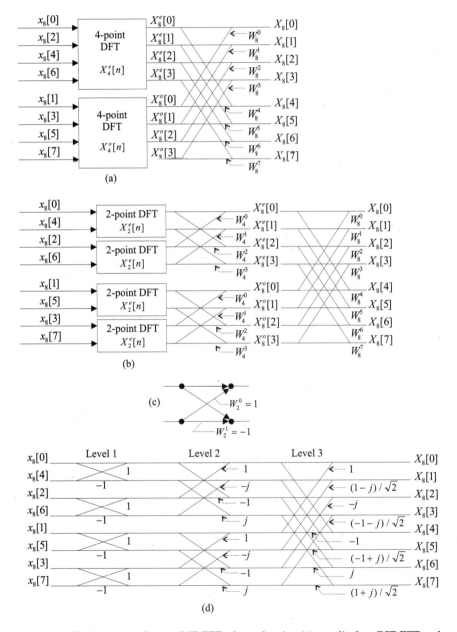

Figure 4.1 Eight-point radix-two DIT FFT where showing (*a*) a radix-four DIT FFT and radix-two FFT sections, (*b*) a radix-two FFT, (*c*) a radix-two butterfly, and (*d*) a detailed schematic of a radix-two eight-point FFT.

$X_2^e[k] = \{x[0], x[4]\}$ and $X_2^o[k] = \{x[2], x[6]\}$, respectively. Each of the two-point time-series can then be processed by the two-point DFTs given by and as shown in Fig. 4.1c. Specifically,

$$X_2[n] = \sum_{k=0}^{1} x_2[k] W_2^{nk} \qquad (4.17)$$

is often referred to as an FFT *butterfly*.

A partitioning similar to that applied to the even indexed samples is used to map $x_4^o[k] = \{x_4^o[0], x_4^o[1], x_4^o[2], x_4^o[3]\} = \{x_8[1], x_8[3], x_8[5], x_8[7]\}$ into the two two-point time-series $x_2^e[k] = \{x_8[1], x_8[5]\}$ and $x_2^o[k] = \{x_8[3], x_8[7]\}$, respectively. The two-point DFTs of four two-point time-series are shown in Fig. 4.1d. The outputs of the two-point DFTs are then combined according to Eq. (4.16) for $N = 4$. The resulting output of the synthesized four-point DFTs (constructed from nested two-point DFTs) would in turn be combined at the output stage to define an eight-point DFT.

Referring to Fig. 4.1d, it can be seen that the input data has been reordered to define an input array which is a permuted version of $x_N[k]$. In particular, $x_N[k] = \{x_N[0], x_N[1], x_N[2], x_N[3], x_N[4], x_N[5], x_N[6], x_N[7]\}$ is permuted into $\{x_N[0], x_N[4], x_N[2], x_N[6], x_N[1], x_N[5], x_N[3], x_N[7]\}$ for $N = 8$. The permutation, or *index shuffle* rule that maps the order $I = \{0,1,2,3,4,5,6,7\}$ into $J = \{0,4,2,6,1,5,3,7\}$ is called *bit-reversing*. For $N = 8$, each index is represented by a three-bit word listed in Table 4.1. The data displayed are the sequential indices and their three-bit bit-reversed permutations. Notice that the order defined by the J column of Table 4.1 is a highly decimated version of the ordinal order, I, also found in Table 4.1.

Table 4.1 $N = 8$ Bit-Reversing Permutation for an $N = 8$ DIT FFT

I (decimal)	I (binary)	Reverse[I(binary)] = J (binary)	J (decimal)
0	000	000	0
1	001	100	4
2	010	010	2
3	011	110	6
4	100	001	1
5	101	101	5
6	110	011	3
7	111	111	7

Returning again to Fig. 4.1d, it can be seen that the $N = 8 = 2^3$ transform consists of three distinct levels. Upon inspection, note that at each level, two input samples are mapped to two output samples. What differentiates the two-point operation at one level from another is the distance separating the input samples and the applied value of W_N^n, which is called a *twiddle factor*. At the first level, adjacent sample values are combined from the permuted input string and weighted by $W_2^n = \{1,-1\}$. At the next level, the inputs are spaced two locations apart and weighted by $W_4^n = \{1,-j,-1,j\}$. At the last level, the inputs are spaced four locations apart and weighted by $W_8^n = \{1,(1-j)/\sqrt{2},-j,$ $(-1-j)/\sqrt{2},-1,(-1+j)/\sqrt{2},j,(1+j)/\sqrt{2}\}$. This pattern can be extended to the case where $N = 2^n$, $n > 3$, in a straightforward manner. Notice that, at the first level, applying the butterfly coefficients $W_2^n = \{1,-1\}$ technically does not require any multiplications but rather an identity operation ($W_2^0 = 1$) or negation ($W_2^1 = -1$).

The radix-two DIT FFT algorithm shown in Fig. 4.1d often appears in what is referred to as the *simplified form* shown in Fig. 4.2. In Fig. 4.2a, a two-point FFT having twiddle adjustments W_N^r and $W_N^{(r+N/2)}$ found throughout Fig. 4.1b is reduced from a two-multiplier butterfly structure to a one-multiplier butterfly architecture defined in terms of W_N^r and scale factor ± 1. Extending this concept to the entire eight-point FFT the architecture shown in Fig. 4.1d can be simplified to the one shown in Fig. 4.2b.

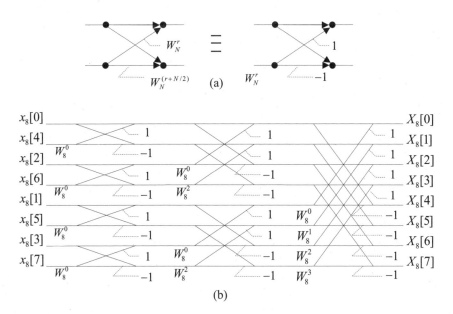

(b)

Figure 4.2 Simplified eight-point DIT FFT.

As a point of nomenclature, the DFT algorithm which reduces a $N = 2^n$-point DFT down to a set of two-point DFTs is called a *radix-two DFT*. If $N = 4^n$, then a radix-four algorithm results and can be considered as an alternative to a radix-two DFT. While having a more restrictive N than in the radix-two case, some designers take advantage of the fact that at the first level, the attendant butterfly coefficients are $W_4^n = \{1, -j, -1, j\}$. These coefficients can be implemented with non-arithmetic operations selectively applied to the real or imaginary data location. Radix-eight DFTs are also commonly found in practice.

One of the remarkable features of the FFT is that computation can be performed *in place*. This means that once a memory array has been allocated to store the input time-series $x_N[k]$, it can be reused *in place*. Data is taken from the array, processed at a given level in accordance with the DFT rules, and then returned to its original location. At the end of an FFT cycle, the data is found where it was originally placed except it has now been transformed into the frequency domain and is in a permuted order. For example, the $N = 8$-point FFT previously considered would be loaded into memory in bit-reversed order, processed using a two-point FFT, twiddle adjusted, and returned to memory. This would be repeated for all three levels and the outcome would be

$$
\begin{bmatrix} x_N[0] \\ x_N[4] \\ x_N[2] \\ x_N[6] \\ x_N[1] \\ x_N[5] \\ x_N[3] \\ x_N[7] \end{bmatrix} \xrightarrow{\text{FFT}} \begin{bmatrix} X[0] \\ X[1] \\ X[2] \\ X[3] \\ X[4] \\ X[5] \\ X[6] \\ X[7] \end{bmatrix} \tag{4.18}
$$

The implementation of a DIT-FFT as a computer program consists of a bit-reversing section and one that performs the individual two-point DFTs in a recursive manner. An example of a C program for implementing a radix-two DIT-FFT is shown below, and can be found in "src\c\chap4\fft.c". The program reads complex data from the standard input (e.g., the keyboard), computes the FFT using the DIT algorithm and double precision arithmetic, and prints the results to the standard output (e.g., the console). While the containing program is only designed to perform a single FFT, the underlying code is designed for efficient FFT processing when more than one FFT is to be

performed. To realize an efficient FFT for bulk processing, two potentially expensive operations must be managed. First, the value of the
complex exponential, W_N^n, is needed many times during the execution
of the FFT. Fortunately, there are only N values of n, so the complex
exponential is easily tabulated. Furthermore, it is possible to take advantage of the fact that $\sin(x) = \cos(x - \pi/2)$ and $W_N^n = \cos(2\pi n/N) -$
$j \sin(2\pi n/N)$, so that only the real (or imaginary) values of the complex
exponential are stored. The second obstacle to an efficient implementation is the need for bit-reversed indexing of data in the FFT. General-purpose computers, and the C language in particular, do not have
a particularly efficient means of performing bit-reversal. The solution
is, again, to compute a lookup table that will allow subsequent bit-reversal operations to occur using Table lookup rather than direct computation. The code contained in the given C implementation
dynamically sizes the cosine and bit-reversal tables to match the largest FFT size used; FFTs of smaller records can be computed with the
same tables used to compute larger FFTs. This approach ensures that
no more memory is consumed with the lookup tables than is necessary
while avoiding unnecessary recompilation of the tables when smaller
length FFTs are desired.

```
/* Program: FFT (c)
   Author: Jon Mellott
   Date: 2/01/97

   Description: Contains a function and supporting functions to perform
   a radix-2 DIT FFT of user defined length.

*/


#include <stdio.h>
#include <math.h>
#include <stdlib.h>
#include <stdio.h>


/* ------------- --------- */
/* Preprocessor Constants */
/* ------------- --------- */

#define PI 3.14159265359
#define PI2 6.28318530718

/* -------- ---------- */
/* Function Prototypes */
/* -------- ---------- */

long lBitReverse(long, int);
```

```
void FFTInit(int);
void FFT(double *, double *, int);

/* -------- ----------- */
/* Constant Definitions */
/* -------- ----------- */

const unsigned long ulExp2[] =
                {1,2,4,8,16,32,64,128,256,512,1024,2048,4096,
                8192,16384,32768,0x10000,0x20000,0x40000,
                0x80000,0x100000,0x200000,0x400000,0x800000,
                0x1000000,0x2000000,0x4000000,0x8000000,
                0x10000000,0x20000000,0x40000000,0x80000000};

/* ------ --------- */
/* Global Variables */
/* ------ --------- */

static int iLgMaxFFTLen = -1;      /* Log_2 of maximum FFT length. */
static double *dpFFTFact = NULL;   /* Pointer to FFT twiddle factor
   table. */
static long *lpBitRev = NULL;      /* Pointer to bit-reversal lookup
   table. */

/* --------- -------- */
/* PROCEDURE DIVISION */
/* --------- -------- */

/* Function: main
   Description:
        Accepts a single command line parameter describing the length
   of the FFT (log format). Calls the DIT FFT routine and prints the
   results. See usage message for more information.

   Parameters
   ----------
   argc -- Number of command line parameters.
   argv -- Array of pointers to command line parameters.
   ----------

   Returns: Zero if successful, a non-zero value if not successful.

*/
int main(int argc, char *argv[])
{
   int iLgLen;
   long lLength,lN;
   double *dpXR, *dpXI;

/* Check command line. */
   if (argc = = 2 && sscanf(argv[1],"%d",&iLgLen)==1 && iLgLen>0 &&
   iLgLen<32)
        lLength = (long)ulExp2[iLgLen];
```

```
else {
        /* Print usage message. */
        fprintf(stderr,"\nUsage: fft [n]"
                "\n Where n is the base-2 log of the length of the "
                "\n FFT to be performed. Program then accepts up to "
                "\n 2^n complex data where each datum is separated by "
                "\n white space and ordered r0, i0, r1, i1,...\n");
    exit(-1); /* Exit abnormally. */
}

/* Allocate complex data array. */
if ((dpXR = (double *)malloc(lLength*sizeof(double))) = = NULL ||
    (dpXI = (double *)malloc(lLength*sizeof(double))) = = NULL) {

    /* Print error message. */
    fprintf(stderr,"\n^^*ABEND: Memory allocation error.\n");

    exit(-2); /* Exit abnormally. */
}

/* Init the data array. */
for (lN = 0; lN<lLength; ++lN) dpXR[lN] = dpXI[lN] = 0.0;

/* Read data from the user. */
for (lN = 0; lN<lLength; ++lN) { /* Assume data is good or EOF. */
    if (scanf("%lg",dpXR+lN) == EOF) break; /* Read real datum. */
    if (scanf("%lg",dpXI+lN) == EOF) break; /* Read imaginary datum. */
}

/* Call the FFT. */
FFT(dpXR,dpXI,iLgLen);

/* Print results. */
for (lN = 0; lN<lLength; ++lN) printf("%lg %lg\n",dpXR[lN],dpXI[lN]);

/* Done. */
return 0;
}

/* Function: lBitReverse

   Description:
        Accepts a long integer of specified width and bit-reverses it
   using a direct computation.

   Parameters
   ----------
   lX        -- Long integer to bit-reverse.
   iWidth    -- Width of integer to bit-reverse. No error checking is
                performed on this parameter.
   ----------
Returns: Bit-reversed version of parameter lX.
*/
```

```
long lBitReverse(long lX, int iWidth)
{

    long lY;
    int iN;

    /*   Initialize result. */
        lY = 0;

    /* Perform bit reversal up to the width of the parameter. */
    for (iN = 0; iN<iWidth; ++iN) lY = (lY<<1)|((lX>>iN)&1);

    /* Done. */
    return lY;

}
/* Function: FFTInit
```

Description:
    Initializes FFT twiddle factor Table and bit-reversal tables
for use by the FFT routine. Tables are initialized according to a
specified maximum FFT size. If tables are already initialized and the
new length is larger, then larger tables are created.

Parameters
----
iLgMaxLen -- Base-two log of maximum FFT length. Must be a value in
[1,31]. No error checking is performed on this parameter.

Returns: Nothing.

```
*/

void FFTInit(int iLgMaxLen)
{
    long lN,lFFTLen;

    if (iLgMaxLen>iLgMaxFFTLen) { /* Check for larger FFT size than
        current. */
    if (dpFFTFact! = NULL) free(dpFFTFact); /* Free old factor array. */
    if (lpBitRev! = NULL) free(lpBitRev); /* Free old bit reversal table.
    */

    /* Compute new maximum FFT length. */
    lFFTLen = (long)ulExp2[iLgMaxLen];
    /* Record length of factor/bit reversal Table array. */
    iLgMaxFFTLen = iLgMaxLen;

    /* Allocate new factor array and bit reversal table. */
    if ((dpFFTFact = (double *)malloc(lFFTLen*sizeof(double)))==NULL ||
        (lpBitRev = (long *)malloc(lFFTLen*sizeof(long))) = = NULL) {

    /* Print error message. */
    fprintf(stderr,"\n***ABEND: Memory allocation error.\n");
```

```
    exit(-2); /* Exit abnormally. */
}

/* Initialize bit reversal and factor array. */
for (lN = 0; lN<lFFTLen; ++lN) {
  lpBitRev[lN] = lBitReverse(lN,iLgMaxLen);
  dpFFTFact[lN] = cos(PI2*(double)lN/lFFTLen);
  }

  /* Make zeros exact. */
      if (lFFTLen> = 4) dpFFTFact[lFFTLen>>2]=
          dpFFTFact[3*(lFFTLen>>2)] = 0.0;
  }
/* Done. */
}


/* Function: FFT
  Description:
      Performs an in-place (results overwrite original data) DIT FFT
  using a radix-two algorithm.

  Parameters
  ----------
  dpXR -- Array of real coefficients.
  dpXI -- Array of imaginary coefficients.
  iLgLen -- Base-two log of length data array on which to perform the
      FFT.
  ----------
  Returns: Nothing.
*/
void FFT(double *dpXR, double *dpXI, int iLgLen)
{
  long lN, lLength, lBStep, lBStride, lT1, lT2, lT3, lT4, lT5;
  int iLevel;
  double dTemp0, dTemp1;

  /* Verify that the tables are big enough for this problem. */
  if (iLgLen>iLgMaxFFTLen) FFTInit(iLgLen);

  /* Initialize length of data array. */
  lLength = (long)ulExp2[iLgLen];

  /* Bit-reverse index original data. */
  for (lN = 0; lN<lLength; ++lN) {
      lT1 = lpBitRev[lN]>>(iLgMaxFFTLen-iLgLen); /* Compute bit
          reversal of lN. */
      if (lT1>lN) { /* Exchange only if it hasn't been done for this
          pair. */
          dTemp0 = dpXR[lN]; /* Exchange real data. */
          dpXR[lN] = dpXR[lT1];
          dpXR[lT1] = dTemp0;

          dTemp0 = dpXI[lN]; /* Exchange imaginary data. */
```

```
                        dpXI[lN] = dpXI[lT1];
                        dpXI[lT1] = dTemp0;
                        }
              }

/* Outer level loop. */
for (iLevel = 1; iLevel< = iLgLen; ++iLevel) {
   lBStep = (long)ulExp2[iLevel-1];        /* Init butterfly step. */
   lBStride = 2*lBStep;                    /* Init butterfly stride. */

   /* Apply twiddle factors, except before first level of butterflys. */
   if (iLevel>1) {
        for (lT1 = 0; lT1<lLength; lT1+ = lBStride) {
             for (lT2 = 0; lT2<lBStep; ++lT2) {
             /* Compute locations of coefficients. */
             lT3 = lT2<<(iLgMaxFFTLen-iLevel);
             lT5 = ((lT2+((3*lBStride)>>2)) & (lBStride-1)) <<
                   (iLgMaxFFTLen-iLevel);
             lT4 = lT1+lBStep+lT2;

        /* Compute twiddled data. */
        dTemp0 = dpXR[lT4]*dpFFTFact[lT3]+dpXI[lT4]*dpFFTFact[lT5];
        dTemp1 = dpXI[lT4]*dpFFTFact[lT3]-dpXR[lT4]*dpFFTFact[lT5];

             /* Save twiddled data. */
             dpXR[lT4] = dTemp0;
             dpXI[lT4] = dTemp1;
        }
   }
}

/* Perform butterflys. */
for (lT1 = 0; lT1< = (lLength-lBStride); lT1+ = lBStride) {
   for (lT2 = 0; lT2<lBStep; ++lT2) {
        lT3 = lT1+lT2;          /* Compute locations of data. */
        lT4 = lT3+lBStep;

        dTemp0 = dpXR[lT3]+dpXR[lT4]; /* Butterfly real data. */
        dTemp1 = dpXR[lT3]-dpXR[lT4];
        dpXR[lT3] = dTemp0;
        dpXR[lT4] = dTemp1;

        dTemp0 = dpXI[lT3]+dpXI[lT4]; /* Butterfly imaginary data. */
        dTemp1 = dpXI[lT3]-dpXI[lT4];
        dpXI[lT3] = dTemp0;
        dpXI[lT4] = dTemp1;
        }
   }
}
/* Done. */
}
```

Another version of the C program given above can be found in "`src\c\chap4\fftbr.c`". The program found in this file is essentially the same as that found in "`fft.c`" except that it stores the cosine Table in bit-reversed order. While there is no apparent advantage to this storage methodology in a C program that is intended to execute on a general-purpose computer, there may be an advantage to this approach when it is used in a DSP microprocessor implementation. Many DSP microprocessors have built-in, direct support for bit-reversed addressing of data in the form of address or index registers that can be manipulated and used directly in bit-reversed mode with no computational overhead.

Example 4.3: FFT Speed Measures

Problem Statement. The DFT is an order N^2 algorithm, and the FFT is an order $N\log_2(N)$ algorithm. Verify the relative performance of the DFT versus the FFT.

Analysis and Conclusions. To test the relative execution speed of a DFT and FFT, consider transforming a 256-sample random signal, $x_{256}[k]$. Depending on the computer used to perform the evaluation, a longer or shorter sequence may be warranted.

Computer Study. The following sequence of Siglab commands will measure the relative speed of the direct DFT and FFT (results will vary depending on the computer).

```
>x = rand(256)      # random time-series
>time xd = dft(x)   # direct
  DFT 00:20.88
>time xf = fft(x)   # FFT
  00:00.49
```

It should be easy to predict the performance of an FFT on a general-purpose computer given the performance of a DFT based upon the fact that the FFT is $O(N\log_2(N))$ and the DFT is $O(N^2)$. However, there are a number of factors that impact the execution of an algorithm. For example, the order of the algorithm is not an exact metric of the number of operations required: the direct DFT requires N^2 complex multiplications and N^2 complex additions while the radix-two FFT requires $2N\log_2(N)$ complex additions but only $(N/2)(\log_2(N)-1)$ complex multiplications. Another aspect of the performance measures is the relative performance of the same algorithm when N is changed. The following sequence of commands will demonstrate the change in execution time for the DFT and FFT when the number of samples is doubled.

```
>x = rand(16384)          # random time-series
>time xd = dft(x[0:255])  # direct DFT
  00:00.66
>time xd = dft(x[0:511])  # direct DFT
  00:02.64
>time xf = fft(x[0:8191]) # FFT
```

```
    00:00.88
>time xf = fft(x)          # FFT
    00:01.81
```

Here, when N is doubled for the DFT the execution time increases by about four times. This is consistent with the fact that the DFT is $O(N^2)$: for every doubling in N, execution time should increase by a factor of four. When N is doubled for the FFT the execution time increases by slightly more than a factor of two. This observation is consistent with the fact that the FFT is $O(N\log_2(N))$. For every doubling in N, the execution time of the FFT is expected to increase by slightly more than a factor of two, depending on the value of N, or more precisely, $2\log_2(N+1)/\log_2(N)$.   ❖

4.4 Decimation-in-Frequency FFT

When the time-series $x_N[k]$ was partitioned into even and odd indexed sequences, the DIT FFT resulted. Consider now the case where the input time-series $x_N[k]$ is partitioned into two sequentially indexed sequences of length $N/2$. From Eq. (4.1), the DFT, one obtains

$$X[n] = \sum_{k=0}^{N-1} x_N[k] W_N^{nk}$$

$$= \sum_{k=0}^{N/2-1} x_N[k] W_N^{nk} + \sum_{k=N/2}^{N-1} x_N[k] W_N^{nk}$$

$$= \sum_{k=0}^{N/2-1} x_N[k] W_N^{nk} + W_N^{(N/2)n} \sum_{k=0}^{N/2-1} x_N[k+N/2] W_N^{nk}$$

$$= \sum_{k=0}^{N/2-1} (x_N[k] + (-1)^n x_N[k+N/2]) W_N^{nk} \tag{4.19}$$

The N harmonics $X[n]$ can be divided into two groups, one consisting of the even and the other the odd harmonics, denoted $X_{N/2}^e[n] = X[2n]$ and $X_{N/2}^o[n] = X[2n+1]$. It then follows that

$$X_{N/2}^e[n] = \sum_{k=0}^{N/2-1} \left(x_N[k] + x_N\left[k+\frac{N}{2}\right] \right) W_N^{2nk}$$

$$= \sum_{k=0}^{N/2-1} \left(x_N[k] + x_N\left[k+\frac{N}{2}\right] \right) W_{N/2}^{nk} \tag{4.20}$$

and

$$X_{N/2}^{o}[n] = \sum_{k=0}^{N/2-1} \left(x_N[k] - x_N\left[k + \frac{N}{2}\right] \right) W_N^{2nk} W_N^k$$

$$= \sum_{k=0}^{N/2-1} \left(x_N[k] - x_N\left[k + \frac{N}{2}\right] \right) W_{N/2}^{nk} W_N^k$$

(4.21)

which is graphically interpreted in Fig. 4.3a. The sequence $X_{N/2}^{e}[n]$ is clearly the DFT of the sequence $x_N[k] + x_N[k + N/2]$ for $k \in \{0,1,2,...,N/2 - 1\}$, while the sequence $X_{N/2}^{o}[n]$ is the DFT of the sequence $(x_N[k] - x_N[k + N/2])$ for $k \in \{0,1,2,...,N/2-1\}$. Upon expanding the $N/2 = 4$-point transform using Eqs. (4.20) and (4.21), an eight-point DFT results as shown in Fig. 4.3b. The $8 = 2^3$-point DFT can be seen to be a three-level process having twiddle factor corrections applied between

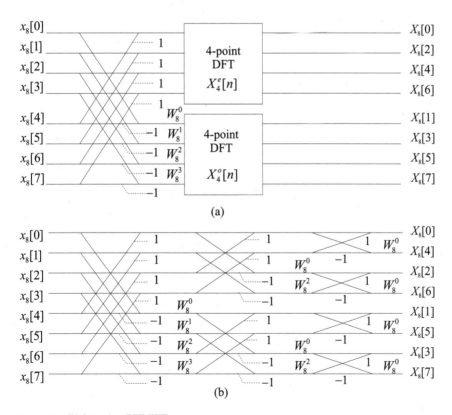

Figure 4.3 Eight-point DIF FFT.

level and at the output level. The primary difference between the DIF FFT shown in Fig. 4.1*d* and the transform reported in Fig. 4.3*b* is the input or output permutation. The FFT algorithm described in Fig. 4.3*b* is seen to decimate the input time-series and produce a sequentially ordered spectrum. As a result it is called a *decimation-in-frequency* or *DIF FFT*.

Example 4.4: Four-Point FFT

Problem Statement. Compute the DFT of a four-point time-series $x_4[k] = \{1,1,0,0\}$ using the DIF FFT algorithm.

Analysis and Conclusions. A four-point DIF FFT is detailed in Fig. 4.4. Algebraically, the DIF FFT algorithm given in Eqs. (4.20) and (4.21). Beginning with Eq. (4.20), it follows that

$$X^e_{N/2}[n] = X[2n] = \sum_{k=0}^{N/2-1} (x_N[k] + x_N[k+N/2])W_N^{2nk}$$

Evaluating for $N = 4$, one obtains

$$X^e_2[n] = \sum_{k=0}^{N/2-1} (x_4[k] + x_4[k+N/2])W_2^{nk}$$

$$= (x[0]W_2^{0n}W_4^0 + x[1]W_2^{1n}W_4^1 + x[2]W_2^{0n}W_4^0 + x[3]W_2^{1n}W_4^1)$$

$$= W_2^{0n} + W_2^{1n}$$

which gives $X^e[0] = X[0] = (1 + 1) = 2$ and $X^e[1] = X[2] = (1 - 1) = 0$ as expected. From Eq. (4.21),

$$X^o_{N/2}[n] = X[2n+1] = \sum_{k=0}^{N/2-1} (x_N[k] - x_N[k+N/2])W_N^{2nk}W_N^k$$

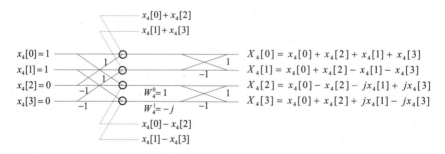

Figure 4.4   Four-point DIF FFT.

For $N = 4$, this becomes

$$X_2^o[n] = \sum_{k=0}^{N/2-1} (x_4[k] - x_4[k+N/2])W_2^{2nk}W_4^k$$

$$= (x[0]W_2^{0n}W_4^0 + x[1]W_2^{1n}W_4^1 - x[2]W_2^{0n}W_4^0 - x[3]W_2^{1n}W_4^1)$$

$$= W_2^{0n} + W_2^{1n}W_4^1$$

which gives $X^o[0] = X[1] = (1 + 1(-j)) = 1 - j$ and $X^o[1] = X[3] = (1 - 1(-j)) = 1 + j$ as expected. ❖

Besides the DIT and DIF FFT algorithm, other basic DFT algorithms are in common use. Those of significance are introduced in the next sections.

4.5 Real FFTs

The FFT has established itself to be a fundamental tool in DSP. Software, firmware, and hardware FFT computing resources are abundant. Recall, however, that the FFT implements a DFT which maps a complex time-series into the complex frequency-domain (i.e., $C^N \leftrightarrow C^N$). It is common, however, to find that the input signal is real valued. If so, some additional efficiencies can be gained. Consider two N sample time series, $x_1[k]$ and $x_2[k]$, which are combined to form a synthetic N sample complex time-series, $x_N[k]$, where

$$x_N[k] = x_1[k] + jx_2[k] \tag{4.22}$$

The DFT of this sequence is given by

$$X[n] = X_1[n] + jX_2[n] \tag{4.23}$$

where $X_1[n]$ and $X_2[n]$ are complex valued. From the properties of a DFT (Table 3.3), it is known that

$$x^*[k] \overset{\text{DFT}}{\leftrightarrow} X^*[-n\bmod(N)] = X_1^*[-n\bmod(N)] + jX_2^*[-n\bmod(N)] \tag{4.24}$$

Combining these two results, the DFTs of $x_1[k]$ and $x_2[k]$ can be extracted from $X[n]$ using the expression

$$X_1[n] = \frac{X[n] + X^*[-n\bmod(N)]}{2} \tag{4.25}$$

and

$$X_2[n] = \frac{X[n] - X^*[-n \bmod(N)]}{2j} \qquad (2.26)$$

This is the *doubling real FFT* (or DFT) algorithm, which states that if there are two N sample real data records, their DFTs may be compute using a single FFT or DFT computation. One record is loaded into the real DFT input array, and the other into the imaginary. Upon completion of an N-point DFT, the individual, complex-valued transforms of the two real time-series can be separated and recovered. For an N sample record, both DFTs may be recovered using a total of only $4N$ real ($2N$ complex) addition/subtraction operations, and $4N$ real scaling operations where the scaling factor is one-half. In conventional binary arithmetic, the scaling by one-half is accomplished by an arithmetic shift right of one bit, which is a very inexpensive operation.

Another variation on this theme also exists. Suppose $x_N[k]$ is a real time-series of length $2N$ samples. Consider subdividing $x_N[k]$ into two real N sample time-series:

$$x_{even}[k] = x_N[2k]$$

$$x_{odd}[k] = x_N[2k+1] \qquad (4.27)$$

for $k \in \{0,1,2,...,N\}$. This is the same decimation in time approach that led to the DIT FFT algorithm. The two real N-sample time series are now packed into an N sample complex array as before in Eq. (4.22), say $z[k] = x_{even}[k] + jx_{odd}[k]$. Suppose further that the DFT of the complex time-series $z[k]$ is $Z[n]$. Then, from Eqs. (4.25) and (4.26), it is known that

$$X_{even}[n] = \frac{Z[n] + Z^*[-n \bmod(N)]}{2} \qquad (2.28)$$

and

$$X_{odd}[n] = \frac{Z[n] - Z^*[-n \bmod(N)]}{2j} \qquad (2.29)$$

From Eq. (4.16), the DFTs of the decimated sequences may be recombined to form the DFT of the original sequence using

$$X[n] = X_{even}[n] + W_{2N}^n X_{odd}[n] \qquad (4.30)$$

for $n \in \{0,1,2,...,2N-1\}$.

This algorithm is called the *packed FFT* for obvious reasons. Like the doubling algorithm, the packed algorithm reduces the computational effort expended performing a DFT or FFT in trade for a relatively small number of additional arithmetic operations. The total computational effort required to use this algorithm to compute the DFT of a real valued record of length $2N$ is one DFT or FFT of a record of length N, $4N$ real addition/subtraction operations, $4N$ real scaling operations (scaling by one-half), and $2N$ complex multiplications and additions.

4.6 Chirp z-Transform

The *chirp z transform*, or CZT, does not have the computational efficiency of the FFT but does offer several unique features. Recall that the DFT,

$$X[n] = \sum_{k=0}^{N-1} x_N[k] W_N^{nk} \qquad (4.31)$$

projects a time-series $x_N[k]$ to discrete frequency locations distributed along the periphery of the unit circle as shown in Fig. 4.5. However, there are times when the interpretability of the spectrum can be improved by choosing a trajectory other than the periphery of the unit circle (i.e., $z = e^{jn\phi_0}$). The spiral trajec-

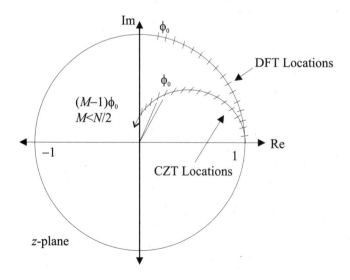

Figure 4.5 Evaluation trajectories for a DFT and CZT.

tory shown in Fig. 4.5 is used by the CZT to evaluate the spectrum of a time-series along the trajectory $z = a^n e^{jn\phi_0}$. Working off the unit circle has a tendency of broadening the spectral peaks of a pure tone relative to that obtained using a DFT. Using the trajectory shown in Fig. 4.5, a modified DFT formula results and is given by:

$$X[n] = \sum_{k=0}^{N-1} a^{-k} x[k] W_N^{nk} \tag{4.32}$$

where $0 < a < 1$. The CZT algorithm is based on Eq. (4.32) and the use of *Bluestein's identity* which states that

$$nk = \frac{1}{2}(n^2 + k^2 - (k-n)^2) \tag{4.33}$$

Substituting Eq. (4.33) into Eq. (4.32) yields

$$X[n] = \sum_{k=0}^{N-1} a^{-k} x[k] W_N^{(n^2 + k^2 - (k-n)^2)/2}$$

$$= W_N^{n^2/2} \sum_{k=0}^{N-1} y[k] W_N^{-(k-n)^2/2} \tag{4.34}$$

where

$$y[k] = a^{-k} x[k] W^{k^2/2} \tag{4.35}$$

The waveform $W^{k^2/2}$ is called an *up chirp* or *chirp* and represents a linearly swept FM signal. The interpretation of Eq. (4.34) is shown in Fig. 4.6 and consists of the following set of operations:

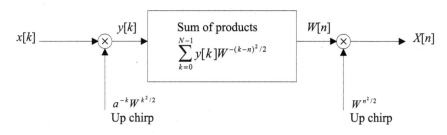

Figure 4.6 CZT architecture.

1. Preprocess $x_N[k]$ with a scaled up chirp $a^{-k} W_N^{k^2/2}$ to form $y[k]$.

2. Evaluate the sum (called a *convolution sum* in Chap. 5)

$$W[n] = \sum_{k=0}^{N-1} y[k] W_N^{-(k-n)^2/2}$$

with respect to a down-chirp.

3. Postprocess $W[n]$ by multiplying by an up chirp $W_N^{n^2/2}$.

The result is the spectrum $X[k]$. For a DFT (i.e., $a = 1$), it is obvious that this operation is far more complex than an FFT. It therefore can be justified only if a technology can be found that implements the three-step CZT process efficiently. One technology often used to implement CZTs is a *charge-transfer device* or CTD (a.k.a., *charge-coupled device* or CCD). While the CTD is not a digital technology, it is a discrete-time approach and represents an important implementation technology for the CZT.

Example 4.5: Chirp z-Transform

Problem Statement. Experimentally investigate the effects of the parameter a in computing the CZT of a broadband signal process.

Analysis and Conclusions. Consider a broadband signal process $x[k]$ which contains frequencies over the normalized range $\varpi \in [0,\pi)$ of essentially equal energy. The CZT of $x[k]$ for $a = 1$ would be equivalent to the DFT of $x[k]$. However, for $0 < a < 1$, the evaluation trajectory moves away from the unit circle to an evaluation point $z_k = a^k e^{j2\pi k/N}$. As a result, the CZT will attenuate the higher frequencies relative to a constant gain DFT spectrum. This is exemplified in Fig. 4.7, which displays the CZT of a broadband signal, having energy of the normalized frequency range $\varpi \in [0,\pi/2)$, for values of $a \in \{1, 0.75, 0.5\}$.

Computer Study. The data presented in Fig. 4.7 is generated using the s-files "chirp.s" and "czt.s." The function chirp creates a swept FM signal having energy in frequency bands from zero to $f_s/4$. The function czt has a format czt(a,x), where a is the attenuation parameter and x is a bandlimited signal having a maximum frequency of $f_s/4$. The following sequence of Siglab commands uses these functions to generate the spectra shown in Fig. 4.7.

```
>include "chap4\chirp.s"
>include "chap4\czt.s"
>x = chirp(501,0,0,.25)
>czt1 = mag(czt(1,x))
>czt2 = mag(czt(0.75,x))
>czt3 = mag(czt(0.5,x))
>ograph(czt1,czt2,czt3)
```

❖

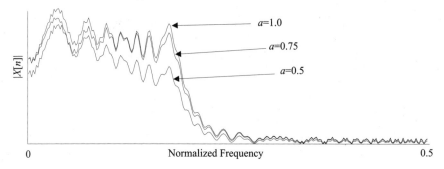

Figure 4.7   Magnitude CZT spectrum of a broadband signal for values of $a \in \{1, 0.75, 0.5\}$.

4.7 Rader Prime Factor DFT

One of the more unusual, yet useful DFT formulas is the *Rader prime factor algorithm* (RPFA). The RPFA is restricted to transforms of length N, where N is prime. The derivation of the RPFA is dependent on an understanding of finite field theory. Finite fields are a well understood topic in number theory and abstract algebra, however, the details and subtleties of this branch of mathematics are beyond the scope of this work.[1] Therefore, only the mechanics of an RPFA will be presented.

The finite field that consists of the set of integers $\{0,1,2,\ldots,p-1\}$, where p is prime, the definition of the addition/subtraction operation on that set (modulo p), and the multiplication operation on that set (again, modulo p) is a *Galois field*. The Galois field is denoted $GF(p)$, and its operations are illustrated in Table 4.2 for $p = 7$. Here, $GF(7) =$

Table 4.2   Additive and Multiplicative Operations in GF(7)[*]

+	0	1	2	3	4	5	6		×	0	1	2	2	3	5	6
0	0	1	2	3	4	5	6		0	0	0	0	0	0	0	0
1	1	2	3	4	5	6	0		1	0	1	2	3	4	5	6
2	2	3	4	5	6	0	1		2	0	2	4	6	1	3	5
3	3	4	5	6	0	1	2		3	0	3	6	2	5	1	4
4	4	5	6	0	1	2	3		4	0	4	1	5	2	6	3
5	5	6	0	1	2	3	4		5	0	5	3	1	6	4	2
6	6	0	1	2	3	4	5		6	0	6	5	4	3	2	1

*Shaded blocks correspond to the additive and multiplicative identities, respectively.

$(\{0,1,2,3,4,5,6\},+,\times)$. Notice that, for the multiplication operation, the product of two non-zero elements is always a non-zero element of $GF(7)$.

One of the remarkable properties of a Galois field is that there always exists an integer α, called a *generator*, such that for each $I \in \{1,2,3,\ldots, p-1\}$, there exists a unique $i \in \{0,1,2,\ldots, p-2\}$ such that $I = \alpha^i$. That is, given a generator α for the Galois field denoted $GF(p)$, there is a group isomorphism (a one-to-one and onto correspondence) between the non-zero elements of $GF(p)$ (i.e., $GF(p)\backslash\{0\}$) and the set $\{0,1,2,\ldots,p-2\}$. This is illustrated for $p = 7$, with generator $\alpha = 3$, in Table 4.3.

One important feature that is evident from the data presented in Table 4.3 is that the elements of $GF(7)\backslash\{0\}$ are generated cyclically, which can be exploited in implementing the DFT. This can be demonstrated by considering a seven-point DFT expressed in matrix-vector form as

$$
\begin{bmatrix} X[0] \\ X[1] \\ X[2] \\ X[3] \\ X[4] \\ X[5] \\ X[6] \end{bmatrix} =
\begin{bmatrix}
1 & 1 & 1 & 1 & 1 & 1 & 1 \\
1 & W_7^1 & W_7^2 & W_7^3 & W_7^4 & W_7^5 & W_7^6 \\
1 & W_7^2 & W_7^4 & W_7^6 & W_7^1 & W_7^3 & W_7^5 \\
1 & W_7^3 & W_7^6 & W_7^2 & W_7^5 & W_7^1 & W_7^4 \\
1 & W_7^4 & W_7^1 & W_7^5 & W_7^2 & W_7^6 & W_7^3 \\
1 & W_7^5 & W_7^3 & W_7^1 & W_7^6 & W_7^4 & W_7^2 \\
1 & W_7^6 & W_7^5 & W_7^4 & W_7^3 & W_7^2 & W_7^1
\end{bmatrix}
\begin{bmatrix} x[0] \\ x[1] \\ x[2] \\ x[3] \\ x[4] \\ x[5] \\ x[6] \end{bmatrix}
\tag{2.36}
$$

Table 4.3   Number Theoretic Logarithms {0,1,...,5}

i	$\alpha^i = 3^i$	$I \in GF(7)\backslash\{0\}$ such that $I \equiv \alpha^i \bmod 7$
0	1	1
1	3	3
2	9	2
3	27	6
4	81	4
5	243	5
6	729	1

Assume that $X[0]$ is separately computed as the sum of $x[0]$ through $x[6]$, the remaining harmonics are represented by the 6×6 matrix DFT equation as

$$
\begin{bmatrix} X[1] \\ X[2] \\ X[3] \\ X[4] \\ X[5] \\ X[6] \end{bmatrix} = \begin{bmatrix} W_7^1 & W_7^2 & W_7^3 & W_7^4 & W_7^5 & W_7^6 \\ W_7^2 & W_7^4 & W_7^6 & W_7^1 & W_7^3 & W_7^5 \\ W_7^3 & W_7^6 & W_7^2 & W_7^5 & W_7^1 & W_7^4 \\ W_7^4 & W_7^1 & W_7^5 & W_7^2 & W_7^6 & W_7^3 \\ W_7^5 & W_7^3 & W_7^1 & W_7^6 & W_7^4 & W_7^2 \\ W_7^6 & W_7^5 & W_7^4 & W_7^3 & W_7^2 & W_7^1 \end{bmatrix} \begin{bmatrix} x[1] \\ x[2] \\ x[3] \\ x[4] \\ x[5] \\ x[6] \end{bmatrix} + \begin{bmatrix} x[0] \\ x[0] \\ x[0] \\ x[0] \\ x[0] \\ x[0] \end{bmatrix} \qquad (4.37)
$$

It can be seen that the exponents of W_7 appear in a chaotic order. Therefore, each element of the result of Eq. (4.37) would have to be computed separately, requiring six complex multiplications per harmonic. What is desired is to put some structure and order into the way the exponents fill the DFT coefficient matrix. Consider permuting the time-series vector $x[k]$ using the permutation rule given in Table 4.3, along with an attendant permutation of the columns of the coefficient matrix to produce

$$
\begin{bmatrix} X[1] \\ X[2] \\ X[3] \\ X[4] \\ X[5] \\ X[6] \end{bmatrix} = \begin{bmatrix} W_7^1 & W_7^3 & W_7^2 & W_7^6 & W_7^4 & W_7^5 \\ W_7^2 & W_7^6 & W_7^4 & W_7^5 & W_7^1 & W_7^3 \\ W_7^3 & W_7^2 & W_7^6 & W_7^4 & W_7^5 & W_7^1 \\ W_7^4 & W_7^5 & W_7^1 & W_7^3 & W_7^2 & W_7^6 \\ W_7^5 & W_7^1 & W_7^3 & W_7^2 & W_7^6 & W_7^4 \\ W_7^6 & W_7^4 & W_7^5 & W_7^1 & W_7^3 & W_7^2 \end{bmatrix} \begin{bmatrix} x[1] \\ x[3] \\ x[2] \\ x[6] \\ x[4] \\ x[5] \end{bmatrix} + \begin{bmatrix} x[0] \\ x[0] \\ x[0] \\ x[0] \\ x[0] \\ x[0] \end{bmatrix} \qquad (2.38)
$$

Using the same logic, now consider permuting the harmonic vector using the same permutation rule, along with an attendant permutation of the rows of the DFT coefficient matrix to produce

$$\begin{bmatrix} X[1] \\ X[3] \\ X[2] \\ X[6] \\ X[4] \\ X[5] \end{bmatrix} = \begin{bmatrix} W_7^1 & W_7^3 & W_7^2 & W_7^6 & \{W_7^4\} & \langle W_7^5 \rangle \\ W_7^3 & W_7^2 & W_7^6 & \{W_7^4\} & \langle W_7^5 \rangle & W_7^1 \\ W_7^2 & W_7^6 & \{W_7^4\} & \langle W_7^5 \rangle & W_7^1 & W_7^3 \\ W_7^6 & \{W_7^4\} & \langle W_7^5 \rangle & W_7^1 & W_7^3 & W_7^2 \\ \{W_7^4\} & \langle W_7^5 \rangle & W_7^1 & W_7^3 & W_7^2 & W_7^6 \\ \langle W_7^5 \rangle & W_7^1 & W_7^3 & W_7^2 & W_7^6 & \{W_7^4\} \end{bmatrix} \begin{bmatrix} x[1] \\ x[3] \\ x[2] \\ x[6] \\ x[4] \\ x[5] \end{bmatrix} + \begin{bmatrix} x[0] \\ x[0] \\ x[0] \\ x[0] \\ x[0] \\ x[0] \end{bmatrix} \quad (4.39)$$

where $\langle \, \rangle$ and $\{ \}$ are notational conveyances used to highlight the spatial relationship among matrix elements. Notice the exponents now have diagonal circular symmetry as indicated by viewing $\langle W_7^5 \rangle$ and $\{W_7^4\}$, how for example, also follows a diagonal path from the location [5,1] to location [1,5] and then wraps around to location [6,6]. A matrix with this type of symmetry is referred to as being *left-circulant*. The consequence of this is graphically interpreted in Fig. 4.8. It can be seen that a circular shift-register and a multiply-accumulator array, having one fixed set of operands, are all that is needed to compute a seven-point DFT. This method can be extended to a DFT of any prime length record. While this approach does not actually reduce the number of arithmetic operations required to perform a direct DFT, it can be leveraged to produce an efficient hardware implementation. Efficiency is derived from the regular, linear arrangement of the storage and arithmetic elements as shown in Fig. 4.8.

Example 4.6: Rader Prime Factor Algorithm

Problem Statement. Experimentally verify the $p = 7$ RPFA DFT formula given in Eq. (4.39) and Fig. 4.8.

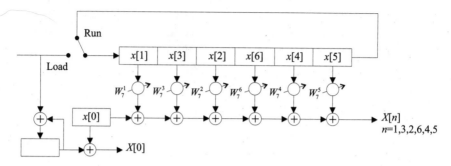

Figure 4.8   Length $p = 7$ Rader prime factor DFT implementation.

Analysis and Conclusions. The test signal $x_7[k] = \delta[k - 2]$ (i.e, delayed impulse), should produce a spectrum having a value of $X_7[n] = W_7^{2n} = e^{-j4\pi n/7}$. Directly interpreting Eq. (4.39), one obtains

$$
\begin{bmatrix} X[1] \\ X[3] \\ X[2] \\ X[6] \\ X[4] \\ X[5] \end{bmatrix} = \begin{bmatrix} W_7^1 & W_7^3 & W_7^2 & W_7^6 & \{W_7^4\} & \langle W_7^5 \rangle \\ W_7^3 & W_7^2 & W_7^6 & \{W_7^4\} & \langle W_7^5 \rangle & W_7^1 \\ W_7^2 & W_7^6 & \{W_7^4\} & \langle W_7^5 \rangle & W_7^1 & W_7^3 \\ W_7^6 & \{W_7^4\} & \langle W_7^5 \rangle & W_7^1 & W_7^3 & W_7^2 \\ \{W_7^4\} & \langle W_7^5 \rangle & W_7^1 & W_7^3 & W_7^2 & W_7^6 \\ \langle W_7^5 \rangle & W_7^1 & W_7^3 & W_7^2 & W_7^6 & \{W_7^4\} \end{bmatrix} \begin{bmatrix} x[1] \\ x[3] \\ x[2] \\ x[6] \\ x[4] \\ x[5] \end{bmatrix} + \begin{bmatrix} x[0] \\ x[0] \\ x[0] \\ x[0] \\ x[0] \\ x[0] \end{bmatrix}
$$

$$
= \begin{bmatrix} W_7^1 & W_7^3 & W_7^2 & W_7^6 & \{W_7^4\} & \langle W_7^5 \rangle \\ W_7^3 & W_7^2 & W_7^6 & \{W_7^4\} & \langle W_7^5 \rangle & W_7^1 \\ W_7^2 & W_7^6 & \{W_7^4\} & \langle W_7^5 \rangle & W_7^1 & W_7^3 \\ W_7^6 & \{W_7^4\} & \langle W_7^5 \rangle & W_7^1 & W_7^3 & W_7^2 \\ \{W_7^4\} & \langle W_7^5 \rangle & W_7^1 & W_7^3 & W_7^2 & W_7^6 \\ \langle W_7^5 \rangle & W_7^1 & W_7^3 & W_7^2 & W_7^6 & \{W_7^4\} \end{bmatrix} \begin{bmatrix} 0 \\ 0 \\ 1 \\ 0 \\ 0 \\ 0 \end{bmatrix} + \begin{bmatrix} 0 \\ 0 \\ 0 \\ 0 \\ 0 \\ 0 \end{bmatrix} = \begin{bmatrix} W_7^2 \\ W_7^6 \\ W_7^4 \\ W_7^5 \\ W_7^1 \\ W_7^3 \end{bmatrix}
$$

as expected. The value of $X[0]$ is the sum of the time series, which is $W_7^{2(0)} = e^{-j4\pi(0)/7} = 1$, as expected.

Computer Study. The s-file "rpdft.s", contains a function, rpdft(x,g), that will compute the RPFA DFT of a time-series $x[k]$ of prime length p with a generator g. A set of harmonic test signals, of normalized frequency $nf_s/7$, where $n \in \{0,1,2,3\}$, are used to test the RPFA using the following sequence of Siglab commands.

```
>include "chap4\rpdft.s"
>mag(rpdft(mkcos(1,0/7,0,7)),3) # 0th harmonic
  2e-16  2e-16  2e-16  7  2e-16  2e-16  2e-16
>mag(rpdft(mkcos(1,1/7,0,7)),3) # 1st and 6th harmonic
  2e-16  1e-16  3.5  1e-16  3.5  2e-16  3e-16
>mag(rpdft(mkcos(1,2/7,0,7)),3) # 2nd and 5th harmonic
  5e-16  3.5  3e-16  1e-16  0  3.5  4e-16
>mag(rpdft(mkcos(1,3/7,0,7)),3) # 3rd and 4th harmonic
  3.5  2e-16  0  0  2e-16  2e-16  3.5
```
❖

4.8 FFT Ordering

The previous sections have focused on developing DFT implementations for the study of signal having $N = r^n$ samples, as in the case of the DIT and DIF FFT. For arbitrary N, a direct DFT or CZT can be used. If N is prime, the RPFA may also be used effectively. A problem of constant interest is designing long DFTs from a suite of shorter DFTs. This process is called *ordering*. By properly ordering and integrating the results from short DFTs, a long DFT can be created.

4.9 Cooley-Tukey FFT

The FFT algorithms developed earlier assumed that $N = r^m$, where $r = 2$ is called the radix. The DIT FFT and DIF FFT algorithms developed are called radix-two Cooley-Tukey FFTs. It is also possible to apply the decimation-in-time and decimation-in-frequency approaches to develop FFTs with radices other than two. For example, suppose that $r = 3$. Then let $N = 3^m$, and let $x[k]$ be a sequence of length N. Now, decimate $x[k]$ in time three ways to produce three sequences,

$$x_0[k] = x[3k]$$
$$x_1[k] = x[3k + 1] \qquad (4.40)$$
$$x_2[k] = x[3k + 2]$$

for $k(\in \{0,1,2,...,N/3-1\}$. Then the DFT of $x[k]$ is

$$X[n] = \sum_{k=0}^{N-1} x[k] W_N^{nk}$$

$$= \sum_{k=0}^{N/3-1} (x_0[k] W_N^{n3k} + x_1[k] W_N^{n(3k+1)} + x_2[k] W_N^{n(3k+2)})$$

$$= \sum_{k=0}^{N/3-1} x_0[k]W_{N/3}^{nk} + W_N^n \sum_{k=0}^{N/3-1} x_1[k]W_{N/3}^{nk} + W_N^{2n} \sum_{k=0}^{N/3-1} x_2[k]W_{N/3}^{nk}$$

$$(4.41)$$

The three summations in Eq. (4.41) are clearly the DFTs of $x_0[k]$, $x_1[k]$, and $x_2[k]$, respectively. The final expression of the DFT in Eq. (4.41) is essentially the same as the radix-two DIT FFT developed in Eq. (4.16). Since, in this case, the length of the original sequence is a power of three, the radix-three DIT approach developed in Eq. (4.41) can be applied repetitively, as was the case with the radix-two DIT FFT, to produce a radix-three FFT. For the radix-three FFT, the basic DFT operation is a length three DFT, which can be thought of as a three-point butterfly.

It is now apparent that the Cooley-Tukey FFT is not limited to radix-two FFTs of sequences of a power of two length. However, the radix-two FFT is considered by most engineers to be *the* FFT. The reason is that the radix-two FFT has a number of implementation advantages, especially when the FFT is performed with digital hardware. First, the radix-two butterfly can be implemented without generalized multiplication, whereas the constituent DFTs of FFTs with radices other than two may require generalized complex multiplication. The savings realized by eliminating generalized complex multiplication in the constituent DFTs are significant. Another aspect of the radix-two FFT that simplifies implementation is the ability to perform the necessary permutations via bit-reversal of array indexes. The hardware implementation of a bit-reversed array index is trivial, whereas the implementation of index permutations required by FFTs having radices other than two may be substantially more complicated.

The FFT algorithm earlier developed assumed that $N = r^m$. A Cooley-Tukey DIT or DIF FFT algorithm can be developed for a general composite integers N (i.e., $N = p_1^{m_1}p_2^{m_2}p_3^{m_3}...p_L^{m_L}$, where $p_i \neq p_j$ if $i \neq j$ for $i,j \in \{1,2,3,...,L\}$ and m_i are positive integers) as well. Consider the N-point DFT given by

$$X[n] = \sum_{k=0}^{N-1} x_N[k]W_N^{nk} \qquad (4.42)$$

Suppose $N = N_1N_2$ and a mapping rule between the sample index k and a two-tuple $[k_1,k_2]$ is given by

$$k = k_1 + N_1k_2 \qquad (4.43)$$

where $k_1 \in \{0,1,2,\ldots,N_1 - 1\}$ and $k_2 \in \{0,1,2,\ldots,N_2 - 1\}$. This corresponds to writing the sample value $x_N[k]$ into a 2-D matrix, as shown in Fig. 4.9, for the case where $N = 15$, $N_1 = 3$ and $N_2 = 5$. The sample value $x_N[k_1,k_2]$ corresponds to a unique value of $x_N[k]$, which is given by

$$x_N[k] = x_N[k_1 + N_1 k_2] = x_N[k_1, k_2] \tag{4.44}$$

where the relation between k, k_1, and k_2 is given by Eq. (4.43). Using a similar argument, consider representing the harmonic index n as the two-tuple $[n_1, n_2]$ using the mapping defined by

$$n = N_2 n_1 + n_2 \tag{4.45}$$

where $n_1 \in \{0,1,2,\ldots,N_1 - 1\}$ and $n_2 \in \{0,1,2,\ldots,N_2 - 1\}$. The harmonic $X[n]$ is uniquely mapped into

$$X[n] = X[N_2 n_1 + N_2] = X[n_1, n_2] \tag{4.46}$$

where the relationship between n, n_1, and n_2 is given by Eq. (4.45). The harmonic database can also be viewed as a 2-D $N_1 \times N_2$ matrix as shown in Fig. 4.9. The index permutations that appear in Fig. 4.9 are the result of the *index shuffling* that is introduced by Eqs. (4.43) and (4.45). Previously, the radix-two DIT FFT was shown to require an index shuffle which corresponds to bit reversal of the index.

Using a two-tuple representation, the DFT formula given by Eq. (4.42) can be expressed as

$$X[n_1, n_2] = \sum_{k_1 = 0}^{N_1 - 1} \sum_{k_2 = 0}^{N_2 - 1} x_N[k_1, k_2] W_N^{nk} \tag{4.47}$$

where nk can be rewritten as

k_1 \ k_2	0	1	2	3	4
0	$x[0]$	$x[3]$	$x[6]$	$x[9]$	$x[12]$
1	$x[1]$	$x[4]$	$x[7]$	$x[10]$	$x[13]$
2	$x[2]$	$x[5]$	$x[8]$	$x[11]$	$x[14]$

n_1 \ n_2	0	1	2	3	4
0	$X[0]$	$X[1]$	$X[2]$	$X[3]$	$X[4]$
1	$X[5]$	$X[6]$	$X[7]$	$X[8]$	$X[9]$
2	$X[10]$	$X[11]$	$X[12]$	$X[13]$	$X[14]$

Figure 4.9 Block decomposition of time- and frequency-domain data for $N = 15$, $N_1 = 3$, $N_2 = 5$

$$nk = (N_2 n_1 + n_2)(k_1 + N_1 k_2)$$

$$= N_2 n_1 k_1 + N_1 k_2 n_2 + N_1 N_2 n_1 k_2 + n_2 k_1$$

$$= N_2 n_1 k_1 + N_1 k_2 n_2 + N n_1 k_2 + n_2 k_1 \qquad (4.48)$$

Therefore, from Eq. (4.47), W_N^{nk} reduces to

$$
\begin{aligned}
W_N^{nk} &= W_N^{N_2 n_1 k_1 + N_1 k_2 n_2 + N n_1 k_2 + n_2 k_1} \\
&= W_N^{N_2 n_1 k_1} W_N^{N_1 k_2 n_2} W_N^{N n_1 k_2} W_N^{n_2 k_1} \\
&= W_{N_1}^{n_1 k_1} W_{N_2}^{k_2 n_2} W_N^{n_2 k_1}
\end{aligned}
\qquad (4.49)
$$

and it follows that Eq. (4.47) can be expressed as

$$
X[n_1, n_2] = \sum_{k_1 = 0}^{N_1 - 1} W_{N_1}^{n_1 k_1} \left(W_N^{n_2 k_1} \sum_{k_2 = 0}^{N_2 - 1} x[k_1, k_2] W_{N_2}^{n_2 k_2} \right) \qquad (4.50)
$$

The generalized form the Cooley-Tukey algorithm can be applied to compute the FFT of a sequence of any length. The anatomy of the algorithm presented in Eq. (4.50) is, from the inside out:

1. Compute the N_2-point FFT

$$
Y[n_1, n_2] = \sum_{k_2 = 0}^{N_2 - 1} x[k_1, k_2] W_{N_2}^{n_2 k_2}
$$

for each k_1.

2. Apply the twiddle factor $W_N^{n_2 k_1}$ to $Y[n_1, n_2]$ to form $Z[n_1, n_2, k_1] = W_N^{n_2 k_1} Y[n_1, n_2]$.

3. Compute the N_1-point FFT $X[n_1 n_2] = \sum_{k_1 = 0}^{N_1 - 1} W_{N_1}^{n_1 k_1} Z[n_1, n_2, k_1]$.

The complexity of a Cooley-Tukey FFT can be analyzed by dissecting Eq. (4.50). If implemented using direct DFT, the first level would require $N_1(N_2)^2$ multiplications. The second level would require $N_2(N_1)^2$ multiplications. There are also N twiddle factor multiplications needed to process the data moving from level one to level two. The total complexity is therefore on the order of $M = N_1(N_2)^2 + N_2(N_1)^2 + N = N(N_1 + N_2 + 1)$. This concept can be extended to multilevel setting. In general, N can be a highly composite number requiring many N_i point FFT and twiddle factor adjustments. For efficiency, however, a Cooley-Tukey is generally defined in terms of the repeated use of a highly optimized, small radix-N FFT operator. Thus, N is normally a composite number of the form $N = r^m$ with a common choice of r being two or four. An advantage can be gained if an efficient radix-r butterfly processor can be used recursively versus the case where butterfly processors of varying sizes are required. This is the secret of success of the radix-two or four FFT, which reuses a simple two- or four-point butterfly processor over and over. Multiple two- or four-point butterfly processors can be interlaced to achieve even higher performance.

Example 4.7: Cooley-Tukey Radix-Two FFT

Problem Statement. In Figs. 4.1 and 4.3, an $N = 8$ FFT was studied. Explain the structure of a radix-two FFT using a Cooley-Tukey ordering.

Analysis and Conclusion. The procedure developed in this section is applied to the case where $N = N_1 N_2$. Initially, let $N_1 = 4$ and $N_2 = 2$, then:

$$k = k_1 + 4k_2; \; k_1 \in [0, 3], k_2 \in [0, 1]$$

$$n = 2n_1 + n_2; \; n_1 \in [0, 3], n_2 \in [0, 1]$$

Now consider subdividing $N_1 = N_{11}N_{12}$ where $N_{11} = N_{12} = 2$. The result should be a radix-two eight-point FFT. The new indexing rules are given by:

$$k = (k_{11} + 2k_{12}) + 4k_2 = k_{11} + 2k_{12} + 4k_2; \; k_{11} \in [0, 1], k_{12} \in [0, 1], k_2 \in [0, 1]$$

$$n = 2(n_{12} + 2n_{11}) + n_2 = n_2 + 2n_{12} + 4n_{11}; \; n_{11} \in [0, 1], n_{12} \in [0, 1], n_2 \in [0, 1]$$

which is graphically interpreted in Fig. 4.10 The vertices of the cube correspond to their indices shuffled order in the context of the chosen decomposition.   ❖

There are times when the length of the time-series suggests that the DFT should be non-radix-two. A calendar year database may consist of 352 sample values. The signal may be padded with 160 zeros to create a 512-sample database which can be processed using a radix-two FFT. However, the padded database is a distorted version of the original signal. A Cooley-Tukey algorithm having a composite value of $N = 352 = 11 \times (2^5)$ can be used to compute an exact 352-point spectrum.

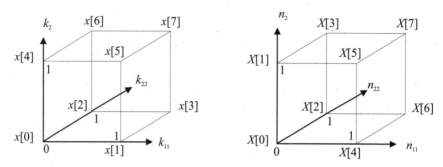

Figure 4.10   Cooley-Tukey index shuffle for a $N = 8$ radix-two DFT.

Example 4.8: Mixed-Radix Cooley-Tukey Transform

Problem Statement. Analyze a $N = 192$-point Cooley-Tukey FFT and experimentally verify its performance.

Analysis and Conclusions. The length $N = 192 = 64 \times 3 = 2^6 4 \times 3$ is a composite length having a corresponding Cooley-Tukey FFT architecture shown in Fig. 4.11. This structure shown is a two level transform. The first level is the set of N_2-point FFTs described in Step 1. The second level is the set of N_1-point FFTs described in Step 3. In between butterflies, correction terms $W_N^{n_2 k_1}$ are

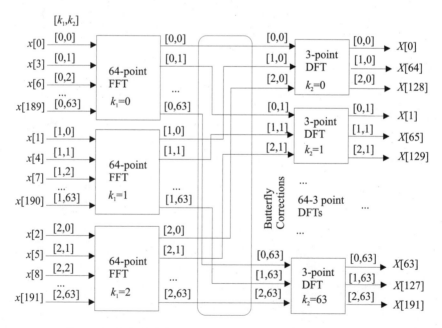

Figure 4.11   Example of an $N = 192$-point ($N_1 = 64$, $N_2 = 3$) Cooley-Tukey FFT.

applied to the data coming from level one heading toward level two, as described in Step 2. Level one, in turn, can be implemented using three 64-point DFTs (radix-two FFT) and level two by 64 three-point DFTs. The indices found at levels one and two are shown in Fig. 4.12.

Computer Study. The s-file "`ctdft.s`" implements a 192-point DFT using Cooley-Tukey ordering. The function is called using "`ctdft(x)`" where x is a 192-sample time-series. The data presented in Fig. 4.13 is created by the following sequence of Siglab commands.

```
>include "chap4\ctdft.s"
>x = mkcos(1,5/192,0,192)
>graph(mag(ctdft(x)))
```
❖

4.10 Good-Thomas FFT

Another ordering policy is attributed to Good and Thomas. The mathematical foundation of the Good-Thomas algorithm is based heavily on

$k=k_1+N_1k_2$

k_1 \ k_2	0	1	2	
0	$x[0]$	$x[1]$	$x[2]$	
1	$x[3]$	$x[4]$	$x[5]$	$N_1=3$
2	$x[6]$	$x[7]$	$x[8]$	$N_2=64$
⋮	⋮	⋮	⋮	
63	$x[189]$	$x[190]$	$x[191]$	

$n=N_2n_1+n_2$

n_1 \ n_2	0	1	2	·
0	$X[0]$	$X[64]$	$X[128]$	
1	$X[1]$	$X[65]$	$X[129]$	
2	$X[2]$	$X[66]$	$X[130]$	
⋮	⋮	⋮	⋮	
63	$X[63]$	$X[127]$	$X[191]$	

Figure 4.12   Cooley-Tukey index shuffle for an $N = 192 = 64 \times 3$-sample DFT.

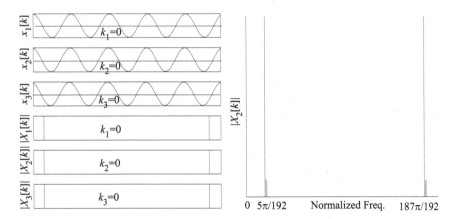

Figure 4.13   192-point Cooley-Tukey order DFT of a pure sinusoid having a frequency $f = 5f_s/192$.

number theory. The full development of this algorithm is beyond the scope of this text. Instead, the *Good-Thomas FFT* will be presented as a procedure.

The length of a sequence suitable for application of the Good-Thomas FFT algorithm is given by

$$N = \prod_{i=1}^{L} N_i \tag{4.51}$$

where N_i and N_j are relatively prime (i.e., $\gcd(N_i,N_j) = 1$) for all $i,j \in \{1,2,3,...,L\}$ where $i \neq j$. The Good-Thomas algorithm can be motivated by studying the case $N = N_1 N_2$. The sample index k is defined by

$$k = \left\{ m_1((m_1^{-1}k_1)\text{mod}N_1) \right\} + m_2 \left\{ (m_2^{-1}k_2)\text{mod}N_2 \right\}\text{mod}N \tag{4.52}$$

where

$$k_i = (k)\text{mod}(N_i) \tag{4.53}$$

each m_i is defined by

$$m_i = N/N_i \tag{4.54}$$

and m_i^{-1}, which is called the *multiplicative inverse* of m_i modulo N_i and is defined to be a number that satisfies

$$m_i m_i^{-1} \equiv 1 \ \text{mod}N_i \tag{4.55}$$

This rather complicated formula is a direct consequence of the *Chinese Remainder Theorem* (CRT), which is an important element of algebraic coding theory and residue arithmetic. Now, the harmonic index n is defined by

$$n = (m_1 n_1 + m_2 n_2)\text{mod}N \tag{4.56}$$

The mapping defined by Eq. (4.56) is *not* the mapping defined by the CRT. It is not the case here that $n_i = n \bmod N_i$. Given the CRT mapping of the index k and the mapping defined in Eq. (4.56) for the index n, consider the complex exponential W_N^{nk},

$$W_N^{nk} = \prod_{i=1}^{L} \prod_{j=1}^{L} W_N^{(m_i n_i)[m_j(m_j^{-1}k_j \bmod N_j)]} = \prod_{i=1}^{L} W_{N_i}^{m_i k_i} \qquad (4.57)$$

which leads to an expression for the DFT of $x[k]$ given by

$$X[n_1, ..., n_L] = \sum_{k_i=0}^{N_1-1} W_{N_1}^{n_1 k_1} \left[... \left[\sum_{k_L=0}^{N_L-1} x[k_1, ..., k_L] W_{N_L}^{n_L k_L} \right] ... \right] \qquad (4.58)$$

Like the Cooley-Tukey FFT, this derivation leads to a series of nested DFT expressions. The Good-Thomas FFT algorithm has arithmetic order

$$M = N \sum_{i=1}^{L} N_i$$

which has a similar growth rate versus N as the Cooley-Tukey FFT.

Example 4.9: Good-Thomas Indexing

Problem Statement. Again consider $N = 15$, $N_1 = 3$ and $N_2 = 5$. Describe the indexing scheme for time- and frequency-domain data.

Analysis and Conclusions. For $N = 15$, $N_1 = 3$ and $N_2 = 5$, it follows that:

$$m_1 = (N_2)^{-1} \bmod(N_1) = 2 \ni (2*5)\bmod(3) = (10) \bmod 3 = 1.$$

$$m_2 = (N_1)^{-1} \bmod(N_2) = 2 \ni (2*3)\bmod(5) = (6) \bmod 5 = 1.$$

Continuing,

$$n_1 = (m_1 n)\bmod(N_1) = (2n)\bmod(3)$$

$$n_2 = (m_2 n)\bmod(N_2) = (2n)\bmod(5)$$

$$k_1 = (k)\bmod(N_1) = (k)\bmod(3)$$

$$k_2 = (k)\bmod(N_2) = (k)\bmod(5)$$

The distribution of time- and frequency-domain data indices is graphically interpreted in Fig. 4.14. ❖

Returning to the case where $N = N_1 N_2$, the Good-Thomas FFT is given by

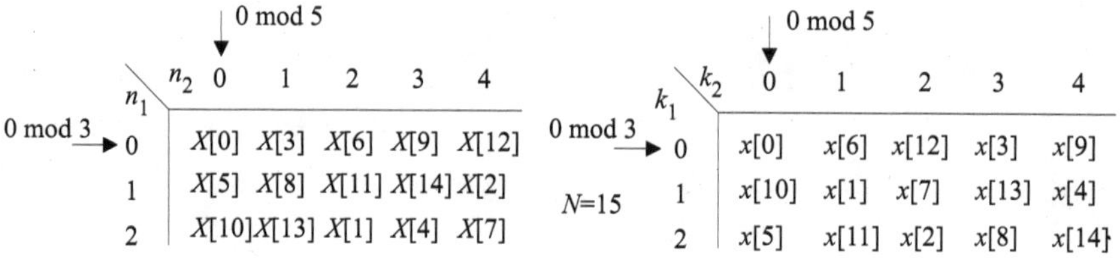

$$0 \bmod 5 \downarrow$$

	n_2 0	1	2	3	4
n_1					
$0 \bmod 3 \rightarrow 0$	$X[0]$	$X[3]$	$X[6]$	$X[9]$	$X[12]$
1	$X[5]$	$X[8]$	$X[11]$	$X[14]$	$X[2]$
2	$X[10]$	$X[13]$	$X[1]$	$X[4]$	$X[7]$

$$0 \bmod 5 \downarrow$$

	k_2 0	1	2	3	4
k_1					
$0 \bmod 3 \rightarrow 0$	$x[0]$	$x[6]$	$x[12]$	$x[3]$	$x[9]$
$N=15$ 1	$x[10]$	$x[1]$	$x[7]$	$x[13]$	$x[4]$
2	$x[5]$	$x[11]$	$x[2]$	$x[8]$	$x[14]$

Figure 4.14   Good-Thomas DFT indexing scheme for $N = 15$, $N_1 = 3$, and $N_2 = 5$.

$$X[n_1, n_2] = \sum_{k_1 = 0}^{N_1 - 1} W_{N_1}^{n_1 k_1} \sum_{k_2 = 0}^{N_2 - 1} x[k_1, k_2] W_{N_2}^{n_2 k_2} \qquad (4.59)$$

From the expression in Eq. (4.59), it is clear that the Good-Thomas FFT can be constructed using N_1 DFTs of length N_2 and N_2 DFTs of length N_1.

Example 4.10: Good-Thomas DFT

Problem Statement. In Example 4.9, an $N = 15$ Good-Thomas index set was studied. Use the Good-Thomas ordering to produce to compute a 15-point DFT.

Analysis and Conclusion. The indexing for an $N = 15$ Good-Thomas DFT is presented in Fig. 4.14. The resulting architecture in shown in Fig. 4.15. It is important to note that there are no butterfly corrections between level one and level two as was the case for the Cooley-Tukey algorithm as shown in Fig. 4.11.   ❖

Example 4.11: Good-Thomas DFT

Problem Statement. Analyze a $N = 192$-point DFT Good-Thomas algorithm. Where, in a 2-D data array, is $x[77]$ and $X[53]$ located?

Analysis and Conclusions. $N = 192 = 64 \times 3$ is a composite number where $N_1 = 64$ and $N_2 = 3$ are relatively prime. The values of m_1' and m_2 found in Eq. (4.54) are $m_1 = 3$ and $m_2 = 64$, respectively. Thus, $x[77] = x[k_1 = (k) \bmod 64, k_2 = (k) \bmod 3] = x[13,2]$ and $X[53] = X[n_1 = (3n) \bmod 64, n_2 = (64n) \bmod 3] = X[31,2]$.   ❖

4.11 Summary

In this chapter, the question of computing a DFT was addressed. The principal contribution was the Cooley-Tukey and Good-Thomas FFT algorithms. Of the two, the Cooley-Tukey ordering algorithm is by far the most popular. If the number of samples to be transformed is composite of the form $N = r^n$, then the classic radix-r FFT results. Other

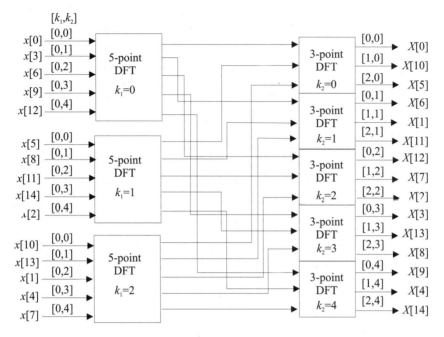

Figure 4.15 Good-Thomas DFT architecture for $N = 15$, $N_1 = 3$, and $N_2 = 5$.

DFT forms were also presented which, while valuable, have only highly specialized uses.

4.12 Self-Study Problems

1. Signal Signature Analysis You are limited to use only a 256-point FFT. Implement a 512-sample FFT using the packing and doubling algorithm. Experimentally verify the performance of each algorithm.

2. Chirp-Z Write a Siglab script file which will evaluate the chirp-z baseband transform (i.e., frequency range [0, $f_s/2$], of a signal of length N, $N \leq 256$, having a format czt(N,a), where a is the chirp-z spiral coefficient. Using a chirp-z transform with $a = 1$, $a = 0.99$, $a = 0.95$, analyze the 64-sample test signal czt.imp and compare the results to what is obtained with a 64-point FFT. Discuss the effect of the parameter a.

3. Cooley-Tukey Ordering Develop a Cooley-Tukey DFT program for a 192-point DFT. Test using the test signal generated by the Siglab command x = mkcos(1,5/192,0,192).

Compare to a 256-point radix-two FFT which has symmetric zero padding of $x[k]$ out to 256-samples.

Compare the execution speed of both solutions.

4. Good-Thomas Ordering Develop a Good-Thomas DFT program for a 192-point DFT. Test using the test signal generated by the Siglab command x = mkcos(1,5/192,0,192). Compare the execution time to that of the 192-point Cooley-Tukey algorithm.

5. Prime-Factor DFT Write a script file that will evaluate an 11-point RPFA. Conduct tests to verify the RPFA DFT.

4.13 References

1. Jacobson, N.A., *Basic Algebra I*. New York: Freeman, 1985.
2. Blahut, R.E., *Fast Algorithms for Digital Signal Processing*. Reading, MA: Addison-Wesley, 1985.
3. Jacobson, N.A., *Basic Algebra I*., New York: Freeman, 1985.

Time-Domain Representation of Discrete-Time Systems

5.1 Introduction

In Chap. 1, discrete-time signals were introduced. In Chap. 2, the z-transform was developed for discrete-time signals. In Chaps. 3 and 4, discrete-time signals were investigated in the frequency-domain using the Fourier transform. In Chap. 5, discrete-time systems will be studied in the discrete-time domain. Systems, it may be recalled, are those entities that modify and manipulate a signal. Examples of discrete-time system models are Shannon's interpolator, digital filters, recursive algorithms, and so forth.

Fundamentally, discrete-time system theory, like continuous-time system theory, can be partitioned into two broad classes called *linear* and *nonlinear*. The differences between these two are profound. Linear systems are based on linear mathematical models and operations that are very familiar to the engineers and scientists. Linear system theory is replete with a myriad of analysis and synthesis tools that are both robust and generally lend themselves to implementation using a general-purpose computer. Nonlinear systems, however, are often mathematically intractable and can frustrate even the most serious mathematician. They often cannot be approached in a straightforward manner and, in such cases, can be explored only by using simulation. The difficulties of working with nonlinear systems are often so severe that engineers and scientists will often overlook reality and assume or approximate the problem as being linear. In some cases, engineers simply give up and use neural nets or statistics in an attempt to predict the performance of a nonlinear system. It is therefore incumbent on the DSP practitioner to fully appreciate what a linear system is and what it is not.

5.2 Linear and Nonlinear Discrete-Time Systems

Linear discrete-time systems exhibit a number of important properties. One of the most important is the superposition principle. The *superposition principle* can be stated as follows.

Superposition Principle

Suppose that there exists a system S, and upon applying an input $x_i[k]$, the output response is measured to be $y_i[k]$, denoted $y_i[k] = S(x_i[k])$. Given a linear combination of L inputs, $x_i[k]$,

$$x[k] = \sum_{k=1}^{L} \alpha_i x_i[k] \tag{5.1}$$

where each $x_i[k]$ has a system output response $y_i[k] = S(x_i[k])$. If

$$y[k] = \sum_{k=1}^{L} \alpha_i S(x_i[k]) = S(x[k]) \tag{5.2}$$

then the system S is said to possess the *superposition property*.

The superposition property is the combination of the system properties *additivity* and *homogeneity*, defined in Sec. 1.3. If a system possessing the superposition property is presented with input $x[k]$, which is also decomposable as the sum of elementary signals, then the output response can be synthesized by simply summing all of the individual system's responses to the individual elementary inputs. This is fortunate, since in Chap. 2 a discipline was presented, based on the z-transform, that represented complicated signals as a linear combination of elementary signals (see Table 3.1). Therefore, it would seem possible that a complicated problem involving a system that has the superposition property can be attacked and solved in a piecemeal manner. This concept is graphically interpreted in Fig. 5.1 in the dis-

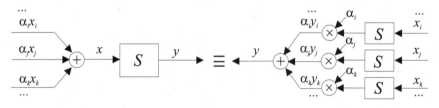

Figure 5.1 Superposition property.

crete-time domain. This property will also be extensively used again in Chap. 6 when transforms are reintroduced for the study of linear, time-invariant systems.

A system is said to be a *linear* system if it exhibits the superposition property. If a system is not linear, then it is said to be *nonlinear*. It was earlier stated the analysis of linear systems, relative to nonlinear, is generally far simpler and more direct. It is therefore important that we establish *a priori* procedures to classify systems. It is tempting to assume that if a system possesses a linear algebraic model it is linear (i.e., if it looks linear, it must be linear). The next example offers a strong caveat to that presumption.

Example 5.1: Superposition

Problem Statement. Consider the discrete-time system S characterized by the linear equation

$$y[k] = ax[k] + b$$

where $b \neq 0$. Is the system linear?

Analysis and Conclusions. Consider the input to be $x[k] = x_1[k] + x_2[k]$. It then follows that

$$y_1[k] = S(x_1[k]) = ax_1[k] + b$$

and

$$y_2[k] = S(x_2[k]) = ax_2[k] + b$$

If the system possesses the superposition property, then it *must* follow that $y[k] = y_1[k] + y_2[k] = S(x_1[k] + x_2[k])$. Calculating $S(x_1[k] + x_2[k])$ produces

$$S(x_1[k] + x_2[k]) = ax_1[k] + ax_2[k] + b$$

but

$$y[k] = y_1[k] + y_2[k]$$

$$= S(x_1[k]) + S(x_2[k])$$

$$= (ax_1[k] + b) + (ax_2[k] + b)$$

$$= ax_1[k] + ax_2[k] + 2b$$

$$\neq S(x_1[k] + x_2[k])$$

Thus, even though the equation defining the system is a linear equation, the system does *not* obey the superposition principle and is therefore classified as nonlinear! ❖

The reason why this apparently benign first system modeled by a linear difference equation failed the superposition test was due to the presence of non-zero initial conditions. This is an often overlooked requirement for linearity, even in academic texts. Systems that are technically nonlinear are often misclassified as being linear simply because engineers and scientists have a natural tendency to *assume* that a system is at rest (zero initial conditions). It should be appreciated that digital signal processing systems are the antithesis of analog signal processors, which tend to lose any residual charge left in the system over time. Digital systems remember! The results from a previous calculation may remain loitering in some digital register and assimilated unwittingly into a DSP algorithm during run-time. It is therefore strongly recommended that, unless it can be absolutely guaranteed that the working registers of a DSP system are zero, they be explicitly initialized to zero.

Example 5.2: Superposition

Problem Statement. Consider again the discrete-time system S studied in Example 5.1, namely

$$y[k] = ax[k] + b$$

except now let $b = 0$. Is the system now linear?

Analysis and Conclusions. Suppose again $x[k] = x_1[k] + x_2[k]$. Then

$$y_1[k] = S(x_1[k]) = ax_1[k]$$

and

$$y_2[k] = S(x_2[k]) = ax_2[k]$$

Calculating $S(x_1[k] + x_2[k])$, one obtains

$$S(X_1[k] + x_2[k]) = a(x_1[k] + x_2[k]) = ax_1[k] + ax_2[k]$$

Finally,
$$y[k] = y_1[k] + y_2[k]$$

$$= S(x_1[k]) + S(x_2[k])$$

$$= (ax_1[k]) + (ax_2[k])$$

$$= ax_1[k] + ax_2[k]$$

$$= S(x_1[k] + x_2[k])$$

Thus the system, in this case, does obey the superposition principle for the given inputs $x_1[k]$, and $x_2[k]$. The moral of these two examples is that due diligence must be observed when determining if a system is linear. ❖

The two examples point out an important observation that even though the equation modeling the system is linear, namely $y[k] = ax[k] + b$, the system itself is *linear only with respect to zero initial conditions*. In other words, given all inputs zero, all outputs should be zero. Because of this systemic weakness in the superposition definition, it is generally assumed that when a system's linearity or nonlinearity is in doubt, the system is tested in an at-rest state or equivalently with zero initial conditions. In later sections, the z-transform will be applied to model systems. A consequence of this application will be a mathematical model called a *transfer function*. Implicit in the transfer function definition will be the assumption that the system is at rest, for reasons made apparent by Examples 5.1 and 5.2.

5.3 Time-Invariant and Time-Varying Systems

A discrete-time system can also be classified in terms of signal propagation. It is generally considered to be a good system attribute if a system behaves consistently every time it is used. Suppose, a causal signal $x[k]$ is presented to a system S and results in a response $y[k]$, namely

$$x[k] \xrightarrow{S} y[k] \tag{5.3}$$

A discrete-time system is said to be *shift-invariant* or *time-invariant* if a shift or delay in the input produces an identical shift or delay in the output. That is, if the previous input $x[k]$ is delayed by m samples, and the output response is identical to what it is was previously, except that it is also delayed by m samples, i.e.,

$$x[k-m] \xrightarrow{S} y[k-m] \tag{5.4}$$

then the system S is said to be time-invariant. This is the test for time-invariance.

Example 5.3: Time-Invariance

Problem Statement. Suppose a discrete-time system is given by $y[k] = (\alpha k + \beta)x[k]$. Is the system time-invariant?

Analysis and Conclusions. To test for time-invariance, substitute for $x[k]$ a shifted time-series, say $x[k - m]$. The output to $x[k-m]$ will be defined to be $w[k] = (\alpha k + \beta)x[k - m]$. If the system is time-invariant, then it must follow that $w[k] = y[k - m]$. However, $y[k - m] = (\alpha(k - m) + \beta)x[k - m] \neq (\alpha k + \beta)x[k - m] = w[k]$ for all $\alpha \neq 0$ and $m \neq 0$. The system is time-invariant only if $\alpha = 0$. ❖

If a system is both linear and time-invariant, then it is said to be *linear time-invariant* (LTI) or, equivalently, *linear shift-invariant* (LSI). LTI systems are of fundamental importance to digital filtering and are generally easy to analyze and behave in a predictable manner. However, it is important to remember that a system can be linear and not be time-invariant (i.e., linear systems with time-varying coefficients).

5.4 Memoryless Systems

The simplest of all systems is one that simply scales a signal by some prespecified constant such as $y[k] = \alpha x[k]$. Such systems are called *memoryless* since they have no knowledge of the past. An example of a memoryless system was given in Example 5.3, in the form of the system modeled by $y[k] = (\alpha k + \beta)x[k]$. Many discrete-time systems, however, generally do contain memory in the form of shift registers or RAM which is used to store past inputs or a system's time history. While any real implementation of a system will have a finite amount of memory (and therefore only be able to store a finite quantity of past input), a digital system that stores state information may have essentially infinite memory. An excellent example of this is an ideal digital integrator, which may be modeled by $y[k] = y[k-1] + x[k]$. An implementation of this system requires only one memory cell to store $y[k-1]$; however, the value stored in that memory is dependent upon *all* previous inputs to the system.

5.5 Causal and Noncausal Systems

A system is said to be *causal*, or *nonanticipatory*, if it cannot produce an output until an input is present and the system is turned on (enabled). An LTI system is causal if and only if its unit impulse response (i.e., $\delta[k]$) is zero for sample values preceding the application of the impulse (i.e., $k < 0$). If a system is not causal, it is said to be *noncausal*, or *anticipatory*.

5.6 LTI Discrete-Time Systems

LTI systems are at the core of most contemporary DSP filter designs. As will be shown in Chap. 11 and beyond, the majority of digital filters currently being designed can be modeled as an Nth order LTI system given by

$$\sum_{m=0}^{N} a_m y[k-m] = \sum_{m=0}^{M} b_m x[k-m] \qquad (5.5)$$

with the appropriate initial conditions specified. If $N \geq M$, the system is said to be *proper*. Equation (5.5) can also be expressed as

$$y[k] = \frac{1}{a_0}\left(-\sum_{m=1}^{N} a_m y[k-m] + \sum_{m=0}^{M} b_m x[k-m] \right) \tag{5.6}$$

If $a_0 = 1$, then the system is said to be *monic*. The solution to Eq. (5.6) can be modeled as the sum of two solutions. The first of these to be considered is the solution to the *homogeneous equation*

$$\sum_{m-0}^{N} a_m y[k-m] = 0 \tag{5.7}$$

The *homogeneous solution* to the linear discrete-time system modeled by Eq. (5.6) shall be denoted $y_h[k]$. Solutions to the homogeneous equation can be defined in terms of roots of the system's *characteristic equation*, which is given as

$$\sum_{m=0}^{N} a_m r^{N-m} = 0 \tag{5.8}$$

There may be up to N distinct solutions to this equation. If a root is unique (i.e., it is not repeated) then it is said to be *distinct* or have *multiplicity* one. If a root is repeated n times then it is said to be *repeated* and has *multiplicity* n. The multiplicity of root r_i will be denoted n_i. The solution to the homogeneous equation, Eq. (5.7), is

$$y_h[k] = \sum_{i=1}^{N} \gamma_i k^{n_i-1} r_i^k \tag{5.9}$$

where the coefficients, γ_i, for all $i \in \{1,2,3,...,N\}$, are constants whose value is dependent upon $N-1$ initial conditions. This defines a system of equations which can be solved for γ_i which, when computed, will define the homogeneous solution to the LTI system given by Eq. (5.6).

Example 5.4: Homogeneous Solution

Problem Statement. Consider the homogeneous LTI system equation given by

$$y[k] - \frac{1}{4}y[k-1] - \frac{1}{8}y[k-2] = 0$$

$$y[-1] = 1$$

$$y[-2] = 1$$

What is the homogeneous solution?

Analysis and Conclusions. Beginning with the characteristic equation, observe that

$$r^2 - \frac{1}{4}r - \frac{1}{8} = 0$$

or

$$\left(r - \frac{1}{2}\right)\left(r + \frac{1}{4}\right) = 0$$

The roots of the characteristic equations are seen to be distinct and are given by $r_1 = 1/2$ and $r_2 = -1/4$ (i.e., $n_1 = n_2 = 1$). Therefore, the homogeneous solution is given by

$$y[k] = \gamma_1 \left(\frac{1}{2}\right)^k + \gamma_2 \left(-\frac{1}{4}\right)^k$$

upon solving for the corresponding values of γ_1 and γ_2. The question, of course, is how to solve the above equation efficiently. To solve a system of two independent equations in two unknowns, two data points need to be specified. The two data points which are known in this case are the given initial conditions $y[-1]$ and $y[-2]$. This information can be used to specify

$$y[-1] = \gamma_1 \left(\frac{1}{2}\right)^{-1} + \gamma_2 \left(-\frac{1}{4}\right)^{-1} = 2\gamma_1 - 4\gamma_2 = 1$$

$$y[-2] = \gamma_1 \left(\frac{1}{2}\right)^{-2} + \gamma_2 \left(-\frac{1}{4}\right)^{-2} = 4\gamma_1 + 16\gamma_2 = 1$$

which, in turn, can be expressed in matrix-vector form as

$$\begin{bmatrix} 2 & -4 \\ 4 & 16 \end{bmatrix} \begin{bmatrix} \gamma_1 \\ \gamma_2 \end{bmatrix} = \begin{bmatrix} 1 \\ 1 \end{bmatrix}$$

The solution to this equation is $\gamma_1 = 5/12$ and $\gamma_2 = -1/24$. For $k \geq 0$, it follows that

$$y[k] = \frac{10}{24}\left(\frac{1}{2}\right)^k - \frac{1}{24}\left(\frac{1}{4}\right)^k \qquad \text{❖}$$

The homogeneous solution, by itself, has limited value. What is of general interest is knowledge of the forced or *nonhomogeneous* system response to an arbitrary input $x[k]$. The nonhomogeneous response of a discrete-time LTI system will be shown to be defined in terms of the system's *impulse response*. The impulse response of a causal LTI system is, as the name implies, the output trajectory of the system with an input $x[k] = \delta[k]$, provided that the system is initially *at rest* (i.e., $y[k] = 0$ for all $k < 0$).

Assume that the response of the system (i.e., $y[k]$) defined in Eq. (5.5) to an impulse is denoted $h[k]$. From Eq. (5.5) (repeated for clarity)

$$\sum_{m=0}^{N} a_m y[k - m] = \sum_{m=0}^{M} b_m x[k - m] \tag{5.10}$$

it follows that

$$\sum_{m=0}^{N} a_m h[k - m] = \sum_{m=0}^{M} b_m \delta[k - m] \tag{5.11}$$

The solution to Eq. (5.11) is called the *impulse response* and can be obtained in a manner similar to that used to compute the homogeneous solution of an LTI [i.e., Eq. (5.7)]. The previous use of Eq. (5.7) was to compute the homogeneous system response in terms of the system's known initial conditions. The nonhomogeneous solution assumes zero initial conditions and therefore must be "tricked" into finding a replacement for the initial conditions. Consider, then, using a set of *pseudo-initial conditions*, say $\{h[0],h[1],...,h[M]\}$, which are the first $M + 1$ samples of output after the system is driven by a unit impulse. The pseudo-initial conditions satisfy the matrix-vector equation

$$\mathbf{Ah} = \begin{bmatrix} a_0 & 0 & \cdots & 0 \\ a_1 & a_0 & \cdots & 0 \\ \vdots & \vdots & \ddots & \vdots \\ a_M & a_{M-1} & \cdots & a_0 \end{bmatrix} \begin{bmatrix} h[0] \\ h[1] \\ \vdots \\ h[M] \end{bmatrix} = \begin{bmatrix} b_0 \\ b_1 \\ \vdots \\ b_M \end{bmatrix} = \mathbf{b} \tag{5.12}$$

This set of values can then be used to "seed" Eq. (5.11), thus producing the nonhomogeneous impulse response, assuming $N \geq M$. The solution to Eq. (5.12) can therefore be defined in terms of the pseudo-initial conditions $h[i]$, and the system coefficients, a_i and b_i, $i \in \{0,1,2,...,M\}$, if the matrix equation $\mathbf{Ah} = \mathbf{b}$ has a solution. Assuming a solution exists, Eq. (5.12) may be solved for $\mathbf{h}$ by

$$\mathbf{h} = \mathbf{A}^{-1}\mathbf{b} \tag{5.13}$$

The values of **h** obtained here serve the role of the initial conditions used in the solution of Eq. (5.7). Once the pseudo initial-condition vector is determined, the analog to Eq. (5.9) can be formed and solved. Specifically, Eq. (5.9) becomes

$$h[k] = \gamma_0\delta[k] + \sum_{i=1}^{N} \gamma_i k^{n_i-1} r_i^k \qquad (5.14)$$

and Eq. (5.7) translates to

$$\sum_{m=0}^{N} a_m h[k-m] = 0 \qquad (5.15)$$

Once the parameters γ_i are computed, the impulse response can be formally expressed in terms of Eq. (5.14).

Example 5.5: Impulse Response

Problem Statement. The at-rest third-order LTI system studied in Example 5.4 was given by

$$y[k] - \frac{1}{4}y[k-1] - \frac{1}{8}y[k-2] = 2x[k] - x[k-1] + x[k-2]$$

What is the system's impulse response?

Analysis and Conclusions. If the input $x[k]$ is replaced by an impulse (i.e., $x[k] = \delta[k]$), then it directly follows that $b_0 x[0] = 2$, $b_1 x[1] = -1$ and $b_2 x[2] = 1$. The impulse response for $k \in \{0,1,2\}$ can be defined in terms of the impulse response vector, **h**, which satisfies the matrix equation $\mathbf{Ah} = \mathbf{b}$, after Eq. (5.12). That is

$$\mathbf{Ah} = \begin{bmatrix} 1 & 0 & 0 \\ -\frac{1}{4} & 1 & 0 \\ \frac{1}{8} & -\frac{1}{4} & 1 \end{bmatrix} \begin{bmatrix} h[0] \\ h[1] \\ h[2] \end{bmatrix} = \begin{bmatrix} 2 \\ -1 \\ 1 \end{bmatrix} = \mathbf{b}$$

Solving the matrix equation results in $h[0] = 2$, $h[1] = -1/2$, and $h[2] = 9/8$. Based on the solution methodology explored in Example 5.4, the assumed form of the impulse response is given by Eq. (5.14), and it is

$$h[k] = \gamma_0\delta[k] + \gamma_1 r_1^k + \gamma_2 r_2^k$$

where r_1 and r_2 are again

$$r_1 = \frac{1}{2} \text{ and } r_2 = -\frac{1}{4}$$

Evaluating the pseudo-initial conditions shown above results in

$$
\begin{bmatrix}
1 & 1 & 1 \\
0 & \dfrac{1}{2} & -\dfrac{1}{4} \\
0 & \left(\dfrac{1}{2}\right)^2 & \left(-\dfrac{1}{4}\right)^2
\end{bmatrix}
\begin{bmatrix}
\gamma_0 \\
\gamma_1 \\
\gamma_2
\end{bmatrix}
=
\begin{bmatrix}
2 \\
-\dfrac{1}{2} \\
\dfrac{9}{8}
\end{bmatrix}
$$

which yields

$$
\gamma_0 = -8, \quad \gamma_1 = \frac{8}{3}, \quad \gamma_2 = \frac{22}{3}
$$

It therefore follows that the impulse response is given by

$$
h[k] = \left(-8\delta[k] + \frac{8}{3}\left(\frac{1}{2}\right)^k + \frac{22}{3}\left(-\frac{1}{4}\right)^k \right)u[k]
$$

Notice also that $h[k]$ agrees with the pseudo-initial conditions at sample instances $k = \{0,1,2\}$.   ❖

Unfortunately, it can also be seen that this method is computationally intensive and may be unsuitable for use with high-order systems. Later, the z-transform will be used to simplify the process of computing an impulse response. Nevertheless, by introducing the impulse response at this time, a gateway is provided to the study of one of the most fundamentally important concepts found in DSP: *convolution*.

5.7 Discrete-Time Convolution

The study of linear continuous-time systems is intimately linked to the concept of convolution. Convolution explains how communications signals are distorted through a channel, and why optical, electronic, and acoustical filters can be designed to be frequency selective. Convolution is also the framework that explains the nature and form of the signals observed in the natural world.

The nonhomogeneous response of an at-rest causal discrete-time LTI system to an arbitrary time-series $x[k]$, described by Eq. (5.16), is defined in terms of a *discrete-time convolution sum*,

$$
y[k] = \frac{1}{a_0}\left(\sum_{m=0}^{M} b_m x[k-m] - \sum_{m=1}^{N} a_m y[k-m] \right) \tag{5.16}
$$

The process suggested by Eq. (5.16) is graphically interpreted in Fig. 5.2. The LTI system is presented with an input time-series $x[k]$

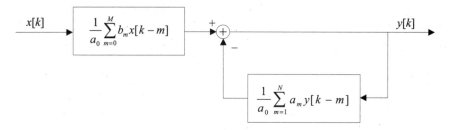

Figure 5.2 LTI system.

and produces an output time-series $y[k]$. The LTI system shown also contains a memory of past inputs (i.e., $x[k-m]$) and outputs (i.e., $y[k-m]$). When combined together in the proper amounts, filters and signal conditioning systems result.

For convenience, assume that the at-rest causal LTI system is *monic* (i.e., $a_0 = 1$). Then the solution to Eq. (5.16) can be expanded as

$$y[0] = b_0 x[0]$$

$$y[1] = b_1 x[0] + b_0 x[1] - a_1 y[0] = [b_1 - a_1 b_0]x[0] + b_0 x[1]$$

$$y[2] = b_2 x[0] + b_1 x[1] + b_0 x[2] - a_1 y[1] - a_2 y[0]$$

$$= [b_2 - (a_1 b_1 + a_2 b_0) + a_1^2 b_0]x[0] + [b_1 - a_1 b_0]x[1] + b_0 x[2]$$

$$\vdots \tag{5.17}$$

which is seen to become unmanageable as k increases. Fortunately the solution shown in Eq. (5.17) can be expressed in terms of a collection of responses defined by replacing the input $x[k]$ with an impulse. Specifically, consider the input $x[k]\delta[k]$, then Eq. (5.17) reduces to

$$y_0[0] = b_0 x[0]$$

$$y_0[1] = b_1 x[0] - a_1 y_0[0] = (b_1 - a_1 b_0)x[0]$$

$$y_0[2] = b_2 x[0] - a_2 y_0[0] - a_1 y_0[1]$$

$$= [b_2 - (a_1 b_1 + a_2 b_0) + a_1^2 b_0]x[0]$$

$$\vdots \tag{5.18}$$

which would define the impulse response of the LTI system over the first few samples. If the input were set to $x[k]\delta[k-1]$, then

$$y_1[0] = 0$$

$$y_1[1] = b_0 x[1] - a_1 y_1[0] = b_0 x[1]$$

$$y_1[2] = b_1 x[1] - a_1 y_1[1] - a_2 y_1[0] = [b_1 - a_1 b_0] x[1]$$

$$\vdots$$

$$(5.19)$$

The response has the mathematical structure identical to that shown in Eq. (5.18) except that the indices are delayed by one sample. Thus, Eq. (5.18) is the delayed impulse response of the system. For the input $x[k]\delta[k - 2]$, it follows that

$$y_2[0] - 0$$

$$y_2[1] = 0$$

$$y_2[2] = b_0 x[2]$$

$$\vdots$$

$$(5.20)$$

and the previous impulse response argument can be repeated. This connection between the nonhomogeneous response of an LTI to an arbitrary input and a delayed impulse response now can be formalized. The nonhomogeneous solution to Eq. (5.16) is defined by what is commonly referred to as the *convolution sum*. The convolution sum is defined in terms of the system's impulse response $h[k]$. The impulse response $h[k]$ is defined by Eq. (5.11) and is seen to be a function of both the a_m and b_m variables found in Eq. (5.16). The *convolution*, or to be more technically correct, *linear convolution*, of an arbitrary time-series $x[k]$ by an LTI system having an impulse response $h[k]$, is denoted $y[k] = h[k]*x[k]$, and is given by

$$y[k] = h[k]*x[k] = \sum_{m=0}^{\infty} h[k-m]x[m] = \sum_{m=0}^{\infty} h[m]x[k-m] \quad (5.21)$$

Exploring Eq. (5.21) in more detail, it can be noted that the values of $y[k]$ are given by:

$$y[0] = h[0]x[0]$$
$$y[1] = h[1]x[0] + h[0]x[1]$$
$$y[2] = h[2]x[0] + h[1]x[1] + h[0]x[2]$$

$$\vdots$$

$$y[k] = h[0]x[0] + h[k-1]x[1] + h[k-2]x[2] + \ldots + h[1]x[k-1] + h[0]x[k]$$

$$(5.22)$$

where the terms in Eq. (5.22), can be placed in correspondence with those found in Eq. (5.17). This pattern continues for all sample instances.

Example 5.6: Convolution

Problem Statement. Suppose $h[k] = 0.9^k u[k]$ is the impulse response of a causal LTI system, and $x[k] = u[k]$ is a unit step. Compute the first few output sample values of $y[k] = h[k]*x[k]$.

Analysis and Conclusions. Applying Eq. (5.21), it follows that:

$$y[0] = h[0]x[0] = 1$$

$$y[1] = h[1]x[0] + h[0]x[1] = 1 + 0.9 = 1.9$$

$$y[2] = h[2]x[0] + h[1]x[1] + h[0]x[2] = 1 + 1.09 + 0.81 = 2.71$$

In the example, the system's output can be seen to be monotonically increasing in time. However, at this point all that is actually known is a few sample values. It is unknown what the actual future values of $y[k]$ will be or what the steady-state step response value will be (i.e., $\lim_{k \to \infty} y[k]$). Therefore, it is generally desirable that the solution be expressed in a closed form.   ❖

Example 5.7: Convolution

Problem Statement. Consider again the system studied in Example 5.6 which possesses an impulse response given by $h[k] = 0.9^k u[k]$, and $x[k] = u[k]$, a unit step. What is $y[k] = h[k]*x[k]$ for all $k \geq 0$?

Analysis and Conclusions. Applying Eq. (5.21) this time, it follows that:

$$y[k] = \sum_{m=0}^{k} h[m]x[k-m] = \sum_{m=0}^{k} h[m] = \sum_{m=0}^{k} (0.9)^m$$

$$= \frac{1 - (0.9)^{k+1}}{1 - 0.9} = 10(1 - (0.9)^{k+1})$$

Notice that from the above closed form expression for $y[k]$ produces,

$$y[0] = 1$$

$$y[1] = 1.9$$

$$y[2] = 2.71$$

which agrees with the previously computed result. The steady-state unit step response is seen to be given by

$$\lim_{k \to \infty} y[k] = 10(1 - (0.9)^{k+1})\big|_{k \to \infty}$$

❖

Example 5.8: Convolution

Problem Statement. In Example 5.4, the homogeneous solution of a second order system is known. Determine the nonhomogeneous unit-step response of that system, given by

$$y[k] - \frac{1}{4}y[k-1] - \frac{1}{8}y[k-2] = 2u[k] - u[k-1] + u[k-2]$$

$$y[-1] = y[-2] = 0$$

Analysis and Conclusions. The impulse response has been previously established to be:

$$h[k] = \left(-8\delta[k] + \frac{8}{3}\left(\frac{1}{2}\right)^k + \frac{22}{3}\left(-\frac{1}{4}\right)^k \right) u[k]$$

Applying Eq. (5.21) this time, it follows that

$$y[k] = \sum_{m=0}^{k} h[k-m]u[k] = \sum_{m=0}^{k} \left(-8\delta[m] + \frac{8}{3}\left(\frac{1}{2}\right)^m + \frac{22}{3}\left(-\frac{1}{4}\right)^m \right)$$

which can be reduced to

$$y[k] = \sum_{m=0}^{k} (-8\delta[m]) + \sum_{m=0}^{k} \left(\frac{8}{3}\left(\frac{1}{2}\right)^m\right) + \sum_{m=0}^{k} \left(\frac{22}{3}\left(-\frac{1}{4}\right)^m\right)$$

$$= -8 + \frac{8}{3}\left(\frac{1 - \left(\frac{1}{2}\right)^{k+1}}{\frac{1}{2}} \right) + \frac{22}{3}\left(\frac{1 - \left(-\frac{1}{4}\right)^{k+1}}{\frac{5}{4}} \right)$$

$$= -8 + \frac{16}{3}\left(1 - \left(\frac{1}{2}\right)^{k+1}\right) + \frac{88}{15}\left(1 - \left(-\frac{1}{4}\right)^{k+1}\right)$$

Notice that the steady-state unit step value is given by

$$\lim_{k \to \infty} y[k] = \frac{48}{15} = 3.2$$

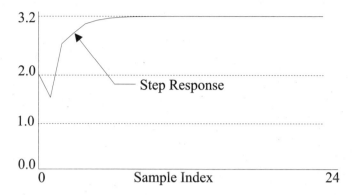

Figure 5.3 Impulse response of a typical second-order system.

The trajectory $y[k]$ over a range of k is reported in Fig. 5.3. Notice that steady state is essentially achieved by the 25-sample sequence generated.

Computer Study. These results can be verified using simulation by executing the following sequence of Siglab commands.

```
>y=zeros(25)        # 25 points of step response will be gener-
                      ated.
>y[0]=2             # Manually enter the initial values for y[k].
>y[1]=(2-1)+0.25*y[0]
>for (k=2:24)       # Compute y[k] for remaining 23 points.
>>y[k]=(2-1+1)+0.25*y[k-1]+0.125*y[k-2]
>end
>graph(y) # Plot results.                                    ❖
```

Consider focusing only on characterizing the response of an at-rest LTI. Since the homogeneous response of an at-rest LTI is zero for all time, the output is completely characterized by the convolution operation given in Eq. (5.21) which, in turn, is defined in terms of the system's impulse responses. The impulse response of an LTI can be classified as being an *infinite impulse response* (IIR) or a *finite impulse response* (FIR). An $(M+1)$th order (i.e., $M+1$ coefficients) FIR system is modeled as

$$y[k] = \sum_{m=0}^{M} b_m x[k-m] \qquad (5.23)$$

Equation (5.23) is seen to be Eq. (5.16) devoid of all the terms associated with coefficients a_m, for $m > 0$ and $a_0 = 1$. The impulse response of the system, defined by Eq. (5.23), is easily verified to be

$$h[k] = \sum_{m=0}^{M} b_m \delta[k-m]$$

$$= b_0 \delta[k] + b_1 \delta[k-1] + \ldots + b_{M-1}\delta[k-M-1] + b_M\delta[k-M] \tag{5.24}$$

and is seen to persist for only $M + 1$ samples. The output of an Mth order FIR system with an N sample input time-series $x[k]$ is given by the finite convolution sum

$$y[k] = \sum_{m=0}^{N-1} h[k-m]x[m] = \sum_{m=0}^{M-1} h[m]x[k\text{-}m] \tag{5.25}$$

where $h[k] = b_k$ is the kth sample of the FIR system's impulse response for $k \in \{0,1,2,\ldots,M-1\}$. The output sample values are given by

$$y[0] = h[0]x[0]$$

$$y[1] = h[1]x[0] + h[0]x[1]$$

$$\vdots$$

$$y[k] = h[k]x[0] + h[k-1]x[1] + \ldots + h[0]x[k]$$

$$\vdots$$

$$y[M+N-2] = h[M-1]x[N-1]$$

That is, the nonhomogeneous response begins with $y[0]$ defined by a single product term, then moves to a middle point where there are either M or N sums-of-products used to define each output sample. The nonhomogeneous response finally terminates at the $k = M + N - 2$ sample, which once again is a single product term. Therefore, the convolution sum, given in Eq. (5.25), may produce a sequence with up to $M + N - 1$ non-zero elements. For all $k > M + N - 2$, the output is zero.

Example 5.9: FIR Convolution

Problem Statement. An Mth order FIR having an impulse response $h[k] = \{1,1,\ldots,1,1\}/M$ is called *moving average* (MA) filter and will be studied in detail in Chap. 11. Specifically,

$$h[k] = \left(\frac{1}{m}\right)\sum_{m=0}^{M-1} \delta[k-m] = \begin{cases} 1/M & \text{for } (k \in \{0, 1, 2, \ldots, M-1\}) \\ 0 & \text{otherwise} \end{cases}$$

If the system begins at-rest, what is its response to a single thirty-two sample triangular wave and a periodic thirty-two sample triangular wave if $M = 9$?

Analysis and Conclusions. If the input $x[k]$ is a length N time-series, then the convolution $x[k]*h[k]$ is a length $(N + M - 1)$ or an $N + 8$ sample time-series. The jth output sample, namely $y[j]$, is equal to the average value of the most recent nine samples. The short-term (nine-sample) average value of a triangle wave is a smoothed signal having the shape shown in Fig. 5.4. Observe that the FIR response to the aperiodic 32-sample triangular wave consists of an initial (transient) period in which energy internally builds up in the filter, a period in which the input fills the filter with sample values, and finally a decay interval in which energy leaves the FIR system. The total length of this process is 40 samples, as predicted. If the input is periodically extended, then after the initial transient build-up, the system response is a periodic sequence.

Computer Study. These results can be verified by simulation using the following sequence of Siglab commands. The sequence of commands shown below is used to simulate a nine sample moving average FIR and produce the data shown in Fig. 5.4.

```
>h=ones(9)/9        # Define FIR impulse response.
>x1=tri(32,32)&zeros(256-32) # Generate single triangle
>x2=tri(32,256)     # Generate eight triangles.
>y1=h$x1            # Produce system response for input x1.
>y2=h$x2            # Produce system response for input x2.
>graph(y1,y2)       # Graph results.                      ❖
```

The previous example demonstrated that the response of an LTI to a periodic input, $x[k]$, is also periodic with the same period as the input. In fact, one of the axioms of a linear system theory is that it linear systems cannot create new frequencies. Instead, an LTI can only scale and rephase (phase shift) a signal. That is, if an LTI system having an impulse response $h[k]$ is presented with a pure sinusoid, say $x[k] = A_i \sin(\omega_i k + \varphi_i)$, then the output $y[k] = h[k]*x[k] = A_0\sin(\omega_0 k + \varphi_0)$ where (A_0, φ_0) may or may not equal (A_i, φ_i) but $\omega_0 = \omega_i$. Nonlinear systems, however, can create new frequencies. Consider a memoryless nonlin-

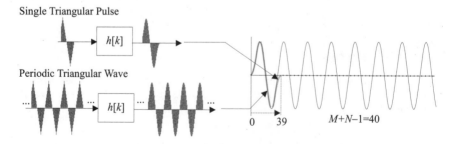

Figure 5.4 Processing a periodic square wave through a length nine moving average filter.

ear system defined by $y[k] = x^2[k]$. The output corresponding to an input $x[k] = A_i\cos(\omega_i k)$ would be $y[k] = (A_i/2)(1 + \cos(2\omega_i k))$, which consists of signals at frequencies of zero and $2\omega_i$.

Another important type of convolution is *circular convolution*. Suppose that $h[k]$ and $x[k]$ are two sequences of length N, defined over $k \in \{0,1,2,...,N-1\}$. The circular convolution of $x[k]$ and $h[k]$ is periodic with period N and is denoted $y[k] = h[k] \otimes x[k]$, and is defined as

$$y[k] = h[k] \otimes x[k]$$

$$= \sum_{m=0}^{N-1} h[(k-m)\mathrm{mod}\,N]x[(m)\mathrm{mod}\,N]$$

$$= \sum_{m=0}^{N-1} h[(m)\mathrm{mod}\,N]x[(k-m)\mathrm{mod}\,N] \qquad (5.26)$$

The processes defined by Eq. (5.26) is called *circular convolution*. The importance of circular convolution will become apparent in the next chapter, where transforms are reintroduced.

Example 5.10: Circular Convolution

Problem Statement. A signal $x[k] = \{1,1,1,1,0,0,0,0\}$ is to be linearly and circularly convolved by a system having a length eight impulse response $h[k] = \{1,1,1,1,0,0,0,0\}$. What is $y[k] = h[k]*(x[k]$ and $y[k] = h[k] \otimes x[k]$?

Analysis and Conclusions. The linear and circular convolution of $x[k]$ and $h[k]$ are summarized in Fig. 5.5. It can be seen that they are identical for $k \in \{0,1,2,...,7\}$ but differ elsewhere. It is sometimes useful to envision the circular convolution operations described in terms of the optical disk analogy also shown in Fig. 5.5. Shown are two coaxial disks, one coded with $x[k]$ and the other with $h[k]$. Wherever $x[k]$ or $h[k]$ is one, a hole is cut into the disk. If zero, the corresponding section of the disk is left opaque. A light source is placed at the center of the disk, and photodetectors are set along the periphery of the outer disk. The inner disk is spun and the amount of light energy collected by the detectors and integrated. At one point, the two disks will perfectly overlap, and the maximum optical energy will be transmitted to the detectors. After progressing π-radians, the hole of one disk overlaps the opaque section of the second, and no light is received by the photodetectors. Continuing, π-radians after that, maximum overlap again occurs. In between, the detected energy linearly increases and decreases.   ❖

5.8 Properties of Discrete-Time Convolution

The discrete-time convolution sum possesses the important properties that are summarized below.

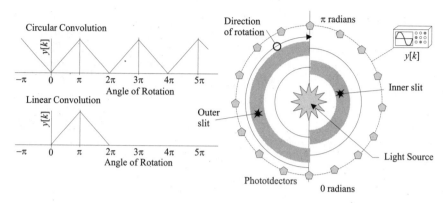

Figure 5.5 Circular and linear convolution of two pulses and an optical circular convolution analogy.

- Commutative Property: $x[k]*y[k] = y[k]*x[k]$
- Associative Property: $(x[k]*y[k])*z[k] = x[k]*(y[k]*z[k])$
- Distributive Property: $x[k]*(y[k] + z[k]) = x[k]*y[k] + x[k]*z[k]$

The distributive property allows higher-order filters to be created by placing lower-order filters in parallel. The commutative and associative properties allows LTI systems to be cascaded together in any order to form more complicated systems, as shown in Fig. 5.6. This can be easily proven. Again, referring to Fig. 5.6,

$$y_1[k] = h_1[k]*x[k] \qquad (5.27)$$

and

$$y[k] = h_2[k]*y_1[k] \qquad (5.28)$$

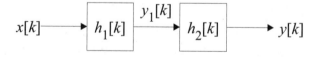

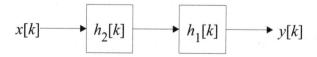

Figure 5.6 Order independence of cascaded LTI systems.

The proof of this fact starts by substitution of Eq. (5.27) into Eq. (5.28) and proceeds using the commutative and associative properties as follows

$$y[k] = h_2[k]*y_1[k]$$

$$= h_2[k]*(h_1[k]*x[k]) \quad \text{(substitution)}$$

$$= (h_2[k]*h_1[k])*x[k] \quad \text{(associativity)}$$

$$= (h_1[k]*h_2[k])*x[k] \quad \text{(commutitivity)}$$

$$= h_1[k]*(h_2[k]*x[k]) \quad \text{(associativity)} \tag{5.29}$$

and so it is shown. Therefore, the two systems shown in Fig. 5.6 are completely equivalent.

5.9 Stability of Discrete-Time Systems

The study of continuous-time systems has historically stressed stability analysis. Continuous-time systems were classified as being stable or unstable. Discrete-time systems can also be stable or unstable. One of the more fundamental definitions of stability is based on the *bounded-input bounded-output* (BIBO) *principle*. A system is said to be BIBO stable if the output for every possible bounded input is likewise bounded.

A discrete-time system with an impulse response $h[k]$ is BIBO stable if and only if

$$\sum_{k=-\infty}^{\infty} |h[k]| < \infty \tag{5.30}$$

That is, to be BIBO stable, the impulse response must be absolutely summable. The condition reported in Eq. (5.30) corresponds to the *worst-case* system response if the input time-series is assumed to have an amplitude bounded by unity (i.e., $|x[k]| \leq 1.0$). The *worst-case input*, up to a scale factor, and assuming bounded input, is denoted $x_w[k]$, and is an input that produces a maximal absolute convolution sum $|y[k]| = |h[k]*x[k]|$. The worst-case input is given by

$$x_w[k] = \frac{h[L-k]}{|h[L-k]|} \tag{5.31}$$

where L is an integer. Note that $|x_w[k]| = 1.0$ (i.e., $x_w[k] = \pm 1$). For some value of N and L, the sample values of $x_w[k]$ will be aligned with those of $h[b - m]$ so that every product $h[k - m]x_w[k]$ has a maximal positive value. This can be seen by substituting Eq. (5.31) into the convolution sum to produce

$$y[k] = \sum_{m = -\infty}^{\infty} h[k - m]x_w[m] = \sum_{m = -\infty}^{\infty} h[k - m]\frac{h[L - m]}{|h[L - m]|} \qquad (5.32)$$

For $k = L$, Eq. (5.32) simplifies to

$$y[k] = \sum_{m = -\infty}^{\infty} \frac{(h[L - m])^2}{|h[L - m]|} = \sum_{m = -\infty}^{\infty} |h[m]| \qquad (5.33)$$

which is also the bound established by Eq. (5.30).

The impulse response of a discrete-time LTI was earlier [Eq. (5.14)] expressed as a linear combination of functions,

$$h[k] = \gamma_0\delta[k] + \sum_{i = 1}^{N} \gamma_i k^{n_i - 1} r_i^k \qquad (5.34)$$

Therefore, in order for $h[k]$ to be absolutely summable, it must follow that the modulus of r_i must satisfy the condition $|r_i| < 1.0$. Therefore, if the moduli of the roots of a discrete-time LTI system's characteristic equation are bounded by unity, then the corresponding LTI system will be BIBO stable.

Example 5.11: Stability of Discrete-Time Systems

Problem Statement. An at-rest discrete-time LTI system given by

$$y[k] - 0.875y[k - 1] + 0.219y[k - 2] - 0.016y[k - 3] = x[k]$$

has a characteristic equation given by

$$r^3 - 0.875r^2 + 0.219r - 0.016 = 0$$

which factors into

$$\left(r - \frac{1}{2}\right)\left(r - \frac{1}{4}\right)\left(r - \frac{1}{8}\right) = 0$$

The roots of the characteristic equation are $r \in \{1/2, 1/4, 1/8\}$ which are bounded in absolute value below unity. Therefore the system is BIBO stable. To explore this claim further, assume that the impulse response of the discrete-time system is given by

$$h[k] = \left(\frac{8}{3}\left(\frac{1}{2}\right)^k - 2\left(\frac{1}{4}\right)^k + \frac{1}{3}\left(\frac{1}{8}\right)^k\right)u[k]$$

Consider using a unit step $x[k] = u[k]$ as a test signal. The step response, given by the linear convolution sum, is equal to

$$y[k] = \left(\frac{8}{3}\sum_{m=0}^{k}\left(\frac{1}{2}\right)^m - 2\sum_{m=0}^{k}\left(\frac{1}{4}\right)^m + \frac{1}{3}\sum_{m=0}^{k}\left(\frac{1}{8}\right)^m\right)$$

The unit step response $y[k]$ monotonically increases from its initial value of $y[0] = 1$ to

$$\lim_{k \to \infty} y[k] = \left(\frac{8}{3}\frac{1}{\left(1-\frac{1}{2}\right)} - 2\frac{1}{\left(1-\frac{1}{4}\right)} + \frac{1}{3}\frac{1}{\left(1-\frac{1}{8}\right)}\right) = 3.04762 < \infty$$

The bounded step response does not mean that the system is necessarily stable for all arbitrary inputs. Since the impulse response $h[k]$ is non-negative, the worst-case input given by Eq. (5.31) is $x_w[k] = u[k]$ in this case. Therefore the step response bound is also the worst-case bound, up to a scale factor. The system is therefore guaranteed to be BIBO stable.

Computer Study. The following sequence of Siglab commands computes the impulse response $h[k]$ as well as the step response for $k<32$. The results are displayed in Fig. 5.7. Note that the steady-state value is nearly achieved after only ten samples.

```
>k=rmp(32,1,32)        # Generate time index 0-31
>z=(8/3)*(1/2)^k-2*(1/4)^k+(1/3)*(1/8)^k # Compute impulse response.
>ograph(z,intg(z))     # Plot impulse, step responses.              ❖
```

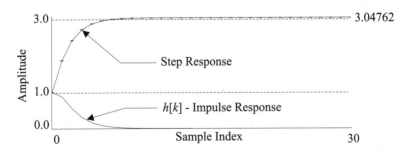

Figure 5.7 Step and impulse response.

5.10 Summary

In Chap. 5, discrete-time systems were introduced. The study focused on linear shift invariant or LTI models which were shown to interact with signals through a process called *convolution*. Two types of convolution were introduced: linear and circular. Linear convolution can be used for arbitrary signals and systems. Circular convolution may be used with periodic (noncausal) signals and/or systems. In the next chapter, transform-domain methods will be used to convert the complicated time-domain implementation of a convolution integral or summation into a set of simple algebraic operations.

Stability was also studied in the context of a bounded-input bounded-output (BIBO) criteria. The stability of an LTI system was seen to be established by the roots of the system's characteristic equation. In the next chapter, this concept will be extended using transform-domain methods.

5.11 Self-Study Problems

1. Homogeneous solution   Consider the Fibonacci rabbit population model given by $y[k] = y[k-1] + y[k-2]$ with initial conditions $y[-1] = y[-2] = 1$ and $k \le 0$. Derive the closed form homogeneous solution for $y[k]$.

2. Convolution   Linearly convolve the following impulse response and signal pairs.

$$h[k] = (0.9)^k u[k]$$
$$h[k] = \delta[k] - \delta[k-10]$$
$$h[k] = u[k] - u[k-10]$$
$$x[k] = \delta[k]$$
$$x[k] = \delta[k-10]$$
$$x[k] = u[k]$$
$$x[k] = u[k-10]$$

3. Stability   Determine the stability of the following LTI system:

$$y[k] - 3.5y[k-1] + 4.85y[k-2] - 3.325y[k-3] + 1.1274y[k-4] - 0.1512y[k-5]$$

$$= x[k-2] - x[k-3] + x[k-4] - x[k-5]$$

What are the roots of the system's characteristic equation?

4. Stability   Determine the stability of Fibonacci's rabbit population described in Problem 1.

6

Transform-Domain Representation of Discrete-Time Systems

6.1 Introduction

In Chap. 5, discrete-time LTI systems were studied in the time domain. The response of a discrete-time LTI system was shown to consist of homogeneous (unforced) and nonhomogeneous (forced) components. The nonhomogeneous solution was found to be given in terms of a convolution sum having a kernel defined by the system's impulse response. It was shown that computing an impulse response in the time domain can be a challenging problem, especially when the system order becomes large. In Chap. 6, transforms will be used to algebraically simplify convolution. It will also be shown that transforms can also be used to determine the stability of an LTI.

6.2 Convolution

Knowledge of an at-rest discrete-time, causal LTI system's input $x[k]$ and impulse response $h[k]$ is sufficient to determine the system's output $y[k]$. In particular, the output time-series $y[k]$ is given by the linear convolution sum

$$y[k] = \sum_{m=0}^{\infty} h[m]x[k-m] \qquad (6.1)$$

Suppose that both $h[k]$ and $x[k]$ possess z-transforms which are given by

$$h[k] \overset{Z}{\leftrightarrow} H(z)$$

$$x[k] \overset{Z}{\leftrightarrow} X(z) \tag{6.2}$$

The z-transform of Eq. (6.1) can then be expressed as

$$Y(z) = Z[y[k]] = Z[h[k]*x[k]] = Z\left(\sum_{m=0}^{\infty} h[m]x[k-m]\right)$$

$$= \sum_{k=0}^{\infty}\left(\sum_{m=0}^{\infty} h[m]x[k-m]\right)z^{-k}$$

$$= \sum_{m=0}^{\infty} h[m]\left(\sum_{k=0}^{\infty} x[k-m]z^{-k}\right) \tag{6.3}$$

Upon substituting $p = k - m$ into Eq. (6.3), it follows that

$$Y(z) = \sum_{m=0}^{\infty} h[m]\left(\sum_{p=0}^{\infty} x[p]z^{-(p+m)}\right)$$

$$= \sum_{k=0}^{\infty} h[m]z^{-m}\left(\sum_{p=0}^{\infty} x[p]z^{-p}\right)$$

$$= H(z)X(z) \tag{6.4}$$

Therefore, the z-transform of a time-domain linear convolution sum $y[k] = h[k]*x[k]$ is mathematically equivalent to the product of the z-transforms of $h[k]$ and $x[k]$ in the z-domain. Equation (6.4) is also known by its popular name, the *convolution theorem* for z-transforms. This theorem provides a bridge between time-domain convolution and transform operations. Formally, the convolution theorem is stated below.

Convolution Theorem for z-Transforms

If $y[k] = h[k]*x[k]$ and $y[k]$, $h[k]$, and $x[k]$ have z-transforms of $Y(z)$, $H(z)$, and $X(z)$, respectively, then

$$y[k] = h[k]*x[k] \overset{Z}{\leftrightarrow} H(z)X(z) = Y(z) \tag{6.5}$$

If the regions of convergence for $X(z)$ and $H(z)$ are R_x and R_h respectively, then the region of convergence of $Y(z)$ is R_y, where $R_y \supset R_x \cap R_h$.

This process is described in Eq. (6.5) and graphically interpreted in Fig. 6.1. The attraction of the convolution theorem is that it replaces a numerically complicated convolution sum computation with a set of simple algebraic operations in the z-domain.

6.3 Transfer Functions

Applying the convolution theorem to the at-rest LTI model found in Eq. (5.5) of Chap. 5 (repeated for clarity),

$$\sum_{m=0}^{N} a_m y[k-m] = \sum_{m=0}^{M} b_m x[k-m] \tag{6.6}$$

produces

$$\sum_{m=0}^{N} a_m Y(z) z^{-m} = \sum_{m=0}^{M} b_m X(z) z^{-m} \tag{6.7}$$

which can be simplified to read

$$\left(\sum_{m=0}^{N} a_m z^{-m} \right) Y(z) = \left(\sum_{m=0}^{M} b_m z^{-m} \right) X(z) \tag{6.8}$$

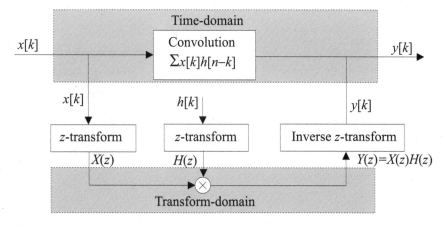

Figure 6.1 Equivalence of time- and transform-domain convolution methods.

The ratio of $Y(z)$ to $X(z)$ is formally called the *transfer function* and is denoted $H(z)$. The transfer function is formally given by

$$H(z) = \frac{Y(z)}{X(z)} = \frac{\left(\displaystyle\sum_{m=0}^{M} b_m z^{-m}\right)}{\left(\displaystyle\sum_{m=0}^{N} a_m z^{-m}\right)} \tag{6.9}$$

and describes how the z-transform of the input signal is manipulated into the z-transform output under linear convolution. Since working with signals in the transform domain has been shown to be a straightforward process, this approach to linear system analysis is preferred by many.

Observe that if $x[k]$ was an impulse, that is $x[k] = \delta[k]$, then $X(z) = 1$. It would then follow from Eq. (6.9) that

$$H(z) = Y_\delta(z) = \frac{\left(\displaystyle\sum_{m=0}^{M} b_m z^{-m}\right)}{\left(\displaystyle\sum_{m=0}^{N} a_m z^{-m}\right)} \tag{6.10}$$

which is, by definition, the z-transform of the at-rest LTI system's impulse response. That is, $h[k] = Z^{-1}\{H(z)\}$. This, in general, provides a simpler method of computing a system's impulse response in comparison to the method developed in Chap. 5.

Example 6.1: RLC Simulation

Problem Statement. Consider the second-order continuous-time system characterized by the ODE equation

$$\frac{d^2 y(t)}{dt^2} + 3\frac{dy(t)}{dt} + 2y(t) = x(t)$$

The continuous-time system's impulse response is known to be given by

$$h(t) = (e^{-t} - e^{-2t})u(t)$$

The discrete-time version of the continuous-time impulse response can be obtained by sampling $h(t)$ at a rate f_s. For a sample period of $T_s = 1/f_s$ seconds, the following time-series results:

$$h[k] = h(kT_s) = (e^{-kT_s} - e^{-2kT_s})u[k] = (a^k - b^k)u[k]$$

where a and b are defined in an obvious manner. What is the unit-step response of the discrete-time system?

Analysis and Conclusions. The z-transform of $h[k]$ is given by:

$$H(z) = \frac{z}{z - a} - \frac{z}{z - b} = \frac{(a - b)z}{(z - a)(z - b)}$$

It is known that if the input $x[k]$ is a unit step, that is $x[k] = u[k] = 1$, $k \geq 0$, then $X(z) = U(z) = z/(z - 1)$. It immediately follows form Eq. (6.9) that

$$Y(z) = X(z)H(z) = \frac{(a - b)z^2}{(z - a)(z - b)(z - 1)}$$

Using previously established methods of inverting a z-transform, namely partial fraction expansion, the inverse of $Y(z)$ can be shown to be the time-series

$$y[k] = \left(\frac{(a - b)}{(1 - a)(1 - b)} - \frac{a^{k+1}}{(1 - a)} + \frac{b^{k+1}}{(1 - b)} \right)u[k]$$

which is also the known step-response of the LTI system.

Computer Study. The following sequence of Siglab commands computes $y[k]$ using linear convolution and transform methods. The calculations are parameterized by the sample period T_s. The results for $T_s = 1/50$ are shown in Fig. 6.2 and are seen to demonstrate agreement between the two methods.

```
>ts=1/50            # Sampling period.
>n=251              # Number of samples.
>t=rmp(n,1,n)       # Sample number ramp.
>a=e^(-ts)          # Impulse response generation coefficients.
>b=a^2
>h=a^t-b^t          # Compute impulse response.
>u=ones(n)          # Setup unit step function.
>y1=h$u             # Compute step response using linear convolution.
>A=((a-b)/((1-a)*(1-b)))*u      # Compute step response using
>B=(-a/(1-a))*a^t               # the inverse z-transform.
```

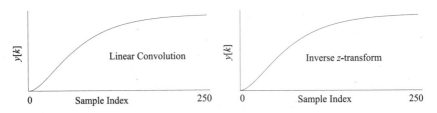

Figure 6.2 Example comparing a convolution sum with the result for the application of the convolution theorem for $T_s = 1/50$.

```
>C=(b/(1-b))*b^t
>y2=A+B+C
>glabel("Linear Convolution",TITLE,0)# Display results
>glabel("Inverse Transform",TITLE,1)
>graph(<t,y1[0:n-1]>,<t,y2[0:n-1]>)
```
❖

Since DSP often is defined in terms of filtering, and filtering is in turn related to convolution, Eq. (6.9) plays an important role in the study of contemporary signal processing. As a result, transfer functions are well studied and documented mathematical tools. Transfer functions may be expressed in a variety of shapes and forms such as the case shown in Eq. (6.11). Here, the transfer function is described as the ratio of two polynomials $N(z)$ and $D(z)$

$$H(z) = \frac{N(z)}{D(z)} = \frac{z^{-M}\left(\displaystyle\sum_{m=0}^{M} b_{(M-m)}z^m\right)}{z^{-N}\left(\displaystyle\sum_{m=0}^{N} a_{(N-m)}z^m\right)} = \frac{\left(\displaystyle\sum_{m=0}^{M} b_m z^{-m}\right)}{\left(\displaystyle\sum_{m=0}^{N} a_m z^{-m}\right)} \qquad (6.11)$$

The roots of the polynomial $N(z) = 0$ are called the system's *zeros* and the roots of the polynomial $D(z) = 0$ are called the system's *poles*. Both play an important role in characterizing and analyzing an LTI system. It can be noted that the poles are also the roots of the system's characteristic equation [i.e., $D(z)$]. Recall that if the order of the numerator is less than or equal to that of the denominator, that is, $N \geq M$, the system is said to be *proper*, and if $N > M$ then the system is said to be *strictly proper*. If the orders are equal, that is $N = M$, $H(z)$ may be represented in the form

$$H(z) = c_0 + \frac{R(z)}{D(z)} \qquad (6.12)$$

where c_0 is a constant, $R(z)$ is called the *remainder polynomial* whose order less than N, the order of $D(z)$. It therefore follows that $R(z)/D(z)$ must be strictly proper. The constant c_0 can be determined using long division or, in the case of a system where $N = M$, $c_0 = b_0/a_0$. The impulse response defined by the inverse z-transform of Eq. (6.12) is given by

$$h[k] = c_0 \delta[k] + Z^{-1}\left(\frac{R(z)}{D(z)}\right) \qquad (6.13)$$

If the order of the numerator equals or exceeds the order of the denominator by an amount $V = M - N > 0$, then the transfer function will be assumed to be factored into the form

$$H(z) = Q(z) + \frac{R(z)}{D(z)} \tag{6.14}$$

where $Q(z)$ is called the *quotient polynomial*. Here $Q(z)$ is of order V, the remainder polynomial $R(z)$ is again of order less than N, and $D(z)$ is of order N, leaving $R(z)/D(z)$ strictly proper. The quotient polynomial can be computed using long division. In particular, if $V > 1$, then

$$H(z) = Q(z) + \left(\frac{R(z)}{D(z)}\right) = c_0 + c_1 z^{-1} + \dots + c_V z^{-V} + \left(\frac{R(z)}{D(z)}\right) \tag{6.15}$$

and the impulse response is given by

$$h[k] = c_0 \delta[k] + c_1 \delta[k-1] + \dots + c_V \delta[k-V] + Z^{-1}\left(\frac{R(z)}{D(z)}\right) \tag{6.16}$$

Discrete-time LTI system models are often defined in terms of a transfer function that has been derived to meet some stated design specification (e.g., a frequency selective digital filter). They can also be determined experimentally by conducting tests on a physical system and then matching the test data to a mathematical model in some acceptable manner. A number of techniques can be used to synthesize a transfer function from observed system data. Some of the more popular modeling techniques belong to a class of estimators called *auto-regressive moving-average* (ARMA) algorithms. Regardless, the production, manipulation, and analysis of transfer functions is fundamentally important to DSP and the DSP engineer.

Due to the convolution theorem, the need to actually compute the impulse response $h[k]$ of an LTI system can often be bypassed. All the information that was possessed by the impulse response is now encapsulated in $H(z)$ in the z-domain. Once a transfer function $H(z)$ is known, and the input transform $X(z)$ is specified, then $Y(z) = H(z)X(z)$ is known. Once $Y(z)$ is defined, the output time-series $y[k]$ can be obtained from an inverse z-transform. The techniques of inverting $Y(z)$ back into a time-domain were developed in Chap. 2. The foundation method was called a *partial fraction expansion*. The methods developed to invert the z-transform of a signal can be directly applied to the study of an LTI. Specifically, if $Y(z)$ is the ratio of polynomials, where the order of the numerator (i.e., M) exceeds the order of the denominator (i.e., N) by an amount $V = M - N > 0$, then $Y(z)$ will be assumed to be factored after Eqs. (6.12) or (6.14) into a quotient and residual term. That is,

$$Y(z) = H(z)X(z) = Q(z) + \frac{R(z)}{D(z)} \tag{6.17}$$

where, as before, the quotient polynomial $Y(z)$ satisfies

$$Y(z) = Q(z) + \left(\frac{R(z)}{D(z)}\right) = c_0 + c_1 z^{-1} + \ldots + c_V z^{-V} + \left(\frac{R(z)}{D(z)}\right) \qquad (6.18)$$

and has an inverse z-transform given by

$$y[k] = c_0 \delta[k] + c_1 \delta[k-1] + \ldots + c_V \delta[k-V] + Z^{-1}\left(\frac{R(z)}{D(z)}\right) \qquad (6.19)$$

The expansion of the strictly proper $R(z)/D(z)$ is accomplished using Heaviside's method as developed in Sec. 2.11. It should be recalled that the partial fraction, or Heaviside expansion, of the rational fraction found in Eq. (6.19) can be accomplished using a general-purpose digital computer. It should also be noted that the formal inversion of $Y(z)$ into $y[k]$ is rarely required. Instead, the design and analysis of an LTI is often conducted entirely in the transform-domain.

Example 6.2: Second-Order System

Problem Statement. Suppose $H(z)$ represents a simple monic second-order system given by

$$H(z) = \frac{z}{(z-0.5)(z+0.5)} = \frac{z}{(z^2 - 0.25)} = \frac{z^{-1}}{(1 - 0.25z^{-2})}$$

and is interpreted as the simulation diagram shown in Fig. 6.3. What is the system's step response?

Analysis and Conclusions. Since $x[k] = u[k]$, $X(z) = z/(z-1)$, it follows that

$$Y(z) = \frac{z^2}{(z-1)(z-0.5)(z+0.5)}$$

Observe that all the poles of the strictly proper $Y(z)$ are nonrepeated (i.e., distinct). The partial fraction expansion of $Y(z)$ would take the form

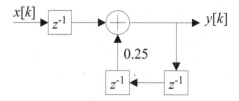

Figure 6.3   Simulation diagram for $H(z)$.

$$Y(z) = a_0 + \alpha_1 \frac{z}{(z-1)} + \alpha_2 \frac{z}{(z-0.5)} + \alpha_3 \frac{z}{(z+0.5)}$$

Recall that, like the expansions found in Sec. 2.11, the above expression is in one-to-one correspondence with the transform pairs cataloged in Table 2.2. As was the case in Sec. 2.11, the Heaviside expansion will be defined in terms of $Y(z)/z$. Formally, $Y(z)/z$ gives

$$\frac{Y(z)}{z} = a_0\left(\frac{1}{z}\right) + a_1\frac{1}{(z-1)} + a_2\frac{1}{(z-0.5)} + a_3\frac{1}{(z+0.5)}$$

Multiplying both sides of the above equation by the poles (i.e., $D(z)$) produces the equality

$$z^2 = a_0(z-1)(z-0.5)(z+0.5) + a_1(z)(z-0.5)(z+0.5) + a_2(z)(z-1)(z+0.5)$$
$$+ a_3(z)(z-1)(z-0.5)$$

Evaluating this equality at the poles of $Y(z)/z$ allows the partial fraction expansion coefficients to be recovered.

$$z = 0 \Rightarrow 0^2 = a_0(-1)(-0.5)(0.5) \Rightarrow \alpha_0 = 0$$

$$z = 1 \Rightarrow 1^2 = \alpha_1(1)(0.5)(1.5) \Rightarrow \alpha_1 = 4/3$$

$$z = 0.5 \Rightarrow (0.5^2) = \alpha_2(0.5)(-0.5)(1) \Rightarrow \alpha_2 = -1$$

$$z = -0.5 \Rightarrow (-0.5)^2 = a_3(-0.5)(-1.5)(-1) \Rightarrow a_0 = -1/3$$

Substituting these coefficients into the expression for the partial fraction expansion gives

$$Y(z) = \frac{4}{3}\frac{z}{(z-1)} - \frac{z}{(z-0.5)} - \frac{1}{3}\frac{z}{(z+0.5)}$$

which defines the system's step response to be

$$y[k] = \left[\frac{4}{3} - (0.5)^k - \frac{1}{3}(-0.5)^k\right]u[k]$$

Computer Study. The s-file "pf.s" performs a partial fraction expansion of $Y(z)/z = z^2/[z(z-1)(z-0.5)(z+0.5)]$, having roots at $z = 0, 1, 0.5$, and -0.5. Notice that one could also have expanded $Y(z)/z = z/[(z-1)(z-0.5)(z+0.5)]$, if the equation is simplified prior to computation. Using the given $Y(z)/z$, the analysis proceeds using the following sequence of Siglab commands.

```
>include "chap2\pf.s"
>n=[1,0,0]            # N(s)=z^2
>d=poly([1,0.5,-0.5,0])   # D(z)= z(z-1)(z-0.5)(z+0.5)
```

```
>d
1  -1  -0.25  0.25  0      # D(z)=z^3-z^2-/4+1/4
>pfexp(n,d)                # Heaviside expansion
pole @ -0.5+0j with multiplicity 1 and coefficients: -0.3333-0j  # a3
pole @ 0.5+0j with multiplicity 1 and coefficients: -1-0j        # a2
pole @ 1+0j with multiplicity 1 and coefficients: 1.33333+0j     # a1
pole @ 0+0j with multiplicity 1 and coefficients: 0+0j           # a0
```

The above results correspond to the manually derived values.   ❖

Example 6.3: Second-Order System

Problem Statement. Suppose $H(z)$ represents a simple second-order system given by

$$H(z) = \frac{3z^2 - 5z + 3}{(z-1)(z-0.5)} = \frac{3z^2 - 5z + 3}{(z^2 - 1.5z + 0.5)} = \frac{3 - 5z^{-1} + 3z^{-2}}{(1 - 1.5z^{-1} + 0.5z^{-2})}$$

and is interpreted as the simulation diagram shown in Fig. 6.4. What is the system's step response?

Analysis and Conclusions. Since $x[k] = u[k]$, $X(z) = z/(z-1)$. Therefore,

$$Y(z) = \frac{3z^3 - 5z^2 + 3z}{(z-1)^2(z-0.5)} = \frac{3z^3 - z^2 + 3z}{z^3 - 2.5z^2 + 2z - 0.5}$$

Observe that all the poles of the proper $Y(z)$ are repeated (i.e., nondistinct). The partial fraction expansion of $Y(z)$ would take the form

$$Y(z) = 0 + \alpha_1 \frac{z}{z-0.5} + \alpha_{21} \frac{z}{(z-1)} + \alpha_{22} \frac{z}{(z-1)^2}$$

Again, as in Example 6.2, the partial fraction expansion of $Y(z)/z$ will be computed instead of $Y(z)$. It therefore follows that

$$\frac{Y(z)}{z} = \frac{a_0}{z} + a_1 \frac{1}{z-0.5} + \alpha_{21} \frac{1}{(z-1)} + \alpha_{22} \frac{1}{(z-1)^2}$$

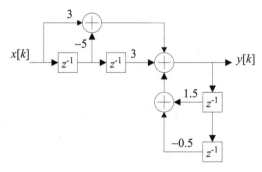

Figure 6.4 Simulation diagram for $H(z)$.

Multiplying both sides of the above equation by the poles (i.e., $D(z)$) produces the equality

$$3z^3 - 5z^2 + 3z = \alpha_1(z-1)^2(z-0.5) + \alpha_1(z)(z-1)^2$$
$$+ \alpha_{21}(z)(z-1)(z-0.5) + a_{22}(z)(z-0.5)$$

Evaluating this equality at the poles of $Y(z)/z$ allows almost all of the partial fraction expansion coefficients to be recovered. Specifically,

$$z = 0 \Rightarrow 0^2 = \alpha_0(1)^2(-0.5) \Rightarrow \alpha_0 = 0$$

$$z = 0.5 \Rightarrow 5/8 = a_1(0.5)(-0.5)^2 \Rightarrow a_1 = 5$$

$$z = 1 \Rightarrow 1 = \alpha_{22}(1)(0.5) \Rightarrow \alpha_{22} = 2$$

All the Heaviside coefficients, save one, are now known. The coefficient a_{12}, unfortunately, cannot be evaluated at $z = 0$, 0.5, or 1. However, a_{12} can be computed at another value of z—say, $z = 2$. Specifically,

$$z = 2 \Rightarrow 10 = a_0(1)^2(1.5) + a_1(2)(1)(1.5) + a_{22}(2)(1.5)$$
$$= 1.5a_0 + 2a_1 + 3a_{21} + 3a_{22} = 0 + 10 + 3a_{21} + 6 \Rightarrow a_{21} = -2$$

Substituting these coefficients into the expression for the partial fraction expansion gives

$$Y(z) = 5\frac{z}{z-0.5} - 2\frac{z}{z-1} + 2\frac{z}{(z-1)^2}$$

which defines the system's step response to be

$$y[k] = [5(0.5)^k - 2 + 2k]u[k]$$

Computer Study. The s-file "**pf.s**" performs a partial fraction expansion of $Y(z)/z$ which has roots at $z = 0$, 0.5, 1 and 1. The analysis proceeds using the following sequence of Siglab commands.

```
>include "chap2\pf.s"
>n=[3,-5,3,0]          # N(z)=3z^2-5z^2+3z
>d=poly([0,1,1,0.5])   # D(z)=(z(z-1)^2)(z-0.5)
>d
1  -2.5  2  -0.5  0    # D(z)=z^3-(2.5)z^2+2z-(0.5)
>pfexp(n,d)            # Heaviside expansion
pole @ 0 + 0j with multiplicity 1 and coefficients 0 + 0j # a0
pole @ 0.5+0j with multiplicity 1 and coefficients: 5+0j # a1
pole @ 1+0j with multiplicity 2 and coefficients: 2+0j -2+0j #
  a22 and a21
```

The results obtained above correspond to the manually derived values. ❖

6.4 Discrete-Time System Frequency Response

It has been established that the response of an LTI consists of an unforced (homogeneous) and forced (nonhomogeneous) time-series. It is normally assumed that the homogenous solution eventually decays to zero in time, or if the system is at rest, remains at zero for all-time. This assumption can be related to the concept of stability, which is addressed in the next section. After the homogeneous solution disappears, all that will be left is the nonhomogeneous response. If the system input is presented with a discrete-time periodic process, then the input signal can be modeled as a suite of sinusoids in the context of a DFT. Recall that a linear system cannot produce any new frequency components. That is, the frequency components found at the output must have been caused by a signal at the input at the very same frequencies. Therefore, the output of an LTI to an input suite of sinusoids will again consist of a collection of sinusoids at the frequency locations found at the input which have been, at most, scaled and phase shifted. The time epoch in which all transient responses have assumed to have left the system is called the *steady-state*. A system's steady-state gives rise to the concept of the *steady-state frequency* response of the LTI.

Formally, the steady-state frequency response of a discrete-time LTI system is determined by evaluating the system's transfer function at frequency locations located along the periphery of the unit circle in the z-domain. The steady-state frequency response is therefore given by

$$\left|H(e^{j\varpi})\right| = \left|H(z)\right|_{z = e^{j\varpi}} \qquad \text{(magnitude frequency response)} \quad (6.20)$$

$$\phi(e^{j\varpi}) = \arg(H(e^{j\varpi})) = \arctan\left(\frac{\text{Im}(H(e^{j\varpi}))}{\text{Re}(H(e^{j\varpi}))}\right) \text{ (phase response)} \quad (6.21)$$

where ϖ is the normalized discrete-time frequency and is restricted to $\varpi \in [-\pi, \pi)$. The system's steady-state frequency response as defined in Eqs. (6.20) and (6.21) is sometimes called the system's frequency response. The magnitude frequency response provides information about the scaling imparted to a periodic input by an LTI at steady-state. The phase response establishes the amount of steady-state phase shift imparted to the periodic input at by the LTI. If $\phi(e^{j\omega})$ is positive, then the system is called a *lead system*. If negative, the system is called a *lag system*.

A discrete-time LTI system will be assumed to have a transfer function representation given by

$$H(z) = K \frac{\displaystyle\prod_{m=1}^{M} (z - z_m)}{\displaystyle\prod_{m=1}^{N} (z - p_m)} \tag{6.22}$$

where z_m and p_m are, in general, the complex zeros and poles of $H(z)$, respectively. Evaluating $H(z)$ along the periphery of the unit circle in the z-domain, namely

$$z = e^{j\omega} \tag{6.23}$$

produces the system's steady-state frequency response at each frequency ω. In particular

$$H(e^{j\omega}) = K \frac{\displaystyle\prod_{m=1}^{M} (e^{j\omega} - z_m)}{\displaystyle\prod_{m=1}^{N} (e^{j\omega} - p_m)} = K \frac{\displaystyle\prod_{m=1}^{M} \alpha_m(j\omega)}{\displaystyle\prod_{m=1}^{N} \beta_m(j\omega)} \tag{6.24}$$

where

$$\alpha_m(j\omega) = \left| \alpha_m(j\omega) \right| e^{j\phi_m}$$

$$\beta_m(j\omega) = \left| \beta_m(j\omega) \right| e^{j\theta_m} \tag{6.25}$$

It then follows that Eq. (6.24) can be expressed more compactly as

$$H(e^{j\omega}) = \left| H(e^{j\omega}) \right| \arg(H(e^{j\omega})) \tag{6.26}$$

where

$$\left| H(e^{j\omega}) \right| = K \frac{\displaystyle\prod_{m=1}^{M} \left| \alpha_m(j\omega) \right|}{\displaystyle\prod_{m=1}^{N} \left| \beta_m(j\omega) \right|} \tag{6.27}$$

and

$$\arg(H(e^{j\omega})) = \sum_{m=1}^{M} \phi_m - \sum_{m=1}^{N} \theta_m + \begin{cases} 0 & \text{if } K > 0 \\ \pi & \text{if } K < 0 \end{cases} \tag{6.28}$$

The values of $|H(e^{j\omega})|$ and $\arg(H(e^{j\omega}))$ for an arbitrary $H(z)$ are graphically examined in Fig. 6.5. Their parametric values can be determined by first locating the point on the periphery of the unit circle corresponding to the desired test frequency, namely $z_0 = e^{j\omega}$. By connecting all the poles and zeros to that point, the magnitude frequency response of the system, $|H(e^{j\omega})|$, can be measured in terms of the rays $\alpha_m(j\omega)$ and $\beta_m(j\omega)$. The constituent phase angles found in $\arg(H(e^{j\omega}))$ can be measured with a protractor. The geometrically obtained data can be used to evaluate Eq. (6.26) at a specific frequency ω. Equation (6.26) also establishes a qualitative relationship between the value of $|H(e^{j\omega})|$ and the proximity of a zero at z_m or pole at p_m to test point $z_0 = e^{j\omega}$ residing on the periphery of the unit circle. It should be apparent that as z_0 approaches z_m, the value of $|H(e^{j\omega})|$ will rapidly converge to zero. As z_0 approaches p_m, the value of $|H(e^{j\omega})|$ will correspondingly move toward infinity. Therefore, system poles should be placed in high-gain regions of the z-plane and zeros placed where attenuation is desired. In between, poles and zeros will interact to produce a gain residing somewhere between these two extremes. This technique is used by experienced designers to manipulate the frequency response of a discrete-time LTI system in order to achieve a desired effect.

Example 6.4: Manual Design

Problem Statement. A classic sixth-order lowpass Chebyshev I filter (see Chap. 13) is designed to have a 250 Hz passband at a sample frequency of 1000 Hz with a transfer function given by

$$H(z) = K \frac{\displaystyle\prod_{m=1}^{6} (z - z_m)}{\displaystyle\prod_{m=1}^{6} (z - p_m)}$$

where $K = 0.009631124$
$z_1 = z_2 = z_3 = z_4 = z_5 = z_6 = 1.0$
$p_1 = 0.00043281697 + j0.939354 = p_2{}^*$
$p_3 = 0.2330879 + j0.766518 = p_4{}^*$
$p_5 = 0.5509034 + j0.335068 = p_6{}^*$.

The pole-zero distribution is shown in Fig. 6.6 along with the filter's frequency response. The passband is seen to be relatively flat.

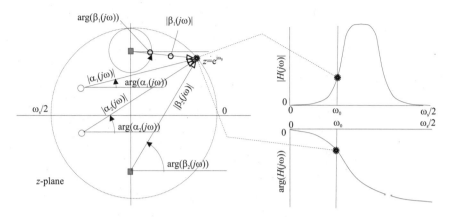

Figure 6.5 Graphical interpretation of an LTI system's steady-state frequency response.

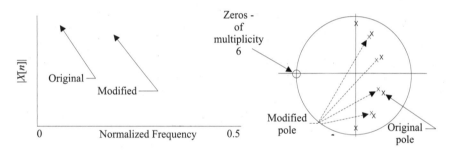

Figure 6.6 Sixth-order Chebyshev I filter magnitude frequency response (left) and the pole-zero distribution (right) along with the modified Chebyshev I lowpass filter response and pole-zero distribution.

Assume that the designer knew that the system was intended to filter signals with a slight 20 percent amplitude "roll-off" near the high end of the pass-band. Desiring a rather flat overall system, or equalized output response, modify the magnitude frequency response to "pre-emphasize" the high-frequency side of the passband. Use manual methods to achieve pre-emphasis.

Analysis and Conclusion. Two options are available. One could move the poles controlling the high-frequency support, located near $\pm\pi/2$ radians in Fig. 6.6, closer to the periphery of the unit circle. Alternatively, the other poles which govern the low frequency behavior of the filter could be moved away from the unit circle (i.e., toward the origin). The latter approach is taken, and using an iterative approach, a filter having the slightly modified pole locations is accepted for the final design. The modified design is given by:

$$K = 0.009631124$$

$$z_1 = z_2 = z_3 = z_4 = z_5 = z_6 = 1.0$$

$$p_1 = 0.0004381697 + j0.939354 = p_2{}^*$$

$$p_3 = 0.22 + j0.75 = p_4{}^* \text{ (changed)}$$

$$p_5 = 0.45 + j0.25 = p_6{}^* \text{ (changed)}$$

The original Chebyshev filter is summarized in Fig. 6.6. The pole-zero distribution of this filter has been modified and is also shown in Fig. 6.6. The pre-emphasis of the passband can be immediately seen. By continuing this process, the shape of the passband can be manipulated to achieve a desired effect.　❖

The graphical method can be tedious if many frequency locations are to be evaluated. In addition, it is at best an approximate method. The frequency response of a system with a known transfer function may also be produced using a digital computer over a dense cover of the normalized frequency interval $\varpi \in [0, \pi)$. Two basic methods can be used and they are

1. DFT/FFT method

2. Polynomial evaluation method

The first method presumes that from a given $H(z)$, the impulse response $h[k]$ is computed. If the impulse response decays to zero (or close to zero) within N samples, then the N sample DFT of the truncated $h[k]$ can be assumed to be a close approximation to $H(e^{j\varpi})$ (see Fig. 6.7). The second method mathematically evaluates Eq. (6.26) at discrete normalize frequencies $\varpi_m \in [0, \pi)$, where $\varpi_m = 2\pi/N$, see Fig. 6.7. In both cases, the larger N becomes, the greater the spectral resolution with an attendant increase in computational complexity.

6.5 Stability of Discrete-Time Systems

The concept of BIBO stability for discrete-time system introduced in Chap. 5 can be further refined. If the homogeneous solution $y[k] \rightarrow 0$ as

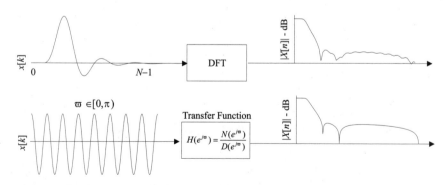

Figure 6.7 Production of approximate steady-state frequency responses.

$k \to \infty$, then the system is said to be *asymptotically stable*. If a discrete-time system is asymptotically stable, then it is also BIBO stable. If the homogeneous response of a discrete-time system can only be guaranteed to be bounded, namely $|y[k]| \leq M < \infty$, for all bound initial conditions, then the system is said to be *marginally* stable or *conditionally* stable. A system is obviously unstable if, for at least one possible initial condition, the homogeneous response becomes unbounded, i.e., $|y[k]| \to \infty$ as $k \to \infty$.

The stability of a discrete-time LTI system can also be determined in the z-transform domain. It was previously noted that there is a correspondence between the roots of the homogeneous equation and the denominator of the corresponding transfer function (giving rise to poles). Since stability in the time-domain was established by the behavior of the system's homogeneous response, a similar conclusion should be obtainable by studying the location of the system poles. In fact, it is known that the system's poles, as defined in Eq. (6.22) are identical to the roots of the characteristic equation, defined in Eq. (5.8). The homogeneous response of the system is

$$y_h[k] = \sum_{i=1}^{N} \gamma_i k^{n_i - 1} p_i^k \qquad (6.29)$$

where each p_i is a pole, n_i is the multiplicity of p_i, and γ_i is a coefficient, for all $i \in \{1,2,3,\dots,N\}$.

Consider an arbitrary element from the string of terms found in Eq. (6.29), namely, ·

$$\phi_i[k] = \gamma_i k^{n_i - 1} p_i^k \qquad (6.30)$$

Since p_i is, in general, a complex pole,

$$\phi_i[k] = \gamma_i k^{n_i - 1} |p_i|^k e^{jk\theta_i} \qquad (6.31)$$

where θ_i is defined in the obvious manner. If $\phi_i[k]$ is to converge asymptotically to zero, then it must be the case that $|p_i|^k \to 0$ as $k \to \infty$. The system is conditionally or marginally stable if $|p_i|^k \leq V$ as $k \to \infty$. Otherwise, the system is unstable. The pole at p_i will cause one of the conditions which are summarized in Table 6.1. It should be noted that if a pole is on the unit circle (i.e., $|p_i| = 1$), it must appear with a multiplicity of one if the system is to remain conditionally stable. Conditional stability, it may be recalled, was associated with a system's homogeneous behavior (i.e., forced by initial.conditions only). If a conditionally stable system is presented with an input signal at the frequency occupied by the conditionally stable pole, instability will results. In this case, a resonant system is being driven at its resonate frequency. Finally, if any pole is unstable, the entire system is unstable. If all the poles are stable, but one or more is conditionally stable, the entire system is conditionally stable. For the system to be asymptotically stable, all the poles must be asymptotically stable. The relationship between pole location and stability is graphically illustrated in Fig. 6.8.

Table 8.1   Pole Stability Conditions

Stability classification	Pole multiplicity	Pole magnitude $|pr|$	BIBO stable
Asymptotic	$\leq N$	<1	yes
Conditional (marginal)	$=1$	$=1$	No
Unstable	>1	$=1$	No
Unstable	$\leq N$	>1	No

Example 6.6: Stability

Problem Statement. Three strictly proper filters having a transfer function $H(z)$, where

$$H(z) = K \frac{\displaystyle\prod_{m=1}^{4} (z - z_m)}{\displaystyle\prod_{m=1}^{5} (z - p_m)}$$

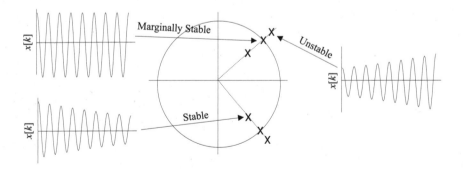

Figure 6.8   z-domain stability cases.

are to be considered. The system transfer functions are listed below. The pole-zero distributions are listed in Table 6.2 and displayed in Fig. 6.9 along with their magnitude frequency, step and harmonic responses. Classify each system in terms of its stability.

$$H_1(z) = \frac{0.002(z^4 + z^3 + 0.25z^2 + 0.25z)}{(z^5 - 3.3z^4 + 5.04z^3 - 4.272z^2 + 2.002z - 0.441)}$$

$$H_2(z) = \frac{0.002(z^4 + z^3 + 0.25z^2 + 0.25z)}{(z^5 - 3.314z^4 + 5.08z^3 - 4.329z^2 + 2.036z - 0.45)}$$

$$H_3(z) = \frac{0.002(z^4 + z^3 + 0.25z^2 + 0.25z)}{(z^5 - 3.304z^4 + 4.886z^3 - 3.869z^2 + 1.572z - 0.308)}$$

Analysis and Conclusions. The pole and zero distribution for each filter is summarized in Table 6.2.

The classification of the system immediately follows from the study of the pole locations and in particular:

$H_1(z)$ = asymptotically (BIBO) stable (all poles interior to the unit circle).

$H_2(z)$ = conditionally stable (two poles on the unit circle at $z = 0.707 \pm j0.707$).

$H_3(z)$ = unstable (three poles exterior to the unit circle at $z = 0.767 \pm j0.767$ and 1.07).

The nonhomogeneous filter responses shown in Fig. 6.9 show that the stability cannot be determined by examining the magnitude frequency response of the transfer function. Note that the step responses tell a different story. The stable system's step response has asymptotically decaying oscillations, while its response to a sinusoid at the resonant frequency results in a bounded harmonic response. The conditionally stable system's step response features a bounded oscillation; however, its harmonic response at the resonant frequency is seen to diverge, but its harmonic response away from the resonant frequency does not diverge. The unstable system's step and harmonic response are both seen to diverge.

Table 6.2   Pole and Zero Distribution.

Filter	K	z_1	z_2	z_3	z_4	p_1	p_2	p_3	p_4	p_5
$H_1(z)$	0.002	0	−1	$j0.5$	$-j0.5$	$0.5 + j0.5$	$0.5 - j0.5$	$0.7 + j0.7$	$0.7 - j0.7$	0.9
$H_2(z)$	0.002	0	−1	$j0.5$	$-j0.5$	$0.5 + j0.5$	$0.5 - j0.5$	$0.707 + j0.707$	$0.707 - j0.707$	0.9
$H_3(z)$	0.002	0	$-j$	$j0.5$	$-j0.5$	$0.5 + j0.5$	$0.5 - j0.5$	$0.767 + j0.767$	$0.767 - j0.767$	1.07

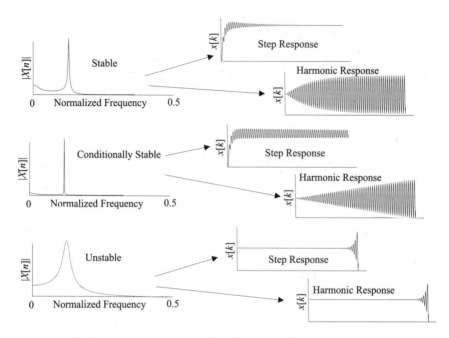

Figure 6.9 Nonhomogeneous responses of an LTI with differing stability properties.

Computer Study. There are three Siglab IIR architecture files, "stable.arc," "marginal.arc," and "unstable.arc," that contain filters with passbands, or resonant frequencies, at $f_s/8$. The following sequence of Siglab commands will subject each of these filters to an impulse, unit step, and a sinusoid at the resonant frequency of the filter. The responses of each of the filters will be plotted, some of which is shown in Fig. 6.9. The harmonic response of the marginally stable filter is also computed with a sinusoid input at $3f_s/8$. Notice that the harmonic response of the marginally stable filter away from the resonant frequency is bounded.

```
>hs=rarc("systems\stable.arc",1)    # Read stable filter arch.
>hm=rarc("systems\marginal.arc",1)  # Read marginal filter arch.
>hu=rarc("systems\unstable.arc",1)  # Read unstable filter arch.
>x=impulse(512)          # Create impulse.
>ys=filt(hs,x); ym=filt(hm,x); yu=filt(hu,x)# Gen impulse responses.
>graph(ys,ym,yu)         # Plot impulse responses.
>x=ones(501)             # Create unit step.
>ys=filt(hs,x); ym=filt(hm,x); yu=filt(hu,x)# Gen step responses.
>graph(ys,ym,yu)         # Plot step responses.
>s=mksin(1,1/8,0,501)  # Create sine at resonant freq.
>ys=filt(hs,s); ym=filt(hm,s); yu=filt(hu,s)# Gen harmonic resps.
# Generate harmonic response of marginally stable system away from
>ym1=filt(hm,mksin(1,3/8,0,501))    # the resonant frequency.
>graph(ys,ym,ym1,yu)                # Plot harmonic responses.  ❖
```

6.6 Summary

In Chap. 6, discrete-time systems were studied using transform domain methods. It was shown that the nonhomogeneous solution can be characterized in terms of a system's transfer function. In general, the qualitative behavior of a linear system was shown to be established by a linear system's pole and zero distribution. The stability of linear systems was also shown to be definable in terms of the system's pole locations.

System stability, based on pole location, was then interpreted in terms of the unit circle in the z-plane (for discrete-time systems). If a discrete-time, stable system is forced by a causal periodic process, it was shown that the total solution would consist of a transient and a periodic steady-state solution. Graphical analysis and display techniques were presented to support steady state system analysis.

The material presented in Chap. 6 will be extended in Chap. 7 to the problem of implementing convolution using a DFT and realizing a system transfer function in hardware. This realization will introduce the concepts of architecture and state variables. This material will provide a foundation upon which systems, having prespecified attributes, can be synthesized and implemented.

6.7 Self-Study Problems

1. Impulse response Find the impulse response of the following discrete-time systems:

$$H_1(z) = \frac{(z+1)^2}{z(z+0.5)(z-0.5)}$$

$$H_2(z) = \frac{(z+1)^4}{z^2(z+0.5)(z-0.5)}$$

$$H_3(z) = \frac{(z+1)^3}{(z+0.5)(z-2z+0.5)}$$

$$H_4(z) = \frac{(z+1)^3}{(z+1)^2(z-z+0.5)}$$

and sketch their frequency responses (magnitude and phase frequency response)

2. Stability Classify the stability of each system defined in Problem 1 and sketch their pole-zero distributions.

3. Nonhomogeneous response   Determine the forced (nonhomogeneous) response of each of the systems defined in Problem 1 to:

$$x_1[k] = u[k]$$

$$x_2[k] = (0.5)^k u[k]$$

$$x_3[k] = \cos(\pi k/16)u[k]$$

4. Graphical response   Graphically compute and sketch the steady-state frequency response of $H(z)$ for

$$H(z) = \frac{z^2}{z^2 - z + 0.8}$$

5. Coherence function   One of the attributes of an LTI is that it cannot create new frequencies. The *coherence function* measures the degree to which a system should be assumed to be linear. For a system having measurable input $x[k]$ and output $y[k]$, the coherence function is given by

$$\delta^2 = \frac{E\left|(G_{yx}(\omega))\right|^2}{E\left|(G_{xx}(\omega))\right| E\left|(G_{yy}(\omega))\right|}$$

where E is the expectation operator and $G_{yx} = Y(\omega)X^*(\omega)$ and is called the *cross-spectral power density*. Similarly $G_{xx} = X(\omega)X^*(\omega)$ $G_{yy} = Y(\omega)Y^*(\omega)$ are called the *input* and *output spectral power densities,* respectively. If a system is linear, then $\delta^2 = 1$ for all frequencies. If there are frequencies appearing at the output due to nonlinearities or noise, then $\delta^2 < 1$ at those frequencies. Test the system

$$y[k] = y^2|k-1| + x[k]$$

in the context of a coherence function. Let $x[k]$ be a chirp and white noise process. Is the system linear?

7

Convolution Methods

7.1 Introduction

In Chap. 5, linear convolution was developed in the time-domain. It was noted that the production rules defined by the convolution sum can become computationally intensive as the order of the system increases. In Chap. 6, convolution was again studied, except in the transform domain. If a signal and system impulse response are both z-transformable, then it has been established that $y[k] = x[k]*h[k] \leftrightarrow Y(z) = X(z)H(z)$. It was noted that transforms can greatly simplify convolution studies but require that the z-transform of both the signal and impulse response be known. Assuming this is the case, the inverse z-transform of $Y(z)$ would still need to be computed to complete the analysis. In Chap. 7, the DFT and the computationally efficient FFT will be used to compute convolutions. Due, however, to the periodic nature of the DFT, linear convolution will have to be carefully interpreted in the context of the properties of a DFT.

7.2 Periodic (Circular) Convolution

In Chap. 5, the linear convolution operation was used to compute the forced response of a system. The convolution of a signal and a causal system was defined to be

$$y[k] = h[k]*x[k] = \sum_{m=0}^{\infty} x[m]h[k-m] = \sum_{m=0}^{\infty} h[m]x[k-m] \qquad (7.1)$$

The limits of the summation are seen to be infinite, which is obviously impossible for computational purposes. It is sometimes the case that the signal to be convolved or the system's impulse response is periodic with period N. It will be shown that this case naturally arises if the signal and system are analyzed using the DFT or FFT. In previous chapters, period N time-series were seen to be characterized by any contiguous N sample record. If this is carried over to convolution, then the infinite sum found in Eq. (7.1) could conceivably be reduced to a finite sum. The case was motivated in Chap. 5.

Assume that $x[k]$ and $h[k]$ are arbitrary time-series. Define periodic sequences $x_p[k + rN] = x[k]$ and $h_p[k + rN] = h[k]$ for $k \in \{0,1,2,...,N-1\}$, and all integers r. The *cyclic convolution* (or *circular convolution*) of $x_p[k]$ and $h_p[k]$ is defined to be

$$y_p[k] = \sum_{m=0}^{N-1} x_p[m]h_p[(k-m)\bmod N] = \sum_{m=0}^{N-1} h_p[m]x_p[(k-m)\bmod N] \quad (7.2)$$

which was denoted $y_p[k] = x_p[k] \otimes h_p[k]$ in Chap. 5. In can be noted that a cyclic convolution is defined over N contiguous samples and is also cyclic with period N (i.e., $y_p[k + rN] = y_p[k]$, for all integers r).

While circular convolution is clearly a useful technique for computing the convolution of a finite sequence and a periodic sequence, it can also be leveraged to compute *linear* convolution. The advantage of this approach will soon become apparent. Consider a sequence, $x[k]$, of length M, and a sequence, $y[k]$, of length N. If each of these sequences are padded with zeros so that their lengths are $L = M + N - 1$ samples, the sequences $x_p[k]$ and $y_p[k]$, of period L, can be formed. In this case, the circular convolution of $x_p[k]$ and $y_p[k]$ is the same as the linear convolution of $x[k]$ and $y[k]$, i.e.,

$$x_p[k] \otimes y_p[k] = x[k]*y[k] \quad (7.3)$$

for $k \in \{0,1,2,...,M + N - 2\}$. Here, only the samples in the range $k \in \{0,1,2,...,M + N - 2\}$ are of interest as the value of $x[k]*y[k]$ for k outside of this range is zero.

Example 7.1: Circular Convolution

Problem Statement. Consider the case where

$$x[k] = h[k] = \begin{cases} 1 & \text{for } k \in \{0, 1, 2, ..., N-1\} \\ 0 & \text{otherwise} \end{cases}$$

What are the results of the linear convolution and circular convolution of these two sequences, as well as the circular convolution of the two sequences after zero-padding to length $2N - 1$?

Analysis and Conclusions. The convolution operations are graphically inter-preted in Fig. 7.1. The linear convolution of $h[k]$ and $x[k]$, given by Eq. (7.1), would produce the triangular wave shown in Fig. 7.1. Suppose, however, that $x[k]$ and $h[k]$ are periodic with period N and represented by $x_p[k]$ and $h_p[k]$, re-spectively. Specifically, consider $x_p[k] = h_p[k] = 1$ for all k. Then the circular convolution of $x_p[k]$ and $h_p[k]$, namely $y_p[k] = x_p[k] \otimes_N h_p[k]$, is described by Eq. (7.2) and also shown in Fig. 7.1 ($\otimes_N$ denotes a period N circular convolution). It can be seen that the linear and circular convolutions do not agree over the sample interval $k \in \{0,1,2,...,N-1\}$ and outside this interval as well.

Consider now that $x_p[k]$ and $h_p[k]$ are zero-padded out to a length of $2N-1$ samples and assumed to be periodic with period $2N-1$. That is,

$$x'_p[k] = h'_p[k] = \begin{cases} 1 & \text{for } k \in \{0, 1, 2, ..., N-1\} \\ 0 & \text{for } k \in \{N, N+1, ..., 2N-2\} \end{cases}$$

Then the circular convolution of $x'_p[k]$ and $h'_p[k]$, namely $y'_p[k] = x'_p[k] \otimes_{2N-1} h'_p[k]$, is given by Eq. (7.2) and is shown in Fig. 7.1. It can be seen that the linear and circular convolutions now agree over the sample interval $k \in \{0,1,2,...,2N-2\}$. However, since $y'_p[k]$ is periodic, with period now equal to $2N-1$, the linear and circular convolutions will not agree outside this inter-val. It can therefore be seen that the circular convolution can be used to re-place linear convolution only with the proper interpretation. ❖

The mechanics of circular convolution can be examined in detail by remembering that $x[(-k) \bmod N] = x[N-k]$ for $k \in \{0,1,2,...,N-1\}$. Sup-pose, for the purpose of discussion, that $x_p[k]$ and $h[k]$ are three sam-ple time-series. The circular convolution of two period-three time-

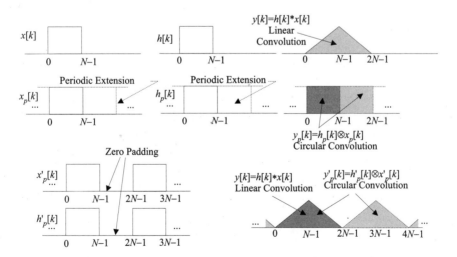

Figure 7.1 Comparison of linear and circular convolution.

series is itself a period-three time-series. The circular convolution operation, defined by Eq. (7.2), provides a means to map $y_p[k] = x_p[k] \otimes_3 h[k]$ into a matrix-vector framework given by

$$
\mathbf{y}_p = \begin{bmatrix} y_p[0] \\ y_p[1] \\ y_p[2] \end{bmatrix} = \begin{bmatrix} h[0] & h[2] & h[1] \\ h[1] & h[0] & h[2] \\ h[2] & h[1] & h[0] \end{bmatrix} \begin{bmatrix} x_p[0] \\ x_p[1] \\ x_p[2] \end{bmatrix} = \mathbf{H}_\otimes \mathbf{x}_p \tag{7.4}
$$

Notice that the elements of $\mathbf{H}_\otimes$ have a circular symmetry—the matrix is right circulant. The linear convolution of $h[k]$ with an arbitrary three sample time-series, $x[k]$, can also be expressed in matrix-vector form. Recall, however, that the linear convolution of two length three time-series is a length five time-series (i.e., $L = 2N-1$). In particular,

$$
\mathbf{y} = \begin{bmatrix} y[0] \\ y[1] \\ y[2] \\ y[3] \\ y[4] \end{bmatrix} = \begin{bmatrix} h[0] & 0 & 0 \\ h[1] & h[0] & 0 \\ h[2] & h[1] & h[0] \\ 0 & h[2] & h[1] \\ 0 & 0 & h[2] \end{bmatrix} \begin{bmatrix} x[0] \\ x[1] \\ x[2] \end{bmatrix} = \mathbf{H}_* \mathbf{x} \tag{7.5}
$$

which, compared to Eq. (7.4), is completely different both in result and structure. Consider, however, what would happen if a five point circular convolution was performed using signals which are zero-padded to fill out a five sample time-series (i.e., $L = 2N - 1$). That is, let $h_z[k] = \{h[k],0,0\}$ and $x_z[k] = \{x[k],0,0\}$. Then $y_z[k] = x_z[k] \otimes_5 h_z[k]$ in matrix vector form would read as

$$
\mathbf{y}_z = \begin{bmatrix} y_z[0] \\ y_z[1] \\ y_z[2] \\ y_z[3] \\ y_z[4] \end{bmatrix} = \left[\begin{array}{ccc|cc} h[0] & h[4] & h[3] & h[2] & h[1] \\ h[1] & h[0] & h[4] & h[3] & h[2] \\ h[2] & h[1] & h[0] & h[4] & h[3] \\ h[3] & h[2] & h[1] & h[0] & h[4] \\ h[4] & h[3] & h[2] & h[1] & h[0] \end{array}\right] \begin{bmatrix} x_z[0] \\ x_z[1] \\ x_z[2] \\ x_z[3] \\ x_z[4] \end{bmatrix} = \mathbf{H}_z \mathbf{x}_z \tag{7.6}
$$

where the elements in $\mathbf{H}_z$ have the diagonal symmetry of the circular convolution matrix shown in Eq. (7.4), scaled to a 5x5 matrix. Since $x_z[3] = x_z[4] = h_z[3] = h_z[4] = 0$, Eq. (7.6) can be reduced to read

$$\mathbf{y}_z = \begin{bmatrix} y_z[0] \\ y_z[1] \\ y_z[2] \\ y_z[3] \\ y_z[4] \end{bmatrix} = \begin{bmatrix} h[0] & 0 & 0 & h[2] & h[1] \\ h[1] & h[0] & 0 & 0 & h[2] \\ h[2] & h[1] & h[0] & 0 & 0 \\ 0 & h[2] & h[1] & h[0] & 0 \\ 0 & 0 & h[2] & h[1] & h[0] \end{bmatrix} \begin{bmatrix} x_z[0] \\ x_z[1] \\ x_z[1] \\ 0 \\ 0 \end{bmatrix} = \mathbf{H}_z \mathbf{x}_z \qquad (7.7)$$

The non-zero coefficients found in right-most 2x5 matrix block of $\mathbf{H}_z$ are seen to be multiplied by $x_z[3] = x_z[4] = 0$ and contribute nothing to the final results. Therefore, Eq. (7.7) produces a result which agrees with the linear convolution results obtained from Eq. (7.5). The difference is that the result obtained in Eq. (7.7) was based on the circular convolution of two time-series which were *zero padded* to fill out a time-series record of at least $L = 2N - 1$ samples in general. By zero padding a time-series of length N out to length $L = 2N - 1$, zero-filled gaps are created along the time-domain axis, which allow the tails of the convolution sum to be computed without error.

It can now be noted that there is a distinction between the convolution of two periodic time-series of length N and the circular convolution of two time-series of length N. In the linear convolution case, periodicity must explicitly exists to produce a periodic output, whereas periodicity is merely assumed in the circular convolution case. This distinction was surfaced when comparing the DTFS and DFT in Chap. 3.

The matrix-vector representation for convolution can also be used to explore both similarity and differences in the linear and circular convolution of two time-series. Suppose that $x[k]$ and $h[k]$ are again considered to be $N = 3$ point time-series. The linear convolution length is again $L = (2N - 1) = 5$. The circular convolution mapping is defined by Eq. (7.4), and the linear convolution mapping is detailed in Eq. (7.5). Comparing the two equations, it can be noted that

$$\mathbf{y}_p = \begin{bmatrix} y_p[0] \\ y_p[1] \\ y_p[2] \end{bmatrix} = \begin{bmatrix} y[0] \\ y[1] \\ y[2] \end{bmatrix} + \begin{bmatrix} y[3] \\ y[4] \\ 0 \end{bmatrix} = \mathbf{H}_\otimes \mathbf{x} \qquad (7.8)$$

where $y_p[k]$ is given by Eq. (7.4) and $y[k]$ is found in Eq. (7.5). Comparing Eq. (7.5) to the zero-filled result found in Eq. (7.7), it is seen that

$$\mathbf{y} = \mathbf{H}_* \mathbf{x} = \begin{bmatrix} y[0] \\ y[1] \\ y[2] \\ y[3] \\ y[4] \end{bmatrix} = \begin{bmatrix} y_z[0] \\ y_z[1] \\ y_z[2] \\ y_z[3] \\ y_z[4] \end{bmatrix} = \mathbf{H}_z \mathbf{x}_z = \mathbf{y}_z \qquad (7.9)$$

That is, $y[k] = y_z[k]$ for $k \in \{0,1,2,\dots,2N - 2\}$. This means that the last $N - 1 = 2$ sample values found at the tail of the linear convolution sum, wrap around modulo N to fill out the matrix $\mathbf{H}_\otimes$. Therefore, linear convolution can always be used to generate a period N circular convolution using a modulo N wrapping of the time-axis (see Fig. 7.2).

7.3 Circular Convolution Using the DFT

The previous example illustrates that there are distinct differences between operations performed on periodic, or presumed periodic, signals and those which are aperiodic. A similar dichotomy was found when the DFT was compared to the DTFT. With the assumption of periodicity it is possible to use the DFT to compute the circular convolution of two sequences. Take two sequences, $x[k]$ and $y[k]$, of length N, with DFTs $X[n]$ and $Y[n]$, respectively:

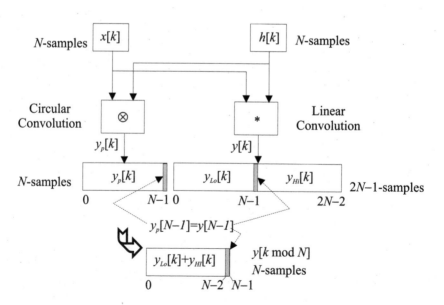

Figure 7.2 Wrapping of a linear convolution into a circular convolution.

$$x[k] \overset{\text{DFT}}{\leftrightarrow} X[n]$$

$$y[k] \overset{\text{DFT}}{\leftrightarrow} Y[n] \qquad (7.10)$$

The circular convolution of $x[k]$ and $y[k]$ was defined in Eq. (5.6), as

$$w[k] = x[k] \otimes y[k]$$

$$= \sum_{m=0}^{N-1} x[(k-m)\text{mod}N]y[(m)\text{mod}N]$$

$$= \sum_{m=0}^{N-1} x[(m)\text{mod}N]y[(k-m)\text{mod}N] \qquad (7.11)$$

The DFT of $w[k]$ as defined above is

$$W[n] = \sum_{k=0}^{N-1}\left(\sum_{m=0}^{N-1} x[(k-m)\text{mod}N]y[(m)\text{mod}N]\right)e^{-j2\pi nk/N}$$

$$= \sum_{m=0}^{N-1}\left(\sum_{k=0}^{N-1} x[(k-m)\text{mod}N]e^{-j2\pi(k-m)n/N}\right)y[(m)\text{mod}N]e^{-j2\pi nm/N}$$

$$(7.12)$$

where the summation in the parentheses is recognized as the DFT of $x[k]$. Since the $k - m$ term is common to the index of x and the complex exponential, the sums may be separated as follows,

$$W[n] = \left(\sum_{k=0}^{N-1} x[k]e^{-j2\pi kn/N}\right)\left(\sum_{m=0}^{N-1} y[m]e^{-j2\pi mn/N}\right) = X[n]Y[n] \quad (7.13)$$

Thus, it is shown that

$$x[k] \otimes y[k] \overset{\text{DFT}}{\leftrightarrow} X[n]Y[n] \qquad (7.14)$$

That is, circular convolution and the DFT can produce an equivalent result in the case where both the signal and system impulse response are assumed periodic. The results shown in Eq. (7.14) are frequently referred to as the *convolution theorem for the DFT*. What is normally desired, however, is to compute the linear convolution of an arbitrary signal and system impulse response. The circular convolution can be leveraged to compute the linear convolution of two finite sequences or a finite sequence and a periodic sequence. The computational advantage of this approach to the computation of convolution comes from the FFT. Suppose that $x[k]$ and $y[k]$ are of the same length, N. Then the computation of the linear convolution of these two sequences is an $O(N^2)$ algorithm. Computation of the linear convolution of these two sequences using the convolution theorem for the DFT proceeds according to the following procedure.

1. Zero pad $x[k]$ and $y[k]$ to length $2N - 1$ ($M + N - 1$ in general, where M and N are the respective lengths of $x[k]$ and $y[k]$), or more, to form $x_p[k]$ and $y_p[k]$. A longer length may be selected to enable the use of an FFT rather than a direct DFT in the subsequent steps.

2. Compute the DFT (or FFT) of $x_p[k]$ and $y_p[k]$, forming $X_p[n]$ and $Y_p[n]$.

3. Compute the inverse DFT (or FFT) of $W[n] = X_p[n]Y_p[n]$, thus forming $w[k] = x_p[k] \otimes y_p[k]$.

By Eq. (7.3), the above algorithm will produce the linear convolution of $x[k]$ and $y[k]$. From a computational perspective, if radix-two Cooley-Tukey FFTs are used, then the total computational requirements are two FFTs of $O(2N\log_2(2N))$, $2N$ complex multiplications, and one inverse FFT of $O(2N\log_2(2N))$. Thus, this whole algorithm is approximately $O(6N\log_2(2N))$. From a computational perspective, this relatively complicated algorithm is inexpensive compared to a direct linear (or circular) convolution implementation. The relative expense of the a direct linear convolution implementation versus an implementation based upon the algorithm detailed above may change if one of the sequences is much shorter than the other.

Example 7.2: Circular and Linear Convolution

Problem Statement. An $N = 32$ sample time-series $x[k]$ is to be convolved with an $M = 9$ sample time-series $y[k]$. Compute the linear and circular convolutions. Where do they agree and disagree?

Analysis and Conclusions. To perform a circular convolution, both the signal and impulse response must be of the same period and length. Therefore, $y[k]$ would need to be zero padded out to length N with $N - M$ zeros. The linear convolution $w[k] = x[k]*y[k]$ is of length $L = N + M - 1 = 40$, whereas the circular convolution is a length $N = 32$ time-series. Suppose, for the purpose of discussion, that $x[k]$ is a random time-series and $y[k]$ is triangularly shaped. Then, the circular and linear convolution results are shown in Fig. 7.3. The linear convolution process actually consists of a buildup transient period of

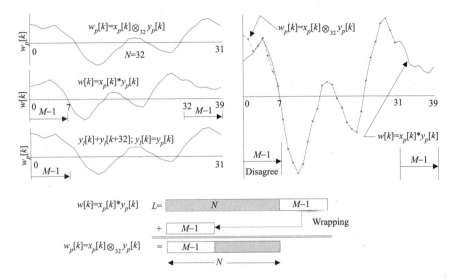

Figure 7.3 Reconciliation between a circular and linear convolution.

length $M - 1 = 8$ samples as the convolver is initialized. At the end of the linear convolution cycle is another eight-sample sequence representing the output leaving the convolver during a build-down phase. Using the argument developed for Eqs. (7.6) and (7.8), upon wrapping the tail of the convolution process modulo N, the circular convolution outcome is shown in Fig. 7.3. The first $M - 1 = 8$ samples are in disagreement with the linear convolution result but the remaining $N - (M - 1) = 24$ exactly matches the $w[k]$ as shown in the overlay graph seen in Fig. 7.3. Therefore, over a subrange of N, the circular convolution outcome $w_p[k]$ equals the linear convolution result $w[k]$.

Computer Results. The following sequence of Siglab commands was used to produce the results reported in Fig. 7.3. Notice that in this case the nine-sample impulse response is zero padded out to 32 samples. In the computation of the circular convolution using the DFT method, the real part of the result is explicitly recovered due to small, non-zero, imaginary artifacts that would interfere with the graphing of the results. The first graph shows, in separate panels, the results of the circular convolution, linear convolution, and the circular convolution reconstructed from the linear convolution. The second graph is an overlay graph of the same set of graphs previously plotted.

```
>x=rand(32)                 # Random input data, N=32.
>y=[1,2,3,4,5,4,3,2,1]      # Triangle input data, M=9.
>xf=fft(x)                  # FFTs of x and y.
>yf=fft(y&zeros(32-9))
>wd=re(ifft(xf.*yf))        # Circular convolution computation.
>w=x$y                      # Linear convolution computation.
>graph(wd,w,w[0:31]+(w[32:39]&zeros(32-8)))   # Graph results.
>ograph(wd,w,w[0:31]+(w[32:39]&zeros(32-8)))  # Graph results. ❖
```

Example 7.3: Linear Convolution Using the Convolution Theorem for DFTs

Problem Statement. Compare a direct linear convolution with a linear convolution implemented using the convolution theorem for DFTs. Demonstrate the correct operation of the algorithm and the performance advantage obtained by using the convolution theorem.

Analysis and Conclusions. Correct operation can be easily verified by convolving two random sequences using direct convolution and the convolution theorem. Performance claims can be demonstrated using two sequences of sufficient length.

Computer Study. The correctness of the convolution algorithm using the convolution theorem for DFTs is verified by a sequence of Siglab commands below. Integrated in this verification of correctness is a demonstration of the performance advantages of the convolution theorem algorithm. The first 100 samples of the linear convolution, computed with both the direct algorithm and the convolution theorem for the DFT, are shown in Fig. 7.4.

```
>x=gn(1024)          # Create test sequences.
>y=gn(1024)
>xp=x&zeros(1024)    # Zero pad sequences to 2^N=2048 points for the
>yp=y&zeros(1024)    # demo of the application of the conv. theorem.
```

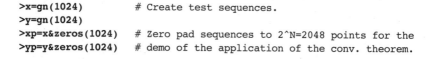

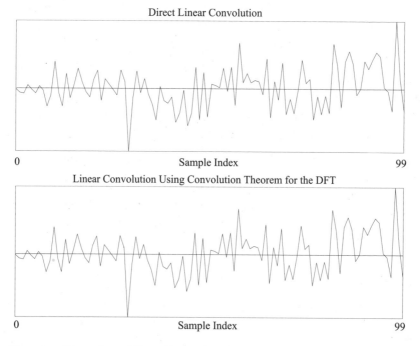

Figure 7.4 Comparison of first 100 samples of linear convolution results computed using the direct algorithm (top) and the convolution theorem for the DFT (bottom).

```
>time dc=x$y          # Compute and time the direct linear conv.
  00:03.85
>time ct=ifft(fft(xp).*fft(yp))      # Compute and time using the
  00:00.49            # convolution theorem for DFTs.
>graph(dc[0:99],re(ct[0:99]))        # Compare first 100 samples.
>sum(mag(dc-ct[0:2*1024-2]))         # Compute the absolute difference
  6.52832e-10         # between the results.
```

The time values will vary greatly, depending upon the computer used. Therefore, it may be necessary to adjust the size of the computation performed. In the case above, the direct linear convolution is more than seven times slower than the DFT implementation. The fundamental sequence length used throughout the above demonstration was 1024 samples; if all instances of 1024 are replaced with a power-of-two value of a different size, the example will continue to function properly. The last Siglab command in the example computes the sum-of-absolute-difference between the linear convolution computed using the direct linear convolution algorithm and the convolution theorem for DFTs. Since random data are being used the actual results may vary, but should be close to zero.   ❖

Example 7.4: Linear Convolution

Problem Statement. Suppose $x[k]$ and $h[k]$ are discrete-time processes given by

$$x[k] = h[k] = \begin{cases} 1 \text{ for } k \in \{0, 1, 2, ..., N-1\} \\ 0 \text{ otherwise} \end{cases}$$

What is $y[k] = h[k]*x[k]$ using a radix-two FFT?

Analysis and Conclusions. Let m be the smallest integer such that $2^m \geq N$. Then $y[k] = h[k]*x[k]$ is a $2N-1 < 2^{m+1}$ sample time-series. The sample values of $y[k]$ can be computed as $y[k] = \text{IDFT}(X(n)H(n))$, where $k \in \{0,1,2,...,2N-1\}$ and $y[k] = 0$ for $k \in \{2N,...,2^{m+1}\}$. If, for example, $N = 64$, then $m = 6$ and $2N-1 = 127 < 128 = 2^7$. Notice that, in this case, 128-point FFTs would be used to compute the convolution result.

Computer Study. The padded data sets are created using the sequence of Siglab commands shown below, along with the production of $y[k] = h[k]*x[k]$ and $y[k] = \text{IDFT}(X(n)H(n))$. The results are displayed in Fig. 7.5. Notice that

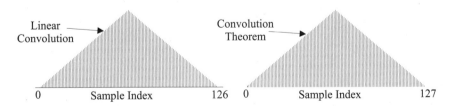

Figure 7.5 Comparison of linear and circular convolution.

the shape of the convolved results are essentially the same with the only difference being the length of the time-series (due to the FFT length being 128 samples) and that $y_{FFT}[128] = 0$.

```
>x = ones(64); xp = x&zeros(64)      # Pad out to 128 samples.
>h = ones(64); hp = x&zeros(64)      # Pad out to 128 samples.
>z = h$x                             # Linear convolution.
>xpf = fft(xp); hpf = fft(hp)        # FFTs.
>yf = xpf.*hpf; y = re(ifft(yf))     # Circular convolution.
>graph(z,y)                          # Graph results.               ❖
```

Example 7.5: Convolution Using DFT Method

Problem Statement. Experimentally verify the use of the convolution theorem for the DFT using audio test records.

Analysis and Conclusion. A chirp is a linearly swept FM signal. A moving average filter (see Example 5.9) was shown to be a smoothing device and, as such, attenuates higher frequencies more severely than lower frequencies. If an $M = $ 9th order moving average filter is convolved with an N sample chirp, the resulting linear convolution would persist for $L = M + N - 1 = N + 8$ samples. If a radix-two FFT is used to implement convolution, then it is reasonable to assume that the FFT length is on the order of $N_{FFT} = N + 8 = 2^n$. Using a chirp of sufficient length to be played through a soundboard, say $N = 2^{14} - 8$ (i.e., $N_{FFT} = 2^{14}$), the chirp would be padded out to $N_{FFT} = 2^{14}$ using eight zeros. The moving average filter, given by $h[k] = (1/9)\{1,1,1,1,1,1,1,1,1\}$, would be padded out to $N_{FFT} = 2^{14}$ using $2^{14} - 9$ zeros. The FFT of the padded signals would be computed, component-wise multiplied, and the inverse FFT of the multiplied spectra computed. The result would be a time-series which equals the linear convolution of the chirp and $h[k]$ over $k \in \{0,1,2,...,L-1\}$. This can be graphically seen in Fig. 7.6, which displays the chirp and linear convolutions computed using the direct (time-domain) and FFT method. It can be seen that the results are identical. It can also be seen the moving average filter acts as a lowpass system.

Computer Study. The program shown below produces the data shown in Fig. 7.6 and plays the original chirp (defined over the frequency range $f \in [0, f_s/4]$) and the convolved result through a 16-bit sound board set to a sampling frequency of 11,025 Hz.

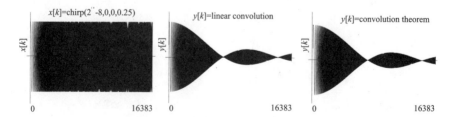

Figure 7.6 Linear convolution of a chirp and moving average filter performed in the time and frequency domains.

```
>include "chap7\chirp.s"
>x=chirp((2^14)-8,0,0,0.25)  # Long chirp.
>h=ones(9)/9                 # Moving average filter.
>y=h$x                       # Direct linear convolution.
>n=len(y); n                 # Length of convolution = 16384
>xf=fft(x&zeros(8))          # FFT of zero padded chirp.
>hf=fft(h&zeros(n-9))        # FFT of zero padded filter.
>yr=(ifft(xf.*hf))           # Convolution theorem for DFT.
>graph(y,yr)                 # Graph results.
>wavout(2^15*x)              # Play original chirp.
>wavout(2^15*y)              # Play filtered chirp.
>wavout(2^15*yr)             # Play chirp filtered using FFT method. ❖
```

7.4 Overlap-and-Add Method

There are times when the input time series is prohibitively long to fit
conveniently into a practical length FFT. The constraint on the FFT
length may come from memory or processor restrictions, or it may
come from an input/output *latency* constraint in a real-time system
(i.e., there may be a constraint on the time lag between an input and
its forced response). In such cases the *overlap-and-add* block convolution model can be used. Suppose that the input time series, $x[k]$, is an
N_1 sample time-series, and the impulse response, $h[k]$, is modeled as
an N_2 sample time-series. Furthermore, it is assumed that $N_1 > N_2$
and that the largest size *input* data block that can be realistically handled is N_0 samples, where $N_0 < N_1$. Consider performing a *block decomposition* of $x[k]$ into data records of N_0 samples each as shown in
Fig. 7.7. The input data block, $x_i[k]$, is defined to be equal to $x[k]$ over
$k \in \{iN_0,...,(i + 1)N_0 - 1\}$ and zero elsewhere. The linear convolution of
an N_0 sample data block and the N_2 sample impulse response is a
time-series of length $L = N_0 + N_2 - 1$. Let the linear convolution of $h[k]$
with the time-series in the ith input data block be denoted $y_i[k]$, where

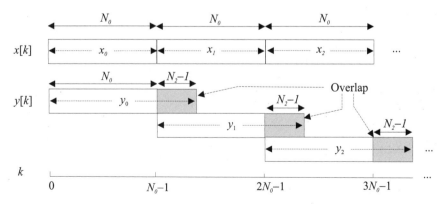

Figure 7.7   Overlap-and-add convolution method.

$$y_i[k] = h[k]*x_i[k] \tag{7.15}$$

which is interpreted in Fig. 7.7. The section of $y_i[k]$ that extends outside the interval $k \in \{iN_0,\ldots,(i+1)N_0 - 1\}$ is seen to invade the next data block or blocks. Usually, the length of the input data block will be selected to be greater than the length of the impulse response, i.e., $N_0 > N_2$. In this case, the section of $y_i[k]$ that extends outside of the range $k \in \{iN_0,\ldots,(i+1)N_0 - 1\}$ will only coincide with the part of the next data block. This portion, called the *overlap*, is added to $y_{i+1}[k]$ to define the convolution sum for $k \in \{(i+1)N_0,\ldots,(i+2)N_0 - 1\}$. In general, if $N_0 < N_2$, then each $y_i[k]$ will coincide with the next $\lceil N_0/(N_2-1) \rceil$ output blocks.

Example 7.6: Overlap-and-Add

Problem Statement. Suppose $x[k] = \{1,2,3,4,5,4,3,2,1\}$ is a length nine ($N_1 = 9$) time-series and $h[k] = \{1,2,1\}$ is of length three ($N_2 = 3$). Their linear convolution is a time-series of length $L = 9 + 3 - 1 = 11$ and is given by the time-series $y[k] = \{1,4,8,12,16,18,16,12,8,4,1\}$. Assume that $N_0 = 3$.

Analysis and Conclusions. The overlap-and-add algorithm proceeds as detailed in the following table.

k	0	1	2	3	4	5	6	7	8	9	10
$y_0[k]$	1	4	8	8	3						
$y_1[k]$				4	13	18	13	4			
$y_2[k]$							3	8	8	4	1
$y[k]$	1	4	8	12	16	18	16	12	8	4	1

If a radix-two FFT is used to compute the convolution block data used in an overlap-and-add algorithm, then L would need to be chosen to be less than or equal to a radix-two number that is an acceptable length for an FFT. The block data sets $x_i[k]$ and $h[k]$ would need to be padded out to the length selected for the FFT. If $N_0 = 6$, then $L = 8$, and the overlap-and-add results computed using the FFT would be given by the following table.

k	0	1	2	3	4	5	6	7	8	9	10	11	12	13
$y_0[k]$	1	4	8	12	16	18	13	4						
$y_1[k]$							3	8	8	4	1	0	0	0
$y[k]$	1	4	8	12	16	18	16	12	8	4	1	0	0	0

Computer Study. The following program generates the tabled data found for the case where $N_0 = 3$ using linear convolution.

```
>x=[1,2,3,4,5,4,3,2,1]; h=[1,2,1] # Input data.
>y=h$x; y                         # Linear convolution.
Columns 0 - 8
0:  1  4  8  12  16  18  16  12  8
Columns 9 - 10
0:  4  1
>x0=[1,2,3]; y0=x0$h; y0          # Block linear convolution.
  1  4  8  8  3
>x1=[4,5,4]; y1=x1$h; y1
  4  13  18  13  4
>x2=[3,2,1]; y2=x2$h; y2
  3  8  8  4  1
```

The following program generates the tabled data found for the case where $N_0 = 6$ using circular convolution and an eight-point FFT.

```
>x=[1,2,3,4,5,4,3,2,1]   # Original data.
>h=[1,2,1]               # Impulse response.
>x0=x[0:5]&zeros(2)      # Extract input blocks of length eight.
>x1=x[6:8]&zeros(5)      # Second block must be padded to length
                           eight.
>H=fft(h&zeros(5))       # Create transformed impulse response,
                           L=8.
>y0=re(ifft(fft(x0).*H)) # Compute output blocks using FFT method.
>y1=re(ifft(fft(x1).*H))
>y0                      # Print output blocks.
  1  4  8  12  16  18  13  4
>y1
  3  8  8  4  1  4e-16  0  -4e-16
>y=(y0&zeros(6)) + (zeros(6)&y1)   # Combine output blocks for
                                     results.
>y                      # Print final results.
Columns 0 - 8
  0:  1  4  8  12  16  18  16  12  8
Columns 9 - 13
  0:  4  1  4e-16  0  -4e-16                          ❖
```

7.5 Overlap-and-Save Method

A variation on the overlap-and-add method is called the *overlap-and-save* convolution model. Again, it is assumed that the input time-series is prohibitively long to conveniently fit into an FFT. The overlap-and-add method was based on decomposing the input into a set of non-overlapped fixed length blocks, each of length N_0. Upon convolving these blocks with an impulse response of length N_2, multiple time-series resulted, each of which overlapped that produced in the next block. From a data management standpoint, it may be desired to eliminate the overlap in the convolution space. Suppose again the input time-series is an N_1 sample sequence and the impulse response is an N_2 sample time-series. Furthermore assume that $N_1 \gg N_2$ and that the largest size *convolution* data block that can be realistically handled is N_0 samples. The *block decomposition* of $x[k]$ into data records of N_0 samples each as shown in Fig. 7.8. In particular, let the linear convolution of $h[k]$ with the time-series in the ith data block be denoted $y_i[k]$, where

$$y_i[k] = h[k]*x_i[k] \qquad (7.16)$$

which is a time-series of length $L = N_0 + N_2 - 1 > N_0$ as shown in Fig. 7.8. Assume that the convolution operation described in Eq. (7.16) is performed using an N_0 sample circular convolution instead of linear convolution. That is, define

$$y_i[k] = h[k] \otimes_{N_0} x_i[k] = \text{IDFT}(\text{DFT}(h[k]) \times \text{DFT}(x_i[k])) \qquad (7.17)$$

Then, the result is periodic with period N_0. The linear and circular convolution will, in fact, disagree, as shown in Fig. 7.3, due to the

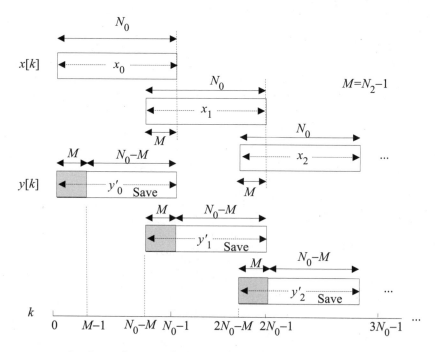

Figure 7.8 Overlap-and-save method where $M = N_2 - 1$.

mechanism developed in Example 7.2. Specifically, $N_2 - 1$ samples of the circular convolution are flawed because of a modulo N_0 wraparound contamination leaving $N_0 - (N_2 - 1)$ values where the circular and linear convolution agree. To account for this pattern of invalid and valid data, the N_0 sample input data records are overlapped by an amount $N_2 - 1$, as shown in Fig. 7.8. The initial $N_2 - 1$ are technically unknown in this process and can either be directly calculated, or the input can be delayed by an amount $N_2 - 1$ to bypass this problem area.

Example 7.7: Overlap-and-Save

Problem Statement. Suppose again that $x[k] = \{1,2,3,4,5,4,3,2,1\}$ is a length nine time-series $(N_1 = 9)$ and $h[k] = \{1,2,1\}$ has length three $(N_2 = 3)$. Then, the linear convolution of x and h is a time-series of length $L = 9 + 3 - 1 = 11$ and is given by $y[k] = \{1,4,8,12,16,18,16,12,8,4,1\}$. Assume that $N_0 = 4 < N_2$.

Analysis and Conclusions. The input is divided into blocks of four samples, each with an overlap of $N_2 - 1 = 2$. To implement circular convolution, $h[k]$ is zero-padded to length $N_0 = 4$ to form $h_p[k]$. This means that in every four sample circular convolution record, two samples are valid, and two samples are flawed. The first two convolution values are manually computed to be $y[0] = x[0]h[0] = 1$ and $y[1] = x[1]h[0] + x[0]h[1] = 4$. This analysis, based on circular convolution, is summarized below:

k	0	1	2	3	4	5	6	7	8	9	10	11
$y_0[k]$	12	8	8	12								
$y_1[k]$			16	14	16	18						
$y_2[k]$					12	16	16	12				
$y_3[k]$							4	8	8	4		
$y_4[k]$									1	2	1	0
$y[k]$	1	4	8	12	16	18	16	12	8	4	1	0

If an FFT is used to compute the data used in an overlay-and-save algorithm, then N_0 would need to be chosen to be a radix-two number, say $N_0 = 4$. The data sets $x_i[k]$ would be blocked into records of length N_0 and $h[k]$ would need to be padded out to $N_0 = 4$. The resulting circular convolutions would produce a result identical to that shown above.

Computer Study. The following sequence of Siglab commands generates the tabled data found for the case where $N_0 = 4$ using circular convolution by computing the linear convolution and wrapping its tail modulo four. Given the usage of block linear convolution, it would be reasonable to use the overlap-and-add algorithm rather than the overlap-and-save algorithm; however, the following demonstration reinforces the principle demonstrated in Eq. (7.8).

```
>x=[1,2,3,4,5,4,3,2,1]; h=[1,2,1]  # Original data.
>hp=[h,0]                          # Zero padded impulse response.
>h$x                               # Direct linear convolution.
 Columns 0 - 8
 0: 1 4 8 12 16 18 16 12 8
 Columns 9 - 10
 0: 4 1
>x0=x[0:3]; x1=x[2:5]; x2=x[4:7]   # Extract overlapping blocks of x.
>x3=x[6:8]&[0]; x4=x[8]&zeros(3)
>y0=hp$x0; y1=hp$x1; y2=hp$x2      # Compute block linear convolution.
>y3=hp$x3; y4=hp$x4
>y0[0:3]+[y0[4:6],0]               # Display block circular convolution.
 12  8  8  12
>y1[0:3]+[y1[4:6],0]               # Display block circular convolution.
 16  14  16  18
>y2[0:3]+[y2[4:6],0]               # Display block circular convolution.
 12  16  16  12
>y3[0:3]+[y3[4:6],0]               # Display block circular convolution.
  4  8  8  4
>y4[0:3]+[y4[4:6],0]               # Display block circular convolution.
  1  2  1  0
```

The following sequence of Siglab commands generates the tabled data found for the case where $N_0 = 4$ by performing circular convolution using a four-point FFT.

```
>x=[1,2,3,4,5,4,3,2,1]; h=[1,2,1]# Original data.
>hp=[h,0]                          # Zero padded impulse response.
>hpf=fft(hp)                       # Transform of padded response.
>x0=x[0:3]; x1=x[2:5]; x2=x[4:7]   # Extract overlapping blocks of x.
>x3=x[6:8]&[0]; x4=x[8]&zeros(3)
>y0=re(ifft(hpf.*fft(x0)))         # Compute block circular convolution.
>y1=re(ifft(hpf.*fft(x1)))
>y2=re(ifft(hpf.*fft(x2)))
>y3=re(ifft(hpf.*fft(x3)))
>y4=re(ifft(hpf.*fft(x4)))
>y0                                # Display convolution results.
 12  8  8  12
>y1                                # Display convolution results.
 16  14  16  18
>y2                                # Display convolution results.
 12  16  16  12
>y3                                # Display convolution results.
  4  8  8  4
>y4                                # Display convolution results.
  1  2  1  0
```

❖

7.6 Building Blocks

Convolution has been presented in linear and circular form. Circular convolution is generally associated with DFT or FFT implementations, particularly of linear convolution. Linear convolution is normally identified with digital filtering. Digital filters are devices which continuously accept time-series samples, $x[k]$, and linearly convolve the input time-series with the filter's impulse response to produce a continuous output time-series which achieves a specified effect. If the desired effect is defined in the context of frequency domain behavior, the filter is often called a frequency selective filter (e.g., lowpass, highpass, bandpass, bandstop). Examples of digital filters fulfilling a time-domain mission are matched filters, waveform generators, and so forth. Digital filters which possess an LTI model, can be classified as finite impulse response (FIR) or infinite impulse response (IIR) filters. In either case the fundamental building blocks used to implement the linear convolution operation, whether in hardware or software, are

- Shift registers
- Scalars/multipliers
- Adders/subtractors

Referring to the linear convolution equation shown in Eq. (7.1), implementations generally take the form

$$y[k] = h[k]*x[k] = K \sum_{m=0}^{L} x[m]h[k-m] = K \sum_{m=0}^{L} h[m]x[k-m] \quad (7.18)$$

where L is finite for an FIR and infinite for an IIR (obviously, an infinite L is impractical for a real implementation). The shift registers are used to store the sample values and coefficients. The scale factor, K, found in Eq. (7.18) is used to adjust the output gain to a desired level. Multipliers are used to combine two variables, such as $x[m]h[k-m]$, and send the results to an adder/subtractor which, with a register, is used as an accumulator. Because multiply followed by accumulation is a core DSP operation, it has be institutionalized as a *multiply-accumulate* (MAC) operation. The MAC operation has been highly optimized in DSP microprocessors. Developers of fixed-point DSP microprocessors have realized that the execution of a MAC operation can have two outcomes. The output can be a valid fixed-point number or dynamic range overflow. The latter will occur when the data exceeds the dynamic range limit of a fixed-point format. Specifically, an M bit, signed fixed-point data format having one sign bit, I integer bits, and $F = N - I - 1$ fractional bits is said to have a $[N{:}F]$ format and represents a number X as

$$X = -(2^I)x_I + \sum_{i=0}^{I-1} x_i 2^i + \sum_{i=1}^{F} x_{-i} 2^{-i} \qquad (7.19)$$

where $x_i \in \{0,1\}$ for $i \in \{-F, -F+1, \ldots, 0, \ldots, I\}$, is the ith bit of binary representation of X. This format is summarized below. The radix point may be inferred to lie between the integer and fractional part.

Sign bit	Integer part			Fractional part		
x_I	x_{I-1}	...	x_0	x_{-1}	...	x_{-F}

The precision of the representation system is denoted Δ, where

$$\Delta = 2^{-F} \qquad (7.20)$$

is commonly measured in volts per bit. The dynamic range of the system defined in Eq. (7.19) is given by

$$-2^I \le X < 2^I \qquad (7.21)$$

Suppose two M bit words X and Y are to be multiplied. Their product $Z = XY$ would have extreme values ranging between

$$-(2^{2I}) \le Z < 2^{2I} \qquad (7.22)$$

For analysis purposes, DSP engineers often consider data to be in an integer form. Defining $X' = 2^F X$, it follows that

$$X' = -(2^{N-1})x_{N-1} + \sum_{i=0}^{N-2} x_i 2^i \qquad (7.23)$$

and

$$-(2^{(N-1)}) \le X' < 2^{(N-1)} \qquad (7.24)$$

The product of two integers having the format shown in Eq. (7.23), namely $Z' = X'Y'$, would have extreme values

$$-(2^{2(N-1)}) \le Z < 2^{2(N-1)} \qquad (7.25)$$

Generally, the full precision product of two N bit data words is considered to be a $2N$ bit result [technically, Eq. (7.25) shows that there is actually only $2(N - 1)$ bits of information due to a redundant sign-bit occupying the $(2N - 1)$th bit location]. Sending full precision outputs to a $2N$ bit accumulator will require that the accumulator register be tested for possible overflow after each accumulation cycle. Using a $2N + M$ bit accumulator, the MAC could accumulate up to 2^M worst-case full precision products without overflowing the accumulator. It is for this reason that most DSP MAC units make use of an extended precision accumulator and a *scaler* at the output of the multiplier. By scaling the product down (i.e., reducing the precision of the product), more bits in accumulator are available for growth of the accumulated value.

A simplified block diagram of the essential components of a typical fixed-point DSP microprocessor is given in Fig. 7.9. The essential components include a MAC unit which is shown in detail assuming a common 16-bit data word size. In this case, the product produced by the multiplier will be either 31 or 32 bits, depending on whether the input operands are signed. The scaler allows the precision of the product to be adjusted, setting the location of the radix point in the accumulator. The extended precision accumulator generally includes an output scaler and a multiplexer to format the output and break it into the native word width of the processor (16 bits, in this case). The MAC unit is connected to memory resources, which may include on-chip RAM, ROM, and register files, as well as off-chip memories, which are typically RAM and ROM. For a typical DSP application, coefficients may be stored either in RAM or ROM, depending on the nature of the application and its implementation, while data is stored in RAM. DSP microprocessors also include memory addressing resources that allow

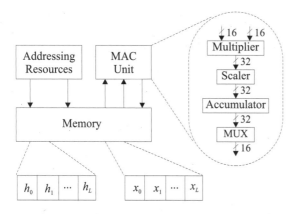

Figure 7.9 Typical general-purpose DSP microprocessor architecture.

addressing modes such as circular and bit-reversed addressing to be implemented with no performance penalty. DSP microprocessors usually have some sort of peripheral support; however, this is not indicated in Fig. 7.9.

7.7 Summary

In Chap. 7, the implementation of convolution was studied. It was shown that there are two distinct types of convolution called *linear* and *circular*. It was shown that they are both similar and dissimilar. Circular convolution was shown to be importance since it can computed using the numerically efficient FFT. By zero-padding a time-series, a circular convolution can be made to agree with a linear convolution. Methods were also developed that will allow linear convolution to be computed using short linear or circular convolutions. Finally, the computational building blocks of convolution were examined in the form of memory, shift-registers, and arithmetic units.

7.8 Self-Study Problems

1. Circular Convolution Circularly convolve a period-eight time series $x[k] = \{1,2,1,0,-1,2,-1,0\}$ with and another period-eight time series $h[k] = \{1,-1,1,-1,1,-1,1,-1\}$.

2. Linear Convolution Linearly convolve the signal from the previous problem using circular convolution operations.

3. Overlap-and-add method Using a maximum size input time series of four-samples, compute the linear convolution of the above time-series using the overlap-and-add method.

4. Overlap-and-save method Using a maximum size output time series of four-samples, compute the linear convolution of the above time-series using the overlap-and-add method.

8

Discrete-Time System Representation

8.1 Introduction

In Chaps. 5 through 7, discrete-time systems were developed in terms of time- and transform-domain concepts. In Chap. 8, the synthesis of discrete-time systems will be pursued as a foundational study. Discrete-time structures will also be examined in terms of *block diagrams* and *signal flow graphs*. These techniques are applicable to the problem of mapping a given discrete-time system structure into a transfer function $H(z)$. The concepts introduced in this chapter will be extended to the study of digital filter architectures in later chapters. At that time, architectures will be developed that related to the implementation of digital filters in hardware and/or software. In Chap. 9, the discrete-time representation list will be extended with the concept of state variables.

8.2 System Block Diagrams

The information paths found within a discrete-time systems can be represented in *block diagram* form. The block diagram representation of a discrete-time systems is often defined in terms of collections of subsystems, say $H_i(z)$, which are expressed in terms of the z-transform delay operator. The implementation of a transfer function $H(z)$ would therefore naturally consist of a collection of interconnected nodes, shift registers/memory, and arithmetic elements. Block diagrams provide a natural graphical means of describing a transfer function in terms of a collection of small, well studied subsystems. To illustrate, suppose two LTI subsystems $H_1(z)$ and $H_2(z)$ are arranged as shown in

Fig. 8.1. At the input-output systems level, the transfer function $H(z)$ has the well known form given by

$$H(z) = \frac{H_1(z)}{(1 - H_1(z))H_2(z)} \tag{8.1}$$

This useful formula can be applied anytime the structure shown in block diagram form, as seen in Fig. 8.1, is encountered. The objective is to selectively combine the various subsystem found in a block diagram description of a system into one block that characterizes the input-output system behavior. This activity is called *block diagram reduction*.

Besides the feedback structure shown in Fig. 8.1, there are other fundamental operations that define how block diagrams can be manipulated and reduced. The two basic block diagram reduction methods shown in Figs. 8.2 and 8.3 are called *forking* and *joining*. Forking occurs at nodes called *pick-off points* or *points-of-bifurcation*. A pick-off point can be moved through a subsystem, or back through a subsystem, as shown in Fig. 8.2. Externally, the two systems shown if Fig. 8.2 are indistinguishable from each other.

Joining operations are performed using adders to combine data from different data paths. Joining operations may be moved as shown in Fig. 8.3. Again, each of the systems shown in Fig. 8.3 retains its transfer function under movement of the join operation as shown.

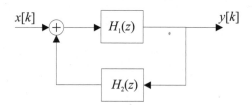

Figure 8.1 Simple feedback structure consisting of two subsystems, $H_1(z)$ and $H_2(z)$.

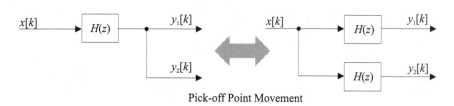

Pick-off Point Movement

Figure 8.2 Pick-off point movement.

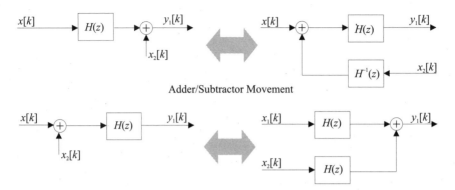

Figure 8.3 Joining operation movement.

Example 8.1: Block Diagram Reduction

Problem Statement. A system consisting of nodes, adders, directed paths, and subsystems is shown in Fig. 8.4. Using block diagram reduction methods, reduce the system to a single input-output relationship.

Analysis and Conclusions. Refer to Fig. 8.4 and follow the illustrated sequential block diagram reduction steps. The final step results in the definition of the transfer function

$$H(z) = d + \frac{c_1 z^{-1} + c_2 z^{-2} + c_3 z^{-3}}{1 + a_1 z^{-1} + a_2 z^{-2} + a_3 z^{-3}} \qquad \diamondsuit$$

While block diagram reduction methods are useful, they can prove to be difficult to manage when the system order becomes large. Therefore, the reduction problem is often avoided in favor of simulation. In this mode, a block diagram representation of a system is created in software, and the inputs and outputs of the subsystems are interconnected. By selecting an input $x[k]$, the evolution of the signal from input to output can be traced and recorded. While useful in some cases, it is also loaded with hazards, since all that is known about the system is what can be deduced from the test signals. If the set of test signals is not sufficiently robust, the analyst may remain oblivious to the system's pathological behavior under certain inputs and/or initial conditions.

8.3 Signal Flow Diagrams

A *signal flow diagram* or *signal flow graph* can also be used to represent the information flow and constraining paths found within a sys-

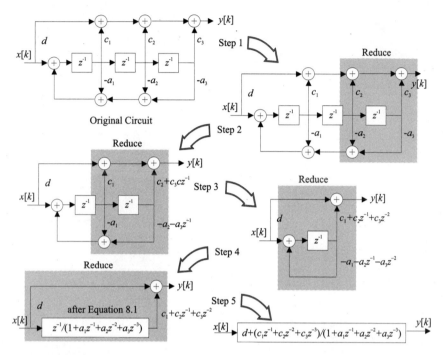

Figure 8.4 Block-diagram reduction example.

tem. The signal flow graph is essentially equivalent to the block diagrams previously discussed. A simple signal flow graph is shown in Fig. 8.5 and has the following elements:

- *Node:* Equivalent to an adder in a block diagram. In the following discussions each node may be numbered to uniquely identify that node (e.g., [1], [2], [3], and [4] in Fig. 8.3).

- *Branch:* A directed line segment connecting two nodes (e.g., [1,2], [3,2] in Fig. 8.3). A branch may have a gain associated with it (e.g., [1,2] has a branch gain of A in Fig. 8.3). Typically, no explicit gain will be indicated when a branch has unity gain.

- *Input node (source):* A node possessing only outgoing branches (e.g., [1] in Fig. 8.3).

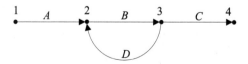

Figure 8.5 Signal flow graph.

- *Output node (sink):* A node possessing only incoming branches (e.g., [4] in Fig. 8.3).

- *Internal node:* A node having both branches directed toward it and branches directed away from it (e.g., [2], [3] in Fig. 8.3).

- *Path:* Any continuously connected set of branches (e.g., [1,2,3,4] in Fig. 8.3).

- *Feedback path (or loop):* A path that originates and terminates on a common node and through which all interconnected nodes are traversed only once (e.g., [2,3,2] in Fig. 8.3).

- *Feedforward path:* A path connecting an input node to an output node in which all interconnected nodes are traversed only once (e.g., [1,2,3,4] in Fig. 8.3).

- *Path gain:* The product of all branch gains along a connected path. (For example, [1,2,3,4] has path gain of ABC in Fig. 8.3.) The path gain could be a simple constant or a complex system with a specified transfer function, $H_i(z)$.

- *Loop gain:* The product of all branch gains along a closed loop. (For example, [2,3,2] has a loop gain of BD in Fig. 8.3.)

Example 8.2: Flow Diagram

Problem Statement. Represent the transfer function $H(z)$ shown in Eq. (8.1) in signal flow diagram form shown in Fig. 8.5.

Analysis and Conclusions. Refer to Fig. 8.5 and note the $A = C = 1$, $B = H_1(z)$, and $D = H_2(z)$, producing the signal flow graph shown in Fig. 8.6.   ❖

8.4 Mason's Gain Formula

Mason's gain formula or *Mason's theorem* can be used to directly map, or reduce, a signal flow diagram corresponding to a discrete-time LTI system into a transfer function. It is a procedure that can be implemented manually or using a general-purpose digital computer.

Mason's gain formula for discrete-time systems is given by

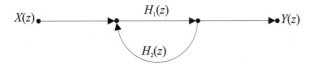

Figure 8.6   Signal flow diagram of a simple feedback system.

$$H(z) = \sum_k \frac{M_k(z)\Delta_k(z)}{\Delta(z)} \qquad (8.2)$$

where $M_k(z)$ is the path gain of the kth feedforward path, and $\Delta(z)$ is called the *system determinant* and is given by

$\Delta(z) = 1 - [\Sigma$ gains from all individual loops]

$\qquad + [\Sigma$ gains from all possible combinations of dual non-touching loops]

$\qquad - [\Sigma$ gains from all possible combinations of triplet non-touching loops]

$\qquad + \ldots \pm$ until all loops of all multiplicity have been taken into account

$\qquad\qquad\qquad\qquad\qquad\qquad\qquad\qquad\qquad\qquad (8.3)$

Equation (8.3) can be more compactly expressed as

$$\Delta(z) = 1 - \left[\sum_m P_{m,1}(z)\right] + \left[\sum_m P_{m,2}(z)\right] - \left[\sum_m P_{m,3}(z)\right] + \ldots \qquad (8.4)$$

where the (m,n)th feedback path (i.e., nth collection of m non-touching feedback loops) has a gain $P_{m,n}(z)$. Finally, $\Delta_k(z)$ is defined by Eq. (8.3) [i.e., $\Delta(z)$] for the part of the graph not touching the kth feedforward path.

Example 8.3: Mason's Gain Formula

Problem Statement. A discrete-time system has a known signal flow diagram shown in Fig. 8.4. What is the system's transfer function $H(z)$?

Analysis and Conclusions. From the signal flow graph, shown in Fig. 8.7, defines the elemental parts of the system in the context of Mason's gain formula. The signal flow diagram is seen to contain of 11 labeled nodes. The input is found at Node 1 and the output at Node 11. It can also be seen that there are three single (i.e., P_{m1}) closed feedback paths which are as shown below.

Path	Path gain
$P_{1,1} = [3,4,7,3]$	$P_{1,1}(z) = -a_1 z^{-1}$
$P_{2,1} = [3,4,5,8,7,3]$	$P_{2,1}(z) = -a_2 z^{-2}$
$P_{3,1} = [3,4,5,6,8,7,3]$	$P_{3,1}(z) = -a_3 z^{-3}$

Notice that all the closed feedback paths share a common node, namely Node 3. Therefore, there are no higher order terms (i.e., $P_{m,j} = 0$ for $j > 1$). Mason's gain formula can now be used to define the system determinant,

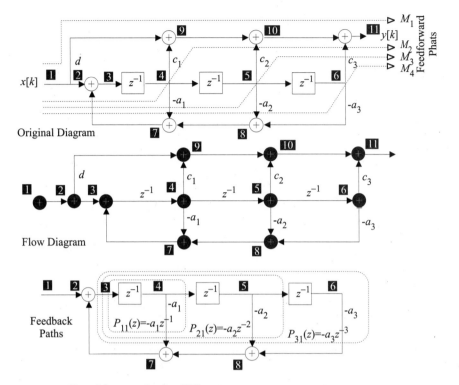

Figure 8.7 Signal flow graph of an LTI system.

$$\Delta(z) = 1 - [P_{11}(z) + P_{21}(z) + P_{31}(z)]$$

$$= 1 - [- a_1 z^{-1} - a_2 z^{-2} - a_3 z^{-3}]$$

$$= 1 + a_1 z^{-1} + a_2 z^{-2} + a_3 z^{-3}$$

A direct search yields four distinct feedforward (input-output) paths:

Path	Path gain	$\Delta_i(z)$
$M_1 = [1,2,9,10,11]$	$M_1(z) = d$	$\Delta_1(z) = \Delta(z)$ (touches no feedback paths)
$M_2 = [1,2,3,4,9,10,11]$	$M_2(z) = c_1 z^{-1}$	$\Delta_2(z) = 1 - (0) = 1$
$M_3 = [1,2,3,4,5,10,11]$	$M_3(z) = c_2 z^{-2}$	$\Delta_3(z) = 1 - (0) = 1$
$M_4 = [1,2,4,5,6,11]$	$M_4(z) = c_3 z^{-3}$	$\Delta_4(z) = 1 - (0) = 1$

From Eq. (8.2), it follows that

$$H(z) = \sum_{k=1}^{3} \frac{M_k(z)\Delta_k(z)}{\Delta(z)} = \frac{d\Delta(z) + c_3 z^{-1} + c_2 z^{-2} + c_1 z^{-3}}{\Delta(z)}$$

which simplifies to

$$H(z) = d + \frac{c_1 z^{-1} + c_2 z^{-2} + c_1 z^{-3}}{1 + a_1 z^{-1} + a_2 z^{-2} + a_3 z^{-3}} \qquad ❖$$

Example 8.4: Mason's Gain Formula

Problem Statement. A discrete-time system has a known signal flow diagram shown in Fig. 8.8. What is the system's transfer function, $H(z)$?

Analysis and Conclusions. From the signal flow graph, define the elemental parts of the system in the context of Mason's gain formula. The signal flow diagram is seen to consist of six labeled nodes. The input is found at Node 1 and the output at Node 6. A direct search yields three distinct input-output paths:

Path	Path gain	$\Delta_i(z)$
$M_1 = [1,2,3,4,5,6]$	$M_1(z) = c_{12} z^{-3}$	$\Delta_1(z) = 1 - (0) = 1$
$M_2 = [1,2,4,5,6]$	$M_2(z) = c_{12} c_{24} z^{-1}$	$\Delta_2(z) = 1 - (0) = 1$
$M_3 = [1,2,5,6]$	$M_3(z) = c_{12} c_{25}$	$\Delta_3(z) = 1 - c_{43} z^{-1}$

It can also be seen that there are three single (i.e., $P_{m,1}$) closed feedback paths, namely:

Path	Path gain
$P_{1,1} = [2,3,2]$	$P_{1,1}(z) = c_{32} z^{-1}$
$P_{2,1} = [3,4,3]$	$P_{2,1}(z) = c_{43} z^{-1}$
$P_{3,1} = [2,4,3,2]$	$P_{3,1}(z) = c_{24} c_{43} c_{32}$

The feedback paths or loops are shown in Fig. 8.9.

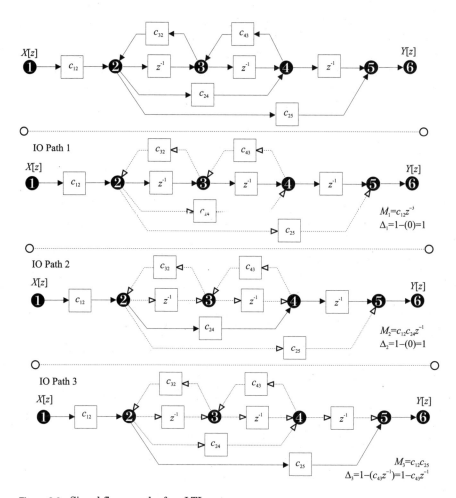

Figure 8.8   Signal flow graph of an LTI system.

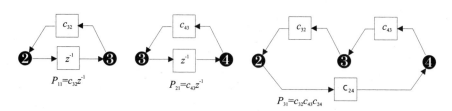

Figure 8.9   Feedback loops.

Notice that all the closed feedback paths share a common node, namely Node 3. Therefore, there are no higher order terms (i.e., $P_{mj} = 0$ for $j > 1$). From these elements, Mason's gain formula can be used to define the system determinant

$$\Delta(z) = 1 - [P_{11}(z) + P_{21}(z) + P_{31}(z)] = 1 - (c_{21} + c_{43})z^{-1} - (c_{24}c_{43}c_{32})$$

From Eq. (8.2), it follows that

$$H(z) = \sum_{k=1}^{3} \frac{M_k(z)\Delta_k(z)}{\Delta(z)} = \frac{1(c_{12}z^{-3}) + 1(c_{12}c_{24}z^{-1}) + (1 - c_{43}z^{-1})(c_{12}c_{25})}{1 - (c_{24}c_{43}c_{32}) - (c_{32} + c_{43})z^{-1}}$$

which simplifies to

$$H(z) = \frac{(c_{12}c_{25}) + (c_{12}c_{24} - c_{43}c_{12}c_{25})z^{-1} + c_{12}z^{-3}}{(1 - c_{24}c_{43}c_{32}) - (c_{32} + c_{43})z^{-1}}$$

$$= \frac{(c_{12}c_{25})z^3 + (c_{12}c_{24} - c_{43}c_{12}c_{25})z^2 + c_{12}}{(1 - c_{24}c_{43}c_{32})z^3 - (c_{32} + c_{43})z^2}$$

❖

Mason's gain formula is a procedure which can be reduced to a computer program. Direct search techniques are used to determine the feedforward and feedback paths, and polynomial arithmetic is relatively straightforward to implement. The program sfg.exe implements these searches and computes the coefficients of a transfer function. The signal flow graph is specified in a source file that takes data in the following format.

```
# comments
BRANCH i j # denotes the definition of a branch from node i to j
NUM      # begin to specify numerator polynomial in descending
           order
 data    # real (c_i) or complex (Re(c_i) Im(c_i)){i.e., real part
         # <space> imaginary part}
END      # end numerator specification
DEN      # begin to specify denominator polynomial in descend-
           ing order
 data    # real (c_i) or complex (Re(c_i) Im(c_i)) {i.e., real part
         # <space> imaginary part}
END      # end denominator specification
```

The keywords are in capital letters in the above format.

Example 8.5: Computer Generated Mason's Gain Formula

Problem Statement. Given

$$H(z) = \frac{z^{-1} + a}{z^{-2} + b} = \frac{z^{-1} + (1 + j2)}{z^{-2} + 2}$$

what is the format of the signal flow graph specification file for use with the sfg program?

Analysis and Conclusions. Consider the discrete-time system represented by a two-node signal flow graph. Let node 1 be the input node and node 2 be the output node. A correct signal flow graph specification is given below.

```
# Path gains
BRANCH 1 2
NUM
1 2        # coefficient az⁰
1          # coefficient z⁻¹
END
DEN
2          # coefficient bz⁰
0          # coefficient 0z⁻¹
1          # coefficient z⁻²
END
```

❖

Example 8.6: Application of Computer-Generated Mason's Gain Formula

Problem Statement. A third-order lowpass filter is given in block diagram form in Fig. 8.10, along with its magnitude frequency response. The coefficients for the filter in the block diagram are given below.

$$K = 0.291371$$

$b_{0,0} = 1.0$	$d_{0,0} = 1.0$
$a_{0,0} = -0.131375$	$c_{0,0} = 0.868625$
$b_{1,1} = 1.0$	$d_{0,1} = 1.0$
$a_{1,1} = 0.022187$	$a_{0,1} = -0.451366$
$c_{1,1} = 0.796895$	$c_{0,1} = 0.548634$

What is $H(z)$?

Analysis and Conclusions. The third-order system is given in signal flow graph form in Fig. 8.11. The program sfg automatically computes a transfer

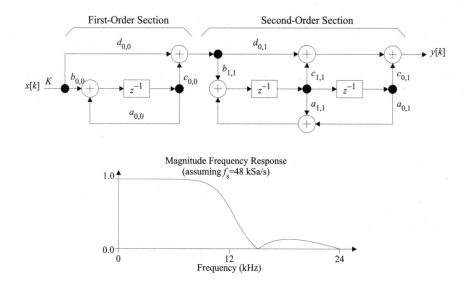

Figure 8.10   Block diagram of a third-order discrete-time system and its magnitude frequency response.

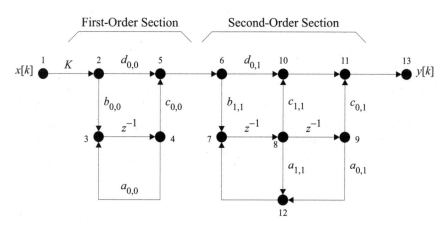

Figure 8.11   Signal flow graph.

function from a list of internode path gains and directed paths. The data used to create the signal flow graph specification file are summarized below.

Unity Gain Paths

$$[2,3] = [2,5] = [5,6] = [6,7] = [6,10] = [10,11] = [11,13] = [12,7] = 1$$

Delay Paths

$$[3,4] = [7,8] = [8,9] = z^{-1}$$

Non-unity Gain Paths

> [1,2] = 0.291371
>
> [4,5] = 0.868625
>
> [4,3] = −0.131375
>
> [8,10] = 0.796895
>
> [8,12] = 0.022187
>
> [9,12] = −0.451366
>
> [9,11] = 0.548634

Computer Study. A valid signal flow graph specification file for the given problem, using the signal flow graph shown in Fig. 8.11 and the given coefficients is shown in Exhibit 8.1.

The signal flow graph specification is found in the file **cas.sfg**, and can be processed using the following command (from DOS command prompt):

```
C:\Chap8>sfg cas.sfg
```

Upon execution, the resulting transfer function is computed using Mason's gain theorem and the following report is generated (abridged where indicated).

```
SFG to Transfer Function Extraction Utility
version 0.0 (2/17/91)
Opening file cas.sfg...
```
                    [Echo input file omitted.]
```
Done reading file.
```
                    [Connection reports omitted.]
```
Signal flow graph transfer function:
------------------------------------
Numerator Polynomial Coefficients
0.291371
0.517098
0.517098
0.291371
Denominator Polynomial Coefficients
0.0592982
0.448451
0.109188
1
------------------------------------
Memory usage: 35900 bytes allocated.
Execution Ends...
```

Exhibit 8.1 Signal flow graph specification file.

```
#--------------------        DEN                     #
# cas.sfg                    1                       # Delays
#--------------------        END                     BRANCH 7 8
# First-order Section        BRANCH 4 3              NUM
# Unity Gain Paths           NUM                     1
BRANCH 2 3                   -0.131375               0
NUM                          END                     END
1                            DEN                     DEN
END                          1                       1
DEN                          END                     END
1                            # -------------------   BRANCH 8 9
END                          # Second-order Section  NUM
BRANCH 2 5                   #                       1
NUM                          # Unity Gain Paths       0
1                            BRANCH 6 7              END
END                          NUM                     DEN
DEN                          1                       1
1                            END                     END
END                          DEN                     #
BRANCH 5 6                   1                       # Non-unity Gain Paths
NUM                          END                     BRANCH 8 10
1                            BRANCH 6 10             NUM
END                          NUM                     0.796895
DEN                          1                       END
1                            END                     DEN
END                          DEN                     1
#                            1                       END
# Delays                     END                     BRANCH 8 12
BRANCH 3 4                   BRANCH 10 11            NUM
NUM                          NUM                     0.022187
1                            1                       END
0                            END                     DEN
END                          DEN                     1
DEN                          1                       END
1                            END                     BRANCH 9 12
END                          BRANCH 11 13            NUM
#                            NUM                     -0.451366
# Non-unity gain paths       1                       END
BRANCH 1 2                   END                     DEN
NUM                          DEN                     1
0.291371                     1                       END
END                          END                     BRANCH 9 11
DEN                          BRANCH 12 7             NUM
1                            NUM                     0.548634
END                          1                       END
BRANCH 4 5                   END                     DEN
NUM                          DEN                     1
0.868625                     1                       END
END                          END
```

From this report, the transfer function can be determined to be

$$H(z) = \frac{0.291371 + 0.517098z^{-1} + 0.517098z^{-2} + 0.291371z^{-3}}{1 + 0.109188z^{-1} + 0.448451z^{-2} + 0.0592982z^{-3}}$$ ❖

8.5 Summary

To this point, it is assumed that a transfer function $H(z)$ is known in one of the following domains:

- Mathematical [i.e., transfer function $H(z)$ or pole-zero description of a filter that meets specific time- or frequency-domain requirements]
- Synthesized [i.e., a transfer function $H(z)$ derived from observed data]
- Block diagram [i.e., $H(z)$ derived from block diagram reduction methods]
- Signal flow diagram [i.e., $H(z)$ derived from Mason's gain formula]

In Chap. 9, another representation form, called *state variables*, will be introduced. Using one or all of these methods, a transfer function $H(z)$ can be mapped into a number of well understood or arbitrary architectures in Chap. 14. Knowledge of the a system's architecture is a prerequisite to its implementation. Once the architecture of a discrete-time has been described, it can be implemented with hardware or software.

8.6 Self-Study Problems

1. Signal flow graph Represent the following discrete-time transfer function in signal flow graph form:

$$H(z) = \frac{1 - \sqrt{2}z^{-1} + z^{-2}}{1 + 1.5z^{-1} + 0.75z^{-2}}$$

Use Mason's gain formula to verify $H(z)$.

2. Block diagram reduction Reduce the system shown in Fig. 8.12 to a transfer function.

3. Mason's gain formula Using Mason's gain formula, derive the four input-output transfer functions associated with the system shown in Fig. 8.13.

4. Non-proper system A non-proper discrete-time system has a transfer function

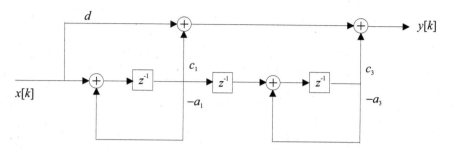

Figure 8.12 Block diagram representation of a discrete-time system.

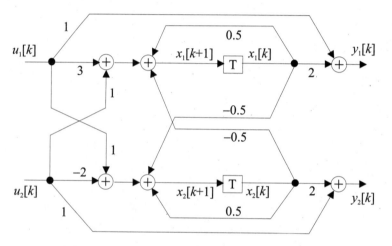

Figure 8.13 Schematic for an example multiple-input multiple-output system.

$$H(z) = \frac{\displaystyle\sum_{m=0}^{4} b_m z^{-m}}{\displaystyle\sum_{m=0}^{3} a_m z^{-m}}$$

Create a signal flow graph representation for $H(z)$. Use Mason's gain formula to verify the validity of the signal flow graph.

9

State-Determined Systems

9.1 Introduction

It has been shown that an LTI system can be modeled as a difference equation or transfer function. In Chap. 8, LTI systems were assumed to have a structure which could be interpreted in terms of a block diagram or signal flow graph. The structure imposed on an LTI by the arrangement of its fundamental building blocks (registers and adders) define its *architecture*. Using block diagram and signal flow graph reduction techniques (e.g., Mason's gain formula), an LTI architecture can be converted into a transfer function $H(z)$. While the transfer function is fundamentally important to the analysis of LTI systems, it should be remembered that transfer functions mask the internal details, or architecture, of the system. A transfer function only quantifies the input-output behavior of an at-rest LTI system. What is desired is a representation methodology that is capable of producing a transfer function while preserving architectural details of the system under study. *State variable* models have this capability. It will be shown that state variables provide a viable and important framework in which to analyze discrete-time systems and architectures.

9.2 State Variables

State variables shall be informally defined as those system parameters which contain sufficient information about the system's state to insure that all future state variables can be computed provided:

1. All past values of state variables are known.

2. The mathematical relationship (sometimes called the *bonds on interaction*) between the system dynamics and state variables is known.

3. All future inputs are known.

State variables, or simply *states,* represent the information stored in the system that quantifies past, present, and future behavior. As such, state variables physically reside at those locations within a system where information is stored. In the case of an analog system, information is stored in capacitors and inductors. For example, the capacitor voltage $v_0(t)$ found in the first-order RC circuit shown in Fig. 9.1 satisfies the requirements to be state variable in that:

1. All past values of state variables are known if the initial condition $v_0(0)$ is known.

2. The mathematical relationship between the system dynamics and state variables is given by $v_0(t) + RCdv_0(t)\dfrac{dv_0(t)}{dt} = v(t)$

3. All future inputs are known if $v(t)$ is known for all $t \geq 0$.

As a consequence, $v_0(t)$ qualifies as the state of the first-order RC circuit. In general, the solution to an Nth-order ODE would contain N state variables, with uniqueness provided by the choice of input and N initial conditions. Similarly, an LTI modeled by an Nth-order difference equation would contain N states, with uniqueness provided by the choice of input and N initial conditions. The states and initial conditions for the discrete-time system would be stored at the register or memory level. This, in fact, is a simple and generally effective way to identify the states in an arbitrary discrete-time system.

In general, a *multiple-input multiple-output* (MIMO) discrete-time system, consisting of P-inputs, R-outputs, and N-states, has a state variable representation given by

state equation: $\mathbf{x}[k + 1] = \mathbf{A}[k]\mathbf{x}[k] + \mathbf{B}[k]\mathbf{u}[k]$ (9.1)

initial conditions: $\mathbf{x}[0] = \mathbf{x}_0$ (9.2)

output equation: $\mathbf{y}[k] = \mathbf{C}^T[k]\mathbf{x}[k] + \mathbf{D}[k]\mathbf{u}[k]$ (9.3)

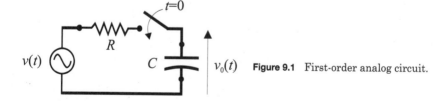

Figure 9.1 First-order analog circuit.

where $\mathbf{A}[k]$ is an $N \times N$ matrix, $\mathbf{B}[k]$ is an $N \times P$ matrix, $\mathbf{C}[k]$ is an $N \times R$ matrix, and $\mathbf{D}[k]$ is an $R \times P$ matrix, $\mathbf{u}[k]$ is an arbitrary $P \times 1$ input vector, $\mathbf{x}[k]$ is an $N \times 1$ state vector, and $\mathbf{y}[k]$ is an $R \times 1$ output vector. Such a system can also be represented by the four-tuple of the form $\mathbf{S}$ = $(\mathbf{A}[k], \mathbf{B}[k], \mathbf{C}[k], \mathbf{D}[k])$. If the discrete-time system is also an LTI, then the state four-tuple consists of a collection of constant coefficient matrices and would have the form $\mathbf{S} = (\mathbf{A}, \mathbf{B}, \mathbf{C}, \mathbf{D})$.

The discrete-time state determined system, based on Eqs. (9.1) through (9.3), is graphically interpreted in Fig. 9.2. In a discrete-time system, state variables are stored in memories (e.g., shift registers). If an Nth-order system can be implemented with N shift-registers, or N states, then the system is said to be *canonic*.

Example 9.1: MIMO State Variable Model

Problem Statement. Consider the two-input, two-output, at-rest MIMO system given by the linear difference equations

$$x_1[k + 1] = 0.5x_1[k] - 0.5x_2[k] + 3u_1[k] + u_2[k]; \; x_1[0] = 0$$

$$x_2[k + 1] = -0.5x_1[k] + 0.5x_2[k] + u_1[k] - 2u_2[k]; \; x_2[0] = 0$$

$$y_1[k] = 2x_1[k] + u_1[k]$$

$$y_2[k] = 2x_2[k] + u_2[k]$$

What is the system's state variable model based on Eqs. (9.1) through (9.3)?

Analysis and Conclusions. Direct application of Eqs. (9.1) though (9.3) produces a state variable model given by

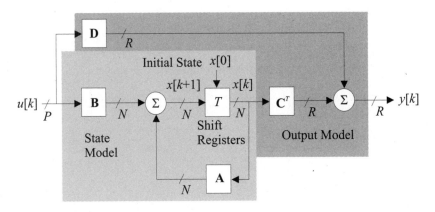

Figure 9.2 Discrete state variable system model.

$$\mathbf{x}[k+1] = \begin{bmatrix} 0.5 & -0.5 \\ -0.5 & 0.5 \end{bmatrix} \mathbf{x}[k] + \begin{bmatrix} 3 & 1 \\ 1 & -2 \end{bmatrix} \mathbf{u}[k]$$

$$\mathbf{x}[k] = \begin{bmatrix} x_1[k] \\ x_2[k] \end{bmatrix}; \quad \mathbf{u}[k] = \begin{bmatrix} u_1[k] \\ u_2[k] \end{bmatrix}; \quad \mathbf{A} = \begin{bmatrix} 0.5 & -0.5 \\ -0.5 & 0.5 \end{bmatrix}; \quad \mathbf{B} = \begin{bmatrix} 3 & 1 \\ 1 & -2 \end{bmatrix}$$

$$\mathbf{x}[0] = \begin{bmatrix} 0 \\ 0 \end{bmatrix}$$

$$\mathbf{y}[k] = \begin{bmatrix} 2 & 0 \\ 0 & 2 \end{bmatrix} \mathbf{x}[k] + \begin{bmatrix} 1 & 0 \\ 0 & 1 \end{bmatrix} \mathbf{u}[k]$$

$$\mathbf{C}^T = \begin{bmatrix} 2 & 0 \\ 0 & 2 \end{bmatrix}; \quad \mathbf{D} = \begin{bmatrix} 1 & 0 \\ 0 & 1 \end{bmatrix}$$

The state determined LTI system is graphically interpreted in Fig. 9.3.   ❖

Many important discrete-time LTI systems are *single-input single-output* (*SISO*) systems which can be modeled by the Nth-order difference equation

$$a_0 y[k] + a_1 y[k-1] + \ldots + a_N y[k-N]$$

$$= b_0 u[k] + b_1 u[k-1] + \ldots + b_N u[k-N] \tag{9.4}$$

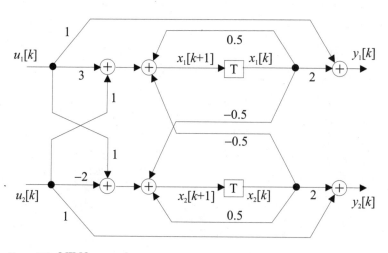

Figure 9.3 MIMO example.

The discrete-time LTI system defined by Eq. (9.4) can also be expressed in terms of a transfer function $H(z)$, where

$$H(z) = \frac{b_0 + b_1 z^{-1} + \ldots + b_N z^{-N}}{a_0 + a_1 z^{-1} + \ldots + a_N z^{-N}}$$

$$= \frac{b_0}{a_0} + \frac{(b_1 - b_0 a_1 / a_0) z^{-1} + \ldots + (b_N - b_0 a_N / a_0) z^{-N}}{a_0 + a_1 z^{-1} + \ldots + a_N z^{-N}}$$

$$= \frac{b_0}{a_0} + \frac{c_1 z^{-1} + \ldots + c_N z^{-N}}{a_0 + a_1 z^{-1} + \ldots + a_N z^{-N}} - d_0 + C(z)\left(\frac{1}{D(z)}\right) \tag{9.5}$$

If $a_0 = 1$, the system is said to be *monic*. The transfer function is seen to consist of three distinct subsystems which include (1) a constant gain path (d_0), (2) an all feedforward system denoted $C(z)$, and (3) an all feedback system denoted $D(z)$.

9.3 Direct II Architecture

For a monic system, the LTI characterized by Eq. (9.5) can be placed into the *direct I* architectural model, shown in Fig. 9.4. The *direct II*

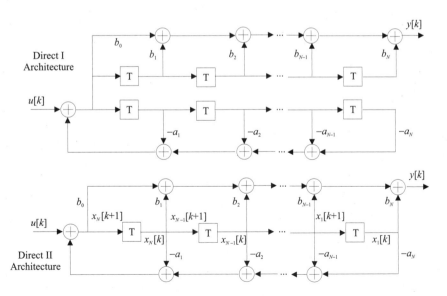

Figure 9.4 Direct I and II architectures.

model is a refinement of the direct I architecture and is also shown in Fig. 9.4. The problem with the non-canonic direct I structure is the need for $2N$ shift registers versus only N for the canonic direct II realization. The direct II state model is defined in terms of an N-element state vector given by

$$\mathbf{x}[k] = \begin{bmatrix} x_1[k] \\ x_2[k] \\ \vdots \\ x_N[k] \end{bmatrix} = \begin{bmatrix} x[k-N] \\ x[k-N+1] \\ \vdots \\ x[k] \end{bmatrix} \qquad (9.6)$$

and the following state assignments

$$\mathbf{x}[k+1] = \begin{bmatrix} x_1[k+1] \\ x_2[k+1] \\ \vdots \\ x_N[k+1] \end{bmatrix}$$

$$= \begin{bmatrix} x_2[k] \\ x_3[k] \\ \vdots \\ -a_N x_1[k] - a_{N-1} x_2[k] + -\ldots -a_2 x_{N-1}[k] - a_1 x_N[k] + u[k] \end{bmatrix}$$

$$(9.7)$$

This result is the *state equation* for a direct II architecture given by

$$\mathbf{x}[k+1] = \mathbf{A}\mathbf{x}[k] + \mathbf{b}u[k] \qquad (9.8)$$

The $N \times N$ coefficient matrix $\mathbf{A}$ is defined to be

$$\mathbf{A} = \begin{bmatrix} 0 & 1 & 0 & \cdots & 0 \\ 0 & 0 & 1 & \cdots & 0 \\ \vdots & \vdots & \vdots & \ddots & \vdots \\ 0 & 0 & 0 & \cdots & 1 \\ -a_N & -a_{N-1} & -a_{N-2} & \cdots & -a_1 \end{bmatrix} \qquad (9.9)$$

where the i,jth element of $\mathbf{A}$ defines the path gain existing between state $x_j[k]$ and $x_i[k+1]$. Since it has been established that the state information resides at the shift register locations of an architecture, equivalently, the i,jth element of $\mathbf{A}$ corresponds to the path gain existing between the output of the jth shift register (i.e., location of state $x_j[k]$) and input of the ith shift-register (i.e., next value of $x_i[k]$, namely $x_i[k+1]$). Continuing, $\mathbf{b}$ is an $N \times 1$ vector of the form

$$\mathbf{b} = \begin{bmatrix} 0 \\ \vdots \\ 0 \\ 1 \end{bmatrix} \tag{9.10}$$

It can be immediately seen that Eqs. (9.9) and (9.10) can be used to construct the feedback system defined by the $1/D(z)$ term of Eq. (9.5).

To this point, however, only the feedback structure of the system has been analyzed and quantified. The output, or *output state equation,* is given by

$$y[k] = \mathbf{c}^T\mathbf{x}[k] + d_0u[k] \tag{9.11}$$

where $\mathbf{c}$ is a length N vector and d_0 is a scalar. Unfortunately the direct II model shown in Fig. 9.3 provides no convenient way to express the output $y[k]$ as a linear combination of only the states $x_i[k]$ and input $u[k]$. The problem point is located at the input of the leftmost shift register, which sends to the output a scaled version of $x_N[k+1]$ (not $x_N[k]$). The architecture shown in Fig. 9.5 is a modification of the direct II architecture shown in Fig. 9.4. The architecture described in Fig. 9.5 will, from this point on, be called the direct II form. The re-engineered architecture can express the system output as a linear combination of states. In particular, the elements of Eq. (9.11) are

$$\mathbf{c}^T = (b_N - b_0a_N \quad b_{N-1} - b_0a_{N-1} \quad \cdots \quad b_1 - b_0a_1) \tag{9.12}$$

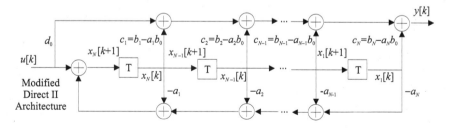

Figure 9.5 Modified direct II architecture.

After Eq. (9.5), d_0 is given by

$$d_0 = b_0 \tag{9.13}$$

Example 9.2: Direct II Architecture

Problem Statement. Many digital filters are specified in terms of a transfer function $H(z)$. Consider, for example, a third-order digital filter having the monic transfer function

$$H(z) = \frac{z^3 - 0.5z^2 - 0.315z - 0.185}{z^3 - 0.5z^2 + 0.5z - 0.25}$$

Realize $H(z)$ as a direct II filter and quantify the state variable model.

Analysis and Conclusions. The transfer function can be directly converted into a state variable model using the established direct II mapping rules. In particular, using long division or Eq. (9.5), $H(z)$ can be reduced to

$$H(z) = \frac{1 - 0.5z^{-1} - 0.315z^{-2} - 0.185z^{-3}}{1 - 0.5z^{-1} + 0.5z^{-2} - 0.25z^{-3}} = 1 + \frac{-0.815z^{-2} + 0.065z^{-3}}{1 - 0.5z^{-1} + 0.5z^{-2} - 0.25z^{-3}}$$

The mapping of $H(z)$ into a direct II architecture results in the filter shown in Fig. 9.6. The LTI system is seen to consist of a line of three shift registers, three feedback, and four feedforward paths. The states of the system reside in the shift registers which are indexed from right to left as shown in Fig. 9.6. The state variable four-tuple, $S = (A,b,c,d)$, for this case, is given by

$$A = \begin{bmatrix} 0 & 1 & 0 \\ 0 & 0 & 1 \\ 0.25 & -0.5 & 0.5 \end{bmatrix}; b = \begin{bmatrix} 0 \\ 0 \\ 1 \end{bmatrix}; c = \begin{bmatrix} 0.065 \\ -0.815 \\ 0.0 \end{bmatrix}; d = 1 \qquad ❖$$

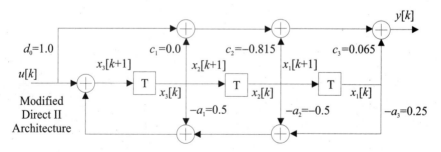

Figure 9.6 Third-order state-determined digital filter reduced to a direct II architecture showing state assignments.

The direct II architecture defined by Eqs. (9.8) and (9.11) sometimes appears in a slightly modified form. Consider again the monic transfer function $H(z) = Y(z)/U(z)$, after Eq. (9.5), namely

$$H(z) = K\left(\frac{b_0 + b_1 z^{-1} + \ldots + b_N z^{-N}}{1 + a_1 z^{-1} + \ldots + a_N z^{-N}}\right) \tag{9.14}$$

where K is a scale factor. The scale factor K is applied to the input $u[k]$ to form an intermediate input $v[k]$, where $v[k] = Ku[k]$. The transfer function of the system receiving a scaled input is denoted $H'(z)$ and is given by

$$H'(z) = \frac{Y(z)}{V(z)} \tag{9.15}$$

Therefore, Eq. (9.14) can then be expressed as

$$H(z) = KH'(z) \tag{9.16}$$

which is graphically interpreted in Fig. 9.7. Sometimes the scaling constant K is required to reduce the input dynamic range to a value that will not cause run-time overflows internal to $H'(z)$. At other times, K is used to amplify a weak input signal to the point where the data has significance inside $H'(z)$. Another use of K is to adjust the overall gain of the system to have a unity gain (0 dB) passband.

The direct II, or scaled direct II, is only one of many architectural manifestations that an LTI system may take. At this time, however, it is sufficient to know that an LTI system can be mapped into at least one state variable architecture, the direct II architecture.

Example 9.3: High-Order Direct II Architecture

Problem Statement. An eighth-order monic LTI system has a transfer function given by

$$H(z) = K\left(\frac{b_0 + b_1 z^{-1} + \ldots + b_8 z^{-8}}{1 + a_1 z^{-1} + \ldots + a_8 z^{-8}}\right)$$

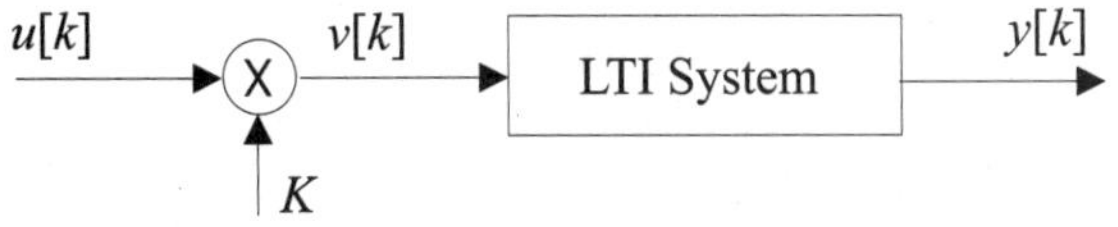

Figure 9.7 Scaled system.

A specific instance of an eighth-order IIR filter can be found in the computer readable file, "chap9-1.rpt". The coefficients for the transfer function as given above are listed as the following (indices reversed).

```
Scale Factor: K = 0.002275075544550193
Numerator Coefficients        Denominator Coefficients
b8 : 1.000000000000000        a8 : 1.000000000000000
b7 : 8.000000000000000        a7 : -1.589136150104044
b6 : 28.00000000000000        a6 : 2.081971254419972
b5 : 56.00000000000000        a5 : -1.530616503476392
b4 : 70.00000000000000        a4 : 0.8682727723194346
b3 : 56.00000000000000        a3 : -0.3186623854575408
b2 : 28.00000000000000        a2 : 0.08195478641299830
b1 : 8.000000000000000        a1 : -0.01222419034039375
b0 : 1.000000000000000        a0 : 0.0008597556308154511
```

If a direct II architecture is assumed, what is the corresponding state variable representation of $H(z)$?

Analysis and Conclusions. The state variable representation for a scaled direct II architecture (found as file chap9-1.rpt), is based on a direct application of Eqs. (9.9) through (9.13), given by scale factor $K = 0.002275075544550193$,

$$
\mathbf{A} = \begin{bmatrix}
0 & 1 & 0 & 0 & 0 & 0 & 0 & 0 \\
0 & 0 & 1 & 0 & 0 & 0 & 0 & 0 \\
0 & 0 & 0 & 1 & 0 & 0 & 0 & 0 \\
0 & 0 & 0 & 0 & 1 & 0 & 0 & 0 \\
0 & 0 & 0 & 0 & 0 & 1 & 0 & 0 \\
0 & 0 & 0 & 0 & 0 & 0 & 1 & 0 \\
0 & 0 & 0 & 0 & 0 & 0 & 0 & 1 \\
-0.000859 & 0.01222 & -0.08195 & 0.31866 & -0.86827 & 1.53061 & -2.08197 & 1.58913
\end{bmatrix},
$$

$$
\mathbf{b} = \begin{bmatrix} 0 \\ 0 \\ 0 \\ 0 \\ 0 \\ 0 \\ 0 \\ 1 \end{bmatrix}, \quad
\mathbf{c} = \begin{bmatrix} 0.9991 \\ 8.0122 \\ 27.9180 \\ 56.3186 \\ 69.1317 \\ 57.5306 \\ 25.9180 \\ 9.5891 \end{bmatrix}, \text{ and } d = 1
$$

Computer Study. The four-tuple $(\mathbf{A}, \mathbf{b}, \mathbf{c}, d)$ is summarized in file "chap9-1.rpt," which is partially listed below. The filter can be generated from the Monarch IIR design menu, and the architecture can be created using the state variable module. The architectural structure can be viewed through the architecture module.

DIGITAL FILTER ARCHITECTURE

TRANSFER FUNCTION
Scale Factor: 0.002275075544550193

Numerator Coefficients	Denominator Coefficients
b8 : 1.000000000000000	a8 : 1.000000000000000
b7 : 8.000000000000000	a7 : -1.589136150104044
b6 : 28.00000000000000	a6 : 2.081971254419972
b5 : 56.00000000000000	a5 : -1.530616503476392
b4 : 70.00000000000000	a4 : 0.8682727723194346
b3 : 56.00000000000000	a3 : -0.3186623854575408
b2 : 28.00000000000000	a2 : 0.08195478641299830
b1 : 8.000000000000000	a1 : -0.01222419034039375
b0 : 1.000000000000000	a0 : 0.0008597556308154511

DIRECT-II STATE VARIABLE FILTER DESCRIPTION
Scale Factor = 0.002275075544550193
A Matrix
Row 1
0.000000000000000   1.000000000000000
0.000000000000000   0.000000000000000
0.000000000000000   0.000000000000000
0.000000000000000   0.000000000000000
Row 2
0.000000000000000   0.000000000000000
1.000000000000000   0.000000000000000
0.000000000000000   0.000000000000000
0.000000000000000   0.000000000000000
Row 3
0.000000000000000   0.000000000000000
0.000000000000000   1.000000000000000
0.000000000000000   0.000000000000000
0.000000000000000   0.000000000000000
Row 4
0.000000000000000   0.000000000000000
0.000000000000000   0.000000000000000
1.000000000000000   0.000000000000000
0.000000000000000   0.000000000000000
Row 5
0.000000000000000   0.000000000000000
0.000000000000000   0.000000000000000
0.000000000000000   1.000000000000000
0.000000000000000   0.000000000000000
Row 6
0.000000000000000   0.000000000000000
0.000000000000000   0.000000000000000
0.000000000000000   0.000000000000000
1.000000000000000   0.000000000000000
Row 7
0.000000000000000   0.000000000000000
0.000000000000000   0.000000000000000
0.000000000000000   0.000000000000000
0.000000000000000   1.000000000000000

Row 8

-0.0008597556308154511	0.01222419034039375
-0.08195478641299830	0.3186623854575408
-0.8682727723194346	1.530616503476392
-2.081971254419972	1.589136150104044

B Vector	C' Vector	D Scalar
0.000000000000000	0.9991402443691846	1.000000000000000
0.000000000000000	8.012224190340394	
0.000000000000000	27.91804521358700	
0.000000000000000	56.31866238545754	
0.000000000000000	69.13172722768057	
0.000000000000000	57.53061650347639	
0.000000000000000	25.91802874558003	
1.000000000000000	9.589136150104043	❖

9.4 State Transition Matrix

The solution to a LTI system consists of two parts, called the *homogeneous* (unforced) and *inhomogeneous solution* (forced). When combined they define the total or complete solution. The solution to the homogeneous state equation is given by

$$\mathbf{x}[k+1] = \mathbf{A}\mathbf{x}[k] \qquad (9.17)$$

with the initial condition $\mathbf{x}[0] = \mathbf{x}_0$ defined. The homogeneous solution can be symbolically represented by the mapping

$$\mathbf{x}[k] = \Phi[k, 0]\mathbf{x}_0 \qquad (9.18)$$

where the $N \times N$ matrix $\Phi[k,m]$ is called the discrete-time *state transition matrix*. The state transition matrix provides a mapping from state at discrete time step m to the state at discrete time step k. The state transition matrix satisfies a number of important properties including

state transition property: $\Phi[k, m] = \Phi[k, j]\Phi[j, m]$ (9.19)

state inversion property: $\Phi[k, m] = \Phi^{-1}[m, k]$ (9.20)

which are graphically interpreted in Fig. 9.8.

For $\mathbf{A}$, an $N \times N$ constant coefficient matrix, the homogeneous solution can be constructed as follows:

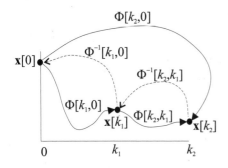

Figure 9.8 Graphical interpretation of a state transition matrix.

$$\mathbf{x}[1] = \mathbf{A}\mathbf{x}[0]$$

$$\mathbf{x}[2] = \mathbf{A}\mathbf{x}[1] = \mathbf{A}^2\mathbf{x}[0]$$

$$\mathbf{x}[3] = \mathbf{A}\mathbf{x}[2] = \mathbf{A}^2\mathbf{x}[1] = \mathbf{A}^3\dot{x}[0]$$

$$\ldots = \ldots$$

$$\mathbf{x}[k] = \mathbf{A}\mathbf{x}[k-1] = \mathbf{A}^2\mathbf{x}[k-2] = \ldots = \mathbf{A}^k\mathbf{x}[0] \qquad (9.21)$$

As a result, the discrete-time transition matrix for an LTI system can be expressed as

$$\Phi[k+m, m] = \Phi[k] = \mathbf{A}^k \qquad (9.22)$$

again, assuming that $\mathbf{A}$ is constant. Therefore, knowledge of $\mathbf{A}$ and the initial state is sufficient to define the homogeneous solution of an LTI system.

Example 9.4: Homogeneous Response

Problem Statement. Let $x_1[k+1] = x_2[k]$, $x_2[k+1] = x_1[k] + x_2[k]$, and $y[k] = x_1[k] + x_2[k]$ with initial conditions given by $x_1[0] = 0$ and $x_2[0] = 1$. What is the homogeneous system solution?

Analysis and Conclusions. The first few samples of the system's homogeneous response can be shown to be $y[k] = \{1,2,3,5,8,13,21,\ldots\}$. Using the direct II architecture and the four-tuple production rules found in Sec. 9.3, the system's state model is given by

$$\mathbf{A} = \begin{bmatrix} 0 & 1 \\ 1 & 1 \end{bmatrix}; \quad \mathbf{b} = \begin{bmatrix} 0 \\ 0 \end{bmatrix}; \quad \mathbf{c} = \begin{bmatrix} 1 \\ 1 \end{bmatrix}; \quad d = 0$$

The homogeneous response is then given by

$$\mathbf{x}[k] = \mathbf{A}^k \mathbf{x}_0 = \begin{bmatrix} 0 & 1 \\ 1 & 1 \end{bmatrix}^k \mathbf{x}_0, \qquad \mathbf{y}[k] = [1\ 1]\mathbf{x}[k]$$

which can be evaluated using a digital computer or using methods developed later in this chapter.

Computer Study. The homogeneous state and output responses, over the first six sample periods, can be computed using the following sequence of Siglab commands.

```
>a={[0,1],[1,1]}
>c=[1,1]
>x=[0,1]'
>for (i=0:5)
>>y=c*a*x
>>x=a*x; y
>end
 2  3  5  8  13  21
```

The results generated here confirm the results produced manually above.   ❖

The direct calculation of $\Phi[k]$ is seen to be intrinsically inefficient. It is therefore important that an effective computational mechanism be defined. The computation of the discrete-time state transition matrix can be accomplished using several computational formulas. One of the most popular methods is called the *Cayley-Hamilton method*.

9.5 Cayley-Hamilton Method

The Cayley-Hamilton method can be used to produce a discrete-time state transition matrix. The method belongs to a branch of mathematics called functions of a square matrix. The *Cayley-Hamilton theorem* states that a matrix satisfies its own characteristic equation. The *characteristic equation* associated with a given square matrix $\mathbf{A}$ is defined to be

$$P(\lambda) = \det(\lambda \mathbf{I} - \mathbf{A}) = \gamma_0 + \gamma_1 \lambda + \gamma_2 \lambda^2 + \ldots + \gamma_N \lambda^N \qquad (9.23)$$

The values of λ which satisfy $P(\lambda) = 0$ are called the *eigenvalues* of $\mathbf{A}$. The Cayley-Hamilton theorem states that, the solutions of the characteristic equation are also solutions of the equation

$$N(\mathbf{A}) = \gamma_0 \mathbf{I} + \gamma_1 \mathbf{A} + \gamma_2 \mathbf{A}^2 + \ldots + \gamma_N \mathbf{A}^N = \mathbf{0} \qquad (9.24)$$

where $\mathbf{0}$ is a $N \times N$ matrix of zeros. From this result, it is possible to produce a compact representation of any matrix polynomial in $\mathbf{A}$.

Given an $N \times N$ matrix $\mathbf{A}$, with characteristic polynomial $P(\lambda)$ as defined in Eq. (9.23), and a polynomial $M(\lambda)$,

$$\frac{M(\lambda)}{P(\lambda)} = Q(\lambda) + \frac{R(\lambda)}{P(\lambda)} \tag{9.25}$$

where the order of $R(\lambda)$ is less than the order of $P(\lambda)$. It follows that

$$M(\lambda) = Q(\lambda)P(\lambda) + R(\lambda) \tag{9.26}$$

Evaluating Eq. (9.26) at the eigenvalues $\mathbf{A}$ results in $P(\lambda) = 0$, so $M(\lambda) = R(\lambda)$. Therefore, by the Cayley-Hamilton theorem, $M(\mathbf{A}) = R(\mathbf{A})$. This result allows a polynomial, $M(\mathbf{A})$, of any order to be expressed as a polynomial of order less than the order of the matrix $\mathbf{A}$.

Example 9.5: Inverse of a Matrix

Problem Statement. Given an $N \times N$ non-singular matrix

$$\mathbf{A} = \begin{bmatrix} 0 & 1 \\ -2 & -3 \end{bmatrix}$$

use the Cayley-Hamilton method to compute $\mathbf{A}^{-1}$.

Analysis and Conclusion. Given an $N \times N$ matrix $\mathbf{A}$, the characteristic equation is

$$P(\lambda) = \det(\lambda \mathbf{I} - \mathbf{A}) = \gamma_0 + \gamma_1 \lambda + \gamma_2 \lambda^2 + \ldots + \gamma_2 \lambda^N = 0$$

The Cayley-Hamilton theorem states that

$$N(\mathbf{A}) = \gamma_0 \mathbf{I} + \gamma_1 \mathbf{A} + \gamma_2 \mathbf{A}^2 + \ldots + \gamma_N \mathbf{A}^N = \mathbf{0}$$

Assuming that the inverse of $\mathbf{A}$ exists, upon multiplying the above equation by $\mathbf{A}^{-1}$, the following formula results,

$$\mathbf{A}^{-1} = -\left(\frac{\gamma_1 \mathbf{I} + \gamma_2 \mathbf{A} + \ldots + \gamma_N \mathbf{A}^{N-1}}{\gamma_0} \right)$$

In terms of the given 2×2 matrix $\mathbf{A}$, the characteristic equation is

$$P(\lambda) = \det(\lambda \mathbf{I} - \mathbf{A}) = 2 + 3\lambda + \lambda^2$$

where $\gamma_0 = 2$, $\gamma_1 = 3$, and $\gamma_2 = 1$. Therefore,

$$\mathbf{A}^{-1} = -\frac{\gamma_1}{\gamma_0} \mathbf{I} - \frac{\gamma_2}{\gamma_0} \mathbf{A} = -\frac{3}{2} \mathbf{I} - \frac{1}{2} \mathbf{A} = \begin{bmatrix} -\frac{3}{2} & -\frac{1}{2} \\ 1 & 0 \end{bmatrix}$$

which can be verified to be the inverse of $\mathbf{A}$.   ❖

The Cayley-Hamilton method also can be used to compute the state transition matrix $\Phi[k] = \mathbf{A}^k$. Consider an $N \times N$ matrix $\mathbf{A}$ and a polynomial $F(\lambda)$. The matrix $\mathbf{A}$ has a characteristic polynomial equation given in Eq. (9.23), and the polynomial $F(\lambda)$ may be written as

$$F(\lambda) = \sum_{i=0}^{\infty} \kappa_i \lambda^i = Q(\lambda)P(\lambda) + R(\lambda) \tag{9.27}$$

as in Eq. (9.26). The polynomial $R(\lambda)$ is a *remainder polynomial*, and is of the form

$$R(\lambda) = r_0 + r_1\lambda + r_2\lambda^2 + \dots + r_{N-1}\lambda^{N-1} \tag{9.28}$$

The remainder polynomial is of order less than that of the order of $P(\lambda)$ {i.e., degree$[R(\lambda)] <$ degree$[P(\lambda)] = N$}. Invoking the Cayley-Hamilton theorem, it then follows that

$$F(\mathbf{A}) = R(\mathbf{A}) = r_0\mathbf{I} + r_1\mathbf{A} + r_2\mathbf{A}^2 + \dots + r_{N-1}\mathbf{A}^{N-1} \tag{9.29}$$

Now, suppose that the eigenvalues of $\mathbf{A}$ are distinct, that is, $\lambda_i \neq \lambda_j$ for all $i \neq j$, $i,j \in \{1,2,3,\dots,N\}$. Then, by Eqs. (9.28) and (9.29),

$$\begin{bmatrix} F(\lambda_1) \\ F(\lambda_2) \\ \vdots \\ F(\lambda_N) \end{bmatrix} = \begin{bmatrix} 1 & \lambda_1 & \lambda_1^2 & \dots & \lambda_1^{N-1} \\ 1 & \lambda_2 & \lambda_2^2 & \dots & \lambda_2^{N-1} \\ \vdots & \vdots & \vdots & \ddots & \vdots \\ 1 & \lambda_N & \lambda_N^2 & \dots & \lambda_N^{N-1} \end{bmatrix} \begin{bmatrix} r_0 \\ r_1 \\ \vdots \\ r_{N-1} \end{bmatrix} \tag{9.30}$$

Equation (9.30) is solved for the unknown r_i values in terms of the known data. The r_i values can then be returned to Eq. (9.29) to compute the matrix function $F(\mathbf{A})$. Equation (9.29), therefore, can be used to compute the state transition matrix $\Phi[k] = \mathbf{A}^k$. Specifically, in this case,

$$F(\mathbf{A}) = \mathbf{A}^k = r_0\mathbf{I} + r_1\mathbf{A} + r_2\mathbf{A}^2 + \dots + r_{N-1}\mathbf{A}^{N-1} \tag{9.31}$$

Example 9.6: State Transition Matrix

Problem Statement. Given a discrete-time LTI system having an $\mathbf{A}$ matrix

$$A = \begin{bmatrix} 0 & 1 \\ \frac{1}{8} & -\frac{1}{4} \end{bmatrix}$$

compute the state transition matrix.

Analysis and Conclusions. The eigenvalues of A are $\lambda_1 = -1/2$ and $\lambda_2 = 1/4$ (distinct). Then Eq. (9.30) states

$$\begin{bmatrix} \left(-\frac{1}{2}\right)^k \\ \left(\frac{1}{4}\right)^k \end{bmatrix} = \begin{bmatrix} 1 & -\frac{1}{2} \\ 1 & \frac{1}{4} \end{bmatrix} \begin{bmatrix} r_0 \\ r_1 \end{bmatrix}$$

It then follows that

$$\begin{bmatrix} r_0 \\ r_1 \end{bmatrix} = \begin{bmatrix} 1 & \left(-\frac{1}{2}\right) \\ 1 & \frac{1}{4} \end{bmatrix}^{-1} \begin{bmatrix} \left(-\frac{1}{2}\right)^k \\ \left(\frac{1}{4}\right)^k \end{bmatrix} = \begin{bmatrix} \frac{1}{3} & \frac{2}{3} \\ -\frac{4}{3} & \frac{4}{3} \end{bmatrix} \begin{bmatrix} \left(-\frac{1}{2}\right)^k \\ \left(\frac{1}{4}\right)^k \end{bmatrix}$$

or

$$r_0 = \frac{1}{3}\left(-\frac{1}{2}\right)^k + \frac{1}{3}\left(\frac{1}{4}\right)^k$$

$$r_1 = \left(-\frac{4}{3}\right)\left(-\frac{1}{2}\right)^k + \frac{4}{3}\left(\frac{1}{4}\right)^k$$

The state transition matrix is obtained by substitution into Eq. (9.31), in particular,

$$A^k = r_0 I + r_1 A = \begin{bmatrix} \left\{\frac{2}{3}\left(\frac{1}{4}\right)^k + \frac{1}{3}\left(-\frac{1}{2}\right)^k\right\} & \left\{\frac{4}{3}\left(\frac{1}{4}\right)^k - \frac{4}{3}\left(-\frac{1}{2}\right)^k\right\} \\ \left\{\frac{1}{6}\left(\frac{1}{4}\right)^k - \frac{1}{6}\left(-\frac{1}{2}\right)^k\right\} & \left\{\frac{1}{3}\left(\frac{1}{4}\right)^k + \frac{2}{3}\left(-\frac{1}{2}\right)^k\right\} \end{bmatrix}$$

As a numerical check, observe that $A^0 = I$ as expected. ❖

If there are repeated eigenvalues of A, then Eq. (9.30) will need to be modified. Assume that the eigenvalue λ_i occurs with a multiplicity $n(i)$. Then, as was the case for a Heaviside expansion of a transfer function containing repeated eigenvalues, $n(i) - 1$ derivatives are required to create $n(i)$ equations with $n(i)$ unknowns. Specifically, $F(\lambda) = (\lambda_i)^k$ produces

$$
\begin{bmatrix} \lambda_i^k \\ \dfrac{d\lambda_i^k}{d\lambda} \\ \vdots \\ \dfrac{d^{n(i)-1}\lambda_i^k}{d\lambda^{n(i)-1}} \end{bmatrix} = \begin{bmatrix} 1 & \lambda_i & \lambda_i^2 & \cdots & \lambda_i^{n(i)-1} \\ 0 & 1 & 2\lambda_i & \cdots & [n(i)-1]\lambda_i^{n(i)-2} \\ \vdots & \vdots & \vdots & \ddots & \vdots \\ 0 & 0 & 0 & \cdots & [n(i)-1]! \end{bmatrix} \begin{bmatrix} r_0 \\ r_1 \\ \vdots \\ r_{n(i)-1} \end{bmatrix}
\tag{9.32}
$$

Example 9.7: Repeated Eigenvalues

Problem Statement. Consider a discrete-time system having a diagonal matrix

$$
A = \begin{bmatrix} \dfrac{1}{2} & 0 & 0 \\ 0 & \dfrac{1}{2} & 0 \\ 0 & 0 & \dfrac{1}{2} \end{bmatrix}
$$

Compute the state-transition matrix.

Analysis and Conclusions. The matrix has eigenvalues $\lambda = 1/2$ with a multiplicity of three. Therefore $n[1] = 3$ and Eq. (9.32) state that

$$
\begin{bmatrix} \left(\dfrac{1}{2}\right)^k \\ k\left(\dfrac{1}{2}\right)^{k-1} \\ k(k-1)\left(\dfrac{1}{2}\right)^{k-2} \end{bmatrix} = \begin{bmatrix} 1 & \dfrac{1}{2} & \left(\dfrac{1}{2}\right)^2 \\ 0 & 1 & 2\left(\dfrac{1}{2}\right) \\ 0 & 0 & 2 \end{bmatrix} \begin{bmatrix} r_0 \\ r_1 \\ r_2 \end{bmatrix}
$$

which produces

$$
r_0 = \left(\frac{1}{2}\right)^k \left(1 - k + \frac{k(k-1)}{2}\right)
$$

$$
r_1 = \left(\frac{1}{2}\right)^{k-1}(2k - k^2)
$$

$$
r_2 = \left(\frac{1}{2}\right)^{k-2}(-k + k^2)
$$

Finally,

$$\Phi[k] = \mathbf{A}^k = r_0\mathbf{I} + r_1\mathbf{A} + r_2\mathbf{A}^2$$

$$= \begin{bmatrix} r_0 + \dfrac{r_1}{2} + \dfrac{r_2}{4} & r_1 + r_2 & r_2 \\[2mm] 0 & r_0 + \dfrac{r_1}{2} + \dfrac{r_2}{4} & r_1 + r_2 \\[2mm] 0 & 0 & r_0 + \dfrac{r_1}{2} + \dfrac{r_2}{4} \end{bmatrix}$$

$$= \begin{bmatrix} \left(\dfrac{1}{2}\right)^k & k^2\left(\dfrac{1}{2}\right)^{k-1} & (k^2 - k)\left(\dfrac{1}{2}\right)^{k-1} \\[2mm] 0 & \left(\dfrac{1}{2}\right)^k & k^2\left(\dfrac{1}{2}\right)^{k-1} \\[2mm] 0 & 0 & \left(\dfrac{1}{2}\right)^k \end{bmatrix}$$

Again, as a numerical check, $\Phi[0] = \mathbf{A}^0 = \mathbf{I}$ as required. ❖

9.6 The z-Transform Method

The Cayley-Hamilton method operates on the data defined in the discrete-time domain. It is also possible to work in the transform domain as well. The z-transform may be applied to the defining state equation [see Eq. (9.1)],

$$\text{state equation: } \mathbf{x}[k+1] = \mathbf{A}\mathbf{x}[k] + \mathbf{B}\mathbf{u}[k] \tag{9.33}$$

which can also be expressed as a transform equation having the form

$$z\mathbf{X}(z) = \mathbf{A}\mathbf{X}(z) + \mathbf{B}\mathbf{U}(z) \tag{9.34}$$

Equation (9.34) can be simplified to solve for $\mathbf{X}(z)$,

$$\mathbf{X}(z) = (z\mathbf{I} - \mathbf{A})^{-1}\mathbf{B}\mathbf{U}(z) \tag{9.35}$$

The state transition matrix may also be determined via the z-transform. In particular, if $\Phi[k] = \mathbf{A}^k$, as defined in Eq. (9.22), the z-transform of $\Phi[k]$ is

$$Z\{\Phi[k]\} = \sum_{n=0}^{\infty} \mathbf{A}^n z^{-n} = z(z\mathbf{I} - \mathbf{A})^{-1} \tag{9.36}$$

The i,jth element of $\Phi[k]$, denoted $\Phi_{i,j}[k]$, is defined in terms of the inverse z-transform of the i,jth element of $z(z\mathbf{I} - \mathbf{A})^{-1}$ and shall be denoted $\Phi_{i,j}(z)$. Using a Heaviside expansion on each $\Phi_{i,j}(z)$, the matrix elements $\Phi_{i,j}[k]$ will result.

Example 9.8 The z-Transform Method

Problem Statement. Refer to Example 9.6 and consider the homogeneous solution to an LTI system defined by the matrix

$$\mathbf{A} = \begin{bmatrix} 0 & 1 \\ \dfrac{1}{8} & -\dfrac{1}{4} \end{bmatrix}$$

Compute the state-transition matrix.

Analysis and Conclusions. The eigenvalues of A are $\lambda_1 = -1/2$ and $\lambda_2 = 1/4$ (distinct) so that

$$z(z\mathbf{I} - \mathbf{A})^{-1} = \frac{z\begin{bmatrix} z + \dfrac{1}{4} & 1 \\ \dfrac{1}{8} & z \end{bmatrix}}{\left(z - \dfrac{1}{4}\right)\left(z + \dfrac{1}{2}\right)}$$

The inverse z-transform of each of the four elements of the matrix $z(z\mathbf{I} - \mathbf{A})^{-1}$ can be computed using a Heaviside expansion, yielding

$$\Phi[k] = \mathbf{A}^k = \begin{bmatrix} \left\{\dfrac{2}{3}\left(\dfrac{1}{4}\right)^k + \dfrac{1}{3}\left(-\dfrac{1}{2}\right)^k\right\} & \left\{\dfrac{4}{3}\left(\dfrac{1}{4}\right)^k - \dfrac{4}{3}\left(-\dfrac{1}{2}\right)^k\right\} \\ \left\{\dfrac{1}{6}\left(\dfrac{1}{4}\right)^k - \dfrac{1}{6}\left(-\dfrac{1}{2}\right)^k\right\} & \left\{\dfrac{1}{3}\left(\dfrac{1}{4}\right)^k + \dfrac{2}{3}\left(-\dfrac{1}{2}\right)^k\right\} \end{bmatrix}$$

which agrees with the previous result. Also note that $\Phi[0] = \mathbf{I}$.

Computer Study. In Chap. 2, Heaviside expansions were performed with the assistance of a general-purpose digital computer. A computer generated analysis of the $\Phi_{1,1}(z) = (z^2 + z/4)/[(z - 1/4)(z + 1/2)]$ term would be performed by expanding

$$\frac{\Phi_{11}(z)}{z} = \frac{z^2 + \dfrac{z}{4}}{z\left(z - \dfrac{1}{4}\right)\left(z + \dfrac{1}{2}\right)}$$

which may be accomplished using the following sequence of Siglab commands:

```
>include "chap2\pf.s"
>n=[1,1/4,0]
>d=poly([0,1/4,-1/2])
>pfexp(n,d)
pole @ 0.5+0j
with multiplicity 1
and coefficients:
0.333333+0j
pole @ 0+0j
with multiplicity 1
and coefficients:
0+0j
pole @ 0.25+0j
with multiplicity 1
and coefficients:
0.666667+0j
```

This defines the Heaviside coefficients for $\Phi_{11}[k]$ to be 1/3 for the eigenvalue -0.5 and 2/3 for the eigenvalue 0.25. Therefore, $\Phi_{11}[k] = (1/3)(-1/2)^k + (2/3)(1/4)^k$. This agrees with the previously stated result. The other terms would be analyzed in a similar manner.   ❖

A fundamental problem with the z-transform method is that a matrix inverse needs to be taken. It should be appreciated that the problem is not like one of inverting a coefficient matrix (such as $\mathbf{A}$), but rather one involving a variable z, namely $(z\mathbf{I} - \mathbf{A})^{-1}$. Fortunately, there are numerical methods which can work with symbolic matrix inversion. *Leverrier's* method is one which can be efficiently applied to the problem of inverting $(z\mathbf{I} - \mathbf{A})^{-1}$ with a general-purpose digital computer.

9.7 Leverrier's Algorithm

The inverse of $(z\mathbf{I} - \mathbf{A})$ is defined to be

$$(z\mathbf{I} - \mathbf{A})^{-1} = \frac{\text{adj}(z\mathbf{I} - \mathbf{A})}{\det(\mathbf{I} - \mathbf{A})} \tag{9.37}$$

where the *adjoint*, $\text{adj}(z\mathbf{I} - \mathbf{A})$, is given by

$$\text{adj}(z\mathbf{I} - \mathbf{A}) = \mathbf{H}_1 z^{N-1} + \mathbf{H}_2 z^{N-2} + \ldots + \mathbf{H}_{N-1} z^1 + \mathbf{H}_N \tag{9.38}$$

The matrices, $\mathbf{H}_i$, are defined by the following algorithm.

$$H_1 = I$$

$$d_1 = -\text{trace}(A)$$

for $i = 2:N$

$$H_i = AH_{i-1} + d_{i-1}I$$

$$d_i = -\frac{1}{i}\text{trace}(AH_i)$$

end   (9.39)

The trace of a matrix [denoted trace(A)] is defined to be the sum of the on-diagonal elements of A. Leverrier's algorithm can be seen to be well suited for computation using a general-purpose digital computer.

Example 9.9: Leverrier's Algorithm

Problem Statement. Compute adj(zI − A), where A is given by

$$A = \begin{bmatrix} 1.75 & -0.875 & 0.125 \\ 1 & 0 & 0 \\ 0 & 1 & 0 \end{bmatrix}$$

Analysis and Conclusions. Using the procedure shown in Eq. (9.39), it follows that

$$H_1 = \begin{bmatrix} 1 & 0 & 0 \\ 0 & 1 & 0 \\ 0 & 0 & 1 \end{bmatrix}; H_2 = \begin{bmatrix} 0 & -0.875 & 0.125 \\ 1 & -1.750 & 0 \\ 0 & 1 & -1.750 \end{bmatrix}; H_3 = \begin{bmatrix} 0 & 0.125 & 0 \\ 0 & 0 & 0.125 \\ 1 & -1.750 & 0.875 \end{bmatrix}$$

Combining these results gives

$$\text{adj}(z\text{I} - A) = H_1 z^2 + H_2 z^1 + H_3$$

and therefore,

$$\text{adj}(z\text{I} - A) = \begin{bmatrix} z^2 - 0.875z + 0.125 & 0.125z \\ z & z^2 - 1.750z & 0.125 \\ 1 & z - 1.750 & z^2 - 1.750z + 0.875 \end{bmatrix}$$

Computer Study. The coefficients of the adjoint matrix can be computed using the s-file "lever.s" and the following sequence of Siglab commands. The results are packed as [H_1, H_2, H_3].

```
>include "chap9\lever.s"
>A = {[14/8,-7/8,1/8],[1,0,0],[0,1,0]}
>lev = lever(A)
>lev
```

1	0	0	0	-0.875	0.125	0	0.125	0
0	1	0	1	-1.75	0	0	0	0.125
0	0	1	0	1	-1.75	1	-1.75	0.875

The results are recognized to be the previously computed values of H_1, H_2, and H_3. ❖

9.8 General Solution

Consider again a single-input single-output Nth order state-determined discrete-time system which can be modeled as:

state equation: $\qquad \mathbf{x}[k + 1] = \mathbf{A}\mathbf{x}[k] + \mathbf{b}u[k]$ $\qquad$ (9.40)

initial condition: $\qquad \mathbf{x}[0] = \mathbf{x}_0$ $\qquad$ (9.41)

output equation: $\qquad y[k] = \mathbf{c}^T \mathbf{x}[k] + du[k]$ $\qquad$ (9.42)

At some sample index $k = k_0$, the next state is given by

$$\mathbf{x}[k_0 + 1] = \mathbf{A}\mathbf{x}[k_0] + \mathbf{b}u[k_0] \qquad (9.43)$$

and at sample $k = k_0 + 1$, the next state satisfies

$$\mathbf{x}[k_0 + 2] = \mathbf{A}\mathbf{x}[k_0 + 1] + \mathbf{b}u[k_0 + 1]$$

$$= \mathbf{A}^2\mathbf{x}[k_0] + \mathbf{A}\mathbf{b}u[k_0] + \mathbf{b}u[k_{0+1}] \qquad (9.44)$$

Continuing to sample $k = k_0 + n - 1$, one obtains

$$\mathbf{x}[k_0 + 1] = \mathbf{A}^n\mathbf{x}[k_0] + \mathbf{A}^{n-1}\mathbf{b}u[k_0] + \mathbf{A}^{n-2}\mathbf{b}u[k_0 + 1]$$

$$+ \ldots + \mathbf{A}\mathbf{b}u[k_0 + n - 2] + \mathbf{b}u[k_0 + n - 1] \qquad (9.45)$$

Equation (9.45) consists of two parts, namely the homogeneous and inhomogeneous solutions. They can be partitioned into the sum

$$\mathbf{x}[k] = \mathbf{A}^{k-k_0}\mathbf{x}[k_0] + \sum_{i=k_0}^{k-1} \mathbf{A}^{k-i-1}\mathbf{b}u[i] \qquad (9.46)$$

where the homogeneous solution is a function only of the initial conditions ($\mathbf{x}[k_0]$) and the inhomogeneous solution is defined by a convolution sum. Upon substituting Eq. (9.46) into the output equation [Eq. (9.42)], the generalized output equation results,

$$y[k] = \mathbf{c}^T\left(\mathbf{A}^{k-k_0}\mathbf{x}[k_0] + \mathbf{c}^T\sum_{i=k_0}^{k-1} \mathbf{A}^{k-1-i}\mathbf{b}u[k]\right) + du[k] \qquad (9.47)$$

From the delay theorem found in Table 2.2, it follows that the z-transform of $\mathbf{x}[k+1]$ is given by $Z(\mathbf{x}[k+1]) = z\mathbf{X}(z) - z\mathbf{x}[0]$. Continuing the z-transform of Eqs. (9.40) and (9.42) results in

$$z\mathbf{X}(z) - z\mathbf{x}_0 = \mathbf{A}\mathbf{X}(z) + \mathbf{b}U(z)$$

$$Y(z) = \mathbf{c}^T\mathbf{X}(z) + dU(z) \qquad (9.48)$$

The homogeneous, or at-rest, solution can be determined by setting $U(z) = 0$. Solving for $\mathbf{X}(z)$ and $Y(z)$ in Eq. (9.48), which simplifies [after Eq. (9.36)] to

$$\mathbf{X}(z) = z(z\mathbf{I} - \mathbf{A})^{-1}\mathbf{x}_0 = \Phi(z)\mathbf{x}_0$$

$$Y(z) = \mathbf{c}^T z(z\mathbf{I} - \mathbf{A})^{-1}\mathbf{x}_0 = \mathbf{c}^T\Phi(z)\mathbf{x}_0 \qquad (9.49)$$

This can be seen to be consistent with the results given in Eq. (9.36). The inhomogeneous solution can be computed by setting $\mathbf{x}_0 = \mathbf{0}$ and solving for $Y(z)$ using

$$\mathbf{X}(z) = (z\mathbf{I} - \mathbf{A})^{-1}\mathbf{b}U(z)$$

$$Y(z) = \mathbf{c}^T(z\mathbf{I} - \mathbf{A})^{-1}\mathbf{b}U(z) + dU(z) \qquad (9.50)$$

Since the transfer function is traditionally defined by the inhomogeneous solution, namely $H(z) = Y(z)/U(z)$ where $U(z) = 1$, it follows that

$$H(z) = \mathbf{c}^T(z\mathbf{I} - \mathbf{A})^{-1}\mathbf{b} + d \qquad (9.51)$$

After some manipulation, it can be shown that

$$H(z) = \frac{\det[z\mathbf{I} - (\mathbf{A} - \mathbf{b}\mathbf{c}^T)] - (d-1)\det(z\mathbf{I} - \mathbf{A})}{\det(z\mathbf{I} - \mathbf{A})} \qquad (9.52)$$

where the poles are seen to be given by roots of $\det(z\mathbf{I} - \mathbf{A})$.

Example 9.10: Transfer Function

Problem Statement. Let an at-rest LTI system be given by $x_1[k + 1] = x_2[k]$, $x_2[k + 1] = x_1[k] + x_2[k] + u[k]$, and $y[k] = x_1[k] + x_2[k] + u[k]$. What is the system's transfer function?

Analysis and Conclusions. The direct II state representation is given by

$$\mathbf{A} = \begin{bmatrix} 0 & 1 \\ 1 & 1 \end{bmatrix}; \quad \mathbf{b} = \begin{bmatrix} 0 \\ 1 \end{bmatrix}; \quad \mathbf{c} = \begin{bmatrix} 1 \\ 1 \end{bmatrix}; \quad d = 1$$

The computation of $(z\mathbf{I} - \mathbf{A})^{-1}$ yields

$$(z\mathbf{I} - \mathbf{A})^{-1} = \frac{1}{z^2 - z - 1} \begin{bmatrix} z - 1 & 1 \\ 1 & z \end{bmatrix}$$

Using Eq. (9.51), the transfer function may be determined to be

$$H(z) = \mathbf{c}^T (z\mathbf{I} - \mathbf{A})^{-1} \mathbf{b} + d$$

$$= \frac{1}{z^2 - z - 1} \begin{bmatrix} 1 & 1 \end{bmatrix} \begin{bmatrix} z - 1 & 1 \\ 1 & z \end{bmatrix} \begin{bmatrix} 0 \\ 1 \end{bmatrix} + 1$$

$$= \frac{z + 1}{z^2 - z - 1} + 1 = \frac{z^2}{z^2 - z + 1}$$

Using Eq. (9.52), the transfer function may be determined to be

$$H(z) = \frac{\det[z\mathbf{I} - (\mathbf{A} - \mathbf{bc}^T)] - (d - 1)\det(z\mathbf{I} - \mathbf{A})}{\det(z\mathbf{I} - \mathbf{A})}$$

$$= \frac{\left\{ \det\left(\begin{bmatrix} z & 0 \\ 0 & z \end{bmatrix} - \begin{bmatrix} 0 & 1 \\ 0 & 0 \end{bmatrix} \right) - 0(z^2 - z + 1) \right\}}{z^2 - z + 1}$$

$$= \frac{z^2}{z^2 - z + 1}$$

This result agrees with the previously computed $H(z)$.   ❖

9.9 Stability

In Chaps. 5 and 6, the stability of an LTI system was studied. It was found that the stability test in the z-domain is defined in terms of the system's poles locations relative to the unit-circle. With the advent of the general-purpose digital computer and root finding subroutines, this is a relatively easy test to perform. Stability, however, can also be defined within a state-variable framework. Consider again the state equation

$$\mathbf{x}[k+1] = \mathbf{A}\mathbf{x}[k] + \mathbf{b}u[k] \tag{9.53}$$

which is often denoted by the two-tuple $\mathcal{X} = (\mathbf{A},\mathbf{b})$. It has been previously established that the system $\mathcal{X} = (\mathbf{A},\mathbf{b})$ is BIBO stable if the eigenvalues of $\mathbf{A}$ are all interior to the unit circle. That is, if for each λ_i an eigenvalue of $\mathbf{A}$, $|\lambda_i| < 1$, then $\mathcal{X}$ is BIBO stable. In such cases, the matrix $\mathbf{A}$ is said to be a *stability matrix*. To determine the stability of a state-determined system the eigenvalues of $\mathbf{A}$ need to be computed. This is again a relatively straightforward process for a general-purpose digital computer. Furthermore, since the poles and eigenvalues are identical, the eigenvalue test is equivalent to the previously discussed unit circle criterion.

The asymptotic stability of an LTI system can be defined in terms of the following Lemma.

Lemma: Stability Matrix

If $\mathbf{A}$ is an $N \times N$ stability matrix, then for any $N \times N$ matrix $\mathbf{W}$, the matrix series

$$\sum_{m=0}^{\infty} \mathbf{A}^m \mathbf{W} (\mathbf{A}^T)^m \tag{9.54}$$

converges.

This represents a well known body of knowledge from control theory and will not be developed any further in this text. The consequence of the lemma is, however, a state-determined definition of stability. Suppose that $\mathcal{X} = (\mathbf{A},\mathbf{b})$ describes the state equations of an LTI system and $\mathbf{Q}$ is an arbitrary positive semidefinite matrix, that is, $\mathbf{x}\mathbf{Q}\mathbf{x}^T \geq 0$ for all $\mathbf{x}$. It is known that there exists a unique positive semidefinite matrix $\mathbf{K}$ (Lyapunov matrix) such that

$$\mathbf{K} = \mathbf{A}\mathbf{K}\mathbf{A}^T + \mathbf{b}\mathbf{Q}\mathbf{b}^T \tag{9.55}$$

if and only if $\mathbf{A}$ is a stability matrix. While Eq. (9.55) is rarely used as a stability test for an LTI system, it will become an important result to be used at a later time when the dynamic range limits of a state-determined system are computed.

Example 9.11: Stability

Problem Statement. Suppose that a first-order (i.e., $N = 1$) homogeneous system is given by $\mathcal{X} = (\mathbf{A},\mathbf{b}) = (a,1)$ where a is an arbitrary constant. Is the system stable?

Analysis and Conclusions. Let $\mathbf{Q} = \mathbf{I} = [1]$, which is obviously positive semidefinite. A matrix $\mathbf{K} = [k]$ is positive semidefinite if $k \geq 0$. Then, from Eq. (9.55),

$$\mathbf{K} = \mathbf{A}\mathbf{K}\mathbf{A}^T + \mathbf{b}\mathbf{Q}\mathbf{b}^T$$

or, in the case of the example problem

$$k = a^2 k + 1$$

Solving for k produces

$$k = \frac{1}{1 - a^2}$$

from which it may be concluded that $\mathbf{K}$ is positive semidefinite if and only if $|a| < 1$, which is the anticipated result.   ❖

9.10 Summary

In Chap. 9, the concept of state variables was introduced. It was shown that state variables represent how information is stored in the shift registers of an Nth-order canonic LTI. The state variable model can be used to capture the structural details of an architecture as well as providing a mechanism for defining the system's transfer function. A system's homogeneous and inhomogeneous solutions were shown to be defined in terms of the state transition matrix, which is definable in terms of a state variable system representation. Because of its versatility and capabilities, state variable modeling is the preferred architectural modeling methodology.

9.11 Self-Study Problems

1. Direct II   Represent H(z) in state-variable form if a direct II architecture is assumed.

$$H(z) = 0.1 \frac{(z - 1)^4}{(z^2 + 0.9)(z^2 + 0.81)}$$

2. State-Transition Matrix   Compute the state-transition matrix for the system analyzed in Question 1, using the Cayley-Hamilton method.

3. State-Transition Matrix   Compute the state-transition matrix for the system analyzed in Question 1 using the z-transform method.

4. Transfer Function   What is the transfer function derived from the state model of the system analyzed in Question 1?

5. Homogeneous Solution   What is the homogeneous system solution, if all initial conditions are unity, for the system analyzed in Question 1?

6. Inhomogeneous Solution   If the system analyzed in Question 1 is assumed to be at rest, what is the system's step response?

10

Introduction to Digital Signal Processing

10.1 Introduction

Digital signal processing (DSP) is a relatively young branch of engineering and one of the more recent derivatives (and beneficiaries) of the semiconductor revolution which has dominated the second half of the 20th century. Beginning in the 1970s, DSP made its initial commercial appearance and, ever since, has been replacing traditional analog signal processing systems at an accelerating pace. Besides serving as a replacement technology, DSP has enabled signal processing solutions that could not have been achieved with other technologies. DSP has, in fact, entered the layperson's vocabulary as evidenced by the acronym's now routine appearance on consumer electronic communication and entertainment equipment. DSP is now considered to be a discipline unto itself, complete with its own mathematics, analysis and synthesis methodology, terminology, and technology. Much of this theory is based on the material found in Chaps. 1 through 9, which is technically applicable to the study of discrete-time signals and systems. This, in some respects, is unfortunate, because discrete-time analysis and synthesis models and techniques can mask many of the unique and challenging problems facing a DSP engineer. Nevertheless, the future of DSP seems to be limited only by human creativity and our ability to successfully apply DSP technology to the many meaningful problems of the 21st century. Some contemporary applications of DSP are shown in Table 10.1.

The driving force behind the DSP explosion, which truly began in the 1980s, has been technology and not algorithms. The breakthrough technology was the *DSP microprocessor*, which was first introduced in

Table 10.1   Contemporary Applications of DSP

General-purpose	Graphics
• Filtering (Convolution)	• Rotation
• Detection (Correlation)	• Image transmission and compression
• Spectral analysis (Fourier transforms)	• Image recognition
• Adaptive filtering	• Image enhancement
Instrumentation	**Control**
• Waveform generation	• Servo control
• Transient analysis	• Disk control
• Steady-state analysis	• Printer control
• Scientific instrumentation	• Engine control
Information systems	• Guidance and navigation
• Speech processing	• Vibration (modal) control
• Audio processing	• Power system monitors
• Voice mail	• Robots
• Facsimile (FAX)	**Others**
• Modems	• Radar and sonar
• Cellular telephones	• Radio and television
• Modulators/demodulators	• Music and sound reproduction
• Line equalizers	• Entertainment
• Data encryption	• Biomedical signal processing
• Spread-spectrum	• Prosthetics
• Digital and LAN communications	• Wireless LANs

the late 1970s. The first-generation DSP microprocessors contained a fixed-point MAC which was far more capable than those possessed by the prevailing microprocessors at the time, plus an on-chip ADC/DAC. The problems with the first generation devices was a lack of memory space and on-chip digital noise that interfered with the ADC/DAC operation. The second fixed-point generation of DSP microprocessor eliminated the on-chip ADC/DAC and added more memory. The third-generation devices are floating-point devices, and fourth-generation chips are multiprocessor systems. The most widely used generation of DSP microprocessor remains the second-generation chip due to its high speed and low cost. Added to this mix is a host of *application specific integrated circuit* (*ASIC*) devices which have been developed for DSP purposes. Some of them use the cores of existing DSP microprocessors, whereas others break new ground with highly optimized special-purpose circuits. Regardless of the form or function, DSP processors of all sorts will be needed to meet the challenge of 21st century, whether the application be pedestrian or exotic.

A typical, end-to-end digital signal processing stream is shown in Fig. 10.1 under the assumption that the input signal is analog. There is an explicit requirement that the signal domain be changed from analog to digital. This is the role of an *analog-to-digital converter* (ADC). Observe that an analog *anti-aliasing filter* has been placed in front of the ADC. This filter ensures that the highest frequency presented to the ADC is bounded below the Nyquist frequency. Following the ADC is the DSP system, which performs filtering and/or transform operations. The output of the system may or may not be converted back into an analog signal depending on the application. A domain shift from digital to analog, if required, is accomplished using a *digital-to-analog converter* (DAC).

10.2 Data Acquisition Systems

Many digital filters accept data that was originally in analog form. The original signal domain may have been acoustic, optical, mechanical, or electromagnetic but has been electronically converted into a signal that can be presented to an ADC device. An n-bit ADC, as shown in Fig. 10.1, consists of an analog *sample and hold* (S/H) front-end followed by a comparison device, called a *quantizer*. The function of the sample and hold circuit is to convert a continuous-time signal into a discrete-time signal. The purpose of the quantizer is to map the discrete-time sample values into one of 2^n possible discrete levels. ADCs, in general, are quantified in terms of:

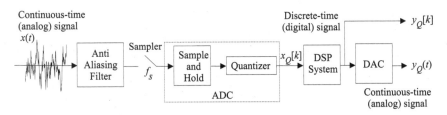

Figure 10.1 Typical DSP signal train.

- *Accuracy.* Absolute accuracy of a conversion relative to a standard voltage or current reference.
- *Aperture uncertainty.* An uncertainty about the sample rate.
- *Bandwidth.* Half-power (3 dB) cutoff frequency of the ADC's analog subsystem.
- *Codes.* Two's complement, one's complement, sign-magnitude, binary coded decimal, and so on.
- *Common-mode rejection.* Measure of the ability to suppress unwanted common-mode input noise.
- *Conversion time.* The maximum conversion rate. (If a separate S/H amplifier is used, the conversion rate is the sum of the ADC and S/H delays.)
- *Full scale.* Dynamic range in volts or amps.
- *Glitch.* Unexpected noise spikes in the output.
- *Linearity.* Measure of linearity across the dynamic range.
- *Offset drift.* Worst-case variation due to parametric changes.
- *Precision.* Repeatability of a measurement.
- *Quantization error.* ±1/2 LSB (least significant bit).
- *Resolution.* n bits for an n-bit converter.

Typical values of some of these parameters are:

- Linearity: ±1/10 LSB
- Thermal drift: ±1/10 LSB
- Resolution: 6–20 bits

Conversion rates range from gigahertz to a few kilohertz. Resolution and speed are typically the biggest factors in determining the cost of the device. Variations on the basic ADC theme are mu-law (μ-law) and A-law companding ADCs (see Chap. 1.6), which provide logarithmic dynamic range compression, and multiplying ADCs or MDACs, which include a general-purpose multiplier for scaling data.

10.3 Fixed-Point Designs

A signal processing stream often begins with an analog-to-digital converter (ADC). From that point on, the signal processing generally follows a floating-point or fixed-point path. *Fixed-point* designs have a natural attraction due to their high speed and low cost, but they have limited precision and dynamic range.

A real number $x \in [-1,1)$, can be represented as

$$x = -b_0 + \sum_{i=1}^{\infty} b_i 2^{-i} \qquad (10.1)$$

where b_i is the ith bit, and b_0 is called the *sign bit*. If the real number $x \in [-V,V)$, then x can be expressed as

$$x = \left(-b_0 + \sum_{i=1}^{\infty} b_i 2^{-i} \right) V \qquad (10.2)$$

A real x can be approximated by $\langle x \rangle_Q$, called the *quantized* value of x, which, for an M-bit representation is given by

$$\langle x \rangle_Q = \left(-b_0 + \sum_{i=1}^{M-1} b_i 2^{-i} \right) V \qquad (10.3)$$

Equation (10.3) is a model that reasonably articulates the precision and range of a *sign-magnitude, one's complement*, and *two's complement* coded number. The binary encoding described in Eq. (10.3) is two's complement. An M-bit sign-magnitude representation of a number reserves one bit for the sign and $M - 1$ bits for the unsigned magnitude (absolute value) of the number. In one's complement, negation of a binary number is accomplished by logical negation of each of its bits. Two's complement negation is the same, except the result is incremented by one. Of these common fixed-point formats, two's complement is by far the most popular fixed-point encoding scheme found in current practice. Sign-magnitude and one's complement have some undesirable properties. Among the most significant is that both sign-magnitude and one's complement have redundant representations for zero.

Two's complement arithmetic is particularly well suited to DSP where sum-of-product calculations are commonplace. An M-bit, two's complement system possesses the "wrap around modulo 2^M" property. Specifically, if S_i denotes the partial sum

$$S_i = \sum_{m=1}^{i} x_m \qquad (10.4)$$

and the actual value of S_L is a valid M-bit, two's complement number (i.e., $-2^{(M-1)} \leq S_L < 2^{(M-1)}$), then S_L will be computed correctly even though the partial sum S_i may overflow an M-bit register during run-time, for $i \in \{2,3,4,\ldots,L-1\}$. That is to say that if the final result is guaranteed to be a valid M-bit two's-complement word, then any overflows that may occur in producing that sum can be ignored. Therefore, a two's complement arithmetic unit need not be polled after each add cycle to determine if a register overflow has taken place. Eliminating such operations, and the attendant overhead, will generally increase the system's computational bandwidth. This is certainly not the case for sign-magnitude or one's complement systems where a register overflow is fatal.

Example 10.1: Two's Complement

Problem Statement. Assume $M = 4$, $V = 1$, and a two's complement code having a single sign-bit and three fractional bits. Compute the sum $S = 7/8 + 7/8 - 1 - 1$, in that order.

Analysis and Results. The two's complement representation of $7/8 \leftrightarrow [0_\Delta 111]$ and $-1 \leftrightarrow [1_\Delta 000]$ where "$_\Delta$" denotes the binary point location. Then $S = 7/8 + 7/8 - 1 - 1 = 7/8 + 7/8 - 8/8 - 8/8 = -2/8$, a valid two's complement four-bit word. A four-bit, two's complement binary adder would produce the following result:

bit location		3	Δ	2	1	0	
		0	Δ	1	1	1	7/8
	+	0	Δ	1	1	1	+7/8
		1	Δ	1	1	0	$= -2/8$ {overflow} $= S_2$
	+	1	Δ	0	0	0	-1
carry out→	1	0	Δ	1	1	0	$= 5/8$ {overflow} $= S_3$
	+	1	Δ	0	0	0	-1
result		1	Δ	1	1	0	$= -2/8 = S_4$

The resulting four-bit word, namely $S_4 \leftrightarrow [1_\Delta 110] \leftrightarrow -2/8$, is the correct result in spite of the fact that two register overflows are encountered during accumulation. This would not be the case if sing-magnitude or one's complement was used. ❖

The M-bit fixed-point representation of x given by Eq. (10.3) covers the dynamic range $\langle x \rangle_Q \in [-V,V)$. The value of each bit level is given by the quantization step-size, denoted Q. The quantization step-size carries units of volts/bit and is given by:

$$Q = \frac{2V}{2^M} = V2^{-(M-1)} \tag{10.5}$$

That is, the analog dynamic range of $2V$ volts is subdivided into 2^M equal intervals. The quantization error is formally defined to be:

$$e_Q = x - \langle x \rangle_Q \tag{10.6}$$

For a *rounding* policy as shown in Fig. 10.2, the maximal quantization error has a value of $e_{Q(\text{max})} = Q/2$. A quantization profile for a truncation is also shown in Fig. 10.2, where the maximal quantization error now has a value of $e_{Q(\text{max})} = Q$. The influence of these errors on the behavior of a system is called *finite wordlength effects*.

It is normally assumed that the value of a real sample $x[k]$ before quantization is completely arbitrary. Therefore, the rounded quantization error $e_Q[k]$ is assumed to be a uniform random process over the quantization interval, so that $e_Q[k] \in [-Q/2, Q/2)$ as shown in Fig. 10.3. For the truncation case, the quantization error $e_Q[k]$ is again assumed to be a uniform random process, except over the interval $e_Q[k] \in [0, Q)$. In the rounded case, the error mean is zero and for truncation, the mean error is $Q/2$. In either case the error variance is given by

$$\sigma^2 = \frac{Q^2}{12} \tag{10.7}$$

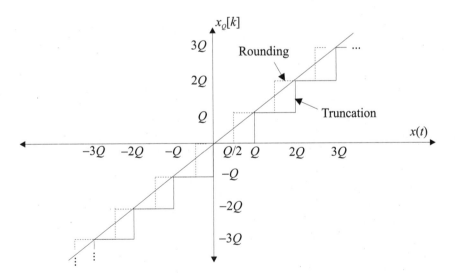

Figure 10.2 Quantization effects showing truncation as a solid line, rounding as a dashed line.

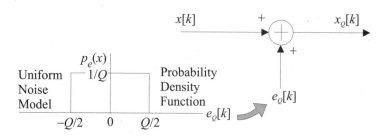

Figure 10.3 Uniform quantization error model.

The quantization noise model provides a mechanism by which finite wordlength effects can be introduced into a mathematical system model as shown in Fig. 10.3. Here, it is assumed that the digital system receives a quantized input signal $x[k]$, which has been quantized into $x_Q[k]$ using the rounding model shown in Fig. 10.2. The quantized signal is mathematically equal to the real (ideal) $x[k]$ plus a random component $e_Q[k]$ from a uniformly distributed population (i.e., $x_Q[k] = x[k] + e_Q[k]$). The expected value of $x_Q[k] = x[k] + e_Q[k]$ is obviously $E(x_Q[k]) = x[k]$ and the error variance is given by $\sigma^2 = Q^2/12$. If the truncation model of Fig. 10.2 had been used, then the expected value of the error would not be zero, but rather $Q/2$, and therefore, $E(x_Q[k]) = x[k] - Q/2$.

The uniform noise model is also useful in that it can be used to represent arithmetic roundoff errors. Suppose that two M-bit words are multiplied, resulting in a $2M$-bit, full precision product. Assume that, after multiplication, the full precision word is rounded to its M most significant bits, leaving the value, or weight, of the least significant bit in the final result to be Q (volts/bit). Here, the roundoff error model can be used to quantify the error associated with reducing the precision of an $2M$-bit double word (full precision product) to that of an M-bit single word. The error can be assumed to be uniformly distributed with zero mean and variance $\sigma^2 = Q^2/12$, if M is sufficiently large.

Example 10.2: Quantization

Problem Statement. A ±15 V input (double-ended), 12-bit ADC is presented with an analog input. The output is formatted as a two's complement word having a sign bit and 11 fractional bits. What is Q and σ measured in volts/bit? Experimentally verify the error model.

Analysis and Results. From the given data, $Q = 2(15)/2^{12} = 15/2^{-11} = 0.007324$ V/b (weight of the LSB). Then $\sigma^2 = Q^2/12 = 4.47035 \times 10^{-6}$, or, as an RMS value, $\sigma = 2.1143 \times 10^{-3}$ V/b. That is, the "statistical precision" of the least significant bit of the word provided by the ADC is approximately 2.1143×10^{-3} V/b.

Computer Study. The statistical precision calculation shown below is normalized based on the assumption that $V = 1$ and displayed in logarithmic form. The following Siglab command generates the graph shown in Fig. 10.4.

```
>include "chap10\ex10-2.s"
```

The graph of the error variance (for $V = 1$), plotted logarithmically as $\log_2(\sigma^2)$ (i.e., variance in bits), as a function of the number of fractional bit is shown in Fig. 10.4 for $V = 1$. It can be seen that the experimental and theoretical error curves are in reasonably close agreement.   ❖

Example 10.3: Audio Quantization Effects

Problem Statement. Experimentally observe the effect of quantization using a pure sinusoid and a soundboard. Create a pure sinusoid and quantize it to eight bits with seven, four, two, and one bit of fractional precision. Evaluate the degree of audio and harmonic distortion.

Analysis and Results. A 275.6 Hz signal was created for playback through an eight-bit soundboard sampling at $f = 11{,}025$ Hz. The signal was quantized to seven, four, two, and one bits of fractional precision. The magnitude FFTs of the signals are shown in Fig. 10.5. It can be observed that the signals having seven and four fractional bits carry a minimal amount of harmonic distortion. The harmonic distortion, however, rapidly increases as the number of fractional bits decreases. This can be substantiated by the audio comparison of the four signals.

Computer Study. The data shown in Fig. 10.5 is generated using the sequence of Siglab commands shown below. The audio record consists of a concatenation of the pure sinusoidal signal quantized to seven, four, two, and one bits of fractional precision. The audio difference between the first two records is slight but becomes pronounced as the quantization error increases.

```
>x=mkcos(1,0.025,0,2^13)        # Create original (8-bit) signal.
>xf=mag(pfft(x))                # Spectrum of original signal.
```

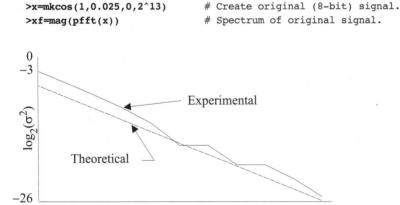

Figure 10.4 Experimental error variance determination.

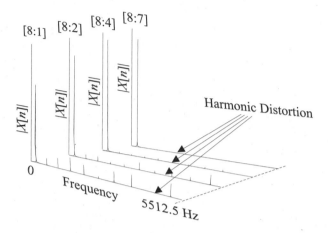

Figure 10.5 Harmonic distortion introduced by quantizing a pure sinusoid to different levels of fractional precision.

```
>wordlen(8,4,1)                    # Set quantization to 4 bits.
>x4=fxpt(x); x4f=mag(pfft(x4))     # Quantize and compute spectrum.
>wordlen(8,2,1)                    # Set quantization to 2 bits.
>x2=fxpt(x); x2f=mag(pfft(x2))     # Quantize and compute spectrum.
>wordlen(8,1,1)                    # Set quantization to 1 bit.
>x1=fxpt(x); x1f=mag(pfft(x1))     # Quantize and compute spectrum.
>wavout(0.7*2^7*(x&x4&x2&x1))      # Play using an 8-bit soundboard.
>yf={xf,x4f,x2f,x1f}               # Plot the results.
>graph3d(yf)                                                        ❖
```

10.4 Other MAC Structures

Floating-point operations and numbers are important in DSP applications requiring high precision and dynamic range. It should be noted that fixed-point will provide a lower cost and higher bandwidth solution to a DSP problem than can be expected form a floating-point solution. Floating-point should be considered, however, when fixed-point *overwhelmingly* fails to provide the needed precision or dynamic range. The DSP application that occasionally needs greater precision or dynamic range than that provided by the native word length of a fixed-point processor is not automatically a candidate for a floating-point processor. It is usually easy to perform extended word length operations on a fixed-point processor. Some DSP microprocessor manufacturers even provide sample assembly language routines to emulate floating-point arithmetic. Examples of applications that overwhelmingly exceed the capabilities of fixed-point DSP microprocessors are long FFTs or highly frequency selective digital filters.

The crossover line between fixed-point implementation and floating-point implementation is currently in flux. Using modern ASIC technology to implement hard-wired DSP solutions, fixed-point arithmetic in 64- to 128-bit word lengths can be a viable alternative to current 32-bit floating-point solutions.

Floating-point arithmetic is an extremely complicated subject. A full exposition on floating-point arithmetic is beyond the scope of this work. Briefly, the problems that arise with floating-point arithmetic are related to the finite, nonuniform sampling of the real number line. Fortunately, most basic DSP operations do not often exercise floating-point pathologies. Nonetheless, any DSP engineer who is depending on floating-point arithmetic should be aware of its potential pitfalls.

A floating-point representation of a real number x is of the form

$$x \approx \pm m_x r^{e_x} \tag{10.8}$$

where m_x is the *mantissa*, r is the numbering system's *radix* (usually 2, but sometimes 16), and e_x is the *exponent*. The mantissa, exponent, and a sign bit are packed into a single or multiple machine words. The M-bit mantissa is frequently represented as a normalized fractional number. For $r = 2$, for example, it is assumed to be bounded according to the inequality $1/2 \leq m_x \leq 1 - 2^{-M}$. Restricting the mantissa to this range will ensure that the mantissa's most significant bit (MSB) will (almost) always have a value of one. Since this bit will always be present, it can be explicitly dropped from the mantissa's binary representation and considered to be present as an "implied" or "hidden" bit. Using a hidden bit increases the precision of the mantissa by one additional bit but does require that the mantissa be renormalized after each floating-point operation. The exponent is generally represented using either two's complement or offset-binary form. The quantized error model for the mantissa is that used for fixed-point quantization. However, the value of the LSB (i.e., Q) is not fixed in a floating-point system but rather scaled by r^{e_x}. It is important to note that there are many floating-point representations in common use, particularly in DSP microprocessors and systems. In comparison, most modern microprocessors use the IEEE 754 standard floating-point representation.

A variation of the floating-point format is called *block floating point*. It sometimes leads to confusion since it suggests the precision of floating-point but is, in fact, a scaled fixed-point representation. In a block floating-point system every number is represented as $x[j] = \pm m_j 2^v$, for a fixed v. Block floating-point is applied to an array of data samples. Assume that v is the minimum value such that the largest element in a K element array is bounded in absolute value according to the inequality $|x[k]| \leq |x_{max}| \leq 2^v$, for all $k \in \{1,2,3,...,K\}$. All numbers would be assigned the same exponent scale factor, namely 2^v, but would have a scaled and rounded mantissa to form $x[j] = \pm m_j 2^v \leq x_{\max}$.

The choice of data format can influence digital arithmetic efficiency. Consider multiplication, which historically is performed using a set of shift and conditional add operations. Specifically, if X and Y are unsigned positive integers having an N-bit representation, then for each multiplication cycle the accumulator operation would be defined to be

$$ACC[j + 1] \leftarrow ACC[j] + XY_j 2^j \qquad (10.9)$$

where $ACC[j]$ is the accumulator contents at iteration j and Y_j is the jth bit of Y. If $Y_j = 0$, then the jth iteration can be bypassed (i.e., the addition may be skipped because $XY_j 2^j = 0$). It is therefore desirable to have Y maximally dense in zeros. In a two's complement or sign-magnitude code, the probability that an individual bit is one or zero is $P_0 = P_1 = 1/2$, assuming a uniform distribution of operands. Therefore, it can be expected that only half the additions can be eliminated. More significantly, in two's complement or sign-magnitude coding, a worst case exists, namely the operand where all bits are one, requiring N additions assuming N-bit operands. The *canonic signed digit* (CSD) code forces a word to be maximally dense in zeros. In the worst case, a CSD representation of a number will have only half of its digits zero. Therefore, the maximum number of additions that must be performed is cut in half compared to the maximum number of additions required when using binary representation. The CSD was developed when computers used vacuum tubes. With this technology, it is possible to produce an efficient adder implementation using ternary arithmetic (i.e., arithmetic with digits $c_i \in \{-1, 0, 1\}$). The CSD representation of a binary number relies upon the fact that a string of contiguous ones in a binary number can be replaced by a single positive one and a single negative one. For example, the CSD representation of $15 = 16 - 1 \leftrightarrow \{1, 0, 0, -1\}$ is a dense-in-zeros array compared to the two's complement representation, $15 = 8 + 4 + 2 + 1 \leftrightarrow \{1, 1, 1, 1\}$. The CSD representation an integer x, having a known M-bit two's complement representation $\{b_{M-1}, b_{M-2}, \ldots, b_0\}$, where

$$x = -b_{M-1} 2^{M-1} + \sum_{i=0}^{M-2} b_i 2^i \qquad (10.10)$$

is $\{c_{M-1}, c_{M-2}, \ldots, c_0\}$, where

$$x = \sum_{i=0}^{M-1} c_i 2^i. \qquad (10.11)$$

with $b_i \in \{0,1\}$ and $c_i \in \{-1,0,1\}$. The CSD digits are chosen so that $c_i c_{i-1}$ = 0 (i.e., at least one digit in a string of two adjacent digits is zero). An encoding table is shown in Table 10.2. For an M-bit integer, the digits b_i and c_i are as given in Eqs. (10.10) and (10.11). To use the conversion table, it is necessary to define $b_M = b_{M-1}$, and $b_{-1} = b_{-2} = 0$. From a logic perspective, it can be seen that the CSD digit c_i is positive one if $\overline{b}_{i+1} \cdot b_i \cdot \overline{b}_{i-1} + \overline{b}_i \cdot b_{i-1} \cdot \overline{b}_{i-2}$, negative one if $b_{i+1} \cdot b_i \cdot \overline{b}_{i-1}$, and zero otherwise.

Table 10.2 Two's Complement to Canonic Signed Digit Conversion Table (x = do not care)

b_{i+1}	b_i	b_{i-1}	b_{i-2}	c_i
×	0	0	×	0
×	0	1	0	0
×	0	1	1	1
0	1	0	×	1
×	1	1	×	0
1	1	0	×	−1

The rationale for the CSD is the creation of operands containing a large number of zeros. In comparison to sign-magnitude or two's complement, which had a probability of a binary digit being zero of $P_0 = 1/2$, the probability that a CSD digit being zero is approximately $P_0 = 2/3$ (assuming a uniform distribution over the valid data range). Each CSD digit requires two bits to store. Therefore the storage of an N-bit two's complement word in the CSD would require $2N$ bits. In those cases where multiplication is to be avoided wherever possible, system constants and coefficients can be chosen to have a CSD representation containing only a few non-zero bits. For example, multiplication of an arbitrary operand X by $Y = 1248 = 1024 + 256 + 32 = 2^{10} + 2^8 - 2^5$ would require shifts of ten, eight, and five bits plus two add/subtract operations (i.e., $X \cdot Y = X \cdot 2^{10} + X \cdot 2^8 - X \cdot 2^5$).

Example 10.4: Canonic Signed Digit Representation

Problem Statement. Represent $x = 3$ and -3 as three-bit CSD numbers.

Analysis and Conclusions. Observe that in two's complement, $3 \leftrightarrow \{0,1,1\}$ and $-3 \leftrightarrow \{1,0,1\}$. Their four-bit sign extended values become $3 \leftrightarrow \{0,0,1,1\}$ and $-3 \leftrightarrow \{1,1,0,1\}$ (remember $b_{-1} = b_{-2} = 0$). Then, from Table 10.2, it follows that for $x = 3$, the CSD representation is $x = 3 = 4 - 1 \leftrightarrow \{1,0,-1\}$. For $x = -3$, it follows that $x = -3 = -4 + 1 \leftrightarrow \{-1,0,1\}$. ❖

Many other data formats exist and can be found in practice. One of the most mathematically elegant is called the *residue number system* or RNS. The RNS generally provides the most compact, highest bandwidth solution when general MAC operations are required.

10.5 General-Purpose Systems

To motivate the differences between general-purpose processors and digital signal processors, the attributes of general-purpose processors need to be fully appreciated. Features found on most general-purpose processors (although not all) include:

- Multiple data types supported by the processor hardware
- Multilevel cache memories
- Paged virtual memory management in hardware
- Support for hardware context management including supervisor and user modes
- Unpredictable instruction execution timing
- Large general-purpose register files
- Orthogonal instruction sets
- Simple or complex memory addressing, depending on whether the processor is RISC or CISC

The most important data types for general-purpose processors are the character type, followed by the integer type. From the viewpoint of market share, the majority of general-purpose processors (at least those used in "computers") will be employed in business applications that involve text and database processing. Floating-point arithmetic is generally not crucial in most applications run on general-purpose computers, although there are niche markets where this is not true (e.g., technical and scientific workstations).

Cache memories have been demonstrated to be a useful enhancement for many general-purpose processors. The inclusion of sometimes substantial cache memories in general-purpose computers is made on the assumption that programs demonstrate instruction or data locality. This assumption will hereafter be referred to as the *cache assumption*. The cache assumption is frequently used to justify the design of shared memory multiprocessing general-purpose computers where the main memory is connected to the processors via a shared bus. If the cache assumption is violated, the performance of single and multiprocessing general-purpose computers is generally degraded. Classic vector supercomputers were assumed, by their designers, to violate the cache assumption for data access and therefore eschew data caches.

Large register files are included in many general-purpose architectures, although there are exceptions (e.g., the Intel x86). Since most general-purpose machines operate on scalar data, and if the cache assumption holds, large register files are generally beneficial. General-purpose registers and orthogonal instruction sets tend to make it easier to write compilers that emit efficient object code, and they are also beneficial to the assembly language programmer. Also, the load-store architectural constraint used in many RISC processors makes larger register files attractive. Since external memory can be accessed only by load and store operations, it is desirable to keep more operands on hand in the register file to enhance performance.

Hardware support for the management of virtual memory and multiple process contexts is desirable in general-purpose computers. Most general-purpose processors support time-shared execution of multiple processes; even single-user desktop computers are generally running many processes in the background. Virtual memory allows programs to run in a degraded manner if their primary memory requirements exceed available resources. The penalty for virtual memory is increased data access latency due to address translation penalties and long page fault latencies. The latter is generally managed by switching the processor context to another process so that the processor does not idle while a page fault is being serviced. Support for multiple process contexts by a general-purpose computer is therefore crucial for optimal utilization of the processor resource among multiple tasks.

Instruction execution timing on general-purpose processors is generally unpredictable. This is a result of a myriad of features designed to enhance the performance of the processor. Cache memory and virtual memory introduce a substantial amount of uncertainty in instruction execution timing. The amount of time required to read or write a particular location in memory will depend on whether a cache hit occurs, at which level of the cache it hits, and whether that virtual address resides in the translation look-aside buffer (TLB). Main memory latency can be affected when a cache fault occurs and the cache needs to be refreshed. Also, variable latencies occur when other processors, such as direct memory access (DMA), cause access contention problems. Various architectural enhancements such as superscalar execution, speculative execution, out-of-order execution, and branch target caches may further confound any attempt to measure the execution time of an instruction.

Another class of general-purpose processor is the microcontroller. Most microcontrollers are derived from successful general-purpose microprocessor designs, although some are original designs. Microcontrollers are targeted at embedded applications that usually do not require significant arithmetic performance. Microcontrollers typically eliminate features such as large cache memories and virtual memory, and, instead, add integrated peripheral interfaces to support the intended embedded applications.

10.6 DSP Systems

Since the 1970s, the semiconductor industry has experienced geometric growth in the number of transistors that can be placed on a chip. One of the principal beneficiaries of the modern high-density, deep submicron capability is DSP. For example, the Texas Instruments TMS320 family of DSPs has logically migrated its fixed-point DSP microprocessor family from the C10 to the larger C20 and then to the more capable C50 in lock-step with CMOS technological advances. Recent semiconductor process improvements have made possible the C80 multiprocessor and the C6x *very long instruction word* (*VLIW*) multiprocessor. For ASIC applications, there are DSP core implementations of popular DSP microprocessors or others which have been developed especially for use in ASIC technologies. All of these solutions incorporate dedicated MAC hardware to form sums of products over the real or complex field (e.g., convolution, correlation, transforms).

10.7 DSP Processor Requirements

DSP processors are typified by the following characteristics (not all of which are always true):

- Only one or two data types supported by the processor hardware
- No data cache memory
- No memory management hardware
- No support for hardware context management
- Exposed pipelines
- Predictable instruction execution timing
- Limited register files with special purpose registers
- Non-orthogonal instruction sets
- Enhanced memory addressing modes
- On-board fast RAM and/or ROM, and possibly DMA

Digital signal processors are designed around a different set of assumptions from those that drive the design of general-purpose processors. First, digital signal processors generally operate on arrays of data rather than scalars; therefore the scalar load-store architectures found in general-purpose RISCs are not essential. The economics of software development for digital signal processors is different from that for general-purpose applications. Digital signal processing problems tend to be algorithmically smaller than, for example, a word processor. In many cases, the ability to use a slower and less expensive digital signal processor by expending some additional software engineering effort is economically attractive: a good return-on-investment may be achieved if five dollars per unit of manufacturing cost can be saved in a product that will ship a million units by expending an extra person-year of development effort. A consequence of these factors is that most serious programming of digital signal processors is done in assembly language rather than high-level languages. In fact, digital signal processors have been architected to allow optimal assembly code to be written quickly to the point that compilers for *standard* high-level languages are unable to produce efficient code. This is essentially the CISC instruction set architecture paradigm.

10.8 DSP Addressing Modes

General-purpose processors have either many addressing modes (CISC processors) or few addressing modes (RISC processors). CISC processors may support addressing modes such as direct, register or memory indirect, indirect indexed, indirect with displacement, indirect indexed with displacement, and the indexed modes may support pre- and post-increment or decrement of the indices. Historically, complex addressing modes have resulted in higher code entropy, which has two consequences. First, the productivity of the assembly language programmer is enhanced and, second, the resulting object code is more compact. A number of factors have contributed to the disappearance of complex addressing modes characteristic of CISC processors. The first is the change in the economics of hardware costs versus software development costs. Thirty years ago, software development was less costly, and hardware was expensive. Thus, hand-coded assembler was commonly used in application programs. Today, hardware is inexpensive relative to software development costs, so most applications are coded exclusively in high-level languages. Another issue is related to the first. It has proven difficult to get compilers to take full advantage of complicated addressing modes and non-orthogonal instruction sets. Another strike against complex addressing modes in general-purpose computers is that the complex addressing modes tend to cause pipeline stalls, due to the complicated data interdependencies produced by the complex addressing modes. Even modern

CISC implementations have been optimized so that better performance results when complex addressing modes are avoided. Eschewing complex addressing modes has led to the adoption of a load-store philosophy that allows functional units to accept issues without stalling due to data dependencies associated with data stored in memory. By moving to a register indirect load-store architecture, all of the more complex addressing operations are performed in software, thus allowing greater flexibility in scheduling instruction issue. A register indirect load-store architecture synthesizes more complicated "addressing modes" with several simple instructions. The compiler is free to statically arrange these instructions with awareness of the impact of adjacent instructions on the scheduling of processor resources. The processor may also elect to rearrange the execution of these simple instructions within the constraints of available resources and data dependencies. In contrast, the classic CISC has the microoperations of each instruction statically scheduled in the microprogram for each instruction.

Digital signal processing applications frequently require nonsequential access to data arrays using modular or bit-reversed addressing. These addressing modes are not easily supported in general-purpose RISC or CISC processors. For maximum performance in digital signal processing applications, it is sensible to add dedicated hardware support for these addressing modes. To summarize, the addressing modes required include:

- Address register indirect
- Address register indirect with unit stride and nonunit stride modular indexing
- Address register indirect with bit-reversed indexing

Most existing DSP architectures are single-issue, so, with the exception of the special modes indicated, the address register file and arithmetic unit would be similar to that found in general-purpose architectures.

10.9 DSP Instruction Set Enhancements

Because execution time in digital signal processing applications is dominated by vector sum-of-product (i.e., MAC) operations, it is sensible to provide instruction set support for executing a loop a fixed number of times. In fact, looping based upon the value of a counter is the most common branching operation in digital signal processors. As a result, many DSP microprocessors have dedicated instructions to implement zero or reduced penalty looping. For example, both the Texas

Instruments TMS320 series processors and Motorola DSP56000 series processors support an instruction that causes the next machine instruction to be repeated a fixed number of times. As a consequence of this, a justification for dedicating substantial resources to a branch-target cache cannot be found. Branch-target caching makes more sense in general-purpose applications, as many of these applications have branching patterns that are difficult to predict at compile time.

Most integer digital signal processors are actually fixed-point arithmetic machines. The fixed-point format is achieved by integrating shifters with the multiplier-accumulator so as to allow pipelined adjustment of operands and results. The multipliers and accumulators included in most fixed-point digital signal processors are oversized to allow transient computations to exceed the normal word length of the processor. For example, the Texas Instruments TMS320C50, a 16-bit fixed-point digital signal processor, has a multiplier that takes two 16-bit operands and produces a 32-bit output as well as a 31-bit accumulator. Likewise, the Motorola DSP56000, a 24-bit fixed-point digital signal processor, has a multiplier that takes 24 bit operands, produces 48 bit outputs, and has a 56-bit accumulator. These processors include architectural support for controlling the rounding and normalization of results.

Since the dominant language for serious programming digital signal processors is assembly, attention has been given to software engineering issues. Multiply-accumulate operations are so common in digital signal processing applications that explicit multiply-accumulate instructions are included in the instruction set of digital signal processors. Exposed pipelines are usually avoided; however, in the quest for higher performance at lower unit cost, some processors (such as the TMS320C50) have switched to exposed pipelines. Exposed pipelines, however, present some challenging programming issues that will probably be exasperated as applications demand deeper pipelines.

10.10 DSP Dataflow Support

Because digital signal processors are designed to support real-time processing of large quantities of sampled data, they generally have support for enhanced dataflow. Modified Harvard architectures are generally applied, particularly with respect to on-chip memories. Some digital signal processors also support modified asymmetric Harvard architectures with respect to external interfaces that support data storage and I/O operations. For example, the TMS320C30 has a 24-bit (16M word) primary addressing space for programs and data, and a second 13-bit (8k word) addressing space for data storage and peripherals.

Some digital signal processors include DMA controllers that are capable of performing memory-memory move operations concurrently with computational tasks. An independent DMA controller would typically be used to load new data into the on-chip memory while some computation is performed. This allows an internal Harvard architecture to be better exploited by keeping the processor busy with computation rather than programmed data I/O. Currently, these DMA resources are managed explicitly by the programmer. To support rapid code development and portability, it is important that the management of the DMA resources be simplified, at least, if not moved completely into the programming tools.

10.11 DSP Instruction-Level Parallelism

As VLSI technology has improved, it has become possible to include additional hardware resources to enhance the performance of general-purpose and application specific processors. To increase throughput in traditional von Neumann machines, additional hardware resources are added to exploit opportunities for instruction-level parallelism. The techniques that have been developed to exploit opportunities for instruction-level parallelism are superpipelining, superscalar architecture, dataflow processors, and very long instruction word architecture. Since software development costs have spiraled upwards, a significant effort has been found in the area of automatic compiler-based optimization of high-level language code. To a certain extent, the ability to automatically optimize code drives general-purpose processor architecture.

The technique of superpipelining has been exploited by processors such as the DEC Alpha and Intel Pentium Pro to achieve high throughput. Superpipelining works by adding pipeline stages so as to achieve a very short machine cycle, thus allowing a high issue rate. While instructions are issued sequentially at a high rate, they take many cycles to complete, so, while one instruction is started, several or many previous instructions may be in various stages of completion. The disadvantage of superpipelining is that it increases latency (the time from when an instruction is issued to when it is completed) and makes pipeline flushes more expensive. From a hardware perspective, the addition of pipeline registers requires significant extra hardware resources. To hide the pipeline from the programmer and/or compiler, the processor must keep track of resources that have been committed to instructions that are in progress in the pipeline. If resource conflicts occur, the pipeline is stalled or bubbles are introduced into the pipeline. Instructions are generally ordered by the compiler or programmer to avoid pipeline stalling whenever possible.

Superscalar processors use multiple functional units to achieve instruction-level parallelism. Examples of modern superscalar processors are the Intel Pentium, which has two integer pipelines and one floating-point pipeline, and the Sun/Texas Instruments Super-SPARC, which also has two integer pipelines and one floating-point pipeline. A high instruction issue rate is achieved by issuing more than one instruction per machine cycle. To do this, the processor must track the resource requirements of each instruction to be sure that it does not conflict with resource requirements of instructions executing on other pipelines. Like superpipelined processors, superscalar processors rely on the programmer or compiler to arrange instructions so as to minimize resource conflicts and thereby maximize the instruction issue rate. Some new processors are capable of changing the order of execution (out-of-order execution) so as to optimize instruction issue; however, this technique is very hardware intensive.

Dataflow processors work by having each instruction indicate which subsequent instructions depend on the results of a particular instruction. With this explicit dependence, information encoded into the instruction stream it is relatively easy to issue instructions so as to achieve an optimal issue rate. Dataflow processors have not found success in the mainstream of general-purpose processors but, rather, in research and application specific processors.

10.12 DSP Multiprocessors

From a commercial point of view, attempts at parallelism for digital signal processing have relied on expensive multiprocessor communications, as in Kung's Warp and iWarp processors and Texas Instruments' TMS320C40 or, alternatively, multiple independently programmed ALUs in Texas Instruments' TMS320C80. The iWarp processor was developed for commercial implementation by Intel but never sold in any volume. The TI TMS320C40 is essentially a TMS320 family floating-point processor with six integrated communications ports, allowing C40 to C40 data I/O. Unfortunately, the C40 has limited appeal due to high cost, largely driven by the die area overhead of the communication ports and 391 pin interstitial ceramic PGA package. Another impairment to widespread use of the C40 was the difficulty in writing code that used the C40's communications features. The newer TMS320C80 combines a RISC floating-point processor with multiple ALUs under independent program control. The ALUs are optimized for pixel processing operations, and the device is optimized for and being marketed toward video processing applications.

One must be aware that most DSP algorithms were developed for uniprocessors and may not translate well to a multiprocessor environment, and that DSPs are designed around a different set of assumptions than those which drive the design of general-purpose processors. As a consequence, most DSP programming is usually done at the assembly language level to optimize run-time performance rather than at the high-level language level.

10.13 Benchmarking

Predicting the performance of a DSP processor in general and applications specific settings is the mission of a benchmark. For a benchmark to be meaningful, it must exercise each element of a processor architecture that affects its performance in the intended application. For example, a DSP processor benchmark that does not test the MAC speed is meaningless for most DSP applications. For most purposes, the best benchmarks are constructed from a suite of representative applications. One selects the elements of a benchmark by choosing applications that exercise differing aspects of an architecture or represent major classes of problems. A typical benchmark suite has been developed by Berkeley Design Technologies and consists of (1) real FIR, (2) complex FIR, (3) real single sample FIR, (4) LMS adaptive FIR, (5) real IIR, (6) vector dot product, (7) vector add, (8) vector maximum, (9) convolution encoder, (10) finite state machine, and (11) radix-two FFT.

For general-purpose computers, there are industry standard benchmarks (such as SPEC) that allow some comparison of different machines. For DSP microprocessors, there is no corresponding industry standard. Most vendors quote peak operation rates and peak arithmetic rates. These numbers may be quite misleading. For example, a vendor might claim that a processor is capable of "400 MOPs" (400 million operations per second); however, that processor might only be capable of 50 million multiply-accumulates per second. The remaining 350 MOPs might be overhead operations such as address arithmetic or DMA. There are innumerable instances of this type of performance inflation, leading to the conclusion that vendor performance claims must be *carefully* scrutinized. Also, peak performance numbers can be very misleading. While in many baseline DSP applications (such as FIR filtering), some DSP microprocessors may be able to come close to sustaining peak operation or arithmetic rates, in many applications, it is not possible to achieve the peak performance rates quoted by the vendor. The bottom line is that there is only one benchmark that really matters: the real application running on the actual hardware that will be used in the application.

10.14 Summary

In Chap. 10, DSP was developed as a discipline and technology. Fixed-point arithmetic remains important to contemporary DSP and DSP applications when cost and/or performance metrics are critical to the overall success of a design. Floating-point arithmetic is generally used when an application has overwhelming dynamic range requirements the preclude the use of fixed-point arithmetic. From an implementation perspective, DSP microprocessors receive the bulk of the attention. However, direct ASIC DSP implementations are growing rapidly in market significance. DSP has also been compared to general-purpose computing, and has been shown to have a number of unique and special attributes.

10.15 Self-Study Problems

1. Number System Representation   Represent $x = -0.1875$ as an eight-bit two's complement and CSD number.

2. Quantization Effects   An eight-bit ADC having a ± 10 V input range is presented an arbitrary signal having a dynamic range likewise bounded by ± 10 V. What is the smallest quantization step size and minimal error variance?

3. Two's Complement   Use a random collection of sixteen signed eight-bit integers given by x = (2^8)*rand(16) to experimentally verify the two's complement property, which states that if the final result is a valid two's complement number, then run-time additive overflows can be ignored. Select a 16-integer string whose sum is representable by a signed 8-bit two's complement number. Perform the same test using a sign-magnitude and one's complement adder.

4. DSP Microprocessors   Select a contemporary DSP microprocessor and analyze in the context of the material presented in Secs. 10.6 through 10.11.

5. DSP Microprocessors   Select two contemporary DSP microprocessors, one fixed-point and one floating-point, that can essentially perform the same number of arithmetic operations (say, multiply-accumulates) per second. How do they compare in price and power dissipation? Consider system integration issues such as the width of the external memory bus and on-chip peripherals. How might these affect system cost?

11

Finite Impulse Response Filters

11.1 Introduction

In Chap. 10, DSP was established as a discipline and technology. The majority of DSP solutions consist of implementing and integrating transforms (e.g., FFT) and/or LTI filters into a DSP solution. Classic LTI digital filters have evolved along two major paths called *finite impulse response* (FIR) *filters* and *infinite impulse response* (IIR) *filters*. FIR filters are purely feed-forward systems that are intrinsically stable, have a simple structure or architecture, and excellent phase management capabilities. Unfortunately, high-order FIR filters are often required to meet aggressive frequency domain specifications. In this chapter, FIR filters and their design processes will be developed.

11.2 Finite Impulse Response Filters

A finite impulse response (FIR) filter, as the name implies, has an impulse response which persists (i.e., is non-zero) for only of a finite number of sample values. The impulse response of an Nth-order FIR filter is given by the sequence

$$h[k] = \{h[0], h[1], ..., h[n-1]\} \tag{11.1}$$

which is graphically interpreted in Fig. 11.1. Observe that if $x[k] = \delta[k]$, the output $y[k]$ is first $h[0]\delta[0] = h[0]$, followed by $h[1]\delta[1-1] = h[1]$, and so forth. The last non-zero output would be $h[N-1]\delta[N-1] = h[N-1]$.

The time-series response of an FIR filter to an arbitrary input $x[k]$ is given formally by the linear convolution sum

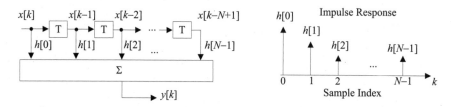

Figure 11.1 Block diagram of (left) an Nth-order FIR filter and (right) impulse response.

$$y[k] = \sum_{m=0}^{N-1} h[m]x[k-m] \tag{11.2}$$

The production of $y[k]$ is also demonstrated in the block diagram found in Fig. 11.1. It can be seen immediately that an FIR filter consists of nothing more than a shift register array of length $N-1$, N multipliers (called *tap weight multipliers*), and an accumulator.

In Chap. 6, discrete-time systems were also expressed in terms of a transfer function. Formally, the z-transform of an FIR filter having the impulse response described by Eq. (11.1) is given by

$$H(z) = \sum_{k=0}^{N-1} h[k]z^{-k} \tag{11.3}$$

Example 11.1: Moving Average and Comb FIR

Problem Statement. The simplest FIR filters to implement are those that are multiplier-free (i.e., contain only shift registers and adders). One such FIR filter is called the *moving average* (MA) filter and is graphically interpreted in Fig. 11.2. Mathematically, a MA filter has a transfer function given by

$$H_{MA}(z) = \frac{1}{N}\sum_{m=0}^{N-1} z^{-m} = \frac{1}{N}\frac{1-z^{-N}}{1-z^{-1}}$$

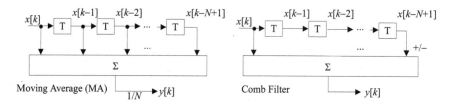

Figure 11.2 Simple multiplier-free FIR filter architectures showing (left) a moving average and (right) comb FIR filter.

Another multiplier-free FIR filter is called a *comb filter*, which is also interpreted in Fig. 11.2. A comb filter has a transfer function given by

$$H_C(z) = 1 \pm z^{-N}$$

What are the frequency responses of the moving average and comb FIR filters?

Analysis and Conclusions. A moving average FIR filter produces an output that represents the average value of the N most recent sample values. The comb filter simply adds or subtracts the value of a delayed sample from that of the current sample. It can be immediately seen that both have a very simple structure and require no multipliers to implement—only shift registers and adders. In addition, the transfer functions of both FIR filters have a similar numerator, namely $N(z) = 1 \pm z^{-N}$, where the N zeros are given by

$$1 + z^{-N} = 0 \rightarrow z_i = e^{j2\pi i/N + j\pi} \text{ for } i \in \{0, 1, 2, ..., N-1\}$$

$$1 - z^{-N} = 0 \rightarrow z_i = e^{j2\pi i/N} \text{ for } i \in \{0, 1, 2, ..., N-1\}$$

It can be seen, however, that the MA filter also contains a pole at $z = 1$ which is not present in the comb filter's transfer function. The pole-zero distributions of the multiplier-less FIR filters are shown in Fig. 11.3 for the case where $N = 7$. Observe that in all cases the zeros are separated by $2\pi/7$ radians and that the MA filter has a pole-zero cancellation at $z = 1$.

The pole-zero annihilation at $z = 1$, in the MA case, gives rise to a finite DC gain (i.e., $H(1) \neq 0$). The comb filter, given by $H(z) = 1 + z^{-7}$, also has a finite DC gain by virtue of the fact that there is no zero present at $z = 1$. However, the filter does have zero gain at the Nyquist frequency due to a zero being located at $z = -1$. The comb filter given by $H(z) = 1 - z^{-7}$ has a zero at $z = 1$ but none at $z = -1$. Therefore, this filter would have zero DC gain and finite gain at the Nyquist frequency.

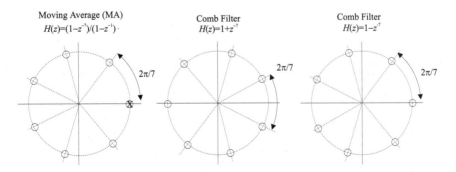

Figure 11.3 Pole-zero distributions of $N = 7$ moving average and comb FIR filters.

Computing the DC gain of the moving average FIR filter provides an interest-ing challenge since $H_{MA}(z)$ evaluated at $z = 1$ is an indeterminate form. The value of $H_{MA}(1)$ can be determine using L'Hôpital's rule as follows:

$$H_{MA}(z)\Big|_{z=1} = \frac{1}{N} \frac{d(1-z^{-N})/dz}{d(1-z^{-N})/dz}\Big|_{z=1} = \frac{1}{N} \left(\frac{Nz^{-(N-1)}}{1}\right)\Big|_{z=1} = 1$$

Computer Study. The pole-zero distributions for a moving average and comb filters (± cases) for $N = 7$ are shown in Fig. 11.3. The magnitude frequency re-sponses, shown in Fig. 11.4, were created using s-file `comb.s` which has a pro-cedure that may be called by "`comb(N,x)`", where N is the order of the multiplier-less FIR filter and $x[k]$ is an input time-series defined for $k \leq 256 - N$. The procedure `comb` linearly convolves $x[k]$ with the impulse responses of a moving average and comb filter. For $x[k]$ defined to be an impulse, and $N = 7$, the following sequence of Siglab commands will produce the results shown in Fig. 11.4.

```
>include "ch12_4.s"
>comb(7,impulse(256-7))
```

The numerator of the transfer function of the moving average and of the comb filters, given by $H(z) = 1 + z^{-N}$, are seen to have a finite DC gain. The other comb filter has a numerator given by $H(z) = 1 - z^{-N}$, resulting in a DC gain of zero. The moving average filter is seen to behave like a lowpass filter while the comb filter functions as a bank of frequency selective subfilters.   ❖

11.3 Stability

If the input to an Nth-order FIR filter and the filter's impulse re-sponse are bounded on a sample-by-sample basis, then the output of the FIR filter is likewise finite and bounded, since

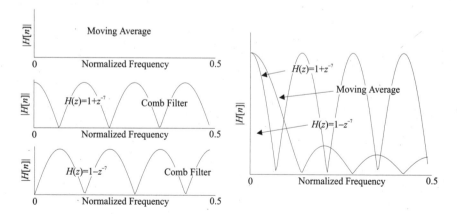

Figure 11.4 Magnitude frequency responses for moving average and comb FIR filters.

$$|y[k]| = \left| \sum_{m=0}^{N-1} h[m]x[k-m] \right| \le \left(\sum_{m=0}^{N-1} |h[m]||x[k-m]| \right) \le \infty \qquad (11.4)$$

The FIR filter is said to be *bounded-input-bounded-output* (BIBO) stable. It can be seen that FIR filters are intrinsically stable. However, with digital filters (FIR or IIR), the issue is rarely stability but rather the filter's run-time dynamic range requirements. This information is essential to establish the binary-point location in a fixed point filter implementation. If the input time-series is bounded by unity (i.e., $|x[k]| \le 1$) on a sample-by-sample basis, then the input which maximizes the FIR output is called the unit bound *worst-case input.* The worst-case input is given by $x[k] = \text{sign}(h[k - L]) = \pm 1$, where L denotes an arbitrary integer delay. The response of an FIR filter to the worst-case input is given by

$$y_{\text{worst case}}[k] = \sum_{m=0}^{N-1} h[k-m]\text{sign}(h[k-m]) = \sum_{m=0}^{N-1} |h[k-m]| \le G_{\text{max}} \le 2^I$$

$$(11.5)$$

The worst-case response is easily computed and can serve as a run-time dynamic range bound on the FIR filter for the purposes of setting the binary-point location in a fixed-point design.

Example 11.2: Half-Band FIR Filter

Problem Statement. A *half-band FIR filter* is a special case that exhibits the magnitude frequency response suggested in Fig. 11.5a. What is the worst-case gain of a 63rd-order half-band FIR filter described in the Computer Study section?

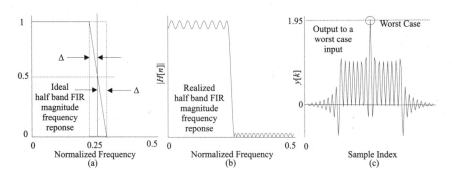

Figure 11.5 Half-band FIR filter: (a) design requirements (b), typical example, and (c) worst-case response.

Analysis and Conclusion. The point of symmetry in the transition band of a half-band FIR filter is found at the normalized frequency $\omega = \pi/2$, which corresponds to half the Nyquist frequency. An important consequence of a half-band FIR filter is shared with the half-wave symmetry property of Fourier series. A periodic signal having half-wave symmetry is known to possess a Fourier series containing only a DC (zeroth harmonic) term and odd harmonics. From the duality principal (i.e., Fourier transform time and frequency properties are interchangeable), symmetry in the frequency domain will result in a time-series which has a non-zero center tap and odd indexed non-zero coefficients. That is, except for the center tap, all even indexed tap weights are zero. This means that about $N/2$ coefficients of an Nth-order half-band FIR filter are zero, which effectively halves the multiply-accumulation burden.

For a 63rd-order half-band FIR filter (designed below), the worst-case output given an unit bounded input $|x[k]| \leq 1$ is $G_{max} = 1.95 < 2^1$ as defined by Eq. (11.5). When the worst-case input, given be $x[k] = \text{sign}(h[-k]) = \pm 1$, is convolved by the half-band FIR filter, the worst-case response is eventually produced as shown in Fig. 11.5c. If this filter is to be implemented using an M-bit fixed-point data format, then at least one integer bit would needed to ensure that the internal registers would not overflow during run-time.

Computer Study. A 63rd-order half-band FIR filter is designed using the following specifications.

Band	Lower edge	Upper edge	Weight	Gain
1	0.00	0.24	1	1
2	0.26	0.50	1	0

Scanning the list of resulting FIR tap weight coefficients shown below, it can be seen that, except for the center-tap position ($N = 31$), beginning with tap weights $N = 29$ (downward) and $N = 33$ (upward), every other tap weight is essentially zero. Notice also that the computed worst-case gain is given by $G = 1.95039$, which is experimentally verified in Fig. 11.5c.

The FIR filter design specifications is saved under the file name "chap11\halfband.fir", which can be viewed after loading it into the FIR design menu. The filter can, at that time, be generated and examined if desired. The engineering details of the resulting FIR filter, saved under the file name "chap11\halfband.rpt", are summarized below.

```
FILTER COEFFICIENTS
H( 0 ) = -0.02055560358601303     = H( 62 )
H( 1 ) = 2.456052766336513e-05    = H( 61 ) ~0
H( 2 ) = 0.006354171958553392     = H( 60 )
H( 3 ) = 1.342026060876857e-05    = H( 59 ) ~0
H( 4 ) = -0.007338967423443881    = H( 58 )
H( 5 ) = 2.678443357822991e-05    = H( 57 ) ~0
```

```
H( 6 ) = 0.008561584503311601      = H( 56 )
H( 7 )  = 1.294058330521374e-05    = H( 55 ) ~0
H( 8 )  = -0.009884130881065349    = H( 54 )
H( 9 )  = 1.949190771700585e-05    = H( 53 ) ~0
H( 10 ) = 0.01151518289891956      = H( 52 )
H( 11 ) = 1.631109925206854e-05    = H( 51 ) ~0
H( 12 ) = -0.01335975854045765     = H( 50 )
H( 13 ) = 1.447941512149825e-05    = H( 49 ) ~0
H( 14 ) = 0.0156814390430 2486     = H( 48 )
H( 15 ) = 2.292061788557374e-05    = H( 47 ) ~0
H( 16 ) = -0.01846566902313194     = H( 46 )
H( 17 ) = 2.639990057454938e-05    = H( 45 ) ~0
H( 18 ) = 0.02211063059077397      = H( 44 )
H( 19 ) = 1.1072705169328550 05    = H( 43 ) ~0
H( 20 ) = -0.02686949701256914     = H( 42 )
H( 21 ) = 1.925958564635403e-05    = H( 41 ) ~0
H( 22 ) = 0.03368516288047901      = H( 40 )
H( 23 ) = 2.744431823231873e-05    = H( 39 ) ~0
H( 24 ) = -0.04412658569115871     = H( 38 )
H( 25 ) = 3.022273599996066e-05    = H( 37 ) ~0
H( 26 ) = 0.06273988425886910      = H( 36 )
H( 27 ) = 1.841193448351651e-05    = H( 35 ) ~0
H( 28 ) = -0.1055005835656519      = H( 34 )
H( 29 ) = 1.201794548504472e-05    = H( 33 ) ~0
H( 30 ) = 0.3181429122318281       = H( 32 )
H( 31 ) = 0.5000148086908596       = H( 31 ) = center tap coefficient
Maximal Gain = 1.95039
```

The simulation of the half-band filter response is produced using the following sequence of Siglab commands.

```
>h = rf("chap11\halfband.imp")  # Load the saved impulse response.
>g = sum(mag(h)); g  # Compute and display worst-case gain Gmax.
1.95039
>worst = sign(rev(h))# Compute the worst-case unit bound input.
>y = h$worst         # Compute the filter response to w.c. input.
>max(mag(y))         # Find maximum output.
1.95039
>graph(y)            # Display result.                          ❖
```

11.4 Linear Phase

The normalized two-sided frequency response of an FIR filter having a transfer function $H(z)$ is given by $H(e^{j\omega})$, which is simply $H(z)$ evaluated along the trajectory $z = e^{j\omega}$ for $\omega \in [-\pi,\pi]$. The frequency response of an FIR filter can be expressed in magnitude-phase form as

$$H(e^{j\omega}) = \left|H(e^{j\omega})\right| \angle \phi(\omega) \qquad (11.6)$$

A system is said to have a *linear phase response* if the measured phase response has the form $\phi(\omega) = \alpha\omega + \beta$ [see Eq. (11.6)]. Linear phase filters are important in a number of applications which are intolerant of a frequency dependent propagation delay. Examples are:

- A phase lock loop (PLL) system used to synchronize data and decode phase modulated signals
- Linear-phase antialiasing filters placed in front of complex (arithmetic) signal analysis subsystems (e.g., DFT)
- Processing phase-sensitive signals such as images

Linear phase filtering is easily achieved with an FIR filter whose tap weight coefficients are symmetrically distributed about the filter's midpoint (see Fig. 11.6). To verify that symmetric FIR filters exhibit linear phase behavior, consider again Eq. (11.3) in a slightly modified form. If N is odd, let $L = (N-1)/2$ and assume that the coefficients are symmetrically distributed about the center tap coefficient $h[0]$ such that $h[i] = \pm h[-i]$. Then,

$$H(z) = z^{-L} \sum_{m=-L}^{L} h[m]z^{-m} = z^{-L}\left(h[0] + \sum_{m=1}^{L} (h[m]z^{-m} \pm h[-m]z^m)\right) \quad (11.7)$$

If N is even, then Eq. (11.7) is suitably modified, and the coefficient $h[0]$ is absent. If the filter has even coefficient symmetry, then, upon substituting $z = e^{j\omega}$ into the Eq. (11.7), one obtains

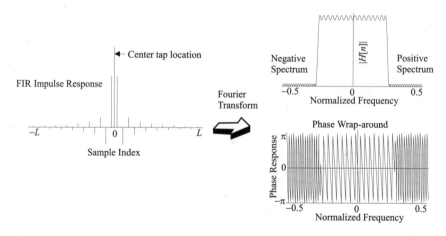

Figure 11.6 Linear phase FIR filter showing impulse response symmetry conditions and modulo(2π) wrapping effect in the phase-plane introduced by the digital computation of atan(ϕ).

$$H(e^{j\omega}) = e^{-jL\omega}\left(h[0] + \sum_{m=1}^{L} (h[m]e^{-jm\omega} \pm h[-m]e^{jm\omega})\right)$$

$$= e^{-jL\omega}\left(h[0] + \sum_{m=1}^{L} 2h[m]\cos(m\omega)\right) = e^{-jL\omega}C(\omega) \tag{11.8}$$

where $C(\omega)$ is seen to be a *real* function of ω. Therefore, the phase of an even symmetry FIR filter is given by

$$\arg[H(e^{j\omega})] = -L\omega + \arg[C(\omega)] \tag{11.9}$$

where $\arg[C(\omega)] = 0$ if $C(\omega) \geq 0$ and $\arg[C(\omega)] = \pm\pi$ if $C(\omega) < 0$. If the filter has odd symmetry, or antisymmetry, then the phase profile has a similar form except that

$$\arg[H(e^{j\omega})] = \frac{\pi}{2} - L\omega + \arg[C(\omega)] \tag{11.10}$$

In either case, Eqs. (11.9) and (11.10) are linear functions of ω and define a linear phase FIR filter. The phase response of a symmetric Nth-order FIR filter is therefore given by $\phi(\omega) = -\lfloor N/2 \rfloor\omega + \lambda$, where λ is a constant and $\lfloor x \rfloor$ is called the floor function and is equal to the integer part of x.

The *group delay* of a symmetric FIR filter is formally given by

$$\tau = -\frac{d[\phi(\omega)]}{d\omega} \tag{11.11}$$

and has a numerical value equal to $\tau = \lfloor N/2 \rfloor$, which corresponds to the midpoint of the FIR filter and is measured in clock delays. If the tap weights are distributed antisymmetrically, the group delay is again equal to $\tau(\omega) = \lfloor N/2 \rfloor$.

Example 11.3: Linear Phase FIR Filters

Problem Statement. Study the phase and group delay response of the 63rd-order half-band FIR filter originally analyzed in Example 11.2.

Analysis and Conclusions. Refer to Fig. 11.7 and note that the half-band filter's impulse response possesses even coefficient symmetry and, therefore, satisfies the conditions for linear phase. The half-band FIR filter impulse response was, as stated, also dense in zeros. The theoretical group delay of a symmetric 63rd-order FIR filter is $\tau = 31$ samples. This is experimentally veri-

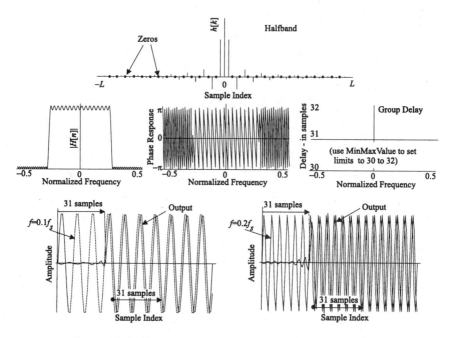

Figure 11.7 Symmetric half-band FIR filter impulse response (dense in zeros), magnitude frequency, phase, and group delay. Shown at the bottom of the figure are input-output responses for sinusoidal inputs at frequencies $f = 0.1f_s$ and $0.2f_s$ demonstrating a 31-sample delay relationship.

fied in Fig. 11.7 (see note regarding the setting the upper and lower magnitude bounds). The implication is that a signal will be delayed 31 samples, or $31T_s = 31/f_s$ seconds traveling through the FIR filter. This can also be experimentally verified using two test signals having frequencies $f = 0.1f_s$ and $0.2f_s$. Measuring the propagation delay through the filter (see Fig. 11.7), it can be seen that the delay difference between the input and output waveforms is 31 samples, as predicted. Notice also that the response to a sinusoidal input goes through a transient buildup period lasting about 31 samples and then transitions into a steady-state mode.

The experiment dramatizes the significance of a constant group delay in the design of systems that require data synchronization. Many systems synchronously recover timing information directly from the signal by sampling. Typically, the timing information is used to allow a signal to be sampled near the middle of a symbol interval or symbol period. For example, a 1 Mbps signal would have a 1 μs symbol period. To decode the message, the signal would be sampled at a 1 MHz rate with the samples taken, ideally, in the middle of each symbol period. If the detector used a linear phase filter, then, regardless of the instantaneous input frequency, the filtered output will always arrive $\lfloor N/2 \rfloor$ samples later. Therefore, the detector sampler can be properly synchronized in both frequency and phase to recover the information embedded in the detected signal process.

Computer Study. The filter's impulse magnitude frequency, phase, and group delay responses are shown in Fig. 11.7. The group delay is seen to be thirty-one delays, which is equal to the theoretical value $[N - 1]/2$. The data displayed in Fig. 11.7 was produced using the following sequence of Siglab commands.

```
>h=rf("chap11\halfband.imp") # Load impulse response.
>pad=h&zeros(512-len(h))     # Increase FFT resolution.
>hf=fft(pad)                 # Compute frequency response.
>graph(mag(hf),phs(hf),grp(hf))# Plot results.
# create test signals and measure group delay
>x=mksin(1,0.1,0,101)        # Test signal.
>y=h$x                       # Convolve test signal with filter.
>ograph(x,y[0:100],ones(31)&zeros(101-31))# I/O and 31 sample
  mark.
>x=mksin(1,0.2,0,101)   # Second test signal.
>y=h$x                  # Convolve test signal with filter.
>ograph(x,y[0:100],ones(31)&zeros(101-31))   # I/O and 31 sample
  mark.                                                      ❖
```

Example 11.4: Phase Delay

Problem Statement. A seventh-order linear phase FIR filter has an impulse response $h[k] = \{1,2,3,4,3,2,1\}$. What phase (propagation) delay will the filter impart to an input $x[k] = \cos(2\pi kn/7)$ for $n = 2$?

Analysis and Conclusions. The theoretical group delay for a seventh-order linear phase FIR filter is $\tau = (N - 1)/2 = 3$ samples. The period of $x[k] = \cos(2\pi kn/7)$ for $n = 2$, shall be denoted to be T_2 where $T_2 = (7/2)T_s$. That is, period of a sinusoid having a frequency 2/7th the sample frequency, or 7/2 clock cycles. Establishing a proportional relationship to determine what portion of a period (ϕ in radians) corresponds to a delay τ, one obtains

$$\frac{T_2}{\tau} = \frac{T_2}{3T_s} = \frac{(7/2)T_s}{3T_s} = \frac{7}{6} = \frac{2\pi}{\phi}$$

or $\phi = (12\pi/7)$ radians.   ❖

11.5 FIR Filter Synthesis Error Criteria

Normally, FIR filters are designed to achieve a prespecified frequency domain response. If $|H_d(e^{j\omega})|$ denotes a desired magnitude frequency response, then the design objective becomes one of deriving an FIR filter with a frequency response $|H(e^{j\omega})|$ which closely approximates $|H_d(e^{j\omega})|$ in some acceptable manner. The metrics used to compare the desired to the realized filter response is generally defined in terms of a frequency dependent error

$$e(\omega) = H(e^{j\omega}) - H_d(e^{j\omega}) \tag{11.12}$$

The criterion of optimization normally takes on one of the following forms:

- *Minimum squared error* (MSE) criterion:

$$\text{minimize } \phi(\omega) = \sum_{\forall \omega} e(\omega)^2 \tag{11.13}$$

- *Minimax error* criterion:

$$\text{minimize}(\text{maximum}(\varepsilon(\omega))); \; \forall \omega \tag{11.14}$$

Both optimization criteria are motivated in Fig. 11.8. The advantage of the MSE method is that it is generally easy to compute a solution when applied to the design of an LTI filter. Examples of design paradigms that use the MSE criterion are *moving average (MA)*, *auto-regressive (AR)*, and *auto-regressive moving average (ARMA)* design methods. The problem with MSE methods is that local errors can be very large, even though the overall MSE error is small. The minimax error criterion is usually more difficult to satisfy but will guarantee that the worst-case error has been reduced to a quantifiable value. This, if it can be achieved, is generally desired.

11.6 Direct Synthesis

The simplest FIR filter synthesis technique is called the *direct method*. It is based on the use of the DFT and IDFT. It involves computing the

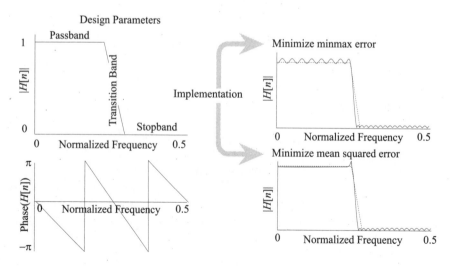

Figure 11.8 FIR filter design objective.

M sample inverse DFT (IDFT) of a specified desired frequency response $H_d(e^{j\omega})$. An Nth-order FIR filter produced using direct synthesis, where $N \le M$, is defined by an N sample time-series extracted from the M sample IDFT, symmetrically oriented about the center tap. The Direct FIR filter design procedure is demonstrated in Fig. 11.9.

Example 11.5: Direct Synthesis FIR Filter Design

Problem Statement. A direct FIR filter design having an ideal magnitude frequency response $|H_d(e^{j\omega})| = \cos(\omega)$, $0 \le \omega \le \pi/2$, can be approximated in a piecewise constant function in the sense of an eight-band 63rd-order direct FIR filter model where the band gains are given by:

$$|H(e^{j\omega})| = \cos(0\pi/14); \quad 0 \le \omega < \pi/8$$
$$|H(e^{j\omega})| = \cos(1\pi/14); \quad \pi/8 \le \omega < 2\pi/8$$
$$|H(e^{j\omega})| = \cos(2\pi/14); \quad 2\pi/8 \le \omega < 3\pi/8$$
$$|H(e^{j\omega})| = \cos(3\pi/14); \quad 3\pi/8 \le \omega < 4\pi/8$$
$$|H(e^{j\omega})| = \cos(4\pi/14); \quad 4\pi/8 \le \omega < 5\pi/8$$
$$|H(e^{j\omega})| = \cos(5\pi/14); \quad 5\pi/8 \le \omega < 6\pi/8$$
$$|H(e^{j\omega})| = \cos(6\pi/14); \quad 6\pi/8 \le \omega < 7\pi/8$$
$$|H(e^{j\omega})| = \cos(7\pi/14); \quad 7\pi/8 \le \omega < 8\pi/8$$

Analysis and Conclusions. An ideal piecewise constant magnitude frequency response can be defined in terms of a dense (high-resolution) complex N sample data array ($N \gg 63$) and presented to an inverse Fourier transform rou-

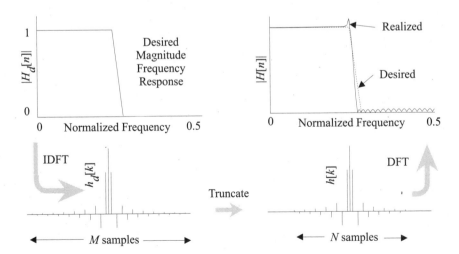

Figure 11.9 Direct synthesis methodology.

tine. The complex data array models, in this case, the ideal magnitude frequency response envelope of the desired FIR filter. If it is assumed that the impulse response is real, then the real part of the computed IDFT is symmetrically truncated to 63 samples and the imaginary part is the spectrum discarded.

Computer Study. The direct FIR filter design process can be investigated by loading the file "chap11\direct.fir" in the direct FIR filter design menu. An engineering report is provided in file "chap11\direct.rpt." This FIR filter design assumes the sample frequency is a normalized $f_s = 1$Hz. The normalized critical frequencies and gains are given below.

Band	f_{lower}	f_{upper}	Gain
1	0.000	0.063	1.00
2	0.073	0.125	0.97
3	0.135	0.188	0.90
4	0.198	0.250	0.78
5	0.260	0.312	0.62
6	0.322	0.375	0.43
7	0.385	0.438	0.22
8	0.448	0.500	0.00

What is interesting about FIR filters is that their critical frequencies scale with respect to the sample frequency f_s. If the filter described above were actually sampled at 11025 Hz, all the FIR filter's critical frequencies would be likewise scaled. In particular, the eighth band would range from 4,939 Hz to 5,512.5 Hz.

When generated, the filter produces the magnitude frequency response shown in Fig. 11.10. Note that the realized response follows the general shape of the desired filter but also exhibits some "ringing" (due to Gibb's phenomenon) locally about each band transition. Here is a case where a low mean squared error is achieved but the realized filter still exhibits a large localized maximum error. Therefore, the direct synthesis method is best applied to those designs that are modeled by a desired filter molded by a continuous function in the frequency domain rather than piecewise constant-magnitude frequency response.

The capability of the designed filter can be explored using a soundboard. The sequence of Siglab commands shown below will create a linearly swept FM signal (i.e., chirp) over the normalized frequency range $f \in [0.0, 0.5]$. It is assumed that a 16-bit soundboard is used with a sampling rate of $f_s = 11,025$ Hz.

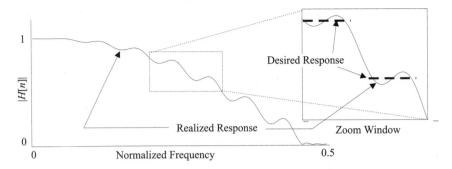

Figure 11.10 Direct FIR filter designed to approximate a piecewise constant magnitude frequency response.

When the chirp is played, the signal will appear to have uniform intensity across the audio spectrum. The chirp, after being convolved with the direct FIR filter, is observed to have lost some of its high-frequency energy. Since the magnitude frequency response of the FIR filter falls gradually, the rate at which the acoustic energy falls is likewise slow.

```
>h=rf("chap11\direct.imp")       # Load direct FIR.
>graph(mag(pfft(h&zeros(2^10)))) # Graph high-resolution FFT.
>include "chap11\chirp.s"        # Chirp signal generator.
>x=chirp(2^14,0,0,0.5)           # Long chirp.
>y=h$x                           # Filter chirp.
>wavout(2^14*x)                  # Play original signal.
>wavout(2^14*y)                  # Play filtered signal.         ❖
```

11.7 Frequency Sampling Method

The *frequency sampling* design technique is a venerable and intuitive FIR filter design method. To motivate the frequency sampling method, consider the ubiquitous N-band hi-fi equalizer, which is used to shape the frequency response envelope of an audio amplifier. Suppose that the owner of the amplifier wishes to override decades of serious engineering to produce a perfectly flat audio amplifier and boost the bass response to the point of audible pain. Assume that the envelope of the altered response is defined by $H_d(e^{j\omega})$. The desired filter can be, in concept, approximated by a FIR filter $H(e^{j\omega})$, which is equivalent to a bank of bandpass FIR filters, each centered about normalized frequencies $\omega = 2\pi n/N$. At the discrete frequency $\omega = 2\pi n/N$, it will be expected that the realized filter, $H(e^{j\omega})$ satisfies

$$H(e^{j2\pi n/N}) = H_d(e^{j2\pi n/N}) = A(n)\angle\phi(n) \tag{11.15}$$

The impulse response of an Nth-order FIR filter, which satisfies Eq. (11.15) for $n \in [0, N-1]$, can be computed using an IDFT of the known $H_d(e^{j2\pi n/N})$, which is formally given by the N harmonic IDFT

$$h[k] = \frac{1}{N} \sum_{n=0}^{N-1} H_d(e^{j2\pi n/N}) e^{j2\pi nk/N} \tag{11.16}$$

It is therefore established that $H(e^{j\omega}) = \text{DFT}(h[k])$ and $H_d(e^{j\omega})$ agree at the discrete frequencies $\omega = 2\pi n/N$. Equivalently, the z-transform representation of the realized FIR filter is given by

$$H(z) = \sum_{k=0}^{N-1} h[k] z^{-k} \tag{11.17}$$

For notational convenience, let $H_d[n] = H_d(e^{j2\pi n/N})$. Upon substitution of Eq. (11.16) into Eq. (11.17), $H(z)$ becomes

$$H(z) = \frac{1}{N} \sum_{k=0}^{N-1} \left(\sum_{n=0}^{N-1} H_d[n] e^{j2\pi nk/N} \right) z^{-k} \tag{11.18}$$

which, when simplified, reduces to

$$H(z) = \frac{1}{N} \sum_{k=0}^{N-1} \sum_{n=0}^{N-1} H_d[n] (e^{j2\pi n/N} z^{-1})^k \tag{11.19}$$

After reversing the order of the summation, one obtains

$$H(z) = \frac{1}{N} \sum_{n=0}^{N-1} H_d[n] \sum_{k=0}^{N-1} (e^{j2\pi n/N} z^{-1})^k \tag{11.20}$$

The sum over the index k in Eq. (11.20) can be expressed as

$$\sum_{k=0}^{N-1} (e^{j2\pi n/N} z^{-1})^k = \frac{1 - z^{-N}}{1 - e^{j2\pi n/N} z^{-1}} \tag{11.21}$$

This allows the FIR filter transfer function to be expressed as

$$H(z) = \frac{(1 - z^{-N})}{N} \sum_{n=0}^{N-1} \frac{H_d[n]}{1 - e^{j2\pi n/N} z^{-1}} \tag{11.22}$$

The preamble term, namely $(1 - z^{-N})$, is seen to be a comb filter which places nulls (zeros) at the reference frequencies $\omega = 2\pi n/N$. The terms in the summation represent a collection of subband filters having poles also at the reference frequencies $\omega = 2\pi n/N$, and operate in parallel as shown in Fig. 11.11. Therefore, there is a natural pole-zero cancellation process going on wherever a subband filter having a nonzero gain is to be located.

Each subband filter functions as a resonator, or sharply tuned filter, and is characterized by

$$H_n(z) = \frac{H_d[n]}{1 - e^{j2n\pi/N}z^{-1}} = \frac{A[n]\angle\phi(n)}{1 - e^{j2n\pi/N}z^{-1}} \tag{11.23}$$

for all $n \in \{0,1,2,\ldots,N-1\}$. The poles of the resonator filters are located along the periphery of the unit circle at locations $z = re^{j2\pi n/N}$. For stability reasons, the distributed poles and zeros are often moved slightly interior to the unit circle by scaling the unit circle poles and zeros back to a radius r (i.e., $z = re^{j2\pi n/N}$), where $r < 1$ but is also close to unity. It can also be observed that the complex poles occur in complex conjugate pairs which allow the filter defined by Eq. (11.22) to be simplified (using Euler's equation) to

$$H(z) = \frac{(1 - r^N z^{-N})}{N}$$

$$\times \left(\frac{H_d[0]}{1 - rz^{-1}} + \sum_{n=1}^{(N-1)/2} \frac{2A[n](\cos(\phi(n)) - r\cos(\phi(n) - 2\pi n/N)z^{-1})}{1 - 2r\cos(2\pi n/N)z^{-1} + r^2 z^{-2}} \right) \tag{11.24}$$

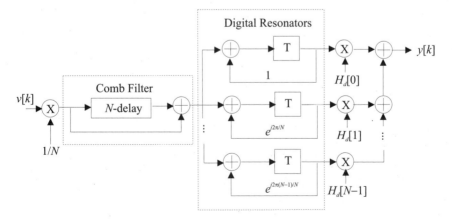

Figure 11.11   Frequency sampling filter system architecture.

if N is odd. If N is even, then

$$H(z) = \frac{(1 - r^N z^{-N})}{N}$$

$$\times \left(\frac{H_d[0]}{1 - rz^{-1}} + \frac{H_d[N-1]}{1 + rz^{-1}} + \sum_{n=1}^{N/2-1} \frac{2A[n](\cos(\phi(n)) - r\cos(\phi(n) - 2\pi n/N)z^{-1})}{1 - 2r\cos(2\pi n/N)z^{-1} + r^2 z^{-2}} \right)$$

(11.25)

While the frequency sampling method has been successfully used for a long time to design FIR filters having an arbitrary magnitude frequency response envelope, it does have acknowledged limitations. A major design problem can occur when a frequency sampling filter is required to model an ideal piecewise constant filter response characterized by $|H(n)| = 1$ and $|H(n + 1)| = 0$ (i.e., abrupt change across the transition band boundary). If N is large, then the frequency $2\pi n/N$ is numerically close to $2(n + 1)\pi/N$, which defines the filter to have a very sharp transition band (i.e., *steep-skirt*). This sharp transition is difficult to model using an IDFT (remember Gibb's phenomenon). The problem can be mitigated somewhat by relaxing the transition band condition so that $|H(n)| = 1 - \alpha$ and $|H(n + 1)| = \alpha$ for $\alpha > 0$. This will reduce the slope of the skirt of the filter so that it can be more easily realized with an IDFT.

A question may arise regarding the elemental filters $H_n(z)$, shown in Eq. (11.23) and Fig. 11.11, which clearly show the presence of feedback. Systems with feedback have the ability to produce an impulse response that can persist forever. How, then, can one guarantee that the filter described by Eq. (11.23) even qualifies as an FIR filter? The device that restricts the impulse response to be of finite length N is the comb prefilter given by $H'(z) = (1 - z^{-N})$. It follows that the impulse response of the prefilter is given by $h'[k] = \delta[k] - \delta[k - N]$. On receipt of the first impulse in $h'[k]$, namely $\delta[k]$, the subfilters $H_n(z)$ will respond with an output time-series (impulse response) that is periodic and resonates at a frequency $2n\pi/N$. This oscillatory behavior would continue forever if not for the second impulse arriving from $h'[k]$, namely $-\delta[k - N]$, which sets up another oscillating waveform, delayed by N samples and 180° out of phase from the first. As a result, the second wavefront perfectly cancels the first, leaving only the first N samples (i.e., finite impulse response) unaltered.

Example 11.6: Frequency Sampling Filter

Problem Statement. Design a 15th-order zero-phase (i.e., $\phi(n) = 0$) frequency sampling FIR filter that emulates an ideal lowpass filter having normalized cutoff frequency located at 0.4π and a normalized transition bandwidth of

0.2π. The frequency sampling filter center frequencies are located at $\omega = 2\pi n/15 = 0.4188n$, for $n \in \{0,1,2,\ldots,14\}$. The target (ideal) filter model requires that there be symmetry in the frequency domain which, in this case, requires

$$H_d(n) = \begin{cases} 1 & n \in \{0, 1, 2, 3, 12, 13, 14\} \\ 0 & n \in \{4, 5, 6, \ldots, 11\} \end{cases}$$

where $n \in \{0,1,2,\ldots,7\}$ correspond to positive frequencies, and $n \in \{9,10,11,\ldots,14\}$ refer to negative frequencies. What is the realized frequency response?

Analysis and Results. The transfer function of the frequency sampling filter, based on the desired filter gain weights and $r = 1$, is given by

$$H(z) = \frac{(1-z^{-15})}{15}\left(\frac{1}{1-z^{-1}} + \sum_{n=1}^{3} \frac{2(1-\cos(2\pi n/15)z^{-1})}{1-2\cos(2\pi n/15)z^{-1}+z^{-2}}\right)$$

The frequency sampling filter can also be defined in terms of relaxed weights which create a smoother transition band. The weights of the relaxed filter are assumed to be $H(3) = H(12) = 0.707$ (i.e., $\alpha = 0.23$) and $H(4) = H(11) = 0.23$. The resulting frequency response of the relaxed filter would be defined again by Eq. (11.24) in terms of the relaxed set of coefficients.

The frequency response of the first (ideal) filter model is shown in Fig. 11.12. Observe that there is local amplitude ripple and overshoot in the vicinity of the transition band. The relaxed filter, also shown in Fig. 11.12, is seen to exhibit less overshoot (ripple) locally about the transition band. It can also be seen that the produced responses are not close replicas of the desired response. To achieve a good fit, a filter order much greater than 15 would be required. However, it should be appreciated that only those filters having nonzero gain at angular frequency $\omega_i = 2\pi i/N$ need be implemented, which can often result in a low-complexity filter solution.

Computer Study. The required first- and second-order filters can be defined in Direct II state variable form. The comb prefilter is implemented using a delay operator. The second-order resonators can be factored as

$$H_i(z) = \left(\frac{2(1-\cos(2\pi n/15)z^{-1})}{1-2\cos(2\pi n/15)+z^{-2}}\right) = 2 + \left(\frac{2(\cos(2\pi n/15)z^{-1}-z^{-2})}{1-2\cos(2\pi n/15)z^{-2}+z^{-2}}\right)$$

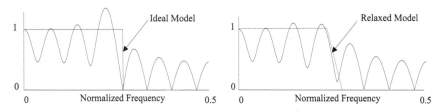

Figure 11.12 Frequency response of a frequency sampling filter modeling (left) an ideal lowpass and (right) relaxed lowpass filter model.

The frequency sampling filter is created using s-file "**freqsam.s**", which defines the function freqsmp(x,h) where x is an input time-series and h is a coarse description of the envelope of the desired filter's magnitude frequency response at the discrete frequencies $\omega_k = 2\pi k/N$. The magnitude frequency response of a 15th-order FIR filter has 15 discrete frequency locations. For an ideal lowpass filter target response, the resulting frequency sampled filter solution is shown in the left panel of Fig. 11.12. A relaxed version is shown in the right panel for $\alpha = 0.23$. The graphs in Fig. 11.12 were produced using the following sequence of Siglab commands.

```
>x=impulse(1024)
>h=[1,1,1,1,0,0,0,0,0,0,0,0,1,1,1]          # Desired response.
>include "chap11\freqsam.s"                 # Freq samp s-file.
>y=freqsamp(x,h)# Impulse response to a piecewise constant model.
>ograph(mag(pfft(y)),(1+sq(512,4/7.5,512))) # Graph results.
>h=[1,1,1,0.77,0.23,0,0,0,0,0,0,0.23,0.77,1,1]# Relaxed model.
>y=freqsamp(x,h)   # Impulse response to a relaxed pw-constant
   model.
>ograph(mag(pfft(y)),(1+sq(512,4/7.5,512))) # Graph results.   ❖
```

The previous frequency sampling example was based on a low-order model and assumed zero-phase shift. In general, FIR filters have a non-zero phase behavior as a function of frequency and, as is often the case, linear phase behavior.

Example 11.7: Complex Frequency Sampling Filter

Problem Statement. Design a frequency sampling filter having linear phase behavior that has a passband from 300 to 700 Hz, and a 12.5 kHz sampling rate. Test the filter.

Analysis and Conclusions. Assume that the resonant filters are to be set on 100 Hz centers which defines $N = 125$. That is, 12.5×10^3 Hz/125 = 100 Hz. Therefore, the impulse response is 125 samples in length. Based on the material presented in Sec. 11.4, assume that the phase changes at a rate of π radians per 100 Hz. This implies that $H(n) = (-1)^n A(n)$. The desired passband and transition band gains are given in the following table.

$f = 200$ Hz	$A(2) = 0.221$	$H(2) = 0.221$
$f = 300$ Hz	$A(3) = 0.707$	$H(3) = -0.707$
$f = 400$ Hz	$A(4) = 1$	$H(4) = 1$
$f = 500$ Hz	$A(5) = 1$	$H(5) = -1$
$f = 600$ Hz	$A(6) = 1$	$H(6) = 1$
$f = 700$ Hz	$A(7) = 0.707$	$H(7) = -0.707$
$f = 800$ Hz	$A(8) = 0.221$	$H(8) = 0.221$

The resulting impulse and magnitude frequency response of frequency sampling filter is shown in Fig. 11.13. Notice that the impulse response has a symmetric shape around the center tap coefficient suggestive of a linear phase filter. The magnitude frequency response is also seen to follow the general shape of the desired filter.

Computer Study. The s-file "freqsmp.s" (similar to freqsam.s used in Example 11.6) may be used to create the data shown in Fig. 11.13 using the following sequence of Siglab commands.

```
>h1=[0,0,0.221,-0.707,1,-1,1,-0.707,0.221]
>h=h1&zeros(63-len(h1))      # Desired response.
>include "chap11\freqsmp.s"
>y=freqsmp(impulse(128),h)   # Impulse response.
>graph(y)                    # Graph impulse response.
>graph(mag(pfft(y))/128)     # Graph magnitude frequency response.❖
```

11.8 Equiripple FIR Filter Design

The direct FIR filter synthesis method can produce a filter having a potentially large local deviation from the desired magnitude frequency response. More specifically, the l_2 error norm of a direct FIR filter maybe small but the maximal error (i.e., l_∞ norm) can be large (e.g., Gibb's phenomenon). Unacceptably large local maximum errors can make a DSP solution completely unusable, even though the overall design is considered to be a successful in a mean-squared error sense. What is often desired is a design paradigm in which the largest error is bounded below some acceptable threshold. This is the role of an equiripple design rule that satisfies the *minimax error criteria* given by

$$\varepsilon_{minimax} = \min\{\max(\varepsilon(\omega))|\omega \in [0, \omega_s/2)\} \quad (11.26)$$

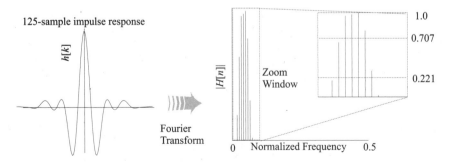

Figure 11.13 Frequency sampling FIR filter (a) impulse response, (b) magnitude frequency, and (c) zoom expansion of the magnitude frequency passband response.

where ε is the error measured in the frequency domain and defined by

$$\varepsilon(\omega) = W(\omega)\left|H_d(e^{j\omega}) - H(e^{j\omega})\right| \tag{11.27}$$

and $W(\omega) \geq 0$ is called the *error weight*. The error $\varepsilon(\omega)$ is the weighted difference between the desired and realized filter's response at frequency ω. For an Nth-order *equiripple FIR filter*, the maximum error occurs at discrete *extremal frequencies* ω_i. One of the interesting properties of an equiripple FIR filter is that *all* the maximal errors, called *extremal errors*, are equal. That is, $\varepsilon_{\text{minimax}} = |\varepsilon(\omega_i)|$ for $i \in \{0,1,2,...,N - 1\}$ where each ω_i is an extremal frequency. In addition, the signs of the maximal errors alternate [i.e., $\varepsilon(\omega_i) = -\varepsilon(\omega_{i+1})$]. For a linear phase FIR filter, the location of the maximum errors can be numerically determined using the *alternation theorem* from polynomial approximation theory. This method was popularized by Parks and McClellan who solved the alternative theorem problem iteratively using the Remez exchange algorithm.[1] The Remez exchange algorithm iteratively adjusts the tentative extremal frequencies ω_i until the minimax criterion is satisfied to within a fixed tolerance. Since all the errors have the same absolute value and alternate in sign, the synthesized FIR filter is generally referred to by its popular name, *equiripple FIR filter*. This method has been used for several decades to design linear phase FIR filters and continues to provide reliable results.

Figure 11.14 graphically interprets a typical equiripple FIR filter approximations to a piecewise constant ideal filter model [i.e., $H_d(e^{j\omega})$]. The design objectives of a equiripple FIR filter can be specified in terms of acceptable passband and attenuation (stop) band deviations (tolerances), denoted δ_p and δ_a, located at the extremal frequencies distributed in the passbands and stopbands.

A logical question to ask is, why is there a need to independently specify the passband and stopband ripple deviation parameters δ_p and δ_a since it is also being claimed that all the minimax (extremal) errors

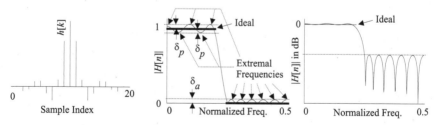

Figure 11.14 Equiripple approximation to a piecewise linear FIR filter magnitude frequency response model showing the equiripple errors at extremal frequency locations.

are equal? This is the role of the error weight $W(\omega)$. Suppose that error deviations in the passband are treated identically to those in the stopband, then $W_{passband}(\omega) = W_{stopband}(\omega)$. The realized filter will produce a magnitude frequency response which satisfies $\delta_p = \delta_a$. If, however, error deviations in the passband are considered to be ten times more serious (undesirable) than those in the stopband, then they would be weighted $W_{passband}(\omega) = 10W_{stopband}(\omega)$. The resulting solutions to the weighted equiripple error minimization problem would produce a solution where $10\delta_p = \delta_a$. That is, the actual filter deviations in the passband will be 1/10th those in the stopband. In this manner, significant to subtle adjustments can be made to an equiripple design.

Example: 11.8 Equiripple FIR Filter

Problem Statement. Design a 51st-order bandpass equiripple FIR filter which meets the specifications:

Sampling frequency f_s = 100 kHz.

Frequency band 1: $f \in [0.0,10]$ kHz; desired gain = 0.0, $W(f) = 1$

Frequency band 2: $f \in [12,38]$ kHz; desired gain = 1.0, $W(f) = 1$

Frequency band 3: $f \in [40,50]$ kHz; desired gain = 0.0, $W(f) = 1$

What is the magnitude frequency response of the realized equiripple filter?

Analysis and Conclusions. The resulting filter is shown in Fig. 11.15. The FIR filter has a measured passband and stopband deviation from the ideal given by $\delta_p = \delta_p \leq -20$ dB or $\delta_p = \delta_p \leq 0.1$. The worst-case gain of the filter is noted to be $G_{max} = 1.999 < 2^1$.

Computer Study. An equiripple FIR filter which meets the posted specifications is shown in Fig. 11.14. The design process can be explored by loading file "equi.fir" into the FIR filter design menu. The engineering details pertaining to the designed filter are saved in file "equi.rpt". The impulse response is seen to have even symmetry and the magnitude frequency response exhibits

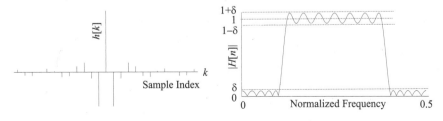

Figure 11.15 Equiripple FIR filter impulse and magnitude frequency response with uniform passband and stopband weights.

equiripple deviation from the ideal in the both the passband and stopband. The measured extremal errors are $\delta_p = \delta_a = 0.0664$ (–23.5 dB). By changing the order to 41, for example, the deviation error will increase. For $N = 61$, the deviation error will decrease. It should be appreciated, however, that filter order is directly correlated to the number of MAC cycles per filter cycle, which in turn is inversely related to the maximum run-time bandwidth of the filter.

The performance of the designed filter can be experimentally evaluated using a soundboard and a chirp swept over the normalized frequency range $f \in [0, 0.5)$. The sequence of Siglab commands shown below assumes a 16-bit soundboard and a sample rate of 11,025 Hz. The input is heard to have a uniform intensity across the passband spectrum but is strong only in the passband.

```
>h=rf("chap11\equi.imp")   # Load equiripple FIR impulse response.
>include "chap11\chirp.s"   # Chirp signal generator.
>x=chirp(2^14,0,0,0.5)      # Create chirp test signal.
>y=h$x                      # Filter chirp test signal.
>wavout(2^14*x)             # Play original chirp.
>wavout(2^14*y)             # Play filtered chirp.            ❖
```

Example 11.9: Weighted Equiripple FIR Filter

Problem Statement. The 51st-order bandpass equiripple FIR filter designed in the previous example has a –23.5 dB passband and stopband deviation. The –23.5 dB stopband attenuation may be inadequate for a particular problem. Suppose that a maximum stopband deviation of –30 dB is desired. Adjust the band weights until the desired stopband deviation is achieved and measure the resulting passband deviation.

Analysis and Conclusions. The required weights were determined experimentally for the an FIR filter having the following design specification.

Sampling frequency $f_s = 100$ kHz:

Frequency band 1: $f \in [0.0, 10]$ kHz; desired gain = 0.0, $W(f) = 4$

Frequency band 2: $f \in [12, 38]$ kHz; desired gain = 1.0, $W(f) = 1$

Frequency band 3: $f \in [40, 50]$ kHz; desired gain = 0.0, $W(f) = 4$

The FIR filter has a measured passband and stopband deviation from the ideal approximately given by $\delta_p \sim -18$ dB and $\delta_a \sim -30$ dB (see Fig. 11.16). While the passband deviation has been relaxed, the stopband attenuation has increased from –23.5 dB to –30 dB. This is often considered to be an acceptable trade-off.

Computer Study. For comparative purposes, the weighted and uniform weighted (Fig. 11.15) magnitude frequency responses are shown in Fig. 11.16. The filter design processes can be studied by loading file "equiw.fir" into the FIR design menu. A detailed engineering report of the designed filter is saved in file "equiw.rpt." The impulse response is seen to have even symmetry, and the magnitude frequency response has an equiripple deviation from the ideal in the both the passband and stopband.

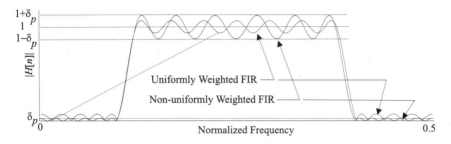

Figure 11.16   Equiripple FIR filter design with weighted passband and stopband.

The performance of the designed weighted filter can be experimentally evaluated using a soundboard and a chirp swept over the normalized frequency range $f \in [0, 0.5]$. The sequence of Siglab commands shown below assumes a 16-bit soundboard and a sample rate of 11,025 Hz. The input is again heard to have a uniform intensity across the spectrum, but the filtered signal is strong only in the passband.

```
>h = rf("chap11\equiw.imp")   # Load equiripple FIR impulse response.
>include "chap11\chirp.s"      # Chirp signal generator.
>x = chirp(2^14,0,0,0.5)       # Create chirp test signal.
>y = h$x                       # Filter chirp test signal.
>wavout(2^14*x)                # Play original chirp.
>wavout(2^14*y)                # Play filtered chirp.          ❖
```

11.9 Hilbert Equiripple FIR Filter

Equiripple filters are commonly used in the design of linear phase lowpass, bandpass, bandstop, and highpass filters. Equiripple filters are also adept at implementing other filter classes such as a *differentiator* of order N, whose frequency responses if given by $H(e^{j\omega}) = (j\omega)^N$. Another important use of the equiripple design paradigm is to construct *Hilbert filters*. Hilbert filters are important elements in many communication systems and have a frequency response given by

$$H(e^{j\omega}) = \begin{cases} j & -\pi/2 \le \omega < 0 \\ -j & 0 \le \omega < \pi/2 \end{cases} \tag{11.28}$$

Observe that a Hilbert filter is essentially an *all-pass filter* in that $|H(e^{j\omega})| = 1$ for all ω but possesses a distinctive quadrature phase shifting property. The phase shifting property of a Hilbert filter is particularly useful in defining single sideband modulators, quadrature amplitude modulation (QAM) communications systems, and in-phase and quadrature phase (I/Q) channel filtering systems.

Example 11.10: Hilbert Transform

Problem Statement. The quadrature phase shifting ability of a Hilbert filter can be viewed in framework of pure cosine and sine waves:

$$\cos(\phi k) = \frac{e^{j\phi k} + e^{-j\phi k}}{2}$$

$$\sin(\phi k) = \frac{e^{j\phi k} - e^{-j\phi k}}{2j}$$

What is the result of Hilbert filtering these sinusoidal sequences?

Analysis and Conclusions. Assuming that the sinusoids are in the passband of the Hilbert filter, the application of Eq. (11.28) produces

$$\cos(\omega k) \overset{F}{\leftrightarrow} \frac{\delta[\omega - k] + \delta[\omega + k]}{2} \overset{\text{Hilbert filter}}{\longrightarrow} \frac{j\delta[\omega - k] - j\delta[\omega + k]}{2} \overset{F}{\leftrightarrow} -\sin(\omega k)$$

$$\sin(\omega k) \overset{F}{\leftrightarrow} \frac{\delta[\omega - k] - \delta[\omega + k]}{2j} \overset{\text{Hilbert filter}}{\longrightarrow} \frac{\delta[\omega - k] + \delta[\omega + k]}{2} \overset{F}{\leftrightarrow} \cos(\omega k)$$

❖

There is a minor problem associated with implementing an equiripple Hilbert filter. In particular, implementing the sharp phase transition from $-90°$ to $90°$ about 0 Hz, and the Nyquist frequency is difficult. This problem can be mitigated by creating a small guardband around DC and $f_s/2$.

Example 11.11: Hilbert FIR Filter

Problem Statement. Design a 63rd-order Hilbert FIR filter that meets the following specifications.

Sampling frequency $f_s = 1$ Hz (normalized)

Frequency band 1: $f \in [0.0, 0.01]$ kHz; desired gain = 0.0, $W(f) = 1$ {guard band}

Frequency band 2: $f \in [0.02, 0.48]$ kHz; desired gain = 1.0, $W(f) = 1$ {passband}

Frequency band 3: $f \in [0.49, 0.50]$ kHz; desired gain = 0.0, $W(f) = 1$ {guard band}

Analysis and Results. The sample frequency was set to unity for convenience. A Hilbert filter was designed with respect to the gain profile shown in Fig. 11.17. The FIR filter has a passband and stopband deviation from the desired of δ_p, which was measured to be approximately -16.8556 dB.

Computer Study. The design of the Hilbert filter can be examined by loading "hilbert.fir" into the FIR filter design menu. An engineering report file

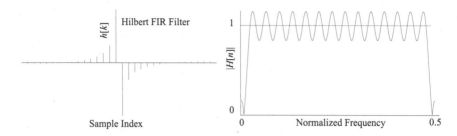

Figure 11.17   Hilbert FIR filter impulse and magnitude frequency response.

was created and saved as "`hilbert.rpt.`" The filter's impulse and magnitude frequency response is shown in Fig. 11.17, and the coefficients are listed below for the purpose of exploring the FIR filter report filter format.

```
FILTER SPECIFICATIONS

Filter Type = HILBERT
Filter Order = 63
Grid Density = 16
```

BAND	LOWER EDGE	UPPER EDGE	WEIGHT	GAIN
1	0	0.01	1	0
2	0.02	0.48	1	1
3	0.49	0.5	1	0

```
Deviation         = 0.143622
                  = -16.8556 dB

FILTER COEFFICIENTS
H(  0 ) = -0.07856705071116282    = -H( 62 )
H(  1 ) = 1.833839954262147e-07   = -H( 61 )
H(  2 ) = -0.01355485416005799    = -H( 60 )
H(  3 ) = 8.674988064002883e-05   = -H( 59 )
H(  4 ) = -0.01359703247490540    = -H( 58 )
H(  5 ) = -1.834823885503183e-05  = -H( 57 )
H(  6 ) = -0.01297411716398010    = -H( 56 )
H(  7 ) = 8.057569734750555e-06   = -H( 55 )
H(  8 ) = -0.01133786816677188    = -H( 54 )
H(  9 ) = 7.506630847580115e-05   = -H( 53 )
H( 10 ) = -0.008588283374313332   = -H( 52 )
H( 11 ) = 6.417112453143428e-05   = -H( 51 )
H( 12 ) = -0.004602441585447758   = -H( 50 )
H( 13 ) = 1.074533931182687e-05   = -H( 49 )
H( 14 ) = 0.0008527606914392194   = -H( 48 )
H( 15 ) = -1.667300267070684e-05  = -H( 47 )
H( 16 ) = 0.008077618281369270    = -H( 46 )
H( 17 ) = -1.934532562832968e-05  = -H( 45 )
H( 18 ) = 0.01752655624011952     = -H( 44 )
H( 19 ) = -1.630281315224673e-05  = -H( 43 )
```

```
H( 20 ) =  0.03001439683049809      = -H( 42 )
H( 21 ) = -7.328714514416189e-06    = -H( 41 )
H( 22 ) =  0.04707275604891830      = -H( 40 )
H( 23 ) =  2.575376140882650e-06    = -H( 39 )
H( 24 ) =  0.07199407577696301      = -H( 38 )
H( 25 ) =  6.355233061357800e-06    = -H( 37 )
H( 26 ) =  0.1134879605121714       = -H( 36 )
H( 27 ) =  8.236755406782923e-06    = -H( 35 )
H( 28 ) =  0.2037848426213617       = -H( 34 )
H( 29 ) =  7.790899568422714e-06    = -H( 33 )
H( 30 ) =  0.6337929424513224       = -H( 32 )
H( 31 ) =  0.000000000000000
Maximal Gain = 2.54035
```
❖

Note the coefficient symmetry typical of a linear phase FIR filter.

11.10 Equiripple Order Estimate

There are several estimation algorithms commonly used to translate the design parameters $(\delta_p, \delta_a, \omega_p, \omega_a)$ into an estimate of the order of an equiripple FIR filter. The most popular estimation formulas are summarized below. The first is given by

$$N_{FIR} \approx \frac{-10\log_{10}(\delta_a\delta_p) - 15}{14\Delta\omega} + 1 \qquad (11.29)$$

where $\Delta\omega$ is the normalized transition bandwidth and is given by $\Delta\omega = (\omega_a - \omega_p)/\omega_s$. The second model is given by

$$N_{FIR} \approx \frac{-10\log_{10}(\delta_a\delta_p) - 13}{14.6\Delta\omega} + 1 \qquad (11.30)$$

These formulas provide an estimate of an equiripple filter's order based on unity band weights [i.e., $W(\omega) = 1$]. The order requirements of a nonunity weighted [i.e., $W(\omega) \neq 1$] equiripple FIR filter need to be determined experimentally.

Steep-skit or narrow-band equiripple FIR filters are known to require a high-order solution. As a "rule of thumb," when the normalized transition band has a value less than 0.04, an equiripple filter is virtually impossible to build. Another problem with high-order FIR filters is that the coefficients found on the tapers (tap weight coefficients found at a distance from the middle tap). Generally, these coefficients have such small values that they cannot be realized with a 16- or 24-bit fixed-point system without introducing serious coefficient roundoff errors. Even when using 32-bit floating point arithmetic, arithmetic anomalies may occur due to underflow or a related pathology known as *catastrophic cancellation*.

Example 11.12: Fixed-Point FIR Filters

Problem Statement. For the purpose of demonstration, consider implementing the 51st-order FIR filter studied in Example 11.8 using an eight-bit microprocessor. Assume that the format consists of one sign bit and seven fractional bits.

Analysis and Results. The maximum difference between that real and fixed-point coefficient can be computed to be 0.0077. The response of the fixed-point filter is compared to the floating-point precision filter response in Fig. 11.18. It can be seen that the fixed-point FIR filter has technically lost its equiripple behavior but still represents a respectable bandpass filter.

Computer Study. The data presented in Fig. 11.18 was produced using the following sequence of Siglab commands and the saved FIR filter impulse reoponac "equi.imp."

```
>h=rf("chap11\equi.imp")   # Read FIR filter impulse response.
>wordlen(8,7,1)            # Set data format to be [8:7],
>hr=fxpt(h)                # and truncate h.
>z=zeros(128)              # Zero padding for high res FFT.
>max(mag(h-hr))            # Compute maximum roundoff error.
  0.0077
>ograph(mag(pfft(h&z)),mag(pfft(hr&z))) # Graph results.
```

Finite wordlength effects can be qualitatively observed using a soundboard. The performance of the designed FIR filter can be experimentally evaluated using a soundboard and a chirp swept over the normalized frequency range $f \in [0,5)$. The Siglab commands shown below assume a 16-bit soundboard and a sampling rate of 11,025 Hz. The input is heard to have a uniform intensity across the spectrum but the filtered signal is strong only in the passband.

```
>h = rf("chap11\equi.imp")   # Read FIR filter impulse response.
>include "chap11\chirp.s"    # Get chirp generator.
```

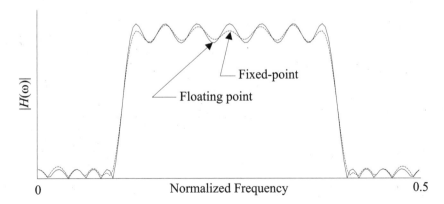

Figure 11.18 Fixed-point and floating-point FIR filter magnitude frequency responses.

```
>x = chirp(2^14,0,0,0.5)   # Generate chirp.
>y = h$x                   # Filter chirp - floating point.
>wordlen(8,2,1)            # Set data format to [8:7].
>hf = fxpt(h)              # Truncate coefficients to [8:7].
>yf = hf$x                 # Filter chirp using truncated coef.
>wavout(2^14*y)            # Play floating point filtered chirp.
>wavout(2^14*yf)           # Play fixed-point filtered chirp.
```

It should be appreciated that the soundboard itself has some intrinsic interpolation abilities which are naturally derived from its analog circuitry. These interpolation abilities tend to mask the finite word length effects illustrated by this example.   ❖

11.11 Mirror and Complement FIR Filters

The design of an FIR filter has been studied in the context of the direct, frequency sampled, and equiripple model. In all cases, a transfer function of an Nth-order FIR filter has the form

$$H(z) = \sum_{m = -M}^{M} h[m]z^{-m} \tag{11.31}$$

where $N = 2M + 1$. It is often desired to develop variations of these filters that maintain a close mathematical relationship to the original but exhibit different frequency domain behavior. Examples of this would be the creation of a filter that mirrors the frequency response of another. If the original, or parent, filter is lowpass having a normalized passband width Δ relative to the sampling frequency, then a *mirror filter* would be a highpass filter also having a bandwidth Δ. This can be achieved simply by modulating the impulse response $h[k]$ by a sinusoid at the Nyquist frequency, namely $c[k] = \cos(2\pi(N/2)k/N) = \cos(\pi k) = (-1)^k$. This action will translate the frequency response of the original FIR filter up to the Nyquist frequency. Therefore, the mirror version of an FIR filter having an impulse response $h[k]$, is simply given by

$$h_{\mathrm{mirror}}[k] = (-1)^k h[k] \tag{11.32}$$

Physically, the only difference between the mirror and the original is the alternating sign on the tap-weight coefficients. The spectral relationship between a mirror and its parent is graphically interpreted in Fig. 11.19.

Another useful filter type is called the *complement filter*. A complement filter is mathematically related to the original odd order $H(z)$ through the equation

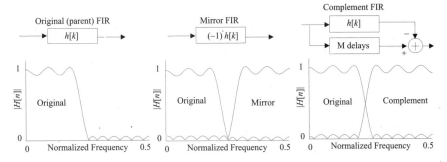

Figure 11.19 Typical mirror and complement FIR filter's relationship to a parent FIR filter.

$$H_{comp}(e^{j\omega}) + H(e^{j\omega}) = 1 \tag{11.33}$$

That is, the original and complement filters, when additively combined, form an all-pass filter. This can be particularly useful in designing banks of filters that must cover a frequency range but must do so with a set of bandlimited filters. The implementation of a complement filter is very simple. Consider again Eq. (11.33) rewritten as

$$H_{comp}(z) = 1 - H(z) \tag{11.34}$$

which implies that the center tap coefficient $h[0]$ of the original FIR filter is subtracted from unity to form the center tap for the complement FIR filter. The other complement FIR filter coefficients are simply the negation of the original. If the initial sample of the complement FIR filter's impulse response is to be indexed at $k = 0$, rather than $h[-M]$ as in Eq. (11.31), then Eq. (11.34) can be expressed as

$$H_{comp\_M}(z) = z^{-M}H_{comp}(z) = z^{-M}(1 - H(z)) = z^{-M} - H_0(z) \tag{11.35}$$

which implies that the compensated filter can be obtained simply with the addition of an M delay shift register, which is a multiplierless operation. This action is interpreted in a block diagram in Fig. 11.19.

Example 11.13: Mirror and Complement FIR Filter

Problem Statement. A 31st-order lowpass FIR filter having a normalized passband ranging over [0.0,0.2] and stopband over [0.225,0.5] is designed using the equiripple method. Create the mirror and complement version of the parent FIR filter.

Analysis and Results. Using Eqs. (11.32) and (11.35), the mirror and complement filters are created and their magnitude frequency response is displayed
in Fig. 11.19. Notice that each FIR filter behaves as predicted, and that each
filter form produces a different interpretation of the original or parent spectrum.

Computer Study. The following program was used to create the data shown
in Fig. 11.19. It is based on the use of a low-order FIR filter saved under the
file name "`loword.fir`" having an impulse response given in the file
"`loword.imp`."

```
>h=rf("chap11\loword.imp")    # Read filter impulse response.
>mirror=zeros(len(h))         # Allocate mirror response.
>for (i=0:len(h)-1)           # Create mirror impulse response.
>>mirror[i]=(-1)^i*h[i]
>>end
>comp=-h                      # Create complement filter.
>center=(len(h)-1)/2          # Compute location of center.
>comp[center]=1+comp[center]  # Adjust center tap.
>hcf=pfft(comp&zeros(500))    # FFT of complement filter.
>hf=pfft(h&zeros(500))        # FFT of original filter.
>hmf=pfft(mirror&zeros(500))  # FFT of mirror filter.
>graph(mag(hf),mag(hmf),mag(hcf)) # Graph magnitude response.
>graph(mag(hf)+mag(hcf))      # Graph original plus complement.  ❖
```

11.12 Windows

In Sec. 3.6, the concept of a *data window* was presented in the context
of improving the appearance of a DFT result. It was shown that windows could smooth a DFT spectrum and suppress local undesirable behavior in the frequency domain. Windows can also be applied to a FIR
filter to suppress large approximation errors at isolated frequencies.
The direct method, for example, was seen to produce filter responses
that can have significant localized overshoot errors at the transition
band edges. This is typical of systems that attempt to represent discrete-time processes requiring an infinitely long time-series with a finite time-series. This problem is pervasive in DSP where, due to finite
memory limitations, arbitrarily long signals may have to be characterized by a time-series containing only a finite number of samples.

Consider the system shown in Fig. 11.20. In this diagram, an arbitrarily long signal $x[k]$ is gated by a finite duration *window function*
$w[k]$. The gated output is defined to be $y[k] = x[k]w[k]$. Assume that
the Fourier transform of $x[k]$ is known to be $X(n)$ and that of the window $w[k]$ has a Fourier transform $W(n)$. Using the *Duality Theorem,* it
can be seen that the Fourier transform of the gated signal $y[k]$ is given
by $Y(n) = X(n)*W(n)$. Ideally, one would like the spectral image of $X(n)$
and $Y(n)$ to be in maximum agreement. However, the spectrum of the

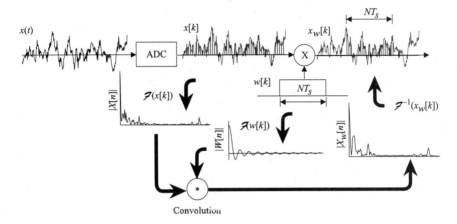

Figure 11.20   Data windows and their effect on signals.

rectangular gating pulse has a $\sin(n)/n = \text{sinc}(n)$ shape. As a result, the spectrum $Y(n) = X(n)*\text{sinc}(n)$ can differ measurably from $X(n)$. An ideal window is one where $Y(n) = X(n)*W(n) = X(n)$. Technically, this would require that $W(n) = \delta(n)$, or $w[k] = 1$ for all time. This obviously is not a window of finite duration and is therefore impractical.

A number of practical windows have evolved that approximate the ideal $W(n) \approx \delta(n)$ to various degrees of accuracy. Some of the more popular of these are the *Blackman, Hamming, Hann,* and *Kaiser* windows. Their discrete-time window specifications are summarized below.

Rectangular Window

A *rectangular window* is given by

$$w[k] = 1 \tag{11.36}$$

for $k \in \{0,1,2,\dots,N-1\}$.

Hann Window

A *Hann window* is given by

$$w[k] = \frac{1}{2}\left[1 - \cos\left(\frac{2\pi k}{N-1}\right)\right] \tag{11.37}$$

for $k \in \{0,1,2,\dots,N-1\}$.

Hamming Window

A *Hamming window* is given by

$$w[k] = 0.54 - 0.46\cos\left(\frac{2\pi k}{N-1}\right) \tag{11.38}$$

for $k \in \{0,1,2,...,N-1\}$.

Blackman Window

A *Blackman window* is given by

$$w[k] = 0.42 - 0.5\cos\left(\frac{2\pi k}{N-1}\right) + 0.08\cos\left(\frac{4\pi k}{N-1}\right) \tag{11.39}$$

for $k \in \{0,1,2,...,N-1\}$.

Kaiser Window

A *Kaiser window* is given by

$$w[k] = \frac{I_0\left[\beta\sqrt{\left(\frac{N-1}{2}\right)^2 - \left(k - \frac{N-1}{2}\right)^2}\right]}{I_0\left(\beta\frac{N-1}{2}\right)} \tag{11.40}$$

for $k \in \{0,1,2,...,N-1\}$, where $I_0(\beta)$ is a zeroth-order Bessel function.

These basic windows can be compared using a number of measures such as main-lobe and maximum sidelobe height as shown in Fig. 11.21 and summarized in Table 11.1. This comparison presumes that the window duration is T seconds, where $T = Nf_s$ for f_s being the sample frequency.

Table 11.1   Windows Parameters

Window	Transition width f_s/N	Highest sidelobe in dB
Rectangular	0.90	−13
Hann	2.07	−31
Hamming	2.46	−41
Blackman	3.13	−58
Kaiser ($\beta = 2.0$)	1.21	−19

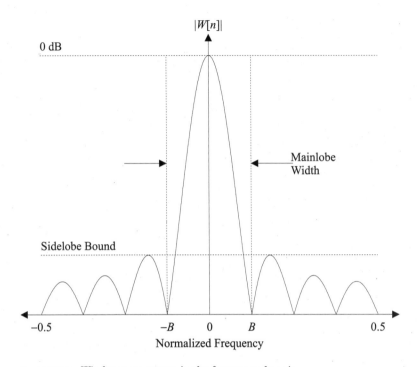

Figure 11.21 Window parameters in the frequency domain.

The effect of applying a window to an FIR filter is to suppress large local deviations from the desired response in the frequency domain. However, it comes at the expense of reducing the steepness of the transition band.

Example 11.14: Effects of Windows

Problem Statement. Many windows have been developed which make subtle trade-offs between the width of the main lobe, the transition bandwidth, and the height of the sidelobes (see Fig. 11.21). Experimentally verify these relationships.

Analysis and Conclusions. Each window can be created numerically and compared in the context of its magnitude frequency response.

Computer Study. The s-file "ch11_14.s" contains the function "*window(M)*" that can be used to create M sample Rectangular, Hann, Hamming, Blackman and Kaiser (β = 2,5) windows. The data is displayed in Fig. 11.22 for M = 51. The data shown were produced using the following sequence of Siglab commands.

```
>include "chap11\ch11_14.s"
>window(51)
```

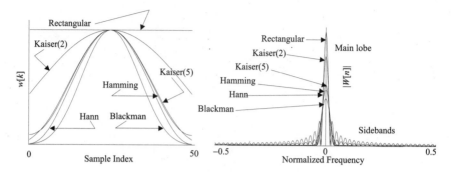

Figure 11.22 Basic windows including a rectangular, Hamming, Hann, Blackman, and Kaiser (β = 2.0, 5.0) along with their magnitude frequency responses.

It can be immediately seen that each window establishes a different trade-off between the main lobe width and the minimum stopband suppression. Using the zoom expansion and crosshair measurement options in Siglab, these differences can be quantified.   ❖

Applying a symmetric window to a symmetric FIR filter will preserve the filter's linear phase response. Since a window is applied to each FIR filter coefficient *a priori*, windowing will not effect the complexity or run-time performance of an FIR filter. The effect of the window is to suppress "ringing" at the expense of widening the transition band.

Example 11.15: Windowing

Problem Statement. In Example 11.5, the direct synthesis method was used to approximate an FIR filter having an ideal piecewise constant magnitude frequency response. The resulting magnitude frequency response, shown in Fig. 11.10, was defined by the $M = 2N + 1$ center samples of $h[k]$ = IDFT($H(n)$). Show that windows can be used to cosmetically improve the appearance of an FIR filter's magnitude response and suppress undesirable large local approximation errors.

Analysis and Conclusions. The direct filter shown in Fig. 11.10 exhibited undesirable ringing. Using the windows reported in Table 11.1, the ringing effect can be suppressed at the cost of broadening the transition band.

Computer Study. The s-file "ch12_9.s" defines the function *winfir(h)*, which accepts an M sample FIR filter impulse response $h[k]$, applies a Blackman window $w[k]$ to $h[k]$, and then displays the two-sided magnitude frequency response of the time-series $h[k]$ and $h[k]w[k]$. The following sequence of commands was used to create the plot shown in Fig. 11.23.

```
>h=rf("chap11\direct.imp") # Read filter impulse response.
>include "chap11\ch11_15.s" # Read function winfir.
>winfir(h) # Compare windowed and nonwindowed mag response.
```

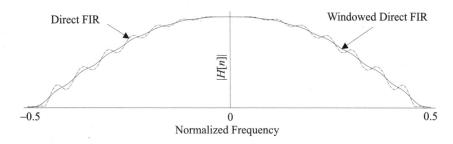

Direct FIR   Windowed Direct FIR

$|H[n]|$

−0.5   0   0.5

Normalized Frequency

Figure 11.23 Effect of windowing on an FIR filter: magnitude frequency responses of the original filter and windowed filter.

The results are displayed in Fig. 11.23. The original filter response exhibits considerable ringing about the values of the specified target filter model. The window is seen to suppress the ringing previously found at the transition band edges. Compared to the nonwindowed FIR filter, the windowed filter response is seen to be in greater agreement with the desired cosine-shaped response.

The capability of the windowed filter can be explored using a soundboard. The sequence of Siglab commands shown below will create a linearly swept FM signal (i.e., chirp) over the normalized frequency range $f \in [0.0, 0.5]$. It is assumed that a 16-bit soundboard is used along with a sampling rate of $f_s = 11,025$ Hz. When the chirp is played, a signal is heard to have uniform intensity across the audio spectrum. After being filtered, the chirp can be heard to have lost energy in the higher frequencies.

```
>h=rf("chap11\direct.imp")    # Read filter impulse response.
>include "chap11\chirp.s"      # Read chirp generator.
>x=chirp(2^14,0,0,0.5)         # Generate test chirp.
>hb=h.*black(len(h))           # Apply a Blackman window.
>y=hb$x                        # Filter the chirp.
>wavout(0.8*2^15*x)            # Play unfiltered chirp.
>wavout(0.8*2^15*y)            # Play filtered chirp.          ❖
```

Applying a window to an FIR filter's impulse response has the effect of "smoothing" the resulting filter's magnitude frequency response. The effect of a window on the steepness of the FIR filter's skirt and stopband suppression can be readily seen in Fig. 11.22. The width of the main lobe of the window controls the amount of smoothing that takes place in the windowed filter's transition band. Which one to use, if any, is an open question. It should be remembered that the precomputed window weights are applied to a signal using sample-by-sample multiplication. Therefore, there is no implementation advantage to be gained by any of the presented nonrectangular windows over another. The choice often is one of personal preference.

11.13 Two-Dimensional FIR Filters

The FIR filter forms developed to this point are 1-D signal processors. There are times, however, when the data has a natural 2-D representation. The classic example of this is image processing. Here, the database is a collection of pixels spatially distributed in two dimensions. Alternatively, the 2-D input can consist of samples taken in time and/or space. The basic image operations are generally considered to be enhancement, manipulation, feature extraction, shape analysis, pattern recognition, scene analysis, and compression/decompression. These are tasks that often lend themselves to 2-D FIR filter solutions. It should be appreciated that 2-D digital filtering is a study unto itself, having its own vocabulary and disciplinarians. It is therefore unrealistic to provide anything more than a cursory overview of 2-D FIR filtering in this chapter.

Two-dimensional filters are often based on general 1-D direct or equiripple FIR filter models extended to 2-D. Two-dimensional FIR filters can also be defined by multiplying two orthogonal 1-D transforms or by "spinning" a 1-D filter about $z_1 = 0$ and $z_2 = 0$ in the z-plane, as was the case in Example 3.9. The basic properties of a 1-D filter can normally be extended to the 2-D case. For example, the linear convolution of an $N_2 \times M_2$ signal with an $N_1 \times M_1$ FIR filter results in a $N_3 \times M_3$ signal where $N_3 = N_2 + N_1 - 1$ and $M_3 = M_2 + M_1 - 1$. Windows (e.g., Hamming) can also be applied to 2-D filters to achieve the same effect they had in the 1-D case.

The output of a causal $N \times M$ 2-D FIR filter having an impulse response $h[k_1,k_2]$ to a 2-D input signal $x[k_1,k_2]$, is given by the *2-D convolution sum*

$$y[k_1, k_2] = \sum_{j_1 = 0}^{N-1} \sum_{j_2 = 0}^{M-1} h[k_1 - j_1, k_2 - j_2] x[j_1, j_2] \qquad (11.41)$$

In the z-transform domain, Eq. (11.41) is given by

$$Y(z_1, z_2) = H(z_1, z_2) X(z_1, z_2) \qquad (11.42)$$

The filter $H(z_1,z_2)$ is said to be *separable* if

$$H(z_1, z_2) = H(z_1) H(z_2) \qquad (11.43)$$

Example 11.16: 2-D Convolution

Problem Statement. A 3×3 uniform 2-D "pedestal" FIR filter has an impulse response given by

$$h[k_1, k_2] = \begin{bmatrix} 1 & 1 & 1 \\ 1 & 1 & 1 \\ 1 & 1 & 1 \end{bmatrix}$$

Suppose the input is a 5×5 database given by

$$x[k_1, k_2] = \begin{bmatrix} 1 & 2 & 3 & 2 & 1 \\ 2 & 3 & 4 & 3 & 2 \\ 3 & 4 & 5 & 4 & 3 \\ 2 & 3 & 4 & 3 & 2 \\ 1 & 2 & 3 & 2 & 1 \end{bmatrix}$$

What is the linear convolution $y = h*x$ and what is the z-transform of $h[k_1,k_2]$?

Analysis and Conclusions. The linearly convolved output is mapped into a 7×7 array. The individual elements are computed to be

$$\begin{bmatrix} 1 & 1 & 1 \\ 1 & 1 & 1 \\ 1 & 1 & 1 \end{bmatrix} * \begin{bmatrix} 1 & 2 & 3 & 2 & 1 \\ 2 & 3 & 4 & 3 & 2 \\ 3 & 4 & 5 & 4 & 3 \\ 2 & 3 & 4 & 3 & 2 \\ 1 & 2 & 3 & 2 & 1 \end{bmatrix} = \begin{bmatrix} 1 & 3 & 6 & 7 & 6 & 3 & 1 \\ 3 & 8 & 15 & 17 & 15 & 8 & 3 \\ 6 & 15 & 27 & 30 & 27 & 15 & 6 \\ 7 & 17 & 30 & 33 & 30 & 17 & 7 \\ 6 & 15 & 27 & 39 & 27 & 15 & 6 \\ 3 & 8 & 15 & 17 & 15 & 8 & 3 \\ 1 & 3 & 6 & 7 & 6 & 3 & 1 \end{bmatrix}$$

where the ijth element of $y[k_1,k_2]$ is simply the sum of the nearest 3×3 samples of $x[k_1,k_2]$ (assuming that x is centered in y). Using a 1-D FIR filter analogy, one should expect that the response of a filter having an impulse response $h[k] = [1,1,1]$ to have a build-up and build-down period. It can be seen that the 2-D output of the 2-D FIR filter consists of a build-up and build-down period found around the periphery of the 7×7 matrix $y[k_1,k_2]$. The maximum output, in fact, is found in the middle of the $y[k_1,k_2]$ matrix.

The z-transform of $h[k_1,k_2]$ is formally given by

$$H[z_1, z_2] = \sum_{k_1=0}^{N-1} \sum_{k_2=0}^{M-1} h[k_1, k_2] z_1^{-k_1} z_2^{-k_2} = \sum_{k_1=0}^{2} \sum_{k_2=0}^{2} z_1^{-k_1} z_2^{-k_2}$$

Computer Study. The s-file "two-d.s" defines a function, $yout(h,x)$, that performs a 2-D linear convolution using a direct sum-of-products approach. Since the linear convolution of an $N_1 \times M_1$ 2-D filter with an $N_2 \times M_2$ 2-D signal is defined over an $N_3 \times M_3$ database, where $N_3 = N_2 + N_1 - 1$ and $M_3 = M_2 + M_1 - 1$, the program zero pads the input signal out to a $N_3 \times M_3$ 2-D array. The s-file "two-dg.s" defines the function $yout(h,x)$ to perform the same 2-D linear convolution as procedure as the function defined in "two-d.s" but displays the input and output time series as 3-D graphs and returns the 2-D convolved result. The format of the function call in either case is given by $yout(h,x)$, where h is the 2-D FIR filter impulse response, and x is a 2-D database. The computed 2-D convolution can be verified by executing the following sequence of Siglab commands.

```
>include "chap11\two-d.s"
>h=ones(3,3)
>x={[1,2,3,2,1],[2,3,4,3,2],[3,4,5,4,3],[2,3,4,3,2],[1,2,3,2,1]}
>yout(h,x) # Convolve h and x.
```

Input data array

0	0	0	0	0	0	0	0	0	0	0
0	0	0	0	0	0	0	0	0	0	0
0	0	0	0	0	0	0	0	0	0	0
0	0	0	1	2	3	2	1	0	0	0
0	0	0	2	3	4	3	2	0	0	0
0	0	0	3	4	5	4	3	0	0	0
0	0	0	2	3	4	3	2	0	0	0
0	0	0	1	2	3	2	1	0	0	0
0	0	0	0	0	0	0	0	0	0	0
0	0	0	0	0	0	0	0	0	0	0
0	0	0	0	0	0	0	0	0	0	0

Convolved output

0	0	0	0	0	0	0	0	0
0	1	3	6	7	6	3	1	0
0	3	8	15	17	15	8	3	0
0	6	15	27	30	27	15	6	0
0	7	17	30	33	30	17	7	0
0	6	15	27	30	27	15	6	0
0	3	8	15	17	15	8	3	0
0	1	3	6	7	6	3	1	0
0	0	0	0	0	0	0	0	0

Notice that the input array is a zero padded version of $x[k_1,k_2]$ and the output agrees with the computed result.   ❖

Eq. (11.42) suggests that the 2-D convolution can be interpreted in the context of 1-D convolution, where it was found to be equivalent to the IDFT of the product of the DFTs of $h[k]$ and $x[k]$. For the 2-D case, the convolved output can be expressed as the 2-D IDFT of the product of the 2-D DFTs of $h[k_1,k_2]$ and $x[k_1,k_2]$. In the frequency domain, the response of the 2-D FIR filter is expressed as

$$H(e^{j\omega_1}, e^{j\omega_2}) = H(e^{j\omega_1})H(e^{j\omega_2}) \qquad (11.44)$$

where ω_1 and ω_2 are normalized frequencies. The frequency response can be represented in magnitude/phase or complex phasor form. One

of the principal functions that a 2-D FIR filter can provide is frequency-selective filtering. As in the 1-D case, 2-D FIR filters can provide precise control over the phase response. This, in some cases, is
not a necessary design objective, but in other cases it is. Video signals,
for example, can be very phase sensitive. A phase shift of 180° of a
monochrome video signal can change the brightest white (e.g., 255) to
the blackest black (e.g., −256). As a result, 2-D image filters are often
designed to have zero phase behavior. A filter $H[z_1,z_2]$ is called *zero
phase* if the frequency response is real. That is,

$$H(e^{j\omega_1}, e^{j\omega_2}) = H^*(e^{j\omega_1}, e^{j\omega_2}) \tag{11.45}$$

Zero-phase FIR filters may be designed using the direct method and a
real spectral model.

A typical use of a 2-D FIR filter is as a data smoother using a 2-D as
a lowpass filter. Consider the 2-D DFT study presented in Example
3.9 and interpreted in Fig. 3.14 (repeated below as Fig. 11.24). The cylindrical 2-D FIR filter impulse response was constructed by taking a
rectangular pulse and spinning it about the center axis. The 2-D DFT
of the FIR filter impulse response is seen to be decidedly lowpass
where most of the spectral energy clustered about $f = 0$.

One-dimensional lowpass filters historically have been used to suppress the effects of broadband noise on a baseband signal. A common
2-D problem is dealing with images that is contaminated by additive
shot or *salt-and-pepper* noise. Shot noise is essentially a high-frequency 2-D random process. If zero mean shot noise $n[k_1,k_2]$ is added
to a signal $x[k_1,k_2]$, then a lowpass 2-D FIR filter will tend to suppress
the effect of the noise in the image plane. However, it should be appreciated that the signal will also smoothed.

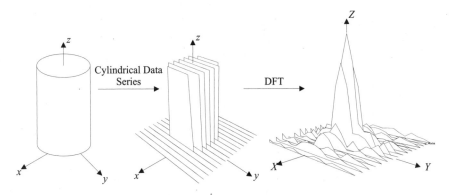

Figure 11.24   2-D DFT (right) of an 8 × 8 cylindrical 2-D FIR filter (left).

Example 11.17: 2-D Lowpass Filter

Problem Statement. The 3×3 uniform 2-D pedestal FIR filter studied in Example 11.16 locally approximates a lowpass FIR filter (i.e., 3×3 moving average filter). Again consider $h[k_1, k_2]$ to be defined as

$$h[k_1, k_2] = \begin{bmatrix} 1 & 1 & 1 \\ 1 & 1 & 1 \\ 1 & 1 & 1 \end{bmatrix}$$

Suppose the input is a simulated 9×9 cross contaminated with additive noise, having the form

$$x[k_1, k_2] = \begin{bmatrix} 0 & 0 & 0 & 0 & 1 & 0 & 0 & 0 & 0 \\ 0 & 0 & 0 & 0 & 1 & 0 & 0 & 0 & 0 \\ 0 & 0 & 0 & 0 & 1 & 0 & 0 & 0 & 0 \\ 0 & 0 & 0 & 0 & 1 & 0 & 0 & 0 & 0 \\ 1 & 1 & 1 & 1 & 1 & 1 & 1 & 1 & 1 \\ 0 & 0 & 0 & 0 & 1 & 0 & 0 & 0 & 0 \\ 0 & 0 & 0 & 0 & 1 & 0 & 0 & 0 & 0 \\ 0 & 0 & 0 & 0 & 1 & 0 & 0 & 0 & 0 \\ 0 & 0 & 0 & 0 & 1 & 0 & 0 & 0 & 0 \end{bmatrix} + n[k_1, k_2]$$

where $n[k_1, k_2]$ is a uniformly distributed random noise. What is the linear convolution $y = h * x$?

Analysis and Conclusions. The linearly convolved output is mapped into a 11 $\times$ 11 array. If $x[k_1, k_2]$ is considered to be noise-free, then the output would be given by

$$\begin{bmatrix} 1 & 1 & 1 \\ 1 & 1 & 1 \\ 1 & 1 & 1 \end{bmatrix} * \begin{bmatrix} 0 & 0 & 0 & 0 & 1 & 0 & 0 & 0 & 0 \\ 0 & 0 & 0 & 0 & 1 & 0 & 0 & 0 & 0 \\ 0 & 0 & 0 & 0 & 1 & 0 & 0 & 0 & 0 \\ 0 & 0 & 0 & 0 & 1 & 0 & 0 & 0 & 0 \\ 1 & 1 & 1 & 1 & 1 & 1 & 1 & 1 & 1 \\ 0 & 0 & 0 & 0 & 1 & 0 & 0 & 0 & 0 \\ 0 & 0 & 0 & 0 & 1 & 0 & 0 & 0 & 0 \\ 0 & 0 & 0 & 0 & 1 & 0 & 0 & 0 & 0 \\ 0 & 0 & 0 & 0 & 1 & 0 & 0 & 0 & 0 \end{bmatrix} = \begin{bmatrix} 0 & 0 & 0 & 0 & 1 & 1 & 1 & 0 & 0 & 0 & 0 \\ 0 & 0 & 0 & 0 & 2 & 2 & 2 & 0 & 0 & 0 & 0 \\ 0 & 0 & 0 & 0 & 3 & 3 & 3 & 0 & 0 & 0 & 0 \\ 0 & 0 & 0 & 0 & 3 & 3 & 3 & 0 & 0 & 0 & 0 \\ 1 & 2 & 3 & 3 & 5 & 5 & 5 & 3 & 3 & 2 & 1 \\ 1 & 2 & 3 & 3 & 5 & 5 & 5 & 3 & 3 & 2 & 1 \\ 1 & 2 & 3 & 3 & 5 & 5 & 5 & 3 & 3 & 2 & 1 \\ 0 & 0 & 0 & 0 & 3 & 3 & 3 & 0 & 0 & 0 & 0 \\ 0 & 0 & 0 & 0 & 3 & 3 & 3 & 0 & 0 & 0 & 0 \\ 0 & 0 & 0 & 0 & 2 & 2 & 2 & 0 & 0 & 0 & 0 \\ 0 & 0 & 0 & 0 & 1 & 1 & 1 & 0 & 0 & 0 & 0 \end{bmatrix}$$

Notice that the cross is now smoothed after being convolved by a 2-D lowpass FIR filter. Because the results near the periphery or boundary of the output space were computed using fewer non-zero sample values than at those points found near the center, the output tapers off rapidly at the edges.

Computer Study. The s-file "two-dg.s" defines the function $yout(h,x)$ which performs a 2-D linear convolution, displays the input and output time series along with their 3-D graphs, and returns the 2-D convolved result. The function is called using the form $yout(h,x)$, where h is the 2-D filter and x is a 2-D database. The 2-D convolution, computed above, can be verified by executing the following sequence of Siglab commands.

```
>include "chap11\two-dg.s"   # Get 2D convolution function.
>h=ones(3,3)                 # Create lowpass filter.
>s=zeros(9,9)                # Allocate image matrix.
>for(i=0:8)                  # Create cross.
>>s[4,i]=1
>>s[i,4]=1
>>end
>x=s + 0.25*rand(9,9)        # Add noise to cross.
>y=yout(h,x)                 # Filter and display results.
Input data array
— data not shown —
Convolved output
— data not shown —
```

The data is graphed in Fig. 11.25. Notice that the convolved response tapers to smaller and smaller values near the periphery of the output array.   ❖

Human visual perception is known to be heavily influenced by the ability to recognize the edges or local boundaries in a complex image. As a result, much attention has been given to developing edge detection and enhancement tools. Edge detection is both critical and often one of the first operations found in an image enhancement signal stream. One class of edge detector is based on gradient methods. Con-

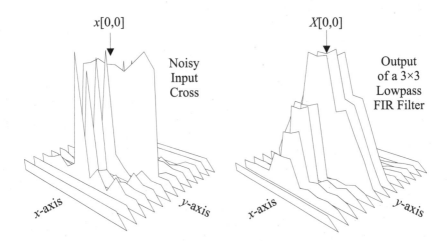

Figure 11.25 Noisy cross time-series (left) convolved with a 3 × 3 lowpass FIR filter (right) (3-D Graph parameters are $x = 18$, $y = 18$, $z = 10$).

sider the data shown in Fig. 11.26, which displays a rising and trailing rounded or smoothed edge and its first and second spatial derivative. The location of the center of edge can be determined by locating the local maxima of absolute value of the first derivative, or from the zero crossing of the second derivative. As a result, edge detection schemes have been developed based on locally computing or estimating the first and/or second derivative of an image.

One of the more popular edge detectors is called the *Sobel edge detector*. Technically, a Sobel edge detector involves taking gradients in the vertical and horizontal direction, computing their square, and a square root. This, for a large image, can be very time consuming. Instead an equivalent technique based on the *Sobel operator* is generally employed. A 3×3 Sobel operator, denoted $S[k_1,k_2]$, is constructed from vertical and horizontal edge detectors, denoted $V[k_1,k_2]$ and $H[k_1,k_2]$ respectively, and are defined in Eq. (11.46).

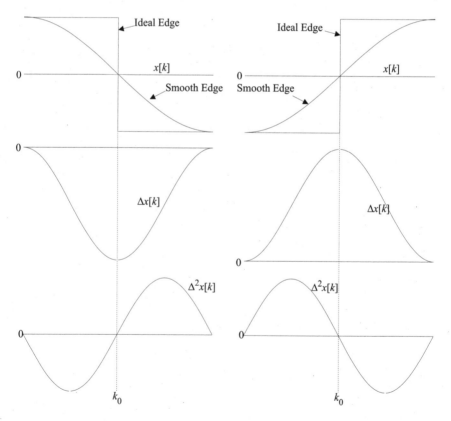

Figure 11.26 A smooth edge $x[k]$ approximation of an ideal edge, first derivative (incremental difference) $\Delta x[k]$, and second derivative (incremental difference) $\Delta^2 x[k]$.

$$V[k_1, k_2] = \begin{bmatrix} 1 & 1 & 1 \\ 0 & 0 & 0 \\ -1 & -1 & -1 \end{bmatrix}$$

$$H[k_1, k_2] = \begin{bmatrix} -1 & 0 & 1 \\ -1 & 0 & 1 \\ -1 & 0 & 1 \end{bmatrix}$$

$$S[k_1, k_2] = V[k_1, k_2] + H[k_1, k_2] = \begin{bmatrix} 0 & 1 & 2 \\ -1 & 0 & 1 \\ -2 & -1 & 0 \end{bmatrix} \tag{11.46}$$

The structure of $S[k_1, k_2]$ implies that the operator is equivalent to first applying a vertical edge detector to an image and then following with a horizontal edge detector.

Example 11.18: 2-D 3 × 3 Sobel Edge Detector

Problem Statement. Apply a 3×3 Sobel edge detector to a 5×5 image $x[k_1, k_2]$ having two level changes,

$$x[k_1, k_2] = \begin{bmatrix} 0 & 0 & -1 & -1 & -1 \\ 0 & 0 & 0 & -1 & -1 \\ 1 & 0 & 0 & 0 & -1 \\ 1 & 1 & 0 & 0 & 0 \\ 1 & 1 & 1 & 0 & 0 \end{bmatrix}$$

Analysis and Conclusions. The convolution of the 5×5 input database with a 3×3 Sobel edge detector is given by the 7×7 data array

$$\begin{bmatrix} 0 & 1 & 2 \\ -1 & 0 & 1 \\ -2 & -1 & 0 \end{bmatrix} \times \begin{bmatrix} 0 & 0 & -1 & -1 & -1 \\ 0 & 0 & 0 & -1 & -1 \\ 1 & 0 & 0 & 0 & -1 \\ 1 & 1 & 0 & 0 & 0 \\ 1 & 1 & 1 & 0 & 0 \end{bmatrix} = \begin{bmatrix} 0 & 0 & 0 & (1) & 3 & 3 & 2 \\ 0 & 0 & -1 & [-1] & (1) & 4 & 3 \\ 0 & -1 & -4 & -4 & [-4] & (1) & 3 \\ (1) & [-1] & -4 & -4 & -4 & [-1] & (1) \\ 3 & (1) & [-4] & -4 & -4 & -1 & 0 \\ 3 & 4 & (1) & [-1] & -1 & 0 & 0 \\ 2 & 3 & 3 & (1) & 0 & 0 & 0 \end{bmatrix}$$

It can be seen that evidence of an edge can be detected by the presence of a sign change across the diagonals of the output matrix (denoted (°) to [°]).

Computer Study. The s-file "two-d.s" used in Example 11.16 can be used to convolve and display the input and output time series. The 2-D convolution

with a 3×3 Sobel operator, computed above, can be verified by executing the
following program

```
>h={[0,1,2],[-1,0,1],[-2,-1,0]}    # Initialize the Sobel operator.
>x=zeros(5,5)                       # Initialize the data matrix.
>x[0,2]=1; x[0,3]=1; x[0,4]=1; x[1,3]=1; x[1,4]=1; x[2,4]=1
>x=-x+x'
>include "chap11\two-d.s"           # Load the 2D convolution function.
>yout(h,x)                          # Filter the data matrix.
Input data array
Columns 0 - 8
— data not shown —
Convolved output
```

0	0	0	0	0	0	0	0	0
0	0	0	0	1	3	3	2	0
0	0	0	-1	-1	1	4	3	0
0	0	-1	-4	-4	-4	1	3	0
0	1	-1	-4	-4	-4	-1	1	0
0	3	1	-4	-4	-4	-1	0	0
0	3	4	1	-1	-1	0	0	0
0	2	3	3	1	0	0	0	0
0	0	0	0	0	0	0	0	0

which agrees with the manually computed results.   ❖

In practice, an image may start out as having sharp edges and, on
transmission, may be received with the edges rounded or smoothed.
The edges can be detected and reconstructed at the receiver using an
edge detection algorithm. Such a situation is diagrammed in Fig.
11.27. The quality of the edge detection is a function of the intrinsic
sharpness of the received edge(s) and the amount of noise contamination.

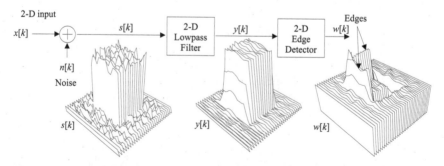

Figure 11.27 Example of typical 2-D image processing and edge detection problem showing a pedestal signal in noise, lowpass smoothing with a 3×3 lowpass filter used in Example 11.17, and a 3×3 Sobel edge detector. (3-D graph parameters are $x = 18$, $y = 18$, $z = 18$).

Example 11.19: Edge Detection

Problem Statement. A 25×25 sample signal having a 13×13 pedestal is convolved with a 3×3 lowpass FIR filter. What is the result of processing the lowpass output with a 3×3 Sobel edge operator?

Analysis and Conclusions. The result of processing the 2-D image through a lowpass and Sobel operator is shown in Fig. 11.27. Notice that the output of the Sobel operator has two sharp positive going ridges and two sharp negative going ridges corresponding to the edges of the original noise-free image.

Computer Study. The s-file "two-dg.s" used in Example 11.17 can be used to convolve and display the input and output time series. The 2-D convolution of the input signal with a lowpass 3×3 lowpass filter produces an output which then convolved with a Sobel operator. These operations are integrated in the s-file "edge.s"

```
>include "chap11\two-dg.s" # Get 2D convolution routine.
>include "chap11\edge.s" # Execute edge script.
— data not shown —
```
❖

The FIR filters considered to this point are constructed using shift registers and multiply-accumulate (MAC) units. By replacing MAC operations with logical operations (AND, OR, EXOR, Exclusive-NOR), other useful 2-D FIR filters can be constructed. One of the most widely used 2-D FIR filters is called a *median filter*. An $N \times N$ median filter interprets an $N \times N$ window in an image database as an array of $M = N^2$ sample values. The median filter sorts through the list of $M = N^2$ values to find the median (middle) value. Such filters can be used successful to remove shot noise from a signal. The median filter works by removing locally extreme samples.

Example 11.20: Median Filter

Problem Statement. Design a 3×3 median FIR filter that maps a 3×3 block of data to the data array's median value.

Analysis and Conclusions. A 3×3 median filter would extract from the nine elements in a local 3×3 regions the fifth element in a sorted list. There are many possible means of sorting a body of data. In general, assuming randomly distributed data, sorting a list of M elements requires at least $M\log_2 M$ comparisons and exchanges. An architectural description of a 3×3 median filter is shown in Fig. 11.28.   ❖

11.14 Summary

In Chap. 11, the concept of an FIR filter was presented. It was shown that an FIR filter can be constructed to exhibit linear phase behavior if the filter's impulse response is symmetric. Stability was shown to be

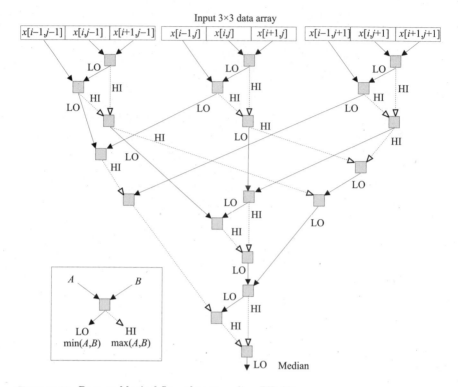

Figure 11.28   Data and logical flow of 3×3 median FIR filter.

easily guaranteed, and a method to ensure that register overflow did not occur during run-time was presented. A number of classic FIR filter design methods were presented, with the most commonly used being the equiripple method. The FIR filter was generally shown to be defined in terms of a collection of multiply-accumulate (MAC) and shift register operations. Fortunately, due primarily to the constant improvements in DSP microprocessors, high-order FIR filters are now a viable option in many applications. As a result, FIR filters have become a fundamentally important element in the DSP engineers inventory.

It should not be forgotten that 2-D FIR filter are also fundamentally important to image processing. They are, in general, simple extensions to the 1-D FIR filters developed in this chapter. Classic examples are 3×3 smoothing filters that compute the weighted average value of nine samples organized in a 3×3 data array. The filter is swept across a 2-D image and the image values replaced by their local average. This will have the effect of denoising an image by suppressing large isolated noise spikes. This field has developed a culture onto itself, and the interested reader is referred to numerous texts on the subject.

11.15 Self-Study Problems

1. Equiripple Design   Design an equiripple FIR filter as specified in Example 11.8, except it is to achieve a –0.1 dB passband ripple deviation. What is the resulting maximum stopband gain?

2. Hilbert Transform   Conduct a set of experiments to verify that the Hilbert filter presented in Example 11.11 achieves a quadrature phase shift at the output. Use the filter as part of the simulation of a phase modulated communication system.

3. Localized Error   Compare the MSE and the minmax error of a direct and equiripple FIR filter design having the specifications presented in Example 11.8. Provide a qualitative analysis with regard to the anticipated performance of each filter in a high-end audio applications (e.g., post-production sound mixing).

4. Frequency Sampling FIR Filter   Design a frequency sampling bandpass filter for use with a 16-bit soundboard sampled at a rate $f = 11025$ Hz. Have the filter achieve a gain of unity at $f \in \{1946, 2162, 2378\}$ Hz and be essentially zero elsewhere. Measure the maximum stopband gain and passband deviation error. Conduct audio tests and subjectively evaluate the quality of the filter. Compare to a 31st-order filter in terms of sound quality and arithmetic complexity.

5. Narrowband Filtering   Use the soundboard to record a section of your favorite music. Add a 60 Hz component to the music to achieve a signal-to-noise ratio of about 0 dB. Design an equiripple FIR filter of order 11, 21, 31, and 61 that removes the unwanted 60 Hz noise. Subjectively evaluate how well your filter serves the role of being a narrowband bandstop filter.

6. Two-Dimensional Filtering   Two-dimensional linear convolution can be performed using sum-of-products of DFTs (i.e., convolution theorem). Revisit Example 11.18 and compare the results obtained using sum-of-products calculation (e.g., s-file "two–dg.s") and a radix-two FFT.

11.16 Reference

1. T.W. Parks and J.M. McClellan, "Chebyshev approximation for nonrecursive digital filters with linear phase." *IEEE Transactions on Circuit Theory*, CT-19, pp. 189–194, March 1972.

12

FIR Filter Implementation

12.1 Introduction

FIR filters have become a popular digital filter methodology due to their intrinsic simplicity and ability to achieve a linear phase response whenever coefficient symmetry was assured. In Chap. 11, the design of a basic FIR filter was developed in the context of direct, frequency sampling, and equiripple synthesis procedures. In any case, it was seen that an FIR filter is implemented using a collection of multiply-accumulate (MAC) operations. In Chap. 12, the mechanics of implementing these basic FIR filter structures will be developed.

12.2 Direct Form FIR

The most common method to implement an Nth-order causal FIR filter having an impulse response given by $h[k] = \{h[0], h[1],...,h[N-1]\}$, with a transfer function

$$H(z) = \sum_{k=0}^{N-1} h[k]z^{-k} \tag{12.1}$$

is the *direct form* FIR filter architecture. The direct form architecture is graphically summarized using a block diagram in Fig. 12.1. It can be seen that a direct form FIR filter consists of a collection of N shift registers with associated tap-weight coefficients h_k, multipliers, and adders. This is equivalent to implementing an inner product of the form $y[k] = a^T x[k]$, where $a \in \Re^N$ (an N-dimensional vector of real constants), $x[k] \in \Re^N$ (an N-dimensional vector of contiguous real samples

Direct Form

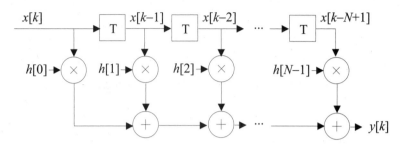

Figure 12.1   Direct Nth-order FIR filter architecture.

$x[k],\dots,x[k-(N-1)])$, "T" denotes the transposition operation, and y is a scalar. For each input sample $x[k]$ and $h_i = h[i]$, a direct **FIR** filter would implement the following set of operations:

For each input sample x[k], do

$$x_0 = x[k]$$

$$y[k] = h_0 x_0 + h_1 x_1 + \dots + h_{N-1} x_{N-1}$$

{update FIFO}

$$x_{N-1} = x_{N-2}$$

$$\dots = \dots$$

$$x_2 = x_1$$

$$x_1 = x_0$$

end do   (12.2)

This simple routine is suited for a sample-by-sample computation. Once initiated, the routine would be repeated over and over during the life cycle of the filter. The designers of modern DSP microprocessors have learned to provide an efficient mechanism for implementing an array of multiply-accumulate (MAC) operations. Most existing DSP microprocessors contain a single MAC unit, which must be efficiently utilized. To implement an FIR filter efficiently, DSP microprocessors generally employ a circular buffers. Data memory is configured as RAM, but coefficients may be stored in either RAM or ROM. In many cases, dual-port memory is used to allow simultaneous supply of two operands to the MAC unit per memory cycle. Data and coefficients are read from memory in a sequential or circular modulo(N) manner, pairwise multiplied to form $h[i]^*x[k-i]$, and sent to the accumulator to create one instance of the partial sum-of-products $y[k]: = y[k] + h[i] x[k-i]$. This operation can, in fact, be performed in a highly pipelined manner. Since DSP microprocessors, in most cases, execute each instruction in a single cycle, the computational latency of a single filter cycle can be estimated to be:

$$T_{FIR\_cycle} \cong (N + 1)T_{inst-cycle} \qquad (12.3)$$

Example 12.1: FIR Filter Performance Prediction

Problem Statement. A DSP microprocessor is assumed to have an instruction cycle time of 50×10^{-9} s (i.e., 50 ns or 20 MIPS) and a data source with a sampling rate of 44.1 kHz. What is the maximum FIR filter order?

Analysis and Results. The FIR filter cycle must be completed in at least $T_s = 1/f_s$ seconds. From Eq. (12.3), $(N + 1) = T_s/T_{inst\_cycle} = 10^{-6}/(44.1 \times 50) \approx 450$. Therefore, the maximum FIR filter order is theoretically bounded by 450, which far exceeds the size of a typical FIR filter having 10 to one 100 taps. As a point of reference, a typical 2-D FIR filter would have a size of 9×9 or fewer taps. The unused cycles, in either case, can be readily used up in performing a host of overhead operations or performing other essential DSP tasks (e.g., FFT). In practice, overhead can consume a significant number of processor cycles. For example, if a typical A/D and D/A converter communicates with the DSP microprocessor using an interrupt-driven protocol, the interrupt service routine (ISR) may consume as many processor cycles as a low-order filter. ❖

As stated, one of the most popular hardware FIR filter implementation technologies is the ubiquitous DSP microprocessor. One hardware alternative to the DSP microprocessor is the DSP *application specific integrated circuit* (ASIC). ASIC implementations of DSP systems may include hardwired DSP primitives and/or DSP microprocessor cores. Another hardware implementation option is to design a DSP system using a *field programmable gate array* (FPGA). The FPGA offers easy programming (and reprogramming), and no nonrecurring engineering (NRE) costs or minimum quantity. However, FPGAs are slower, smaller (in terms of available equivalent gate count), and more expensive than ASICs. Function blocks or application specific standard parts (ASSPs) are standard parts that implement useful DSP operations. Functions such as FIR filters and FFTs are available as ASSPs. ASSPs offer a better cost/performance ratio than DSP microprocessors, involve no NRE costs, and may be obtained in low volume. For low-performance problems, general-purpose microprocessors may be used. Regardless, DSP microprocessors are a principal FIR filter implementation technology and are noted for their powerful MAC units

and an instruction set that is optimized for SAXPY ($s = ax + y$) operations. These operations are at the core of the FIR filter algorithm. An
Nth-order FIR filter would, in general, be expected to perform up to N
SAXPY operations per filter cycle.

Example 12.2: DSP FIR Filter Implementation

Problem Statement. Implement a typical 11th-order equiripple FIR filter using a Texas Instruments TMS320C50 fixed-point DSP microprocessor.

Analysis and Conclusions. A direct FIR filter implementation of the 11th-or-
der FIR filter is shown in Fig. 12.2. For a TMS320C50, the FIR filter main
routine for a direct FIR filter has the following form:

```
        .mmregs            ; Symbols for memory mapped registers.
DATA    .set 02000h        ; Location of data in data memory space.
COEF    .set 02000h        ; Location of coefs in program memory space.
ORDER   .set 11            ; Filter order.
FIRFILT MAR *,AR0          ; Use AR0 for indirect addressing.
        LAR AR0,#DATA + ORDER-1 ; Init data address register.
        RPTZ #ORDER-1      ; Clear P[roduct] register, accumulator
                           ; & repeat the next instr. ORDER times.
        MACD COEF,*-       ; Compute MAC and move (shift) data.
        APAC               ; Perform final accumulation.
        RET                ; Return from routine.
```

The FIR filter implementation is configured as a subroutine that is called
must be called each time an output sample is required. Like most assembly
code, what is accomplished by the above code is not intuitively obvious. The label FIRFILT is the entry point for the routine. The first two instructions in
the routine initialize the address register that will be used to index the data.
The next instruction, RPTZ, clears the multiplier-accumulator pipeline and
causes the next instruction to be executed (in this case) eleven times, with no
looping overhead, except for the initial RPTZ instruction. The MACD instruction
is where the routine will spend the majority of its time. The combination of
the RPTZ and MACD instruction will cause the following events each cycle for
eleven cycles:

- A MAC-step operation will occur using operands obtained by an index
  started at COEF, and the location pointed at by AR0.

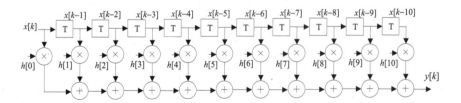

Figure 12.2 Eleventh-order direct FIR filter architecture.

- The data pointed at by AR0 will be shifted "down" in memory.
- The data address register, AR0, will be decremented.
- The coefficient index (which was initialized to COEF on the first MACD execution) will be incremented.

The sequence of events described is essentially a software simulation of the system block diagram shown in Fig. 12.2. The instruction that follows the loop, APAC, finishes the last accumulation of the sequence of MAC operations performed and is necessary, since the C50's MAC unit is pipelined. The result is left in the accumulator register, and the subroutine returns to the caller.

It is clear from the preceding description of the execution of the direct form FIR filter that the C50 microprocessor is highly optimized to execute this type of MAC-intensive algorithm. It should also be appreciated that the routine shown above is one of many possible implementations, all of which have similar performance but differ in the way that data is manipulated.   ❖

12.3 Symmetric FIR Filter Architectures

Most baseline FIR filters are linear phase filters and therefore possess either even or odd coefficient symmetry. Coefficient symmetry allows the direct form FIR filter architecture to be modified as shown in Fig. 12.3. The advantage of this architecture is a reduced multiplication budget. The multiplication complexity, in fact, can be seen to be reduced by nearly a factor of two. If the implementation technology allows additions to be performed at a rate faster than that for multiplication (such as with ASICs), then symmetry can produce a much higher real-time bandwidth (filter cycles per unit time) than an

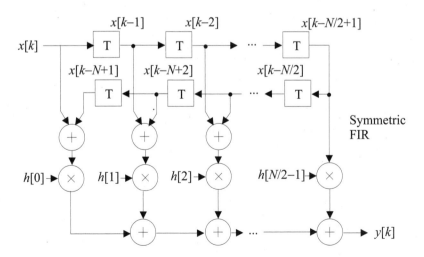

Figure 12.3 Reduced multiplication symmetric FIR filter architecture.

equivalent direct form FIR. In some pipelined hardware systems, the multiply and add rates are identical, in which case the architecture in Fig. 12.3 may only provide a small advantage or even be at a disadvantage compared to a direct form implementation.

Suppose that an Nth-order FIR filter, with even coefficient symmetry, N odd and $h[m]$ being the middle (center) filter coefficient [i.e., $m = (N - 1)/2$], is to be implemented with a symmetric architecture. For each input $x[k]$, an even symmetry FIR filter routine be would implemented by the following set of operations:

For each input sample x[k] and $h_i = h[i]$, do

$$x_0 = x[k]$$

for $(i = 0{:}m-1)$

$$d_i = x_i + x_{N-1-i}$$

end;

$$y[k] = h_0 d_0 + h_1 d_1 + \ldots + h_{m-1} d_{m-1} + h_m d_m$$

{update FIFO stack}

$$x_{N-1} = x_{N-2}$$

$$\ldots = \ldots$$

$$x_2 = x_1$$

$$x_1 = x_0$$

end do   (12.4)

The modification to an even order and/or antisymmetric (odd) FIR filter is straightforward.

Example 12.3: Implementation of a Symmetric FIR

Problem Statement. Consider the implementation of an 11th-order even symmetric FIR using a reduced multiplication approach in microprocessor and ASIC technologies. What are the costs and benefits?

Analysis and Conclusions. A reduced multiplier FIR filter implementation of an 11th-order even symmetric FIR filter is shown in Fig. 12.4. For DSP microprocessors, this implementation approach provides no advantage. DSP microprocessors such as the TMS320C50 and Motorola's 56000 series have an optimized multiplier-accumulator structure that is designed to generate a di-

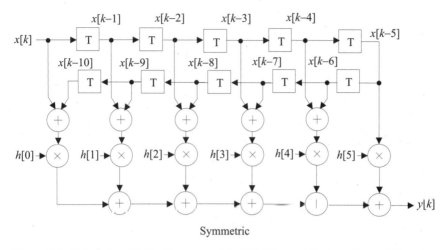

Symmetric

Figure 12.4 Reduced multiplication symmetric FIR filter architecture for an 11th-order FIR.

rect sum-of-products as suggested in Fig. 12.2. The expense of the add, multiply, and multiply-accumulate operations is the same for these DSP microprocessors. Furthermore, because of the MAC structure in the DSP microprocessor, the pre-addition of data before the MAC operation is not efficient. The best that could be achieved with this implementation approach on most DSP microprocessors is a break-even with respect to a direct implementation. However, for most general-purpose RISC and CISC microprocessors, the reduced approach may yield some substantial implementation advantage.

In an ASIC implementation of the architecture shown in Fig. 12.4, the benefits are obvious. If the input data and coefficients are M bits, then adders have a "cost"[*] proportional to M, while multipliers have a cost between $M\log_2 M$ and M^2. In ASIC technologies, the multiplication operation is much more expensive than the addition operation. The number of adders and storage elements are the same in Figs. 12.2 and 12.4, however, the number of multipliers is approximately halved. Therefore, the reduced multiplication symmetric FIR architecture provides substantial benefits in ASIC technologies.   ❖

12.4 Lattice FIR Filter Architecture

The direct form FIR filter is the most often used architecture used to implement basic FIR filters with or without symmetry. Another impor-

---

[*] From a high-level perspective, it can be difficult to exactly quantify how a design decision affects cost. Subsystems cannot be considered in isolation—the entire system must be taken into consideration. The ultimate cost metrics in integrated circuits are die area and packaging.

tant form is called the *lattice* architecture. An Nth-order lattice FIR filter is shown in Fig. 12.5. The coefficients, k_i, are called *PARCOR (partial correlation)* coefficients and will be related to the filter's impulse response in this section. It is seen that the lattice architecture seems to be more complicated than a direct FIR, and in fact it requires $2N$ multiplies per filter cycle versus N for a direct FIR. Nevertheless, a lattice FIR filter is often preferred over a direct FIR filter due to its ability to better suppress coefficient roundoff and finite wordlength arithmetic effects. In addition, lattice structures are important to the design of adaptive filters.

A lattice FIR filter consists of a collection of coefficients, denoted k_i, and are called *PARCOR* or *reflection coefficients*. A lattice architecture can be used to implement a monic FIR filter of the form

$$H(z) = 1 + \sum_{j=1}^{N-1} a_j z^{-j} \qquad (12.5)$$

where the use of the a_j coefficient notation is purposeful to differentiate the coefficients of Eq. (12.5) from Eq. (12.1), which is defined in terms of coefficient $h[j]$. Observe that the filter described by Eq. (12.5) is, in general, nonsymmetric and therefore does not define a linear-phase FIR filter. Furthermore, whereas a typical linear phase FIR filter's impulse response begins with small coefficients, builds in the middle, and then tapers back to small values at the end, no generalizations can be made about lattice filters.

To more fully appreciate the architecture shown in Fig. 12.5, consider the second order lattice filter shown in Fig. 12.6 in the transform-domain. The input-output relationships can be expressed as

$$Y(z) = (1 + k[0]z^{-1} + k[1]z^{-2})X(z)$$

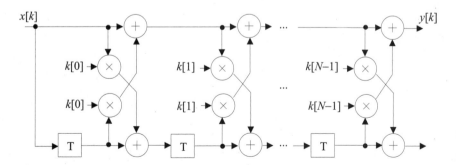

Figure 12.5 Lattice FIR filter architecture.

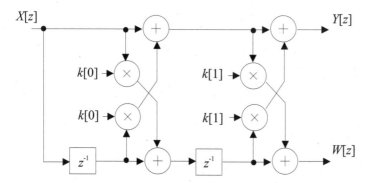

Figure 12.6   Second-order lattice filter.

$$W(z) = (k[1] + k[0]z^{-1} + z^{-2})X(z) \qquad (12.6)$$

It can be seen that $Y(z)$ carries the form suggested in Eq. (12.5). Furthermore, $W(z)$ is related to $Y(z)$ through a similar structure but with coefficients appearing in a permuted (reversed) order.

The mechanics of implementing Eq. (12.5) as a lattice filter are shown below. For an iteration index $i \in \{1,2,...,N-1\}$, define

$$H_i(z) = \left[1 + \sum_{m=1}^{i} a_m^{[i]} z^{-j} \right] \qquad (12.7)$$

The PARCOR coefficients $k[i]$ are related to $a_m^{[i]}$ by

$$a_i^{[i]} = k[i]$$

$$a_m^{[i]} = a_m^{[i-1]} - k[i]a_{i-m}^{[i-1]}$$

$$a_m = a_m^{[N-1]} \qquad (12.8)$$

Reversing the process, for $j \in \{N-1, N-2,..., 2, 1\}$, it follows that

$$k[N-1] = a_{N-1}^{[N-1]}$$

$$k[j] = a_j^{[j]}$$

$$a_m^{j-1} = (a_m^{[j]} + k[j]a_{j-m}^{[j]})/[1 - k^2[j]] \qquad (12.9)$$

Example 12.5: Lattice FIR

Problem Statement. A simple FIR filter has a transfer function $H(z) = 1 - 0.9z^{-1} + 0.64z^{-2} - 0.575z^{-3}$. Compute the PARCOR coefficients for a lattice FIR filter and return the model back to a direct FIR filter form.

Analysis and Results. Directly applying Eqs. (12.8) and (12.9), yields the following:

$$h_{direct} = \{1.0, -0.9, 0.64, -0.575\}$$

$$h_{lattice} = \{-0.672, 0.183, -0.575\}$$

The direct and lattice filters are interpreted in Fig. 12.7. Manually computing the output of both at the first two sample instances produces the following data.

Architecture	$y[0]$	$y[1]$
Direct	1	-0.9
Lattice	1	$-0.672 - 0.674 \times 0.183 + 0.183 \times -0.575 = -0.9$

The data produced using the direct and lattice architectures are seen to be in agreement.

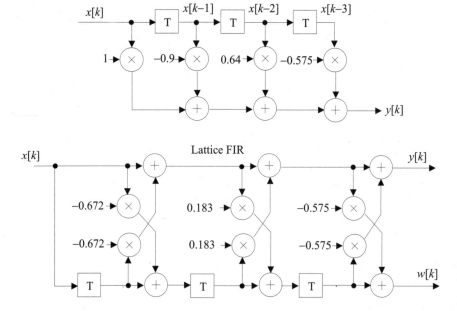

Figure 12.7   Direct and lattice low-order FIR filter architectures.

Computer Study. The s-file "`lattice.s`" defines a function *lattice(x,c)*, where *x* is a coefficient array, *c* = 0 corresponds to a direct to lattice FIR filter conversion, and *c* = 1 corresponds to a lattice to direct FIR filter conversion. To convert $H(z) = 1 - 0.9z^{-1} + 0.64z^{-2} - 0.575z^{-3}$ to a lattice filter and back again, use the following sequence of Siglab commands.

```
>include "chap12\lattice.s"
>a = [1,-0.9,0.64,-0.575]
>b = lattice(a,0)              # 0 => direct to lattice
>b
 -0.672 0.183 -0.575
>lattice(b,1)                  # 1 => lattice to direct
 1 -0.9 0.64 -0.575
```
❖

12.5 Distributed Arithmetic Implementation

Implementing an FIR filter requires that the primitive MAC operation be efficiently implemented. Currently, general purpose DSP microprocessors and commercial ASIC FIR filter chips are commonly used in this role. However, upon close inspection of Eq. (12.1), one notes that the FIR filter coefficients h_i are generally real and known *a priori*. As a result, the partial product terms $h[i] x[k-i]$, found in a convolution sum,

$$y[k] = \sum_{n=0}^{N-1} h[n]x[k-n] \qquad (12.10)$$

are technically defined by a process known as *scaling* and not multiplication.[*] Since scaling rather than multiplication is required, alternatives to traditional MACs can be considered. One of the most attractive alternatives is to replace a traditional multiplier with high-bandwidth *memory table lookup* (TLU) operations. A TLU approach simply emulates how youths were taught to multiply using multiplication tables that are committed to memory. The difference is that now semiconductor memory stores the precomputed values of partial products, which are read out as needed. This is the concept on which *distributed arithmetic* (DA) filters are based.

A DA filter assumes that data is coded as an *M* bit two's-complement word, specifically,

$$x[k] = -x[k{:}0] + \sum_{i=1}^{M-1} x[k{:}i]2^{-i} \qquad (12.11)$$

---

[*] Scaling refers to the multplicative combination of a variable and a constant. Multiplication is defined to be multplicative combination of two variables.

where $x[k{:}i]$ is the ith-bit of sample $x[k]$. Substituting Eq. (12.11) into Eq. (12.10), one obtains

$$y[k] = \sum_{r=0}^{N-1} h[r]\left(-x[k-r{:}0] + \sum_{i=1}^{M-1} x[k-r{:}i]2^{-i}\right)$$

$$= \sum_{r=0}^{N-1} h[r]x[k-r{:}0] + \sum_{r=0}^{N-1}\sum_{i=1}^{M-1} h[r]x[k-r{:}i]2^{-i} \qquad (12.12)$$

If the order of the double summation is reversed, then[*]

$$y[k] = \sum_{r=0}^{N-1} h[r]x[k-r{:}0] + \sum_{i=1}^{M-1} 2^{-i}\sum_{r=0}^{N-1} h[r]x[k-r{:}i] \qquad (12.13)$$

Suppose that a 2^N word memory lookup table, denoted θ, contains a pre-programmed mapping of an array (vector) of one bit values, namely $\underline{x}[k{:}i] = \{x[k{:}i], x[k-1{:}i],..., x[k-N-1{:}i]\}$ into a P bit word (typically $P = M$) given by Eq. (12.14). Specifically, for the elements of $\underline{x}[k{:}i] = \{x[k{:}i], x[k-1{:}i],..., x[k-N-1{:}i]\}$ are taken from the ith-common bit location of $x[k-s]$, $s\in\{0,1,2,...,N-1\}$, $\theta[\underline{x}[k]{:}i]$ is given by

$$\theta[\underline{x}[k]{:}i] = \sum_{r=0}^{N-1} h[r]x[k-r{:}i] \qquad (12.14)$$

where $x[i{:}s]\in\{0,1\}$. If, for example, $N = 8$ and $\underline{x}[k{:}i] = \{10110101\}$, then $[\underline{x}[k]{:}i] = h[0] + h[2] + h[3] + h[5] + h[7]$.

Referring to Fig. 12.8, note that the vector of binary values $[\underline{x}[k]{:}i]$ is presented to the table θ beginning at the common least significant-bit location moving to the most significant. The first $(M-1)$ binary-valued vectors correspond to positive weight bit locations. However, the vector $[\underline{x}[k]{:}i+1]$ has a weight of two relative to vector $[\underline{x}[k]{:}i]$. This explains the scaling factor 2^{-i}, found in Eq. (12.13), implemented using a shift adder as shown in Fig. 12.8. The last, or $(M-1)$st binary-valued vector, is taken from the common sign-bit (negation) location which requires that the output of table θ be subtracted from the accumulator.

---

[*] The redistribution of the order of summation has given this filter class its name, distributed filter.

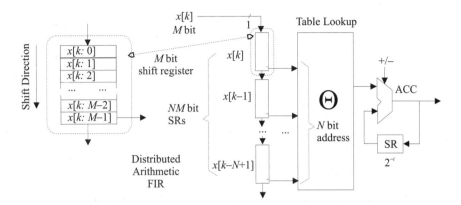

Figure 12.8 Distributed arithmetic architecture.

The advantage of a distributed arithmetic filter implementation is twofold. First, a DA implementation can often provide a definite real-time bandwidth advantage. Suppose an $M = 16$ bits $N = 12$th-order FIR filter is to be implemented using a DSP microprocessor and a distributed filter. The DSP microprocessor will be assumed to have a 100 ns MAC cycle and the distributed arithmetic filter will be based on a $2^{12} \times 16$-bit memory table having a cycle time of 10 ns. A filter cycle for a typical DSP microprocessor would be on the order of

$$T_{FIR\_cycle} = NT_{MAC\_cycle} \qquad (12.15)$$

which results in a 1200 ns cycle time, or a real-time rate of 833 kHz. A distributed arithmetic filter would run the same filter with a latency of

$$T_{FIR\_cycle} = MT_{Memory\_cycle} \qquad (12.16)$$

which results in a filter cycle of 160 ns or a real-time rate of 6250 kHz, which is 750 percent faster. This is done without an appreciable increase in hardware and, in many cases, is less complex than designs based on DSP microprocessors.

The second reason for using a DA implementation is the additional precision gained by forming all of the bit-wise partial sums-of-products before writing them to the table. This can reduce the roundoff error encountered in an implementation. Roundoff error will be developed further in later chapters.

Example 12.6: Distributed Arithmetic FIR Filter

Problem Statement. The simple FIR filter studied as a lattice filter has a transfer function $H(z) = 1 - 0.9z^{-1} + 0.64z^{-2} - 0.575z^{-3}$. Determine the table contents and compute $y[3]$ if the causal input is a four-bit time-series $x[k]$ = {1,0,–1,0,...}.

Analysis and Results. The worst-case input is $x[k]$ = {1,–1,1,–1}, which would produce an output of $3.115 < 2^2$. Therefore, at least two bits must be reserved for the integer field of the output adder. The largest possible table lookup output is $1.64 < 2^1$, which means that the data word presented by the table must have at least one integer bit. Assume, then, for exemplary purposes, that the memory table contents be coded into an eight-bit word consisting of a sign bit, one integer bit, and six fractional bits. The 16 possible table contents of the $2^4 \times 8$ lookup table are computed (e.g., if $[\underline{x}[k]:i]$ = [0,0,1,1] $\rightarrow <0.64 - 0.575>_6$ = 0.0625, where $<x>_6$ denotes rounding x to six fractional bits of precision), and shown in Table 12.1.

The four 4-bit input sample values at sample index three are $x[3]$ = 0 $\leftrightarrow$ $[0_\Delta 000]$ = 0, $x[2]$ = –1 $\leftrightarrow$ $[1_\Delta 000]$ = –1, $x[1]$ = 0 $\leftrightarrow$ $[0_\Delta 000]$ = 0, and $x[0]$ = 1 $\leftrightarrow$ $[0_\Delta 111]$ = 7/8, where Δ denotes the binary point location. The real output $y[3]$ is given by $y[3] = h[0]x[3] + h[1]x[2] + h[2]x[1] + h[3]x[0] = h[1]x[2] + h[3]x[0]$ =

Table 12.1   DA Table Contents

$v[k-0:i]$	$v[k-1:i]$	$v[k-2:i]$	$v[k-3:i]$	Table
0	0	0	0	0
0	0	0	1	–0.5781
0	0	1	0	0.6406
0	0	1	1	0.0625
0	1	0	0	–0.9062
0	1	0	1	–1.4687
0	1	1	0	–0.2656
0	1	1	1	–0.8281
1	0	0	0	1.0000
1	0	0	1	0.4218
1	0	1	0	1.6406
1	0	1	1	1.0625
1	1	0	0	0.0937
1	1	0	1	–0.4687
1	1	1	0	0.7343
1	1	1	1	0.1718

0.425. Using four-bit words, computing $y[3]$, using a four bit ALU (three fractional bits) would result in $y_{ALU}[3] = <-0.9>_4 \times <-1>_4 + <-0.575>_4 \times <1>_4 = -1 \times -1 + -0.5625 \times 0.875 = 0.453125$, which is in error by 0.028125. The four-tap distributed filter will execute the sequence of operations show in Table 12.2.

At the conclusion of the distributed filter cycle, the accumulator holds 0.40039, which is closer to the desired result. The error is -0.02460, which is smaller that obtained using an general-purpose ALU.   ❖

Higher-order distributed filters can be constructed from lower-order distributed filters using a tree architecture as shown in Fig. 12.9. Higher-order designs are those whose order exceeds that address space of a single lookup table. In such cases, an FIR filter of order N is spread across L tables having an address space of n-bits each. In particular, $L = \lceil N/n \rceil$ where $\lceil ° \rceil$ denotes the ceiling function. This particular design paradigm is also gaining popularity in implementing 2-D FIR filters and transforms having coefficients that are known a-priori.

Example 12.7: 2-D Distributed Arithmetic Implementation

Problem Statement. A 2-D Laplacian mapping is often used in image processing to enhance edges. The coefficients of a 2-D Laplacian filter are given by:

$$L = \begin{bmatrix} -16 & -7 & -13 & -7 & -16 \\ -7 & -1 & 12 & -1 & 7 \\ -13 & 12 & 160 & 12 & -13 \\ -7 & -1 & 12 & -1 & -7 \\ -16 & -7 & -13 & -7 & -16 \end{bmatrix}$$

Implement the 2-D Laplacian filter as a DA FIR.

Analysis and Conclusion. The ij coefficient in the Laplacian matrix L operates on a data element $x_{i,j}$ which is taken from a 2-D image space. In addition, it can be seen that the Laplacian matrix L matrix exhibits considerable coefficient symmetry, which can be exploited in a DA design as shown in Fig. 12.10.   ❖

Table 12.2   Distributed Arithmetic Filter Execution Table

i	Address vector	Table $<\phi>_6$	ACC	$ACC = ACC \pm 2^{-i}\phi$
0	[0001]	−0.5781	0	0+(−0.5781/8) = −0.07226
1	[0001]	−0.5781	−0.07226	0.07266 + (−0.5781/4) = −0.21679
2	[0001]	−0.5781	−0.21679	−0.21679 + (−0.5781/2) = −0.50585
3	[0100]	−0.9062	−0.50585	−0.50585 − (−0.9062/1) = 0.40039

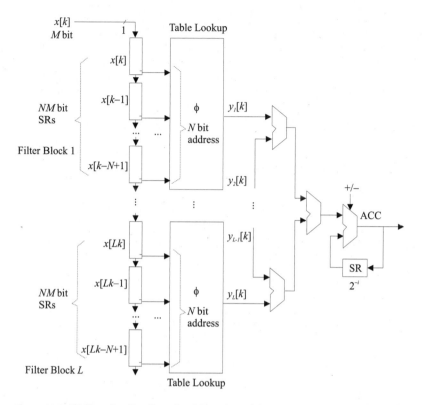

Figure 12.9 High-order distributed arithmetic architecture.

12.6 Canonic Signed Digit

In Chap. 10, the canonic signed digit (CSD) system was introduced. The CSD was employed by a first-generation DSP chip (Intel 2920) in an attempt to accelerate multiplication. Since FIR filters are MAC-intensive, the CSD continues to be considered in special cases as a potentially viable technology. The most common manifestation of the CSD is one which approximates FIR filter coefficients with a highly sparse ternary (i.e., digits in the set $\{-1,0,1\}$) number (i.e., maximally dense in zeros). If a coefficient c_i having an N bit binary representation can be represented with an acceptable error as

$$c_i = C_j 2^j + C_k 2^k + C_m 2^m$$

where $C_i = \pm 1$, then an N-bit word can be approximated by a one-, two-, or three-digit number. As a result, the multiplication $x[k]c_i$ can be performed using only a few weighted shift-adds in comparison to up to N shift-adds in the original case.

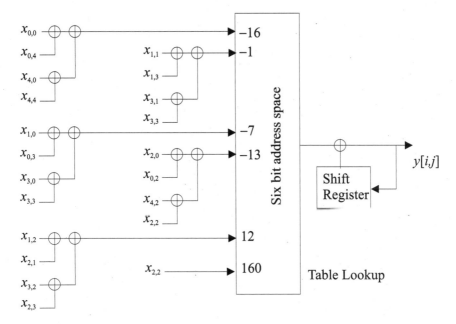

Figure 12.10 Distributed arithmetic implementation of a 2-D image filter.

Example 12.8: Canonic Signed-Digit FIR Filter Implementation

Problem Statement. Assume that an element in the signal processing stream has a magnitude frequency response of the form

$$\left| H(e^{j\omega}) \right| = \frac{\sin(\omega)}{\omega}$$

which is illustrated in Fig. 12.11. Design a 15th-order CSD FIR filter that will compensate for the $\sin(x)/x$ roll-off.

Analysis and Conclusions. Using trial and error, and noting that FIR filters having alternating coefficients generally have a highpass shape, define the 15th-order FIR filter to be $h[k]$ = {−1,4,−16,32,−64,136,−352,1312,−352,136, −64,32,−16,4,−1}, where the filter coefficients have the following CSD representations.

Coefficient	CSD	Number of binary digits
1312	1024 + 256 + 32	3
−352	−256 − 64 − 32	3
136	128 + 8	2

The remaining coefficients only require one binary digit to encode. An FIR filter cycle could then be implemented using 2×5 (one digit coefficients) + 2×1 (for $h[\pm 2]$) + 3×1 (for $h[\pm 1]$) + 3×1 (for $h[0]$) = 18 shift-add operations. The uncompensated, compensator, and compensated magnitude frequency responses are shown in Fig. 12.11. It can be seen that the low-complexity FIR filter does improve the overall system magnitude frequency response, but not to the degree that a carefully designed, but far more complex, equiripple compensator could.

Computer Study. The following sequence of Siglab commands was used to produce the data shown in Fig. 12.11.

```
>w=pi*rmp(512,1/511,512)
>h=ones(1)&((sin(w[1:511]))./(w[1:511]))
>hfir=[-1,4,-16,32,-64,136,-352,1312,-352,136,-64,32,-16,4,-1]
>hf=mag(pfft(hfir&zeros(1024-len(hfir))))
>k=sum(hfir);k
  790
>ograph(h,hf/k,h.*hf/k)                                        ❖
```

12.7 Multiplierless FIR Filters

In the previous section, an FIR filter with low multiplier complexity was presented. The reduced complexity was achieved by approximating a set of FIR filter coefficients as a set of numbers that have a simple implementation using only a few shift and add operations. This is achieved at the expense of introducing additional coefficient rounding errors to the output. The extent of the errors is case dependent. There were, however, multiplier-free FIR filters presented in Chap. 11. Called *moving average* and *comb* filters, they can be trivially implemented using an array of shift registers and distributed adders. Such

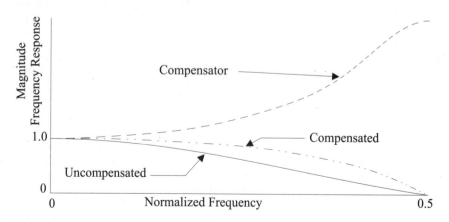

Figure 12.11   CSD implementation of a 15th-order FIR.

filters lend themselves well to ASIC implementation. More complex multiplier-free FIR filters are also in use.

12.8 Two-Dimensional FIR Filters

Two-dimensional FIR filters are generally simple extensions of 1-D architectures. If coefficient symmetry is present, as in often the case in 2-D FIR filters, then a reduced multiplication architecture can be used if a custom design in warranted. Otherwise, the direct architecture is generally employed.

12.9 Finite Wordlength Effects

In Example 11.12, the effects of coefficient roundoff errors was experimentally studied. In Chaps. 15 through 17, a formal examination of rounding and truncation (called *finite wordlength effects*) is presented. This analysis, admittedly, will concentrate on quantifying the finite wordlength effects in IIR filters. The fact that three chapters are needed to address this problem speaks to the difficulty of quantifying finite wordlength effects in IIRs. FIR filters, however, due to their simple structure, are somewhat easier to analyze.

It may be recalled that the worst-case FIR filter response can be trivially computed [Eq. (11.5)]. It shall be assumed that the input has been properly scaled so that no run-time overflow is possible. Therefore, errors due to finite wordlength effects manifest themselves in two forms:

- Coefficient roundoff errors

- Arithmetic errors

Coefficient roundoff errors, as the name implies, corresponds to induced filter imprecision attributable to rounding real coefficients to F bits of fractional precision. Coefficient roundoff errors in IIRs will initially be studied in the context of a sensitivity (partial derivative) analysis (see Chap. 15). It will be found, at that time, that such mathematical methods apply only to cases where the roundoff errors are infinitesimal. In general, simulation is used to experimentally measure the effects of coefficient rounding in complicated systems. Due to the intrinsic simplicity of an FIR filter, it will be shown that simulation, and to some degree, analytical methods can be used to investigate FIR filter coefficient rounding.

Assume that rounding results in a mapping of an ideal coefficient having a value of $h[i]$ to a rounded value of $h[i] + \Delta_i$, where $|\Delta_i| \le 2^{-(F-1)}$ (i.e., $\pm$LSB/2). The size of the error is seen to be a function of F, the number of fractional bits used in the finite wordlength representation of $h[i]$. For an FIR filter having the form

$$y[k] = \sum_{m=0}^{N-1} h[m]x[k-m] \qquad (12.17)$$

the response of an FIR filter implemented using rounded coefficients is given by

$$y'[k] = \sum_{m=0}^{N-1} (h[m] + \Delta_m)x[k-m] = y[k] + \sum_{m=0}^{N-1} \Delta_m x[k-m] \qquad (12.18)$$

The error due to coefficient rounding is therefore

$$e[k] = (y'[k] - y[k]) = \sum_{m=0}^{N-1} \Delta_m x[k-m] \qquad (12.19)$$

Notice that the error budget is a scaled by the individual values of $x[k]$. If the input has a mean value of zero and the FIR filter coefficients are random, then the roundoff error will also have a mean value of zero. Computing the variance, however, is far more challenging, since it is a function of the input signal power. If the input is assumed to be an impulse, then Eq. (12.19) simplifies to

$$e[k] = (y'[k] - y[k]) = \sum_{m=0}^{N-1} \Delta_m \delta[k-m] = \Delta_k \qquad (12.20)$$

where Δ_k comes from a uniformly distributed random population defined over the interval $[-2^{-F}/2, 2^{-F}/2)$. The error is seen, in this case, to have a mean value of zero and a variance $\sigma^2 = 2^{-2F}/12$ (i.e., $Q^2/12$ after Table 1.1). This is of little practical value, however, since the input signal is rarely an impulse. If the input is random and arbitrary, then the error can be modeled as

$$\sigma_e^2 = \frac{N2^{-2F}\sigma_x^2}{12} \qquad (12.21)$$

where σ_x^2 is the signal variance (power). That is, the coefficient roundoff error variance of an Nth-order FIR filter is essentially the roundoff error power of each rounding (i.e., $Q^2/12$) scaled by the signal power and filter order.

Example 12.9: Coefficient Roundoff Error

Problem Statement. Experimentally verify the FIR filter roundoff error models.

Analysis and Results. The error models can be examined using an impulse and random signal. Assume that the FIR filter is modeled by a Nth-order Hamming window. The worst-case gain of the FIR filter is then given by

$$y[k] = \sum_{m=0}^{N-1} |h[m]| \le 2^I$$

If the input is bounded by unity on a sample-by-sample basis, then an M-bit fixed-point FIR filter implementation would require that at least I bits be reserved for the integer data field to prevent run-time overflow. This would leave $F = M - I - 1$ bits of fractional precision (i.e., $Q = 2^{-F}$).

Computer Study. The s-file "coef.s" defines a function that uses a Hamming window as an FIR. The function is called using the format $coef(N,F,x)$ where N is the order of the Hamming FIR filter, F is the number of fractional bits of precision, and x is the input signal. The program generates a rounded Hamming FIR filter and computes the impulse and forced responses. The error variance for both processes are computed. The sequence of Siglab commands shown below uses a 32nd-order Hamming FIR filter with 8 fractional bits of precision, and a 200 sample Gaussian distributed random test signal. The worst-case gain is computed to be 16.82, which technically would require $I = 5$ bits of integer precision. If the FIR filter is executed using saturating arithmetic, it is quite likely that four bits of integer precision would produce satisfactory run-time performance in this case. A comparison of the ideal (floating-point) and rounded fixed-point forced responses are shown in Fig. 12.12. It can be seen that for eight bits of fractional precision, the rounded and ideal filter responses are essentially identical.

```
>include "chap12\coef.s"
>coef(32,8,gn(200))
```

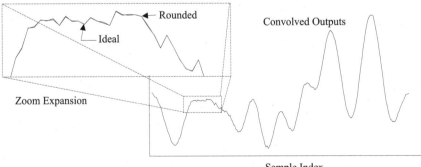

Figure 12.12 Coefficient roundoff error effects.

```
16.82              # Worst case gain.
0.000234375        # Impulse response mean error ~0.
1.32826e-06        # Impulse response error variance.
1.27157e-06        # Error variance model, Eq. (12.20).
-0.000948111       # Forced response mean error ~0.
3.70135e-05        # Forced response error variance.
4.06901e-05        # Error variance model, Eq. (12.21).
```

It can be seen that the roundoff error models closely predict the experimentally produced error variances. ❖

Arithmetic errors can occur when partial products of the form $x[k - m]h[m]$ are computed using finite-wordlength multiplier-accumulators. Arithmetic errors in FIR filters are covered in detail in Chap. 17. For an Nth-order FIR filter, there are N MAC operations, each of which consists of presenting two M-bit words, each having F fractional bits of precision, to a multiplier. The result, in general, is a $2M$-bit full precision product having $2F$ fractional bits of precision. For F sufficiently large ($F > 6$ bits), it is generally assumed that rounding the full precision product back to F bits of fractional precision introduces an error having zero mean and variance $Q^2/12$ into the output, where $Q = 2^{-F}$. It is normally assumed that each error source is statistically independent from the others. Under this assumption, the arithmetic error variance will be shown to be $\sigma^2 = NQ^2/12$ [see Eq. (12.2) and Sec. 17.3]. If the N MAC operations present a full-precision product to an extended-precision accumulator (typically $2M + 3$ bits), then the precision found at the MAC's output is at least equal to the precision of the input data. The error in rounding the extended precision accumulator register to F bits after all N partial products have been computed and accumulated is the now familiar $\sigma^2 = Q^2/12$.

Distributed arithmetic FIR filters were developed in Sec. 12.5. One of the advantages of a DA implementation over a direct FIR filter implementation is a reduced finite wordlength error variance. The statistical roundoff error for an Nth-order direct form FIR filter can be modeled as $\sigma^2 = NQ^2/12$, where Q represents the precision of the least significant bit in volts/bit. For a distributed arithmetic FIR filter, the statistical error budget is $\sigma^2 = Q^2/9$. Thus, the DA precision advantage can be significant, especially when N become large.

It can be noted in Example 12.9 that the worst case gain was $G_{max} = 16.82 < 2^5$. Intuitively, it would make sense to actually have the worst-case gain fall on a power-of-two value, say 2^n, or in the case of Example 12.9, 2^5. This can be achieved by scaling the FIR filter by a factor k where

$$k = \frac{2^n}{G_{max}} \qquad (12.22)$$

By multiplying each filter coefficient $h[i]$ by k, the new FIR filter would have a maximum gain of $G'_{max} = 2^n$. This action would be performed offline and not affect the filter's run-time bandwidth. It should be appreciated, however, that the scaled filter's magnitude frequency response will be increased by a factor k. While the absolute magnitude frequency response will be changed, the relative gain will be left unaffected.

Scaling an FIR filter by a factor will not appreciably affect the realized roundoff error variance, assuming that there are at least six bits or more of fractional precision. What scaling by k will do, however, is change the output signal strength (power) by a factor of k^2. As a result, the *signal-to-noise* ratio (*SNR*), given by

$$SNR = \frac{\sigma_{y[k]}^2 \quad \text{(signal power)}}{\sigma_{n[k]}^2 (\text{roundoff noise power})} \tag{12.23}$$

measured at the output of the FIR filter, will be changed by scaling. The signal power is a function of signal amplitude and can be expressed in terms of G_{max} as

$$\sigma_{y[k]}^2 = G_{max}^2 \alpha^2 \tag{12.24}$$

This is easy to see if the SNR of an unscaled FIR filter has an SNR given by SNR_0, where

$$SNR_0 = \frac{G_{max}^2 \alpha^2}{\sigma_{n[k]}^2} \tag{12.25}$$

Upon scaling by k, given by Eq. (12.22), the SNR for the scaled FIR filter, say SNR_+, is given by

$$SNR_+ = \frac{2^{2n}\alpha^2}{\sigma_{n[k]}^2} = \frac{k^2 G_{max}^2 \alpha^2}{\sigma_{n[k]}^2} = k^2 SNR_0 \tag{12.26}$$

That is, the SNR ratio is improved by a factor k^2. For example, the filter studied in Example 12.9 could be scaled by $k = 32/G_{max} \sim 1.9$, which would result in an improvement SNR of $k^2 = 3.61$.

Scaling can take place in a downward direction as well. Logically, a scale factor $k = 2^{n-1}/G_{max}$ can also be considered. This scale factor k can be used to adjust the gains downward to the nearest power-of-two value. The immediate concern is that the signal power is decreased by a factor of k^2. However, it should be appreciated that number of fractional bits which can be committed to a solution can be increased by one-bit. This means that the noise power decreases by a factor of four. As a consequence, the new SNR ratio, say SNR_-, satisfies

$$SNR_- = \frac{2^{2n-2}\alpha_2}{\left(\dfrac{\sigma_{n[k]}^2(\text{roundoff noise power})}{2^2}\right)} = \frac{2^{2n}\alpha^2}{\sigma_{n[k]}^2} \qquad (12.27)$$

and would result in the same SNR advantage reported in Eq. (12.26).

Example 12.10: Gain Scaling

Problem Statement. Experimentally verify the SNR gains predicted in Eqs. (12.26) and (12.27).

Analysis and Conclusions. A Gaussian random process $x[k]$ can be used as an ideal signal that is rounded to F fractional bits of precision. The scale factors k found in Eqs. (12.26) and (12.27) can be defined in terms of the dynamic range of $x[k]$ before it is rounded. The SNRs can then be compared. The results of rounding $N = 1024$ sample Gaussian random time-series using $F = 8$ is shown in Fig. 12.13. The value of G_{max} is show to be bound within a radix-two range from four to eight. The unscaled and scaled up [see Eq. (12.26)] error variances are seen essentially the same. When the signal is scaled down [Eq. (12.27)], the roundoff error variance is seen to be about one bit less than the unscaled roundoff error variance (decreased in power by four). Since these parameters support the assumptions made in developing the SNR Eqs. (12.26) and (12.27), the conclusions regarding the SNR improvements follow.

Computer Study. The data presented in Fig. 12.13 was produced using s-file "round.s" which defines a function, $bits(F,M)$, where F is the number of fractional bits to be retained of a 1024 sample Gaussian time-series. The experiment is repeated M times. The following sequence of Siglab commands was used to create the data shown in Fig. 12.13.

```
>include "chap12\round.s"
>bits(8,100)                                              ❖
```

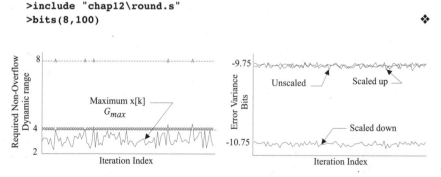

Figure 12.13 Roundoff noise vs. the two FIR filter scaling policies.

12.10 Multiple MAC Architectures

As technology provides more powerful and compact arithmetic devices, multiple MAC architectures are becoming a viable option. FIR
filters have typically required multiple MAC operations be performed
per filter cycle. This provides motivation for implementing multiple
MAC FIR filter architectures. Suppose M MACs are available for implementing an Nth-order FIR filter convolution sum given by

$$y[k] = \sum_{m=0}^{N-1} h[m]x[k-m] \qquad (12.28)$$

where $x[k]$ is a bounded input time-series. The filter tap-weight coefficients are denoted $h[k]$ and are assumed to be real (however, they may
be complex, in general). Upon presenting an input to the FIR, the output time-series is given by

$$y[0] = h[0]x[0]$$

$$y[1] = h[1]x[1] + h[0]x[1]$$

$$\dots = \dots$$

$$y[N-1] = h[N-1]x[0] + h[N-2]x[1] + \dots + h[0]x[N-1] \qquad (12.29)$$

For sample indices $k \geq N-1$, N actual multiplications must be performed per filter cycle. These multiplies can be performed sequentially
or concurrently. If M MACs are available for concurrent use, and N is
divisible by M, then a $K = N/M$-fold speed up can be realized. Suppose,
however, that M does not evenly divide N, so that $N = KM + K_0$,
$K_0 \in \{1,2,3,\dots,N-1\}$, then $K+1$ MAC cycles would be required where
the first K cycles would use all M MACs concurrently and the last cycle would use only K_0 of the available MACs. The efficiency of this action can be mathematically represented by Δ

$$\Delta = \frac{\text{actual speed-up}}{\text{ideal speed-up}} = \frac{L/M}{\lceil L/M \rceil} \qquad (12.30)$$

where $\lceil \circ \rceil$ denotes the ceiling function. The value of Δ is interpreted in
Fig. 12.14 for other cases. Notice that the efficiency improves as L increases which simply reflects that fact the overhead is reduced when a
large number of MAC need to be performed.

The multiple MAC filter shown in Fig. 12.15 is called a *horizontal
architecture,* since the MACs are spread horizontally across the convolution space. It is assumed that the coefficients are loaded into registers and spatially attached to each MAC.

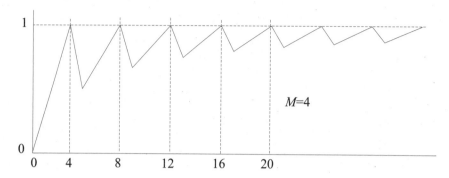

Figure 12.14 Speed-up potential of a four-multiplier FIR filter as a function of filter order.

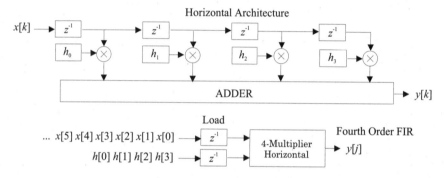

Figure 12.15 Example of a four-multiplier horizontally architected FIR filter.

Each one of the M MACs could be dedicated to computing one sample instances of $y[k]$, as shown in Eq. (12.28). This from is called a *vertical architecture* and is graphically interpreted in Fig. 12.16. Again, the efficiency of the multi-MAC architecture is a function of the relationship between the filter length N and the number of MACs M.

The execution of the first several clock instances of a four-multiplier vertically architected FIR filter is shown in Table 12.3.

12.11 Summary

In Chap. 12, FIR filter implementation techniques were developed. The most popular architecture, by far, is the direct FIR filter form. However, the lattice architecture is the preferred scheme in many cases, including adaptive filters. The other options presented in this chapter—namely, reduce multiplier, distributed, canonic signed digit, and multiplierless sections—presume a custom implementation.

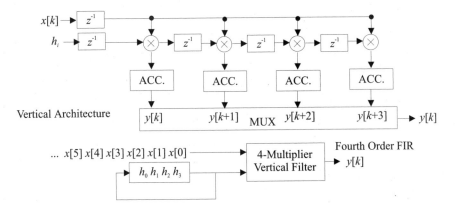

Figure 12.16 Example of a four-multiplier vertically architected FIR filter.

Table 12.3 Four-Multiplier Vertically Architected FIR Filter Execution

Clock	Cell 0	Cell 1	Cell 2	Cell 3	Sum/clear
0	$+h_3x[0]$	0	0	0	0
1	$+h_2x[1]$	$h_3x[1]$	0	0	0
2	$+h_1x[2]$	$+h_2x[2]$	$h_3x[2]$	0	0
3	$+h_0x[3]$	$+h_1x[3]$	$+h_2x[3]$	$h_3x[3]$	Cell 0 = $y[3]$
4	$h_3x[4]$	$+h_0x[4]$	$+h_1x[4]$	$+h_2x[4]$	Cell 1 = $y[4]$
5	$+h_2x[5]$	$h_3x[5]$	$+h_0x[5]$	$+h_1x[5]$	Cell 2 = $y[5]$
6	$+h_1x[6]$	$+h_2x[6]$	$h_3x[6]$	$+h_0x[6]$	Cell 3 = $y[6]$
7	$+h_0x[7]$	$+h_3x[7]$	$+h_2x[7]$	$h_3x[7]$	Cell 0 = $y[7]$

Compared to a general-purpose DSP microprocessors, a custom implementation will generally have a higher real-time bandwidth but may require a longer development cycle, lack robustness, and have higher nonrecurring engineering costs. Custom ASIC solutions are considered for stable high-volume applications where a better price-performance point can be achieved compared to that obtainable with a general-purpose DSP microprocessor. The size, power requirements, and economics of all these design options will vary from case to case.

12.12 Self-Study Problems

1.   Direct and Reduce-Multiplier Design   Implement the half-band FIR filter studied in Example 11.2 as a direct and reduced-multiplier FIR filter. Conduct tests to verify that their frequency responses are identical.

2.   Lattice FIR   Implement the 11th-order equiripple FIR filter specified in "chap12\ch12-2.fir" (passband defined over $f \in [0, 0.226757]$ f_s and stopband over $f \in [0.272109, 0.5]$ f_s), with coefficients listed below, as a lattice FIR filter. Conduct tests to verify that the frequency response is identical to that of a direct FIR filter implementation. Conduct tests using an assumed eight-bit MAC. Compare the statistical error performance of each FIR filter design.

$$H(\ 0\ ) = 0.1181327927118295 \quad = H(\ 10\ )$$
$$H(\ 1\ ) = -0.004271890168912273 = H(\ 9\ )$$
$$H(\ 2\ ) = -0.1075254844085444 \quad = H(\ 8\ )$$
$$H(\ 3\ ) = -0.002083352052220443 = H(\ 7\ )$$
$$H(\ 4\ ) = 0.3143951357614121 \quad = H(\ 6\ )$$
$$H(\ 5\ ) = 0.4956540140664078 \quad = H(\ 5\ )$$

3.   Distributed Arithmetic FIR   Implement the 11th-order FIR filter studied in the previous exercise as an 8-bit distributed arithmetic FIR. Conduct tests to verify that the frequency response is comparable to the floating point frequency response. Conduct tests using an assumed direct implementation with an 8-bit MAC to that of an 8-bit DA filter. Compare the statistical error performance of each FIR filter design.

4.   Canonic Signed Digit FIR   Implement the 2-D Laplacian filter presented in Example 12.7 as a CSD filter having, at most, three coefficients. Compute the magnitude frequency response of the 2-D filter. Conduct tests using an assumed 8-bit data word. Compare the statistical error performance of the CSD implementation and direct FIR filter implementation.

5.   Finite Wordlength Effects in FIR Filtering   Implement the half-band FIR filter studied in Example 11.2 as a direct and reduced-multiplier FIR. Conduct tests to verify that their frequency responses are identical. Conduct tests using an assumed 8-bit MAC and 8-bit coefficients. The input is bounded by unity in absolute value. Use a worst-case analysis to set the location of the binary point in your design. Assume, in one case, that the MAC results are rounded immediately to 8 bits and, in the other case, all products are sent as 16-bit full-precision results to a 21-bit accumulator which is then rounded to 8 bits at the end of each filter cycle. Compare the statistical error performance of each FIR filter design. Isolate coefficient rounding errors from arithmetic errors.

13

Infinite Impulse Response Filters

13.1 Introduction

In Chap. 11, the concept of FIR filtering was introduced. In Chap. 12, FIR filters were shown to have a very regular dataflow and simple architecture. Frequency-selective FIR filtering, unfortunately, generally evokes high-order FIR filter solutions, which can carry a serious implementation and real-time bandwidth penalty. For frequency selective filtering, alternative structures may be required. Fortunately, there are a host of well known classical low-order frequency-selective analog filter models that can be assimilated into digital filters. The origins of these design options can be traced back to the analog electrical and electronic filters used in the design of early radios. The purpose of these radio filters was to selectively process information in the frequency domain. In this capacity, filters would remove unwanted signal energy and amplify desired signal attributes based on their frequency location. What was characteristic of these radio filters was the presence of feedback. It may be recalled that FIR filters are intrinsically feedforward devices devoid of feedback. Therefore, developing a class of discrete-time filters that include feedback may result in a level of performance unobtainable with an FIR system.

Before the advent of the general-purpose digital computer, the design of analog filters was based on the use of carefully assembled standardized tables, charts, and graphs. Because the production of these tables was labor intensive, only a few important filters types were ever reduced to a standardized database. From this short menu, the radio engineer would convert a set of filter specifications into an analog filter definition. The analog filter would then be implemented in analog hardware consisting of a collection of amplifiers, capacitors, inductors, and resistors. These filters have been carried over to a class of digital devices called *infinite impulse response* (IIR) filters and represent a class called *classical IIR* filters.

13.2 IIR Filters

A causal IIR filter, as the name implies, has an impulse response that may persist forever. Such a system can be modeled in terms of a transfer function of the form

$$H(z) = \sum_{k=0}^{\infty} h_k z^{-k} \tag{13.1}$$

In some cases, the transfer function can be reduced to closed form

$$H(z) = \sum_{k=0}^{\infty} h_k z^{-k} = \frac{\displaystyle\sum_{m=0}^{M} b_m z^{-m}}{\displaystyle\sum_{m=0}^{N} a_m z^{-m}} = \frac{z^{-m} \displaystyle\sum_{m=0}^{M} b_{(M-m)} z^{m}}{z^{-m} \displaystyle\sum_{m=0}^{N} a_{(N-m)} z^{m}} = \frac{N(z)}{D(z)} \tag{13.2}$$

where $N(z)$ and $D(z)$ are Mth- and Nth-order polynomials, respectively. $H(z)$ is said to be *monic* if the coefficient multiplying the highest power in the polynomial $D(z)$ is unity (i.e., $a_N = 1$). It is generally assumed that $N \geq M$, in which case, $H(z)$ is said to be a *proper system*. If $N < M$, then the polynomial quotient is said to be *improper*. It is possible to convert an improper polynomial quotient to the sum of a polynomial and a proper polynomial quotient of the form

$$H(z) = \frac{N(z)}{D(z)} = \sum_{i=0}^{M-N-1} c_i z^{-i} + \frac{\displaystyle\sum_{m=0}^{N} b'_m z^{-m}}{\displaystyle\sum_{m=0}^{N} a_m z^{-m}} = C(z) + \frac{N'(z)}{D(z)} \tag{13.3}$$

where the coefficients of $C(z)$ and the remainder polynomial $N'(z)$ can be computed using long division. That is,

$$\begin{array}{r} C(z) \\ D(z)\overline{\smash{\big)}N(z)} \\ \underline{C(z)D(z)} \\ N'(z) \end{array}$$

$$\tag{13.4}$$

The inverse z-transform of $H(z)$ would then contain elements from $C(z)$ which take an FIR form. Since FIR systems have been discussed in previous chapters, unless otherwise stated, only proper systems will be considered from this point on.

A causal IIR system having a transfer function $H(z)$ can also be represented in factored form as

$$H(z) = \frac{\displaystyle\sum_{m=0}^{M} b_m z^{-m}}{\displaystyle\sum_{m=0}^{N} a_m z^{-m}} = K \frac{\displaystyle\prod_{i=0}^{M-1} (z - z_i)}{\displaystyle\prod_{i=0}^{N-1} (z - p_i)} \tag{13.5}$$

where z_i are called the *zeros* of $H(z)$, p_i are called the *poles*, and $K = b_M/a_N$. Because IIR filters contain feedback, they have the ability to sharply shape the filter's frequency response. The frequency response of an IIR filter can be determined by evaluating $H(z)$ along the trajectory $z = e^{j\varpi}$. That is,

$$H(e^{j\varpi}) = \frac{\displaystyle\sum_{m=0}^{M} b_m e^{-jm\varpi}}{\displaystyle\sum_{m=0}^{N} a_m e^{-jm\varpi}} = K \frac{\displaystyle\prod_{i=0}^{M-1} (e^{j\varpi} - z_i)}{\displaystyle\prod_{i=0}^{N-1} (e^{j\varpi} - p_i)} \tag{13.6}$$

where ϖ may be evaluated over the normalized frequency range $-\pi \leq \varpi \leq \pi$.

An IIR filter can generally meet very demanding frequency response specification with a filter whose order is lower than that required of an equivalent FIR filter [see Eq. (11.29) or (11.30)]. IIR filters, therefore, provide a potential implementation advantage over FIR filters when filter cost is evaluated in terms of the "silicon horsepower" required to implement a design. However, there are disadvantages to IIR filters. Because of their feedback structure, IIR filters are more likely to encounter register overflow during run-time than are FIR filters. If this condition is not properly managed, it can cause great havoc. The implementation details associated with an IIR filter design are therefore considerably more challenging than for an FIR filter. A second potential problem with an IIR filter is the loss of linear phase behavior, a strength of FIR filters. For an IIR filter to possess linear phase behavior, complicated all-pass *phase compensators* are needed. As a general rule of thumb, it would take an IIR filter phase compensator of order

$2N$ to correct for the phase distortion introduced by an Nth-order IIR filter.

Classical filters are defined in terms of an ideal magnitude-frequency lowpass, highpass, bandpass, and bandstop model. Much is known about such filters thanks to nearly a century of radio communication and control system engineering. Throughout the 20th century, engineers have elevated the design of analog frequency selective filters to a fine art. The same filter models are extensively used today to design digital IIR filters. Whereas the radio engineer used tables and nomographs to determine the parameters of a filter, the designer of digital filters relegates that task to a general-purpose digital computer.

Example 13.1: Comparison of IIR and FIR Filters

Problem Statement. Using techniques developed in Chap. 11 and those to be developed in Chap. 13, compare the design of a typical FIR filter and IIR filter having the following specifications:

$$f_s = 8 \text{ kHz}$$

$$f_p = 1.75 \text{ kHz}$$

$$f_a = 2 \text{ kHz}$$

$$\delta_p \leq -40 \text{ dB}$$

and

$$\delta_a \leq -0.0875 \text{ dB}$$

Analysis and Conclusions. The FIR filter order needed to meet the posted specification is 63. The specifications are also satisfied by a sixth-order elliptic IIR filter (see Sec. 13.7). Their responses are compared in Fig. 13.1. What is noteworthy is that the IIR filter is one-tenth the order of the FIR filter, which carries with it a potential tenfold increase in run-time throughput. However, it can also be noted that the IIR filter phase and group delay are highly non-linear, especially locally about the transition band.

Computer Study. The design of the FIR filter and IIR filters can be viewed by loading the files "firiir.FIR" and "iirfir.IIR" into the FIR and IIR design menus respectively. The following sequence of Siglab commands is used to produce the data shown in Fig. 13.1.

```
>fir = rf("chap12\firiir.imp")      # Read FIR impulse response.
>iir = rarc("chap12\iirfir.arc",1)  # Read IIR filter arch file.
>include "chap12\ch13_1.s"          # Get plotting routine.
>firiir(fir,iir)                     # Generate analysis.
```

The analysis can be extended using a soundboard. Using a chirp having a frequency range $f \in [0, f_s/4]$, the audio response of each filter can be examined. As-

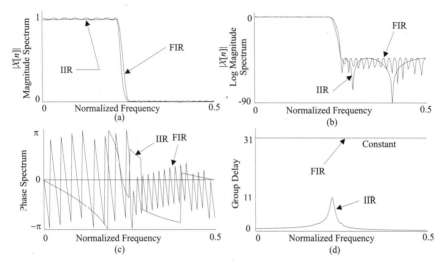

Figure 13.1 Comparison of an FIR and IIR filter showing (*a*) linear magnitude frequency response, (*b*) logarithmic magnitude frequency response, (*c*) phase response, and (*d*) group delay.

suming a 16-bit soundboard is used with a sampling rate of $f = 11,025$ Hz, the following sequence of Siglab commands can be used to the test both the FIR and IIR filters. The signal length is chosen so that a sufficiently long play period can be achieved. There can be heard to be little difference between the FIR filter and IIR filter responses. Both fall off at high frequencies, as expected.

```
>include "chap13\chirp.s"
>fir=rf("chap12\firiir.imp")          # Read FIR impulse response.
>iir=rarc("chap12\iirfir.arc",1)      # Read IIR filter arch file.
>x=chirp(10000,0,0,0.25)              # Generate test signal.
>wavout(0.8*2^15*x)                   # Play unfiltered signal.
>wavout(0.8*2^15*fir$x)               # FIR filtered signal.
>wavout(0.8*2^15*filt(iir,x))         # IIR filtered signal.      ❖
```

13.3 Classic Analog Filters

The design of a classic analog filter, in many respects, remains today as it was practiced during the early days of radio. The design objective of the radio engineer was one of shaping the frequency spectrum of a received or transmitted signal using modulators, demodulators, and frequency-selective filters. Filter designs were based, to various degrees, on ideal lowpass, highpass, bandpass, bandstop, and all-pass models. The frequency response of an analog filter system having a transfer function $H(s)$ was defined by $H(s = j\omega)$, where ω is called the *analog frequency* and is measured in radians per second. The mathematical specifications of classic ideal filters are summarized below (see Fig. 13.2):

Ideal lowpass: $|H(j\omega)| = \begin{cases} 1 \text{ for } \omega \in [-\omega_p, \omega_p] \\ 0 \text{ otherwise} \end{cases}$ (13.7)

Ideal highpass: $|H(j\omega)| = \begin{cases} 0 \text{ for } \omega \in [-\omega_p, \omega_p] \\ 1 \text{ otherwise} \end{cases}$ (13.8)

Ideal bandpass: $|H(j\omega)| = \begin{cases} 1 \text{ for } \omega \in [-\omega_H, -\omega_L] \cup [\omega_L, \omega_H] \\ 0 \text{ otherwise} \end{cases}$ (13.9)

Ideal bandstop: $|H(j\omega)| = \begin{cases} 0 \text{ for } \omega \in [-\omega_H, -\omega_L] \cup [\omega_L, \omega_H] \\ 1 \text{ otherwise} \end{cases}$ (13.10)

Ideal all-pass: $|H(j\omega)| = 1$ (13.11)

The task of the analog filter designer is to create an analog filter whose magnitude frequency response emulates that of an ideal filter. Historically, the design of a classic analog filter has been based on the use of well known models attributed to Bessel, Butterworth, Chebyshev, and Cauer (elliptic). To standardize the design procedure, a set of normalized lowpass filter models were agreed upon and reduced to tables, charts, and graphs. These models, called *analog prototypes*, were based on an assumed –1 dB or –3 dB passband loss from DC out to

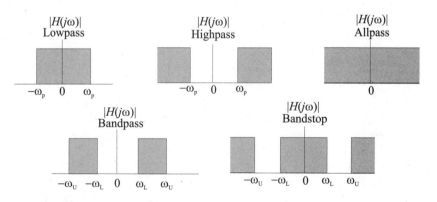

Figure 13.2 Classic ideal filter models.

1 rad/s. In a classical design environment, the analog prototype, denoted by $H_p(s)$, is read from prepared set of tables, charts, and graphs. The desired analog filter, however, generally has a different bandwidth than that possessed by the prototype filter whose cutoff frequency has to be standardized to 1 rad/s. Therefore, the prototype filter $H_p(s)$ needs to be mapped, or converted into a filter $H(s)$ having the desired response. The mapping rule that maps an analog prototype into its final form is called a *frequency-frequency transform*. They are summarized in Table 13.1 and graphically interpreted in Fig. 13.3.

Table 13.1 Frequency-Frequency Transforms

Nth-order prototype	Frequency-frequency transform	Order of $H(s)$
Lowpass to lowpass	$s = s/\omega_p$	N
Lowpass to highpass	$s = \omega_p/s$	N
Lowpass to bandpass	$s = (s^2 + \omega_H\omega_L)/[s + (\omega_H - \omega_L)]$	$2N$
Lowpass to bandstop	$s = [s + (\omega_H - \omega_L)]/(s^2 + \omega_H\omega_L)$	$2N$

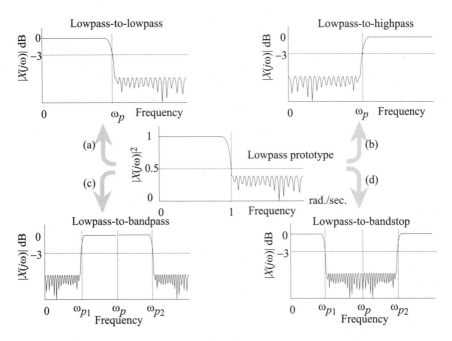

Figure 13.3 Frequency-frequency transforms associated with the mapping of a prototype lowpass to (*a*) lowpass, (*b*) highpass, (*c*) bandpass, and (*d*) bandstop mappings.

Example 13.2: Frequency-Frequency Mapping

Problem Statement. The −3 dB analog prototype circuit shown in Fig. 13.4 has a transfer function given by

$$H_p(s) = \frac{1}{s^3 + 2s^2 + 2s + 1}$$

What is $H(s)$ if the −3 dB frequency is 1 kHz?

Analysis and Conclusions. The gain at $s = j\omega = j1$ rad/s is easily verified to be $|H(j1)| = |1/(1 + j1)| = 0.707 \rightarrow -3$ dB. The poles of $H_p(s)$ are located at $s = -1$ and $-0.50 \pm j0.866$. The −3 dB frequency for the circuit shown in Fig. 13.4(a) has a critical frequency of 1 rad/s. The lowpass prototype can be converted into desired lowpass filter using a lowpass-to-lowpass frequency-frequency transformation. In particular,

$$H(s) = H_p(s)\Big|_{s \to s/\omega_p} = \frac{1}{s^3 + 2s^2 + 2s + 1}\Big|_{s \to s/\omega_p}$$

where $\omega_p = 2\pi \times 1000$ (rad/s)/ 1 (rad/s) = 6280. Upon substituting $s = s/\omega_p$, the final filter $H(s)$ becomes

$$H(s) = \frac{1}{4.03 \times 10^{-12} s^3 + 5.07 \times 10^{-8} x^2 + 3.18 \times 10^{-4} s + 1}$$

where the poles are located at $s = -6280$ and $s = -3140 \pm j5438$ (i.e., poles of $H_p(s)$ are scaled by ω_p). The frequency response of the analog filter, and its attendant circuit elements, are shown in Fig. 13.4(b). The critical frequency for $H(s)$ is 1 kHz or 6280 rad/s.   ❖

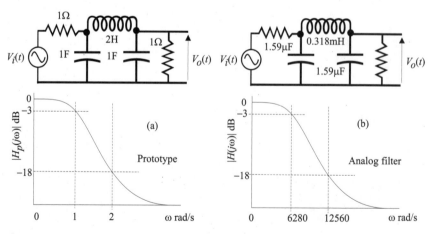

Figure 13.4 Lowpass model showing (a) a prototype and (b) the frequency transform filter.

13.4 Lowpass Prototype Filters

An analog prototype's magnitude squared frequency response, measured at $\omega = 1$ rad/s, is defined to be

$$|H(s)|^2_{s\,=\,j1} = \frac{1}{1 + \varepsilon^2} \tag{13.12}$$

If $\varepsilon^2 = 0.5$, the prototype is said to be a -3 dB filter. Refer to Fig. 13.5 and observe that the analog filter is to be mapped into an analog filter having targeted *critical frequencies* ω_p, ω_a, ω_{p1}, ω_{p2}, ω_{a1}, and ω_{a2}. The passband and stopband gains are specified in terms of the parameters ε and A. The steepness of the filter skirt is measured in terms of the *transition gain ratio* which is given by $\eta = \varepsilon/(A^2-1)^{1/2}$. The *frequency transition ratio*, denoted k_d, measures the normalized width of the transition bandwidth. The value of k_d is given by the following expressions:

$$\text{Lowpass:} \quad k_d = \frac{\omega_p}{\omega_a} \tag{13.13}$$

$$\text{Highpass:} \quad k_d = \frac{\omega_a}{\omega_p} \tag{13.14}$$

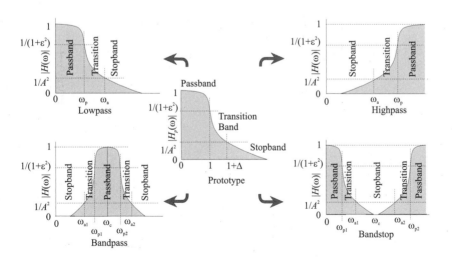

Figure 13.5 Frequency-frequency transforms for lowpass prototype filters.

$$\text{Bandpass:} \quad k_d = \begin{cases} k_1 \text{ if } \omega_u^2 \geq \omega_v^2 \\ k_2 \text{ if } \omega_u^2 < \omega_v^2 \end{cases} \tag{13.15}$$

$$\text{Bandstop:} \quad k_d = \begin{cases} 1/k_1 \text{ if } \omega_u^2 \geq \omega_v^2 \\ 1/k_2 \text{ if } \omega_u^2 < \omega_v^2 \end{cases} \tag{13.16}$$

where

$$\omega_c = \frac{\omega_{p_1} \omega_{p_2}}{\omega_{p_2} - \omega_{p_2}} \tag{13.17}$$

$$k_1 = \left(\frac{\omega_{a_1}}{\omega_c}\right) \frac{1}{\left(1 - \dfrac{\omega_{a_1}^2}{\omega_v^2}\right)} \tag{13.18}$$

$$k_2 = \left(-\frac{\omega_{a_2}}{\omega_c}\right) \frac{1}{\left(1 - \dfrac{\omega_{a_2}^2}{\omega_v^2}\right)} \tag{13.19}$$

$$\omega_u^2 = \omega_{a_1} \omega_{a_2} \tag{13.20}$$

$$\omega_v^2 = \omega_{p_1} \omega_{p_2} \tag{13.21}$$

From $\delta^2 = A^2 - 1$, the *gain ratio* is defined to be $d = \delta/\varepsilon$. From these parameters, the order and transfer function of classical analog prototype filters can be determined.

13.5 Classic Butterworth Filters

An Nth-order *Butterworth filter* has a magnitude squared frequency response given by

$$|H(s)|^2 = \frac{1}{1 + \varepsilon^2 s^{2N}} \tag{13.22}$$

The order is estimated to be

$$N = \log\left(\frac{A^2 - 1}{\varepsilon^2}\right)\frac{1}{2\log\left(\frac{1}{k_d}\right)} \qquad (13.23)$$

The order $2N$ transfer function squared $|H(s)|^2$ can be factored into $|H(s)|^2 = |H(s)H(-s)|$ where $H(s)$ is called the *realizable filter* and $H(-s)$ is called the *unrealizable filter*. The poles of $|H(s)|^2$ are located on a circle of radius $r = 1/\varepsilon^{1/N}$ and separated π/N radians from each other. Specifically, the poles are known to be located at $s_k = re^{-j(2k-1)\pi/2N}$, $k \in \{0,1,2,\ldots,2N-1\}$. Therefore, N of the poles reside in the stable left-hand plane and belong to the realizable part of $|H(s)|^2$, namely $H(s)$. The other N poles belong to the unstable filter $H(-s)$ as shown in Fig. 13.6. A typical magnitude frequency response for a lowpass Butterworth filter is shown in Fig. 13.6.

Example 13.3: Butterworth Filter Design

Problem Statement. Derive the transfer function of a -1 dB lowpass Butterworth filter with a 35 Hz passband and -60 dB stopband beginning at 100 Hz.

Analysis and Conclusion. The lowpass filter shown in Fig. 13.7 has a stopband bounded by $A = 1000$ (-60 dB) and a passband bounded by $\varepsilon = 0.508$ (-1 dB). The remaining design parameters are given by $\eta = \varepsilon/(A^2-1)^{1/2} = 0.508 \times 10^{-3}$ and $k_d = 35/100 = 0.35$. The filter order is estimated to be

$$N = \log\left(\frac{10^6 - 1}{0.258064}\right)\frac{1}{2\log(2.857)} = 7.18 < 8$$

The prototype Butterworth filter is, therefore, given by

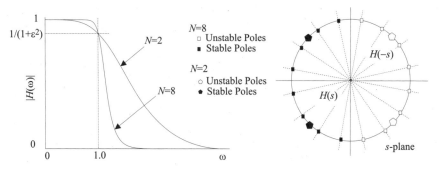

Figure 13.6 Typical Butterworth lowpass filter showing the magnitude frequency response and pole distribution.

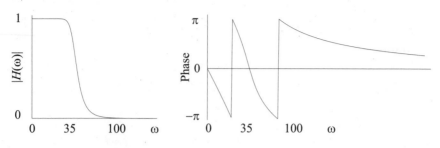

Figure 13.7 Butterworth filter showing the magnitude frequency and phase responses.

$$H_p(s) = \frac{1}{s^8 + 5.125s^7 + 13.137s^6 + 21.846s^5 + 25.688s^4 + 21.846s^3 + 13.137s^2 + 5.125s + 1}$$

Next, the prototype needs to be converted into an analog filter using a lowpass-to-lowpass frequency-frequency transform of the form $s \rightarrow s/\omega_p = s/(2\pi35)$. The resulting transfer function is

$$H(s) = \frac{1.1 \times 10^{19}}{\left(\begin{array}{c} s^8 + 1230s^7 + 7.53 \times 10^5 s^6 + 3.00 \times 10^8 s^5 + 8.42 \times 10^{10} s^4 + \\ 1.72 \times 10^{13} s^3 + 2.47 \times 10^{15} s^2 + 2.31 \times 10^{17} s + 1.1 \times 10^{17} \end{array} \right)}$$

The magnitude frequency and phase response of the lowpass Butterworth IIR filter is shown in Fig. 13.7.   ❖

13.6 Classic Chebyshev Filters

An Nth-order lowpass *Chebyshev-I filter* has a magnitude squared frequency response given by

$$|H(s)|^2 = \frac{1}{1 + \varepsilon^2 C_N^2(s)} \tag{13.24}$$

where $C_N(s)$ is a Chebyshev polynomial of order N. The first few Chebyshev polynomials are $C_1(s) = s$, $C_2(s) = 2s^2 - 1$, $C_3(s) = 4s^3 - 3s$, and so forth. The order of a Chebyshev-I filter is estimated to be

$$N = \frac{\log[(1+\eta)]}{\log\left[\frac{1}{k_d} + \sqrt{\left(\frac{1}{k_d}\right)^2 - 1}\right]} \tag{13.25}$$

The $2N$ poles of $|H(s)|^2 = |H(s)H(-s)|$ lie on an ellipse whose geometry is determined by ε and N. Once again, N of the poles

$|H(s)|^2$ belong to the stable left-hand plane and are assigned to $H(s)$. In particular, the left-hand plane poles are located at $s_k = -\sin(x[k])\sinh(y[k]) + j\cos(x[k])\cosh(y[k])$ where $x[k] = (2k-1)\pi/2N$, and $y[k] = \pm\sinh^{-1}(1/\varepsilon)/N$. The other N poles belong to $H(-s)$.

A variation on the Chebyshev-I model is called the *Chebyshev-II filter* and is given by

$$|H(s)|^2 = \frac{1}{1 + [\varepsilon^2 C_N^2(j\omega_a/s)]} \qquad (13.26)$$

The order estimation formula is the same as that for the Chebyshev-I model and is given by Eq. (13.25).

A typical magnitude frequency response for a Chebyshev-I and Chebyshev-II filter is shown in Fig. 13.8. The Chebyshev-I filter is seen to exhibit ripple in the passband and have a smooth transition into the stopband. The Chebyshev-II filter is seen to have ripple in the stopband and smooth transition into the passband, which is relatively flat.

Example 13.4: Chebyshev Lowpass Filter

Problem Statement. Derive a Chebyshev-I and a Chebyshev-II filter that meet the specifications provided by Example 13.3.

Analysis and Conclusion. Again, the objective is to design a lowpass filter with a stopband bounded by $A = 1000$ (-60 dB) and a passband bounded by $\varepsilon = 0.508$ (-1 dB). The remaining design parameters are given by $\eta = \varepsilon/(A^2 - 1)^{1/2} = 0.508 \times 10^{-3}$ and $k_d = 35/100 = 0.35$. The filter order is estimated to be

$$N = \frac{\log\left[\left(\dfrac{1 + \sqrt{1 - 0.258 \times 10^{-6}}}{0.508 \times 10^{-3}}\right)\right]}{\log[2.857 + \sqrt{7.6163}]} = 4.83 < 5$$

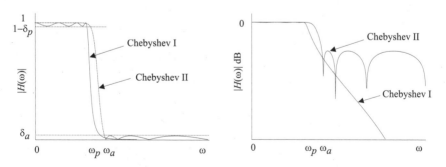

Figure 13.8 Typical Chebyshev-I and Chebyshev-II lowpass filter magnitude frequency responses in linear and logarithmic (dB) units.

The prototype Chebyshev-I and Chebyshev-II filters are given by

$$H_{CI}(s) = \frac{0.122}{s^5 + 0.94s^4 + 1.69s^3 + 0.97s^2 + 0.58s + 0.122}$$

and

$$H_{CII}(s) = \frac{0.0142*(s^4 + 32.65s^2 + 213.25)}{s^5 + 4.00s^4 + 7.89s^3 + 7.52s + 3.05}$$

and after applying a lowpass-to-lowpass frequency-frequency transform, the final version of the Chebyshev-I and Chebyshev-II filters are

$$H_{CI}(s) = \frac{6.32 \times 10^{10}}{s^5 + 206s^4 + 81700s^3 + 1.04 \times 10^7 s^2 + 1.35 \times 10^9 s + 6.32 \times 10^{10}}$$

and

$$H_{CII}(s) = \frac{3.14(s^4 + 1.58 \times 10^6 s^2 + 5.00 \times 10^{11})}{s^5 + 874s^4 + 3.82 \times 10^5 s^3 + 1.04 \times 10^8 s^2 + 1.76 \times 10^{10} s + 1.56 \times 10^{12}}$$

The frequency responses of the designed Chebyshev-I and Chebyshev-II filters are shown in Fig. 13.9.   ❖

13.7 Classic Elliptic Filters

The attenuation of an Nth-order *elliptic filter* is given by the solution to an elliptic integral equation. The order of an elliptic filter is estimated by

$$N \geq \frac{\log(16D)}{\log(1/q)} \tag{13.27}$$

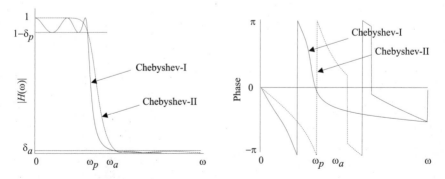

Figure 13.9 Chebyshev-I and Chebyshev-II filters from Example 13.3 showing magnitude frequency and phase responses.

where

$$k' = \sqrt{(1 - k_d^2)} \tag{13.28}$$

$$q_0 = 1 - \frac{\sqrt{k'}}{2(1 + \sqrt{k'})} \tag{13.29}$$

$$q = q_0 + 2q_0^5 + 15q_0^9 + 15q_0^{13} \tag{13.30}$$

$$D = d^2 \tag{13.31}$$

The typical magnitude frequency response of an elliptic lowpass filter is shown in Fig. 13.10. It can be seen that an elliptic filter has ripple in both the passband and stopband.

Example 13.5: Elliptic Filter

Problem Statement. Using the design parameters given in Example 13.3, design an elliptic lowpass filter

Analysis and Conclusions. Again the stopband is bounded by $A = 1000$ (–60 dB) and the passband bounded by $\varepsilon = 0.508$ (–1 dB). The remaining design parameters are given by $\eta = \varepsilon/(A^2 - 1)^{1/2} = 0.508 \times 10^{-3}$ and $k_d = 35/100 = 0.35$. The filter order is computed using the parameters

$$k' = \sqrt{(1 - 0.35)} = 0.9367$$

$$q_0 = \frac{1 - \sqrt{0.9367}}{2(1 + \sqrt{0.9367})} = 0.01631$$

$$q \approx q_0 = 0.01631$$

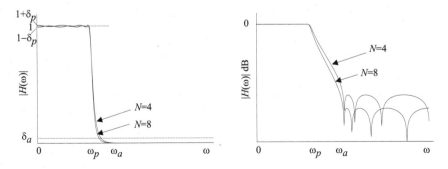

Figure 13.10 Typical elliptic filter magnitude frequency response.

$$D = 1968^2$$

to produce

$$N \approx \frac{\log(16D)}{\log(1/q)} = 3.68 < 4$$

The fourth-order elliptic prototype filter is given by

$$H_p(s) = \frac{0.000461(s^4 + 22.502s^2 + 65.237)}{s^4 + 0.553s^3 + 0.498s^2 + 0.149s + 0.0337}$$

After applying the lowpass-to-lowpass frequency-frequency transform, the final version of the elliptic filter emerges and is given by

$$H(s) = \frac{0.000461(s^4 + 3.229 \times 10^6 s^2 + 1.342 \times 10^{12})}{s^4 + 2.096 \times 10^2 s^3 + 7.148 \times 10^4 s^2 + 8.148 \times 10^6 s + 6.961 \times 10^8}$$

The magnitude frequency response and phase response are reported in Fig. 13.11. The zeros are found at $s = \pm j1654.7$ and $s = \pm j700.3$, while the poles are located at $s = -75.76 \pm j93.21$ and $s = -29.065 \pm j217.7$.   ❖

13.8 Other IIR Filter Forms

Analog filter structures, other than the classic Butterworth, Chebyshev, and elliptic filter models, are also routinely encountered. Filters with an arbitrary magnitude frequency response can be defined by the invention of an engineer or synthesized from measured data using spectral estimation tools such as *auto-regressive* (AR) or *auto-regressive moving-average* (ARMA) models. In all cases, the objective of the

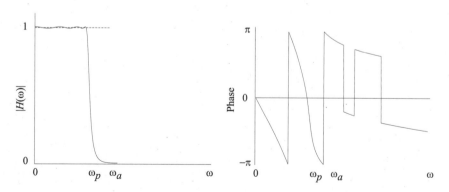

Figure 13.11 Magnitude frequency and phase responses for the example elliptic filter.

designer is to obtain an Nth-order analog filter transfer function of the form

$$H(s) = \frac{\displaystyle\sum_{m=0}^{M} b_m s^m}{\displaystyle\sum_{m=0}^{N} a_m s^m} \tag{13.32}$$

Experienced analog filter designers also work directly in the s-plane. By adjusting the location of the poles and zeros, relative to the $j\omega$ axis, the frequency response of an analog filter can be shaped until it satisfies the designer's requirements.

Regardless of the method used to obtain an analog transfer function, the designer of a digital filter replacement for an analog system must convert $H(s)$ into a discrete-time filter model $H(z)$. The basic domain conversion methods (i.e., $H_a(s) \rightarrow H(z)$) found in common use are

1. Impulse-invariant method

2. Bilinear-z transform method

The first method preserves the time-domain behavior of the system, which is important for some applications, such as in linear control systems. The second transform method preserves the magnitude frequency response of the analog system in the discrete-time domain.

13.9 Impulse-Invariant Transform

The impulse-invariant filter design method produces a discrete-time system which is based on a continuous-time system model. The impulse response of a discrete-time, impulse-invariant system, denoted $h_d[k]$, is related to the sampled values of continuous-time system's impulse response $h_a(t)$, through the defining relationship

$$h_d[k] = T_s h_a(kT_s) \tag{13.33}$$

That is, the discrete-time and continuous-time impulse responses agree, up to a scale factor T_s, at the sample instances. The standard z-transform possesses the *impulse-invariant property* in that it guarantees that the analog and discrete-time impulse responses agree at the sample instances. This can be of significant importance in some applications areas, such as control, where knowledge of the envelope of a signal in the time domain is of more importance than knowledge of its frequency response.

Consider an analog filter having a known impulse response $h_a(t)$ with a known transfer function in the form of Eq. (13.32). The z-transform of $h_a(t)$ can be expressed as

$$h_a(t) \overset{Z}{\leftrightarrow} H_a(z) = \sum_{m=0}^{\infty} h_d[m]z^{-m} \overset{Z^{-1}}{\leftrightarrow} \sum_{m=0}^{\infty} h_d[m]\delta[k-m] = h_d[k] \quad (13.34)$$

where $h_d[k]$ is the sampled value of $h_a(t)$ at the sample instances $t = kT_s$. Since $h_a(t = kT_s) = h_d[k]$ at the sample instances, the z-transform is said to be *impulse invariant*.

For the sake of development, consider the Nth-order system described by the transfer function $H_a(s)$ [from Eq. (13.32)] to have N distinct poles. The frequency response of the system $H_a(j\omega)$, for $\omega \in (-\infty, \infty)$, is shown in Fig. 13.12. Upon taking the impulse-invariant z-transform of $H_a(s)$, the following results:

$$h_a(t) \Leftrightarrow H_a(s) = \sum_{i=1}^{N} \frac{\alpha_i}{(s + p_i)} \overset{Z}{\leftrightarrow} \frac{1}{T_s} \sum_{i=1}^{N} \frac{\alpha_i}{(1 + e^{p_i T_s} z^{-1})}$$

$$= \frac{1}{T_s} H(z) \Leftrightarrow \frac{1}{T_s} h[k]$$

$$(13.35)$$

As a direct consequence of Eq. (13.35), the frequency response of the discrete-time system can be computed to be

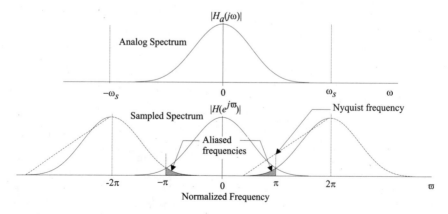

Figure 13.12   Spectrum of an impulse-invariant filter.

$$H(e^{j\omega}) = \frac{1}{T_s} \sum_{k=-\infty}^{\infty} H_a\left[j\left(\frac{\omega}{T_s} - \frac{2\pi k}{T_s}\right)\right] \qquad (13.36)$$

Equation (13.36) states that under the z-transform, the frequency response of the transformed discrete-time system, namely $H(e^{j\omega})$, is directly related to the frequency response of the analog filter, specifically $H_a(j\omega)$. It should be noted, however, that $H(e^{j\omega})$ is constructed using copies of $H_a(j\omega)$ repeated on centers separated by $2\pi/T_s = 2\pi f_s$ radians (see Fig. 13.12) in the frequency domain. Observe that the spectral energy coming from any one band, centered about $\omega = k\omega_s$, can overlap another copy of the analog spectrum centered about $\omega = m\omega_s$, where $m \neq k$. Thus, the spectrum found in one frequency band can leak as aliased energy into a neighboring band.

When the z-transform was introduced in Chap. 2, the fact that a discrete-time signal periodically repeated its spectrum modulo (f_s) was not an issue, since it was assumed that the signal being transformed was sampled above the Nyquist rate. Therefore, there is an assumed absence of signal energy at frequencies above the Nyquist frequency. Thus, no aliased energy exists outside the baseband to leak into the baseband spectrum. Unfortunately, analog filters generally have a frequency response that can persist (i.e., is non-zero) for all frequencies and therefore *can* introduce aliasing errors for any finite sampling frequency. Refer again to Fig. 13.12 to more fully appreciate the deleterious effect of aliasing. Notice that the system's frequency response contains aliased energy at the boundaries $\omega_k = k2\pi/T_s$ of the impulse-invariant discrete-time filter.

Example 13.6: First-Order Impulse-Invariant System

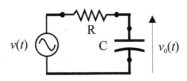

Figure 13.13 *RC* circuit.

Problem Statement. Consider the at-rest first-order *RC* circuit show in Fig. 13.13. Implement a discrete-time impulse-invariant equivalent to the *RC* circuit for a given arbitrary sampling frequency.

Analysis and Results. The relationship between the input forcing function $v(t)$ and voltage developed across the capacitor, namely $v_o(t)$, is defined by the solution to the ordinary differential equation

$$v_0(t) = RC\frac{dv_0(t)}{dt} = v(t)$$

The transfer function associated with the *RC* circuit model is given by

$$H(s) = \frac{1}{1 - RCs}$$

It immediately follows that the impulse response is given by

$$h(t) = \frac{1}{RC}e^{-t/RC}u(t)$$

For a given periodic sampling period of T_s, the resulting sampled impulse response is given by

$$h_d[k] \Leftrightarrow T_s h(kT_s)$$

or, for $k \geq 0$,

$$h_d[k] = \frac{T_s}{RC}e^{-kT_s/RC} = \frac{T_s}{RC}\alpha^k$$

where $\alpha = e^{-kT_s/RC}$. Referring to Table 2.1, a table of standard z-transforms, it can be seen that $h_d[k]$ corresponds to a system with a discrete-time transfer function given by

$$H(z) = \left(\frac{T_s}{RC}\right)\frac{1}{1 - \alpha z^{-1}} = \left(\frac{T_s}{RC}\right)\frac{z}{z - \alpha}$$

which results in the impulse-invariant filter shown in Fig. 13.14. The frequency response of the impulse-invariant filter is given by

$$H(e^{j\omega}) = \frac{T_s}{RC}\left(\frac{e^{j\omega}}{e^{j\omega} - \alpha}\right)$$

which is periodic with period 2π. That is, $H(e^{j\omega + k2\pi}) = H(e^{j\omega})$.

Computer Study. The s-file "std_z1.s" defines a function that simulates the magnitude frequency response of $H(s)$ and its impulse-invariant version, $H(z)$, for the parameters RC and f_s using the format "rcckt(rc,fs)." The function computes both the analog and discrete-time frequency responses of the RC circuit model. The comparison will be performed with two sets of parameters,

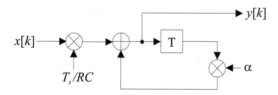

Figure 13.14 Discrete-time system model.

$$\text{(i) } RC = 10^{-2}, f_s = 10^3$$

$$\text{(ii) } RC = 10^{-2}, f_s = 250$$

The responses computed using the above parameters are reported in Fig. 13.15 and are obtained by executing the following sequence of Siglab commands:

```
>include "chap13\std_z1.s"
>rcckt(1e-2,1e3)
>rcckt(1e-2,250)
```

It can be seen that the continuous-time RC system behaves as a lowpass filter. The difference between the analog and impulse-invariant discrete-time filter, out to the sampling frequency, is minor for high sample rates. The analog and impulse-invariant responses, however, begins to diverge when the sample rate is reduced. This is a result of aliasing.   ❖

The problem associated with maintaining an agreement between the frequency response of an analog and impulse-invariant digital filter will be further explored using a second-order example.

Example 13.7: Second-Order Impulse-Invariant System

Problem Statement. Design a second-order −1 dB impulse-invariant Butterworth lowpass filter having a 1 kHz passband and sample frequency of 5 kHz.

Analysis and Results. The analog prototype second-order analog filter has a magnitude squared frequency response given by

$$|H(s)|^2 = \frac{1}{1 + \varepsilon^2 s^4}$$

where ε is chosen to produce a filter gain of −1 dB at ω = 1 radians per second. That is,

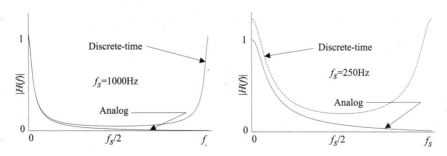

Figure 13.15 Simulated responses of a continuous-time and discrete-time first order RC circuit.

$$\log_{10} \frac{1}{1+\varepsilon^2 s^4}\Bigg|_{s=j1} = -1 \text{ dB}$$

or

$$\frac{1}{1+\varepsilon^2} = 10^{-0.1}$$

Solving for $\varepsilon^2 = 0.2589$, the analog prototype is given by

$$H(s) = \frac{1}{1+0.2589 s^4}$$

The poles of $|H(s)|^2 = H(s)H(-s)$ are located at (see Fig. 13.16)

$$s_k = \frac{1}{\sqrt[4]{0.2589}} e^{j\frac{(\pi+2k\pi)}{4}} = 1.4 e^{j\frac{(\pi+2k\pi)}{4}}$$

for $k \in \{0,1,2,3\}$. Collecting the stable roots together, the analog prototype can finally be computed,

$$H_p(s) = \frac{1}{(0.71335)^2} \frac{1}{(s-s_2)(s-s_3)} = \frac{1.965}{s^2 - \frac{1.414}{0.71335}s + \frac{1}{0.71335^2}} = \frac{1.965}{s^2 - 1.9822s + 1.9652}$$

The prototype is then mapped into the analog filter model using a lowpass-to-lowpass frequency transform given by

$$s = \frac{s}{\omega_p} = \frac{s}{2\pi \times 1000}$$

which results in an analog filter with a transfer function given by

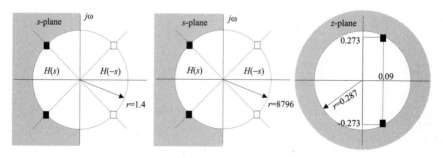

Figure 13.16 Pole-zero distribution for a second-order Butterworth analog prototype and filter model, plus discrete-time impulse-invariant design.

$$H(s) = \frac{1.965}{\dfrac{s^2}{(2000\pi)^2} - 1.9822\dfrac{s}{2000\pi} + 1.9652} = \frac{7.757 \times 10^7}{s^2 - 1.2454 \times 10^4 s + 7.757 \times 10^7}$$

having poles located at $s = -6.228 \times 10^3 \pm j6.228 \times 10^3$ (see Fig. 13.16). The transfer function is seen to have the general form

$$e^{-at}\sin(\omega_0 t)u(t) = (-0.5je^{(-a+j\omega_0)t} + 0.5je^{(-a-j\omega_0)t})u(t) \overset{L}{\leftrightarrow} \frac{\omega_0}{(s+a)^2 + \omega_0^2}$$

$$= \frac{-0.5j}{s + (a - j\omega_0)} + \frac{0.5j}{s + (a + j\omega_0)}$$

The impulse-invariant method, it may be recalled, maps the Laplace transform of $H(s)$ into a discrete-time system using the standard z-transform. After some manipulation, the equivalent discrete-time model of a damped sinusoid impulse response can be shown to satisfy

$$\alpha^k \sin(\beta kT_s) \overset{L}{\leftrightarrow} \frac{2\alpha z \sin(\beta T_s)}{T_s(z^2 - 2\alpha z \cos(\beta T_s) + \alpha^2)}$$

where $T_s = 1/f_s$, $\alpha = e^{-6228T_s}$, and $\beta = 6228$. For example, if $f_s = 5$ kHz, then the poles of the impulse-invariant system are located at

$$z = e^{sT_s} = e^{-(6228(1 \pm j1)/5000)} = 0.0919 \pm j0.272686$$

(See Fig. 13.16). Thus,

$$H(z) = \frac{Kz}{(z - 0.0919)^2 + (0.272686)^2}$$

where the value of K is chosen so that the DC gain, namely $H(1)$, is unity.

Computer Study. The s-file "imp_inv.s" defines a function, "impinv(fs)," that simulates the magnitude frequency response of $H(s)$ and its impulse-invariant version $H(z)$. The function simulates, plots, and compares the baseband frequency response of both systems out to f_s for $f_s = 5$ kHz, $f_s = 10$ kHz, and $f_s = 50$ kHz (see Fig. 13.17). It can be seen that the continuous-time system behaves as a lowpass filter. The difference between the analog and impulse-invariant discrete-time filter, from DC out to the Nyquist frequency, is small for high sampling rates. The responses, however, are divergent when the sample rate is reduced. This, again, is due to aliasing.

```
>include "chap13\imp_inv.s"
>impinv(5e3)
>impinv(10e3)
>impinv(50e3)
```

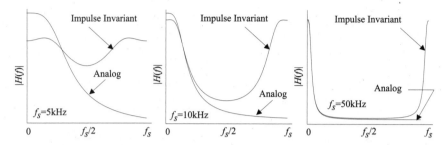

Figure 13.17 Simulated responses of a second-order continuous-time and discrete-time Butterworth filter for different sampling frequencies.

Example 13.8: Higher-Order Example

Problem Statement. In Example 13.5, a fourth-order elliptic lowpass filter was derived having a transfer function

$$H(s) = \frac{0.000461(s^4 + 3.229 \times 10^6 + 1.342 \times 10^{12})}{s^4 + 209.6s^3 + 7.148 \times 10^4 s^2 + 8.148 \times 10^6 s + 6.961 \times 10^8}$$

Convert $H(s)$ into a digital filter using the impulse-invariant transform.

Analysis and Conclusion. From Eq. (13.35), it is known that

$$h(t) \overset{s}{\Leftrightarrow} H(s) = \sum_{i=1}^{N} \frac{\alpha_i}{s + p_i} \overset{z}{\Leftrightarrow} \frac{1}{T_s} \sum_{i=1}^{N} \frac{\alpha_i}{(1 + e^{p_i T_s} z^{-1})} = \frac{1}{T_s} H(z) \overset{z}{\Leftrightarrow} \frac{1}{T_s} h[k]$$

The parameter α_i is historically called the *Heaviside coefficient*. A comparison of the analog and discrete-time filters' frequency responses is shown in Fig. 13.18.

Computer Study. The partial fraction expansion function defined in the s-file "pf.s" can be used to compute the Heaviside coefficients from a given polyno-

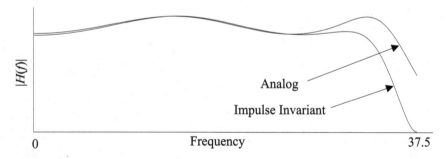

Figure 13.18 Comparison of a fourth-order lowpass elliptic analog and impulse-invariant digital filters' magnitude frequency responses.

mial description of a transfer function $H(s)$. For the given $H(s)$, the following sequence of Siglab commands can be used to factor the polynomial quotient.

```
>include "chap2\pf.s" # Include partial fraction expansion function.
>n=0.000462*[1,0,3.2283e6,0,1.34271e12]     # Numerator polynomial.
>d=[1,2.0966e2,7.1481e4,8.14933e6,6.9608e8] # Denominator polynomial.
>pfexp(n,d) # Perform partial fraction expansion.
Quotient polynomial:
0.000462
pole @ -75.7665-93.207j
with multiplicity 1
and coefficients:
13.8147 + 77.8455j
pole @ -75.7665 + 93.207j
with multiplicity 1
and coefficients:
13.8147-77.8455j
pole @ -29.0635 + 217.716j
with multiplicity 1
and coefficients:
-13.8631 + 26.9737j
pole @ -29.0635-217.716j
with multiplicity 1
and coefficients:
-13.8631-26.9737j
```

The Heaviside expansion of $H(s)$ is therefore computed to be

$$H(s) = k\left(\frac{\beta_1}{s - \alpha_1} + \frac{\beta_1^*}{s - \alpha_1^*} + \frac{\beta_2}{s - \alpha_2} + \frac{\beta_2^*}{s - \alpha_2^*} \right)$$

where $\beta_1 = 13.8147 + j77.8455$, $\alpha_1 = -75.7665 + j93.207$, $\beta_2 = -13.8631 - j26.9737$, $\alpha_1 = -29.0635 + j217.716$, and $k = 0.000462$. Continuing,

$$h(t) = k\left(\beta_1 e^{\alpha_1 t} + \beta_1^* e^{\alpha_1^* t} + \beta_2 e^{\alpha_2 t} + \beta_2^* e^{\alpha_2^* t} \right) u(t)$$

which corresponds to

$$H(z) = \frac{k}{T_s}\left(\frac{\beta_1 z}{z - e^{\alpha_1 T_s}} + \frac{\beta_1^* z}{z - e^{\alpha_1^* T_s}} + \frac{\beta_2 z}{z - e^{\alpha_2 T_s}} + \frac{\beta_2^* z}{z - e^{\alpha_2^* T_s}} \right)$$

which represents the impulse-invariant model of $H(s)$.

The s-file "spec.s" defines a function, "spec(fs)," that simulates the responses of an analog and discrete-time fourth-order lowpass elliptic filter having a 35 Hz passband. The function is called using the format "spec(fs)," where f_s is the chosen sampling frequency. Based on the previous examples, it can be conjectured that if the analog filter's magnitude frequency response persists beyond the Nyquist frequency, aliasing will result. Selecting a sampling frequency of 75 Hz, the filter may be simulated using the following sequence of Siglab commands.

```
>include "chap13\spec.s"
>spec(75)
```

The resulting analog and normalized discrete-time filter spectra (i.e., maximum passband gain set to unity) are shown in Fig. 13.18. It can be seen that there is some disagreement between the analog and discrete-time filters' responses, which is attributable to leakage.   ❖

It can be seen through analysis and simulation that the impulse-invariant design method, while preserving the integrity of the impulse response, has definite limitations when it comes to emulating the frequency response of a parent analog filter. If filter specifications are defined in the frequency domain, it follows that an impulse-invariant design may not be the preferred design approach. The next transform method, called the *bilinear z-transform*, will be shown to overcome the limitations of the impulse-invariant design method.

13.10 Bilinear z-Transform

Recall that the Laplace transform is applicable to modeling continuous-time systems, such as those defined by an ordinary differential equation (ODE). The fundamental ODE operator, used in programming an analog computer simulation of an ODE, is an integrator having a transfer function of the form $H(s) = 1/s$. It would, therefore, make sense to attempt to develop a discrete-time transform based on an integrator model. Specifically, the input-output transfer function of a continuous-time integrator is given by

$$H(s) = \frac{1}{s} \tag{13.37}$$

The discrete-time model of a Riemann integrator is given by

$$y[k+1] = y[k] + \frac{T_s}{2}(x[k] + x[k+1]) \tag{13.38}$$

which, in the z-domain, defines the input-output transfer function model of an integrator to be

$$H(z) = \frac{Y(z)}{X(z)} = \frac{T_s(z^{-1}+1)}{2(z^{-1}-1)} = \frac{T_s(z+1)}{2(z-1)} \tag{13.39}$$

Equations (13.36) and (13.39) define a relationship between the continuous-time and discrete-time interpolators which is given by

$$s = \frac{2(z-1)}{T_s(z+1)} \tag{13.40}$$

or, as an inverse mapping,

$$z = \frac{\left(\dfrac{2}{T_s}\right) + s}{\left(\dfrac{2}{T_s}\right) - s} \tag{13.41}$$

The mapping, defined by Eqs. (13.40) and (13.41), is called a *bilinear z-transform*. One factor that immediately stands out is that the bilinear z-transform is a simple algebraic mapping in contrast with a more complicated exponential mapping (i.e., e^{sT_s}) associated with the standard z-transform. However, this advantage is, by itself, insufficient to consider using the bilinear z-transform to be a replacement to the standard z-transform. The problem associated with the standard z-transform was the leakage (aliasing) that occurred when an analog filter was mapped into the discrete frequency domain. Overcoming this problem would be sufficient justification for using the bilinear z-transform as a replacement for the standard z-transform.

13.11 Warping

Assume that the frequency response of an analog filter is defined by $H_a(j\omega)$. The frequency response of a discrete-time filter is given by $H(e^{j\varpi})$, where $\varpi \in [-\pi,\pi]$. The discrete-time frequency axis can be calibrated with respect to the actual sampling frequency using $\omega \leftarrow \varpi/T_s$. The problem at hand is to construct a mapping of the $j\omega$-axis in the s-plane to the unit circle in the z-plane, in a manner which eliminates aliasing. The bilinear z-transform is capable of achieving this objective. To demonstrate this claim, consider evaluating Eq. (13.40) for a given $s = j\omega$ and $z = e^{j\varpi}$, namely

$$j\omega = \left(\frac{2}{T_s}\right)\frac{e^{j\varpi} - 1}{e^{j\varpi} + 1} = \left(\frac{2}{T_s}\right)\frac{j\sin(\varpi/2)}{\cos(\varpi/2)} = \left(\frac{2}{T_s}\right)j\tan(\varpi/2) \tag{13.42}$$

which, upon simplification, becomes

$$\omega = \left(\frac{2}{T_s}\right)\tan(\varpi/2) \qquad (13.43)$$

or

$$\varpi = \left(\frac{2}{T_s}\right)\tan^{-1}(\omega/2) \qquad (13.44)$$

which are interpreted in Fig. 13.19.

Equation (13.44) is called the *warping equation*, and Eq. (13.43) is the *prewarping equation*. It can be seen that the relationship between the analog and discrete (normalized) frequency axes is nonlinear. As a result, the critical frequencies of an analog filter model will not, in general, align themselves exactly with the critical frequencies of a bilinear z-transformed discrete-time filter. This potential distortion will be dealt with momentarily. What is of significant importance, however, is that the bilinear z-transform overcomes the leakage or aliasing problem associated with the impulse-invariant design method. It can be seen that as $\omega\to\infty$ along the continuous-frequency axis, $\varpi\to\pi$ along the discrete-frequency axis (likewise, as $\omega\to-\infty$, $\varpi\to-\pi$). That is, the entire analog frequency axis is mapped onto the unit circle. As a result, no leakage is possible in a bilinear z-transformed spectrum. Because of this, the bilinear z-transform is well suited to convert a

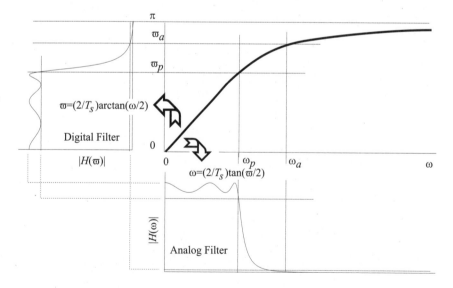

Figure 13.19 Relationship between the continuous and discrete (normalized) frequency axes under the bilinear z-transform.

classic analog filter into a discrete-time IIR model that preserves the shape of the magnitude frequency response of the parent analog system. The design paradigm, unfortunately, is not a straightforward process, due to the fact that the analog prototype is defined in terms of a set of analog frequencies which differ from those possessed by the discrete-time filter. The difference is governed by the warping equations. The possible distortion between the frequency axes, however, can be accounted for by using the procedure outlined below and summarized graphically in Fig. 13.20. The five steps are:

1. Specify the desired discrete-time filter requirements and attributes.

2. Prewarp the discrete time critical frequencies into corresponding analog frequencies and estimate analog filter order.

3. Design an analog prototype filter from the given continuous-time parameters.

4. Convert the analog prototype into an analog filter $H(s)$ using frequency-frequency transforms.

5. Design a digital filter $H(z)$ using a bilinear z-transform of $H(s)$ that warps the frequency axis that has previously been prewarped.

While this method may initially seem to be complicated, it is a simple sequential procedure that can be reduced to a digital computer program. To demonstrate the use of the bilinear z-transform, several simple examples are offered.

Example 13.9: Bilinear z-Transform Design Method

Problem Statement. Design a discrete-time second-order Butterworth filter which meets or exceeds the following specifications:

Maximum passband attenuation = −3 dB

Passband $f \in [0,1]$ kHz

Minimum stopband attenuation = −10 dB

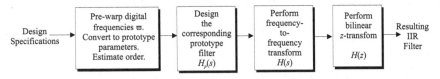

Figure 13.20 Design of a discrete-time IIR filter from an analog model using a bilinear z-transform.

Stopband $f \in [2,5]$ kHz

Sampling frequency $f_s = 10$ kHz

Create a digital IIR filter using the bilinear z-transform.

Analysis and Results. Once the discrete-time filter has been specified, the design parameters are prewarped into their equivalent continuous-time values. In particular, the continuous-time critical frequencies are given by

$$\omega_p = \frac{2}{T_s}\tan(\varpi_p/2) = 20 \times 10^3 \tan(0.1\pi) = 6498 r/s \rightarrow 1.0345 \text{ kHz}$$

and

$$\omega_a = \frac{2}{T_s}\tan(\varpi_a/2) = 20 \times 10^3 \tan(0.2\pi) = 1453 r/s \rightarrow 2.312 \text{ kHz}$$

The second-order prototype Butterworth model is known to satisfy

$$H_p(s) = \frac{1}{s^2 + 1.414s + 1}$$

which can be carried into an analog filter using a lowpass-to-lowpass frequency-frequency transform $s = s/\omega_p$. The resulting analog filter becomes

$$H_a(s) = \frac{4.3 \times 10^7}{s^2 + 9200s + 4.3 \times 10^7}$$

The final step is to apply the bilinear z-transform to $H(s)$ to obtain $H(z)$

$$H(z) = \frac{0.0676(z+1)^2}{z^2 - 1.142z + 0.412}$$ ❖

Example 13.10: Higher-Order Bilinear z-Transformed Filter

Problem Statement. Design an eighth-order elliptic lowpass IIR filter meeting the following specifications.

Sampling frequency $f_s = 100$ kHz

Allowable passband deviation = 1 dB, passband range $f \in [0, 20]$ kHz

Minimum stopband attenuation = 60 dB, stopband range $f \in [22.5, 50]$ kHz

Create a digital IIR filter using the bilinear z-transform.

Analysis and Conclusions. Using a computer (see Computer Study), the elliptic IIR filter shown in Fig. 13.21 was developed. The passband deviation for the digital filter is 1 dB, and the minimum stopband attenuation is 69.97 dB,

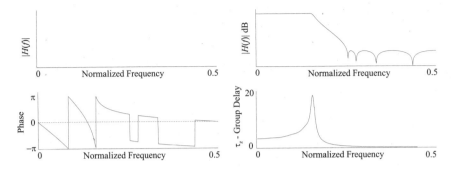

Figure 13.21   Response of an eighth-order elliptic lowpass filter.

which exceeds the 60 dB specification. Returning to Fig. 13.20, the following steps would need to be sequentially completed:

1. Specify the desired discrete-time filter requirements and attributes.

   Sampling frequency $f_s = 100$ kHz

   Allowable passband deviation = 1 dB, Passband range $f \in [0, 20]$ kHz

   Minimum stopband attenuation = 60 dB, Stopband range $f \in [22.5, 50]$ kHz

2. Prewarp the discrete-time critical frequencies into corresponding analog frequencies and estimate filter order.

$$\omega_p = \frac{2}{T_s}\tan(\varpi_p/2) = 2 \times 10^5 \tan(0.2\pi) = 1.45308 \times 10^5 \, r/s \rightarrow 23.1265 \text{ kHz}$$

$$\omega_a = \frac{2}{T_s}\tan(\varpi_p/2) = 2 \times 10^5 \tan(0.225\pi) = 1.70816 \times 10^5 \, r/s \rightarrow 27.1862 \text{ kHz}$$

Order $N = 8$ (given)

3. Design an analog prototype filter from the given continuous-time parameters. The coefficients for the filter are listed below.

```
Scale Factor = 0.0003171790179146282
Numerator Coefficients    Denominator Coefficients
b8: 1.000000000000000      a8: 1.000000000000000
b7: 0.000000000000000      a7: 0.8432321549557102
b6: 22.78212198717270      a6: 2.324679877013270
b5: 0.000000000000000       a5: 1.518496611851494
b4: 100.3988955650298      a4: 1.792686987736361
b3: 0.000000000000000      a3: 0.8384140160892136
b2: 155.0010233324510      a2: 0.4937712826502375
b1: 0.000000000000000      a1: 0.1336234536436084
b0: 78.71438390690363      a0: 0.02801293094358584
```

$$H_p(s) = \frac{0.000317(s^8 + 22.7s^6 + 100.3s^4 + 1.550s^2 + 78.7)}{s^8 + 0.843s^7 + 2.324s^6 + 1.518s^5 + 1.792s^4 + 0.838s^3 + 0.493s^2 + 0.133s + 0.028}$$

4. Convert the analog prototype into an analog filter $H(s)$ using frequency-frequency transforms. The resulting coefficients are listed below.

```
Scale Factor = 0.0003171790179146282
Numerator Coefficients        Denominator Coefficients
b8: 1.000000000000000         a8: 1.000000000000000
b7: 0.000000000000000         a7: 132848.6438766790
b6: 565475906758.6661         a6: 57700966666.66305
b5: 0.000000000000000         a5: 5938048306328768.
b4: 6.185414395968560e+22     a4: 1.104445605602088e+21
b3: 0.000000000000000         a3: 8.137825288365668e+25
b2: 2.370251136743972e+33     a2: 7.550672368678664e+30
b1: 0.000000000000000         a1: 3.219233339889335e+35
b0: 2.987678290671110e+43     a0: 1.063257075570912e+40
```

5. Design a digital filter $H(z)$ using a bilinear z-transform of $H(s)$ which warps the frequency axis which has previously been prewarped. The resulting coefficients are listed below.

```
Scale Factor = 0.006580053908657626
Numerator Coefficients Denominator Coefficients
b8: 1.000000000000000         a8: 1.000000000000000
b7: 1.726394249837218         a7: -3.658329148486950
b6: 3.949504363016771         a6: 7.945038529817918
b5: 4.936136278362346         a5: -11.43292389819922
b4: 5.923576026393579         a4: 11.90622526260542
b3: 4.936136278362346         a3: -8.996275762206013
b2: 3.949504363016771         a2: 4.845038440452235
b1: 1.726394249837218         a1: -1.711127747265849
b0: 0.9999999999999999        a0: 0.3175496992717872
```

These are the coefficients for a digital IIR filter transfer function given by

$$H(z) = \frac{0.00658(z^8 + 1.726z^7 + 3.949z^6 + 4.936z^5 + 5.923z^4 + 4.936z^3 + 3.949z^2 + 1.726z + 1)}{z^8 - 3.658z^7 + 7.495z^6 - 11.432z^5 + 11.906z^4 - 8.996z^3 + 4.845z^2 - 1.711z + 0.317}$$

The frequency response of the eighth-order elliptic filter is plotted in Fig. 13.21, where it can be seen to exhibit the classic ripple in the passband and stopband. Observe also that most of the phase variability is concentrated in the passband and transition band, and early stopband. This is verified by viewing the group delay which indicates that a delay of about 20 samples occurs at the transition-band frequency.

Computer Study. The design of the eighth-order elliptic filter can be viewed by loading file "ellip8.IIR" into the IIR design menu. The filter coefficients shown above were taken from the report file "ellip8.rpt" which was produced during the design process. The data shown in Fig. 13.21 is produced using the sequence of commands shown below.

```
>h = rarc("chap13\ellip8.arc",1) # Load elliptic IIR filter.
>y = filt(h,impulse(1024)) # Compute impulse response of filter.
```

```
>yfp = pfft(y)              # Generate and plot frequency response.
>graph(mag(yfp),20*log(mag(yfp)),phs(yfp),grp(yfp))        ❖
```

Example 13.11: Classic IIR Filter Design

Problem Statement. Design a lowpass Butterworth, Chebyshev-I, Chebyshev-II, and elliptic IIR filters that meet or exceed the following specifications.

Desired Passband Attenuation = 1.0 dB

Desired Stopband Attenuation = 40.0 dB

Sampling Frequency = 10000.0 Hz

Passband Edge = 1500.0 Hz

Stopband Edge = 2500.0 Hz

Implement the digital filter using the bilinear z-transform.

Analysis and Conclusions. Using the stated IIR filter design paradigm and computer-aided design tools, the Butterworth, Chebyshev-I, Chebyshev-II, and elliptic IIR filters were generated. The results from the generation of these filters are shown below.

```
Filter Type: Butterworth
Actual Stopband Attenuation: 40.9855 dB
Order: 8
H(z)
Scale Factor = 0.0005705414218556547
Numerator Coefficients Denominator Coefficients
   b8: 1.000000000000000      a8: 1.000000000000000
   b7: 8.000000000000000      a7: -2.829054555430326
   b6: 28.00000000000000      a6: 4.309045376441891
   b5: 56.00000000000000      a5: -4.092118976022213
   b4: 70.00000000000000      a4: 2.619609039717636
   b3: 56.00000000000000      a3: -1.132089946694619
   b2: 28.00000000000000      a2: 0.3203684065766790
   b1: 8.000000000000000      a1: -0.05378515271965319
   b0: 1.000000000000000      a0: 0.004084412125652116


Filter Type: Chebyshev Type I
Order: 5
Actual Stopband Attenuation: 44.35666 dB
H(z)
Scale Factor = 0.002020169397617657
Numerator Coefficients Denominator Coefficients
   b5: 1.000000000000000      a5: 1.000000000000000
   b4: 5.000000000000000      a4: -3.162364647736195
   b3: 10.00000000000000      a3: 4.760700364549023
   b2: 10.00000000000000      a2: -4.052794082948058
   b1: 5.000000000000000      a1: 1.934390525887203
   b0: 1.000000000000000      a0: -0.4152867390282085
```

```
Filter Type: Chebyshev Type II
Order: 5
Actual Stopband Attenuation: 44.3566 dB
H(z)
Scale Factor = 0.05239987594174734
Numerator Coefficients Denominator Coefficients
   b5: 1.000000000000000      a5: 1.000000000000000
   b4: 2.073170731707318      a4: -1.205636106513167
   b3: 3.170731707317074      a3: 1.213622639361258
   b2: 3.170731707317074      a2: -0.4983685124449977
   b1: 2.073170731707318      a1: 0.1583280351022219
   b0: 1.000000000000000      a0: -0.01358662911081143

Filter Type: Elliptic
Order: 4
Actual Stopband Attenuation: 51.1447 dB
H(z)
Scale Factor = 0.02099872314742810
Numerator Coefficients Denominator Coefficients
   b4: 1.000000000000000      a4: 1.000000000000000
   b3: 1.567841933416378      a3: -2.346138658587813
   b2: 2.196319333663452      a2: 2.690894615786962
   b1: 1.567841933416378      a1: -1.585086766991654
   b0: 1.000000000000000      a0: 0.4130798064394767
```

The frequency domain behavior of the four IIR filters is shown in Fig. 13.22 in the context of their magnitude and log magnitude frequency responses, phase responses, and group delays. Notice that, in each case, the specifications are exceeded in the stopband by 1 to 11 dB. The magnitude frequency responses vary subtly in shape. The greatest comparative differences occur in the phase and group delay responses. Notice that there is a nonlinear frequency dependent delay ranging from near 0 to 16 sample delays. Furthermore, the most egregious phase or group delay is found in the region of the transition band in the frequency domain for all four filter types. This nonlinear phase distortion, in general, is difficult to correct with a phase compensating filter. As noted be-

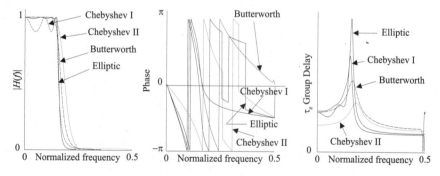

Figure 13.22 The pole-zero distribution of four IIR filters along with their comparative magnitude frequency, phase, and group responses.

fore, as a general rule of thumb, the nonlinear group delay equalization of an Nth-order IIR filter would require another IIR filter of order at least $2N$.

Computer Study. The IIR filter design process can be viewed by loading the files "butt8.IIR," "cheby1_5.IIR," "cheby2_5.IIR," and "ellip4.IIR" into the IIR design menu. The filters, when generated, produced engineering reports saved as "butt.RPT," "chby1_5.RPT," "cheby2_5.RPT," and "ellip4.RPT." The following sequence of Siglab commands produces the data reported in Fig. 13.22.

```
>bb=rarc("chap13\butt8.arc",1)          # Load IIR filter architectures.
>hc1=rarc("chap13\chebl_5.arc",1)
>hc2=rarc("chap13\cheby2_5.arc",1)
>he=rarc("chap13\ellip4.arc",1)
>x=impulse(1024)                        # Generate test impulse.
>yb=fft(filt(hb,x))                     # Generate frequency response.
>yc1=fft(filt(hc1,x))
>yc2=fft(filt(hc2,x))
>ye=fft(filt(he,x))
>y1=yb[512:1023]; y2=yc1[512:1023]      # Extract positive spectra.
>y3=yc2[512:1023]; y4=ye[512:1023]
>ograph(mag(y1),mag(y2),mag(y3),mag(y4))    # Plot positive mag spec-
                                              trum.
>ograph(20*log(mag(y1)),20*log(mag(y2)),20*log(mag(y3)),20*log(mag(y4)))
>ograph(phs(y1),phs(y2),phs(y3),phs(y4))    # Plot positive phase
                                              spectrum.
>ograph(grp(yb),grp(yc1),grp(yc2),grp(ye))  # Plot 2-sided group delay.
```

The analysis can be extended using a soundboard. Using a chirp having a frequency range $f \in [0, f_s/4]$, the audio response of each filter can be tested. Assuming a 16-bit soundboard is used with a sample rate of $f = 11{,}025$ Hz, the following sequence of Siglab commands can be used to test the IIR filters (assuming the filter architectures are loaded as above). The signal length is chosen so that a sufficiently long play period can be achieved. There can be heard to be little to no difference between the IIR filter responses. All fall off at high frequencies as expected.

```
>include "chap13\chirp.s"        # Get chirp generator.
>x=chirp(10000,0,0,0.25)         # Generate chirp.
>wavout(0.8*2^15*x)              # Play unfiltered signal.
>wavout(0.8*2^15*filt(hb,x))     # Butterworth IIR filtered signal.
>wavout(0.8*2^15*filt(hc1,x))    # Chebyshev I IIR filtered signal
>wavout(0.8*2^15*filt(hc2,x))    # Chebyshev II IIR filtered signal
>wavout(0.8*2^15*filt(he,x))     # Elliptic IIR filtered signal   ❖
```

It can be seen that, if maintaining general agreement between the envelope of the magnitude frequency response of a parent analog model [i.e., $H(j\omega)$] and its discrete-time counterpart [i.e., $H(e^{j\omega})$] is important, the bilinear z-transform is the transform of choice. It should be noted, however, that the bilinear z-transform is not an impulse-invariant transform as was the case with the standard z-transform.

Example 13.12: Impulse Invariance and Noninvariance

Problem Statement. A signal $h(t) = tu(t)$ can be uniformly sampled and transformed using the standard and bilinear z-transform. Test the impulse invariance of each transform for a sample rate of $f_s = 1$ Hz.

Analysis and Results. The impulse-invariant model $H(z)$ is defined in terms of the standard z-transform of $H(s) = 1/s^2$. The bilinear z-transform of $H(s)$ is a straightforward mapping process.

$$H_1(z)\Big|_{\text{standard-}z} = \frac{z^2}{z^2 - 2z + 1} = \frac{1}{1 - 2z^{-1} + z^{-2}}$$

$$H_2(z)\Big|_{\text{bilinear-}z} = 0.25\left(\frac{z^2 + 2z - 1}{z^2 - 2z + 1}\right) = 0.25\left(\frac{1 + 2z^{-1} - z^{-2}}{1 - 2z^{-1} + z^{-2}}\right)$$

The short-term impulse responses can be computed using long division. In particular, consider the following cases.

Standard z-transform:

$$z^{-2} - 2z^{-1} + 1 \overline{\smash{\big)}\ z^{-2}} \quad \overset{\displaystyle 1 + 2z^{-1} + 3z^{-2} + \text{ramp time – series}}{}$$

$$\underline{z^{-2} - 2z^{-1} + 1}$$
$$\text{continue}$$

Bilinear z-transform:

$$z^{-2} - 2z^{-1} + 1 \overline{\smash{\big)}\ 0.25z^{-2} + 0.5z^{-1} + 0.25} \quad \overset{\displaystyle 0.25 + 1z^{-1} + 2z^{-2} + \text{others}}{}$$

$$\underline{z^{-2} - 2z^{-1} + 1}$$
$$\text{continue}$$

It can be seen that the standard z-transform produces an impulse response which agrees with $h(t)$ at the sample instances (i.e., impulse-invariant). The bilinear z-transform, however, produces an impulse response which does not agree with the value of $h(t)$ at the sample instances and is therefore not an impulse-invariant mapping. ❖

13.12 Direct Design Method

To this point, LTI transfer functions have been assumed to be defined using classical filter formulas (e.g., Butterworth) or derived from observed filter behavior (e.g., ARMA modeling). Another alternative is possible. It had been earlier stated that analog filter designers often choose to design a filter $H(s)$ by directly specifying the system's pole-zero distribution. The same methodology can also be used to design a discrete-time filter. The steady-state performance of a filter, up to a scale factor, can be deduced from the system's pole-zero distribution. The steady state frequency response of a discrete-time LTI system has a magnitude frequency response given by

$$\left|H(e^{j\varpi})\right| = \left|H(z)\right|_{z = e^{j\varpi}} \tag{13.45}$$

The phase response satisfies

$$\phi(e^{j\varpi}) = \arg[H(e^{j\varpi})] = \tan^{-1}\left\{\frac{\text{Im}[H(e^{j\varpi})]}{\text{Re}[H(e^{j\varpi})]}\right\} \tag{13.46}$$

for $\varpi \in [-\pi,\pi]$. Both the magnitude and phase responses are seen to be specified by the filter's pole-zero distribution up to a constant of proportionality. For an M-zero, N-pole IIR filter design, where $M \le N$, the frequency response of the filter can be quantified as

$$H(e^{j\varpi}) = \frac{\displaystyle\sum_{m=0}^{M} b_m e^{-j\varpi m}}{\displaystyle\sum_{m=0}^{N} a_m e^{-j\varpi m}} = K\frac{\displaystyle\prod_{i=1}^{M}(e^{j\varpi} - z_i)}{\displaystyle\prod_{i=1}^{N}(e^{j\varpi} - p_i)} \tag{13.47}$$

where the poles and zeros of the filter have values z_i and p_i respectively. Evaluating Eq. (13.47) along the periphery of the unit circle, namely, $z = e^{j\varpi}$, it follows that

$$H(e^{j\varpi}) = K\frac{\displaystyle\prod_{i=1}^{M}\alpha_i(j\varpi)}{\displaystyle\prod_{i=1}^{N}\beta_i(j\varpi)} = K\frac{\displaystyle\prod_{i=1}^{M}(e^{j\varpi} - z_i)}{\displaystyle\prod_{k=1}^{N}(e^{j\varpi} - p_i)} \tag{13.48}$$

where

$$\alpha_i(j\varpi) = \left|\alpha_i(j\varpi)\right|e^{j\varphi_i} \tag{13.49}$$

and

$$\beta_i(j\varpi) = |\beta_i(j\varpi)|e^{j\phi_i} \qquad (13.50)$$

Therefore, Eq. (13.48) can be redefined as

$$|H(e^{j\varpi})| = K\frac{\displaystyle\prod_{i=1}^{M}\alpha_i(j\varpi)}{\displaystyle\prod_{i=1}^{N}\beta_i(j\varpi)} \qquad (13.51)$$

so that the phase is computed to be

$$\phi(e^{j\varpi}) = \arg[H(e^{j\varpi})] = \sum_{i=1}^{M}\varphi_i - \sum_{i=1}^{N}\phi_i + \psi \qquad (13.52)$$

where $\psi = 0$ if $K \geq 0$ and $\psi = \pi$ otherwise.

The magnitude and phase of α_i and β_i are directly influenced by their proximity of the unit circle as shown in Fig. 13.23. That is, a pole that is in close proximity to the periphery of the unit circle will cause locally a high gain in proximity of pole. Similarly, a zero that is in close proximity to the periphery of the unit circle will cause locally a null to appear in the magnitude frequency response in proximity of zero.

Example 13.13: Direct Design of an IIR Filter

Problem Statement. Direct IIR filter design methods, based on pole-zero placement, can begin with a clean sheet of paper on which designer places

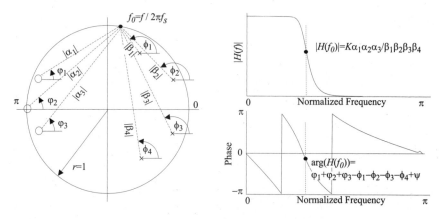

Figure 13.23 Relationships between pole-zero distribution and frequency response based on Eqs. (13.51) and (13.52).

poles within the unit circle in areas where locally high gain is desired. Equivalently, to accentuate the local gain of an existing IIR filter design, the poles can be moved closer to the periphery of the unit circle. Consider the classic fifth-order Chebyshev-II IIR filter studied in Example 13.11 (file "cheby2_5.IIR"). The frequency response shown in Fig. 13.24 exhibits a nearly classic piecewise constant magnitude frequency domain behavior. Modify $H(z)$ to create a new filter that can compensate for an assumed 20 percent roll-off from an analog system placed in front of the ADC.

Analysis and Conclusion. The fifth-order Chebyshev-II IIR filter shown in Fig. 13.24 is given by

$$H(z) = K \frac{\displaystyle\prod_{i=1}^{5}(z - z_i)}{\displaystyle\prod_{i=1}^{5}(z - p_i)}$$

where $K = 0.00963$, and the pole and zero locations are tabulated in Table 13.2. Several design options are available. First, the poles located near the edge of the transition band could be move slightly closer to the periphery of the unit circle to locally elevate the gain in this region. Alternatively, the poles that support the low-frequency range can be pushed back from the periphery of the unit circle to depress the gain over this range. This is the option taken.

The resulting magnitude frequency response of the original and modified IIR filter are interpreted in Fig. 13.24. It can be seen that the response of the modified system emphasizes the high-frequency response of the new system.

Computer Study. The Chebyshev-II filter was designed using computer-aided-design tools and the pole-zero design menu to alter the pole positions. The original classic Chebyshev-II filter is defined in file "cheby2_5.IIR," which produces an engineering report saved as "cheby2_5.RPT." The filter can imported into the Pole-Zero menu and the pole-locations changed as discussed above.

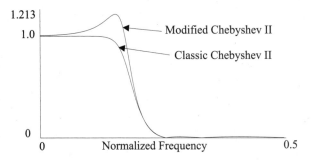

Figure 13.24 Classic fifth-order Chebyshev-II IIR filter showing the magnitude frequency responses of the original and manually modified IIR filter.

Table 13.2   Pole-Zero Distribution of a Chebyshev-II IIR Filter and Modified
Pole-Zero Distribution

Unmodified H(z)	Modified H(z)
$z_1 = z_2^* = -0.051 + j0.988$	$z_1 = z_2^* = -0.051 + j0.988$ (unchanged)
$z_3 = z_4^* = -0.486 + j0.873$	$z_3 = z_4^* = -0.486 + j0.873$ (unchanged)
$z_5 = -1.0$	$z_5 = -1.0$ (unchanged)
$p_1 = p_2^* = 0.191 + j0.379$	$p_1 = p_2^* = 0.191 + j0.379$ (unchanged)
$p_3 = p_4^* = 0.352 + j0.715$	$p_3 = p_4^* = 0.40 + j0.75$ (changed)
$p_5 = 0.119$	$p_5 = 0.119$ (unchanged)

The modified filter is saved under the filename "cheby2_a.ARC." The modified
filter can then be generated and the response displayed. The filter responses
are compared using the following sequence of Siglab commands.

```
>hc2=rarc("chap13\cheby2_5.arc",1)   # Read original filter arch.
>hca=rarc("chap13\cheby2_a.arc",1)   # Read modified filter arch.
>x=impulse(1024)                     # Create impulse test signal.
>yc2=pfft(filt(hc2,x))               # Generate frequency responses.
>yca=pfft(filt(hca,x))
>ograph(mag(yc2),mag(yca))           # Plot positive magnitude spectrum.
```

The analysis can be extended using a soundboard. Using a chirp defined over
the frequency range $f \in [0, f_s/4]$, the audio response of each filter can be tested.
Assuming a sixteen bit soundboard is used with a sample rate of $f = 11,025$
Hz, the following sequence of Siglab commands can be used to test the IIR filters. The signal length is chosen so that a sufficiently long play period can be
achieved. There can be heard to be little to no difference between the two IIR
filters. The subtle changes occur in the higher frequency range.

```
>include "chap13\chirp.s"          # Load chirp generator function.
>x=chirp(10000,0,0,0.25)           # Generate chirp test signal.
>wavout(0.8*2^15*x)                # Play original unfiltered signal.
>wavout(0.8*2^15*filt(hc2,x))      # Play Chebyshev-II filtered signal.
>wavout(0.8*2^15*filt(hca,x))      # Play modified filtered signal.      ❖
```

Example 13.14: All-Pass IIR Filter

Problem Statement. Design an all-pass filter having a magnitude frequency
response $|H(e^{j\omega})| = 1$ for all $\omega \in [-\pi, \pi]$.

Analysis and Conclusions. The transfer function of an LTI IIR filter is given by

$$H(z) = K \frac{\displaystyle\prod_{i=1}^{N} (z - z_i)}{\displaystyle\prod_{i=1}^{N} (z - p_i)}$$

The poles and zeros of an all-pass filter have a reciprocal relationship given by

$$z_i = \frac{e^{j\omega_i}}{r}$$

$$p_i = re^{j\omega_i}$$

for each $\omega_i \in [-\pi, \pi]$. It then follows that

$$\left| H(e^{j\omega}) \right| = K \frac{\displaystyle\prod_{i=1}^{N} \left(e^{j\omega} - \frac{e^{j\omega_i}}{r} \right)}{\displaystyle\prod_{i=1}^{N} (e^{j\omega} - re^{j\omega_i})}$$

For the purpose of stability, choose $r < 1$.

The design parameters for three all-pass filters are listed in Table 13.3. The magnitude and phase responses of all three filters are shown in Fig. 13.25. It can be seen that, in all three cases, the designs result in an all-pass magnitude frequency response. What differentiates one from the other is the passband gain (set by r) and the phase response.

Computer Study. The s-file "allpass.s" defines a function, "allpass(r,ω_1,ω_2,n)," that computes the frequency response of an all-pass filter, where r, ω_1, and ω_2 are found in Table 13.3, and n is the length of the simulated impulse response. The following sequence of Siglab commands generates the plots shown in Fig. 13.25.

Table 13.3 All-Pass Design Objectives

Filter	r	ω_0	ω_1	ω_2	ω_1^*	ω_2^*
$H_1(z)$	0.975	0	$\pi/6$	$\pi/3$	$-\pi/6$	$-\pi/3$
$H_2(z)$	0.900	0	$\pi/6$	$\pi/3$	$-\pi/6$	$-\pi/3$
$H_3(z)$	0.975	0	$\pi/3$	$2\pi/3$	$-\pi/3$	$-3\pi/3$

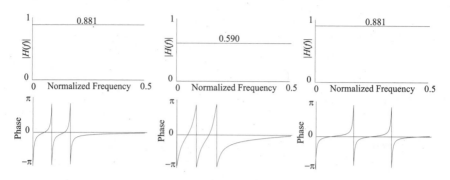

Figure 13.25 Pole-zero distribution influence on the response of an all-pass IIR filter.

```
>include "chap13\allpass.s"        # Get allpass function.
>h=allpass(0.975,pi/6,pi/3,501)    # Compute frequency response.
>graph(mag(h),phs(h))             # Plot magnitude, phase response.
>h=allpass(0.9,pi/6,pi/3,501)      # Compute frequency response.
>graph(mag(h),phs(h))             # Plot magnitude, phase response.
>h=allpass(0.975,pi/3,2*pi/3,501)  # Compute frequency response.
>graph(mag(h),phs(h))             # Plot magnitude, phase response. ❖
```

13.13 Summary

In Chap. 13, classic IIR filters were designed. The classic IIR digital filters are based on classic analog filters. It was shown that the digital IIR filter can be created using either an impulse-invariant or a bilinear z-transform. The impulse-invariant design method produced filters that have limited but, nevertheless, important applications. The most popular and useful IIR filter transform method is the bilinear z-transform method, which produces filters having a magnitude frequency response that closely approximates the shape of the magnitude frequency response of an analog filter model or template. The templates are typically assumed to be Butterworth, Chebyshev types I and II, and elliptic filters. The result, whether using impulse-invariant or bilinear z-transform design methods, is a transfer function. What remains is the challenging part of IIR filter design, namely implementation. Implementation requires that an architecture be chosen along with supporting analysis.

13.14 Self-Study Problems

1. Impulse-Invariant IIR Filter Design Design, using the impulse-invariant method, a third-order Butterworth lowpass filter having a –3 dB passband over $f \in [0,1000]$ Hz and a stopband that begins at $f = 2000$ Hz. Compute the realized stopband attenuation of the discrete-time IIR filter. Compute the frequency response of the analog and discrete-time filter if the sampling frequency is f_s = 5512.5 Hz and f_s = 11,025 Hz. How do they compare at f = 1000 Hz and f = 2000 Hz?

2. Bilinear z-Transform IIR Filter Design Design, using the bilinear z-transform method, a third-order Butterworth lowpass filter having a –3 dB passband over $f \in [0,1000]$ Hz and a stopband that begins at f = 2000 Hz. Compute the realized stopband attenuation of the discrete-time IIR filter. Compute the frequency response of the analog and discrete-time filter if the sampling frequency is f_s = 11,025 Hz. How do they compare at f = 1000 Hz and f = 2000 Hz?

3. Bilinear z-Transform IIR Filter Design Design, using the bilinear z-transform method, a fourth-order Butterworth bandpass filter having a –3 dB passband over $f \in [1500, 2500]$ Hz with the first stopband ending at f = 1000 Hz and the second stopband beginning at f = 3000 Hz. What is the realized transition bandwidths of the analog bandpass filter? Compute the frequency response of the analog and discrete-time filter if the sampling frequency is f_s = 4000 Hz. How do they compare at f = 1000 Hz, f = 2000 Hz and f = 3000 Hz?

4. Butterworth Bilinear z-Transform Design Design a lowpass Butterworth IIR filter that meets or exceeds the following specifications:

Desired Passband Attenuation = 3.0 dB

Desired Stopband Attenuation = 20.0 dB

Sampling Frequency = 10,000.0 Hz

Passband Edge = 2000.0 Hz

Stopband Edge = 2500.0 Hz

What is the required filter order? Sketch the magnitude frequency response and estimate the passband and stopband gains. For those having the professional version of Monarch, plot the magnitude and phase frequency response and the group delay.

5. Chebyshev-I Bilinear z-Transform Design Design a lowpass Chebyshev-I IIR filter that meets or exceeds the specifications given in Problem 13.4. What is the required filter order? Sketch the magnitude frequency response and estimate the passband and stopband gains. For those having the professional version of Monarch, plot the magnitude and phase frequency response and the group delay.

6. Chebyshev-II Bilinear z-Transform Design Design a lowpass Chebyshev-II IIR filter that meets or exceeds the specifications given in Problem 13.4. What is the required filter order? Sketch the magnitude frequency response and estimate the passband and stopband gains. For those having the professional version of Monarch, plot the magnitude and phase frequency response and the group delay.

7. Elliptic Bilinear z-Transform Design   Design a lowpass elliptic IIR filter that meets or exceeds the specifications given in Problem 13.4. What is the required filter order? Sketch the magnitude frequency response and estimate the passband and stopband gains. For those having the professional version of Monarch, plot the magnitude and phase frequency response and the group delay.

8. Manual IIR Filter Design   Consider again the Butterworth filter from Problem 13.2. Manually adjust the poles and/or zeros to increase the passband gain locally about 2000 Hz by 1 to 3 dB for $f_s = 11,025$ Hz. Evaluate the performance of both designs using a soundboard.

14

IIR Filter Implementation

14.1 Introduction

In Chap. 13, the IIR filter was developed. A number of design procedures were presented that can translate a set of frequency domain specifications into an IIR filter transfer function. The classic filters of Butterworth, Chebyshev, and Cauer (elliptic) were discussed in detail and shown to be easily characterized using computer-aided design (CAD) tools. IIR filters, unlike FIR filters, do contain feedback paths that can create potentially disastrous run-time problems. One of the most severe events that can occur is *register overflow*. Others include aggressive roundoff and arithmetic errors that can render a filter useless. Overflow performance and precision requirements are intimately related to the internal structure, or *architecture,* of an IIR filter. Architecture, recall, refers to how the basic building blocks of a filter, such as shift registers, multipliers, and adders are interconnected. Whereas an FIR filter had but a few architectural choices, IIR filters possesses many. In Chap. 14, a collection of IIR filter architectures are presented and analyzed using state variables as the integrating framework.

14.2 IIR Filter Architecture

A monic transfer function $H(z)$ possesses a number of factorizations. A common representation for a proper transfer function $H(z)$ is as the ratio of polynomials, namely

$$H(z) = \frac{N(z)}{D(z)} = K \frac{\displaystyle\sum_{m=0}^{M} b_m z^{-m}}{1 + \displaystyle\sum_{m=1}^{N} a_m z^{-m}} \qquad (14.1)$$

The roots of the polynomial $N(z)$, found in Eq. (14.1), define the location of the filter's *zeros,* which are assumed to be located at z_i, $i \in \{1,2,3,...,M\}$. The roots of the polynomial $D(z)$ define the location of the filter's *poles* which are assumed to be located at λ_i, $i \in \{1,2,3,...,N\}$. (Note that, for purposes of notational convenience, poles will temporarily be denoted λ_i as instead of p_i as was the previous case.) Specifically, the filter's poles and zeros allow $N(z)$ and $D(z)$ to be written as

$$N(z) = \prod_{i=0}^{M-1} (z - z_i)$$

$$D(z) = \prod_{i=0}^{N-1} (z - \lambda_i) \tag{14.2}$$

An alternative representation of $H(z)$ is as the product of low-order polynomial quotients,

$$H(z) = K \prod_{i=1}^{Q} H_i(z) \tag{14.3}$$

where $Q \leq N$. Here, Q will be used to represent the number of first-order or second-order subsystems that can be realized from the factorization of $H(z)$. First-order and second-order systems provide the design framework for many digital filters. The real poles of $H(z)$ will be used to define first-order subsystems and the complex poles of $H(z)$ define the second-order sections. Specifically, if λ_i is a real pole, then the ith first-order subsystem $H_i(z)$ is assumed to satisfy

$$H_i(z) = \frac{q_{i0} + q_i z^{-1}}{1 - \lambda_i z^{-1}} = q_{i0} + \frac{r_i z^{-1}}{1 - \lambda_i z^{-1}} \tag{14.4}$$

where r_i is called the *remainder* and is given by

$$r_i = q_{i1} + \lambda_i q_{i0} \tag{14.5}$$

where q_{i0} and r_i real coefficients. The structure of $H_i(z)$ is reported in Fig. 14.1.

If λ_i is complex and appears with a complex conjugate pair, then the resulting second-order transfer function can be expressed as

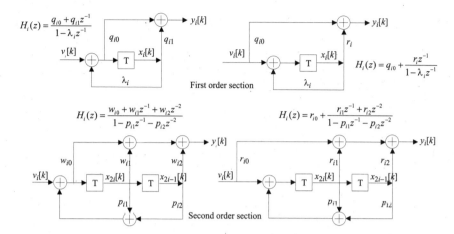

Figure 14.1 Implementation of a first-order section over a real pole, and a second-order section over a complex conjugate pair of poles, using real arithmetic.

$$H_i(z) = d_{i0} + \frac{n_i(z)}{1 - \lambda_i z^{-1}} + \frac{n_i^*(z)}{1 - \lambda_i^* z^{-1}} \tag{14.6}$$

The problem with implementing a second-order filter as two first-order sections having complex coefficients is one of efficiency. It should be appreciated that each complex multiply actually requires four real multiplies and two adds. That is, the product rule for a complex multiply is given by

$$(a + ib)(c + id) = (ac - bd) + i(ad + bc) \tag{14.7}$$

It can be seen that Eq. (14.7) is far more arithmetically complicated than its real counterpart. It is for this reason that DSP solutions generally attempt to avoid implementing complex multiplication if at all possible. The added complexity associated with each complex multiplication operation can, however, be mitigated by combining the two complex first-order sections into one second-order section defined with respect to real coefficients only. That is, Eq. (14.6) can be expressed as

$$H_i(z) = d_{i0} + \frac{n_i(z)}{1 - \lambda_i z^{-1}} + \frac{n_i^*(z)}{1 - \lambda_i^* z^{-1}}$$

$$= d_{i0} + \frac{[n_i(z) + n_i^*(z)] - [n_i(z)\lambda_i^* z^{-1} + n_i^*(z)\lambda_i z^{-1}]}{(1 - \lambda_i z^{-1})(1 - \lambda_i^* z^{-1})}$$

$$= d_{i0} + \frac{2\,\text{Re}[n_i(z)] - 2\,\text{Re}[\lambda_i^* n_i(z)]z^{-1}}{[1 - 2\,\text{Re}(\lambda_i)z^{-1} + |\lambda_i|^2 z^{-2}]} \tag{14.8}$$

The second-order section shown in Eq. (14.8) therefore can be rewritten as

$$H_i(z) = d_{i0} + \frac{r_{i1}z^{-1} + r_{i2}z^{-2}}{1 - p_{i1}z^{-1} - p_{i2}z^{-2}} = \frac{w_{i0} + w_{i1}z^{-1} + w_{i2}z^{-2}}{1 - p_{i1}z^{-1} - p_{i2}z^{-2}} \tag{14.9}$$

where all coefficients are now real and are given by

$$(z - \lambda_i)(z - \lambda_i^*) = z^2 - p_{i1}z - p_{i2}$$

$$d_{i0} = w_{i0}$$

$$r_{i1} = w_{i1} + p_{i1}w_{i0}$$

$$r_{i2} = w_{i2} + p_{i2}w_{i0} \tag{14.10}$$

The filter section defined by Eq. (14.9) is called a *biquadratic* section or a *biquad*. Block diagrams of the biquads defined in Eq. (14.9) are shown in Fig. 14.1. It should be appreciated that the biquad section is defined in terms of two complex conjugate poles and zeros and is realized using only real coefficients.

A logical question to pose is how to pair the zeros from a field of M zeros defined by the roots of $N(z)$ with the N poles obtained by $D(z)$. As a general rule, zeros are paired with the closest poles as suggested in Fig. 14.2. Here, the poles and zeros are paired so that the maximum distance between all the poles and zeros is minimized. The *proximity pairing* strategy will generally result in filter sections that have similar steady-state dynamic range requirements. This is usually desirable in that it will balance out the gains distributed within a system. Other pairing strategies will generally result in a few subsystems having large dynamic range requirements. Protecting these few high-gain sections against run-time register overflow will compromise the overall precision of the entire digital filter.

Example 14.1: Pole-Zero Pairing

Problem Statement. Consider a Chebyshev-II bandpass filter having the following specifications.

   Bandpass attenuation = 1 dB

   Stopband attenuation = 30 dB

Sampling frequency = 100 kHz

Stopband 1 = [0,5] kHz

Passband = [15,30] kHz

Stopband 2 = [35,50] kHz

The resulting Chebyshev-II IIR filter has eight poles and zeros. Pair them using the proximity rule.

Analysis and Conclusions. The poles and zeros of the filter are listed below and displayed in Fig. 14.2.

Zeros	Poles
$z_1 = z_5^* = -0.623 + j0.782$	$\lambda_1 = \lambda_5^* = -0.153 + j0.490$
$z_2 = z_6^* = 0.795 + j0.606$	$\lambda_2 = \lambda_6^* = 0.357 + j0.448$
$z_3 = z_7^* = -0.899 + j0.436$	$\lambda_3 = \lambda_7^* = -0.324 + j0.795$
$z_4 = z_8^* = 0.949 + j0.314$	$\lambda_4 = \lambda_8^* = 0.652 + j0.678$

The poles and zeros are combined using the proximity rule as shown in Fig. 14.2.

Computer Study. The bandpass filter is saved under the file name "ch14.IIR." The filter's pole and zero locations are saved in the engineering

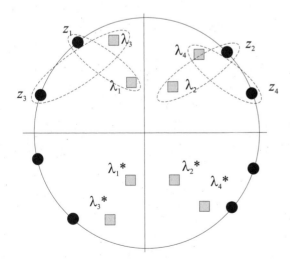

Figure 14.2 Pole-zero pairing strategy for a typical eight-order IIR filter.

report file "ch14.RPT". During the IIR filter design process, the transfer function is factored into a collection of second-order sections with pole-zeros assigned by the proximity paring rule. The resulting second-order sections are reported to be:

```
Section 1 (z₁,z₁*,p₁,p₁*)
b2 = 1.000000000000000          a2 = 1.000000000000000
b1 = 1.246346135098095          a1 = 0.3056833059715465
b0 = 1.000000000000000          a0 = 0.2637713127044173

Section 2 (z₂,z₂*,p₂,p₂*)
b2 = 1.000000000000000          a2 =  1.000000000000000
b1 = -1.590086942029127         a1 = -0.7146996856339551
b0 =  0.9999999999999999        a0 =  0.3279952700473907

Section 3 (z₃,z₃*,p₃,p₃*)
b2 = 1.000000000000000          a2 = 1.000000000000000
b1 = 1.799729700501794          a1 = 0.6475571518393545
b0 = 0.9999999999999999         a0 = 0.7367598807523005

Section 4 (z₄,z₄*,p₄,p₄*)
b2 =  1.000000000000000         a2 =  1.000000000000000
b1 = -1.898930749412855         a1 = -1.124043330876328
b0 =  0.9999999999999999        a0 =  0.7754970370224432
```

The following sequence of Siglab commands to evaluates the four second-order sections for various pairings of poles and zeros. Upon executing the commands, the magnitude frequency response of each second-order section is plotted. The individual second-order section magnitude frequency responses for the pairing shown above are reported in Fig. 14.3. The pairing shown above is seen to have the smallest maximal magnitude frequency response section gain. This pairing would therefore have the lowest dynamic data range requirement, which is generally desirable.

```
>h=rarc("chap14\ch14.arc",2)                    # Read filter architecture.
>freq=<zeros(64),pi*rmp(64,1/64,64)>            # Create frequency sweep.
>for (i=0:3)                                    # Cycle through combinations.
>>z1=h[0,0:1];z2=h[0,2:3];z3=h[0,4:5];z4=h[0,6:7]    # Extract zeros.
>>p1=h[1,0:1];p2=h[1,2:3];p3=h[1,4:5];p4=h[1,6:7]    # Extract poles.
>>v1=polyval(poly(z1),freq)./polyval(poly(p1),freq)  # Evaluate combo.
>>graph(mag(v1))                                # Graph results.
>>v2=polyval(poly(z2),freq)./polyval(poly(p2),freq)  # Evaluate combo.
>>graph(mag(v2))                                # Graph results.
>>v3=polyval(poly(z3),freq)./polyval(poly(p3),freq)  # Evaluate combo.
>>graph(mag(v3))                                # Graph results.
>>v4=polyval(poly(z4),freq)./polyval(poly(p4),freq)  # Evaluate combo.
>>graph(mag(v4))                                # Graph results.
>>max(mag([v1,v2,v3,v4]))           # Print worst case gain.
>>h[1,]=cshift(h[1,],2)             # Rotate for next group of permutations.
>>end
6.43795 # pairing z₁ and λ₁, z₂ and λ₂, z₃ and λ₃, z₄ and λ₄
7.17338 # pairing z₁ and λ₂, z₂ and λ₃, z₃ and λ₄, z₄ and λ₁
7.67403 # pairing z₁ and λ₃, z₂ and λ₄, z₃ and λ₁, z₄ and λ₂
7.92014 # pairing z₁ and λ₄, z₂ and λ₁, z₃ and λ₂, z₄ and λ₃    ❖
```

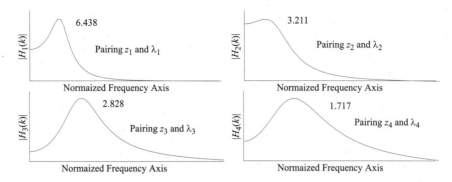

Figure 14.3 Magnitude frequency responses of second-order sections for optimal pole-zero pairing.

The architectural structure suggested by Eq. (14.3) has a product form. Another factorization of $H(z)$ is as a sum of sections, or Heaviside form, which is given by

$$H(z) = K \sum_{i-1}^{Q} H_i(z) \qquad (14.11)$$

where, again, $Q \leq N$, and Q is the number of first- or second-order systems found in $H(z)$. These factorizations are but a few of the possible forms that $H(z)$ can take, each resulting in a different architecture.

The formal study of architecture begins with a generic discrete-time model for a single-input single-output system given by the *monic* Nth-order difference equation

$$y[k] + a_1 y[k-1] + \ldots + a_N y[k-N] = b_0 u[k] + b_1 u[K-1] + \ldots + b_N u[k-N] \qquad (14.12)$$

which corresponds to a transfer function $H(z)$, where

$$\begin{aligned}
H(z) &= \frac{b_0 + b_1 z^{-1} + \ldots + b_N z^{-N}}{1 + a_1 z^{-1} + \ldots + a_N z^{-N}} \\[2mm]
&= b_0 + \frac{(b_1 - b_0 a_1)z^{-1} + \ldots + (b_N - b_0 a_N)z^{-N}}{1 + a_1 z^{-1} + \ldots + a_N z^{-N}} \\[2mm]
&= d_0 + \frac{c_1 z^{-1} + \ldots + c_N z^{-N}}{1 + a_1 z^{-1} + \ldots + a_N z^{-N}}
\end{aligned} \qquad (14.13)$$

This model was examined in Section 9.3 and was referred to as a direct-II filter. This architecture, as well as other basic structures, will be formally developed in this section.

14.3 Direct-II Architecture

The system characterized by Eq. (14.13) can be placed in the direct-II form shown in Fig. 14.4 by using the simple assignment rule developed below. The state-variable model is defined in terms of an N dimensional state vector $\mathbf{x}[k]$. More specifically, for

$$\mathbf{x}[k+1] = \mathbf{A}\mathbf{x}[k] + \mathbf{b}u[k]$$

$$y[k] = \mathbf{c}^T\mathbf{x}[k] + d_0 u[k] \tag{14.14}$$

let

$$\mathbf{x}[k+1] = \begin{bmatrix} x_1[k+1] \\ x_2[k+1] \\ \vdots \\ x_N[k+1] \end{bmatrix}$$

$$= \begin{bmatrix} x_2[k] \\ x_3[k] \\ \vdots \\ -a_N x_1[k] - a_{N-1}x_2[k] - \cdots - -a_2 x_{N-1}[k] - a_1 x_N[k] + u[k] \end{bmatrix} \tag{14.15}$$

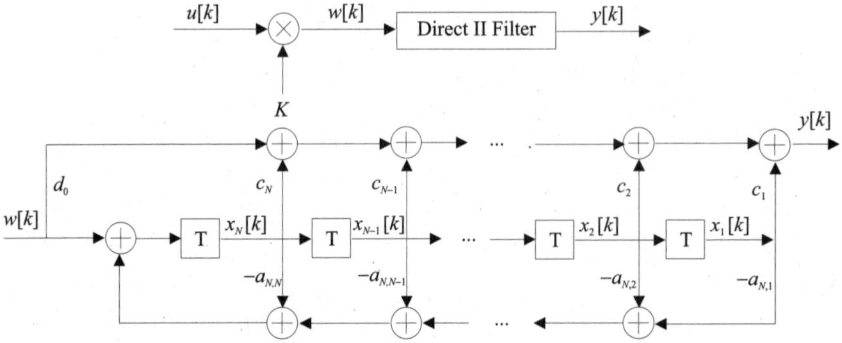

Figure 14.4 Direct-II realization of an Nth-order IIR filter.

It follows that the state four tuple $\mathbf{S} = (\mathbf{A}, \mathbf{b}, \mathbf{c}, d_0)$ is given by

$$
\mathbf{A} = \begin{bmatrix}
0 & 1 & 0 & \cdots & 0 \\
0 & 0 & 1 & \cdots & 0 \\
\vdots & \vdots & \vdots & \ddots & \vdots \\
0 & 0 & 0 & \cdots & 1 \\
-a_N & -a_{N-1} & -a_{N-2} & \cdots & -a_1
\end{bmatrix}
\tag{14.16}
$$

$$
\mathbf{b} = \begin{bmatrix}
0 \\
0 \\
\vdots \\
1
\end{bmatrix}
\tag{14.17}
$$

$$
\mathbf{c}^T = (b_N - b_0 a_N, \, b_{N-1} - b_0 a_{N-1}, \, \ldots, \, b_1 - b_0 a_1)
\tag{14.18}
$$

$$
d_0 = b_0
\tag{14.19}
$$

It can be seen that the construction rules are very straightforward and are developed in detail in Section 9.3. A direct-II implementation of an Nth-order system is seen to require for each filter cycle at most N multiplies from the term $\mathbf{Ax}$, N multiplies from the term $\mathbf{c}^T\mathbf{x}$, and none[*] from the terms $\mathbf{b}u$ and possibly one for $d_0 u$, for a total complexity measure of $M_{\text{multiplier}} \leq 2N + 1$. It will be shown, however, that not every non-zero or non-unity valued parameter found in the state-variable four-tuple $\mathbf{S} = (\mathbf{A}, \mathbf{b}, \mathbf{c}, d_0)$ defines a real coefficient multiply. The actual physical coefficients needed to implement a filter for a given architecture are called *DSP coefficients*. In addition, the scale factor K [see Eq. (14.1)] is often separately supplied, sometimes employed, sometimes ignored. If used, the scale factor can optionally be applied to the input or output, which will result in a scaled filter response. If applied to the input, the dynamic range requirement will be K-fold less than the filter applying the scale factor to the output. Care must be taken, however, to ensure that scaling the input does not render the input impotent. That is, scaling an eight-bit input signal by $K < 0.004$ would completely destroy the signal. If the scale factor is used, the resulting filter passband gain will generally be scaled to unity (0 dB). If the scale factor is not employed, the output will be $(1/K)$ times larger than it would be had the scaling factor been used.

---

[*] Scaling by ± 1 or zero is not counted as a general multiplication.

The implementation of a direct-II IIR filter on a modern DSP micro-processor is a straightforward processes. The following TI TMS320C50 code is typical of a direct-II implementation of a filter of order N.

```
.mmregs              ; Symbols for memory mapped registers.
DATA .set 02000h     ; Location of data in data memory space.
COEFC .set 02000h    ; Location of coefs in program memory space.
COEFA .set 02040h    ; Location of coefs in program memory space.
ORDER .set 11        ; Filter order.
NORMF .set 3         ; Normalization factor.
DIRECT2 MAR *,AR0        ; Use AR0 for indirect addressing.
; Compute the filter output.
LAR AR0,#DATA+ORDER      ; Init data address register.
RPTZ #ORDER          ; Clear P[roduct] register, accumulator
                     ; & repeat the next instr. ORDER+1 times.
MAC COEFA,*-         ; Compute MAC.
APAC                 ; Perform final accumulation.
SACB                 ; Store result in the accumulator buffer.
; Compute the feedback term, shift and update DATA.
LAR AR0,#DATA+ORDER+1    ; Init data address register.
RPTZ #ORDER             ; Clear P register, accumulator, & repeat
                        ; the next instruction ORDER+1 times.
MACD COEFC,*-           ; Compute MAC and move (shift) data.
ADRK #1                 ; Adjust AR0==DATA.
APAC                    ; Perform final accumulation
NOP                     ; Protect the pipeline for AR0.
SACH *,NORMF            ; Store new, normalized x_N.
RET                     ; Return from routine.
```

The given code segment is the heart of the direct-II IIR filtering operation. Storage locations for the filter's state and input data (DATA), the coefficients COEFC ($-c_i$ in Fig. 14.4), and the coefficients COEFA ($a_{N,i}$ in Fig. 14.4) are specified, as is the filter's order. The coefficients must be stored as indicated below.

COEFC					
Offset	0	1	2	...	ORDER
Data	d_0	c_N	c_{N-1}	...	c_1

COEFA					
Offset	0	1	2	...	ORDER
Data	1	$-a_{N,N}$	$-a_{N,N-1}$	...	$-a_{N,1}$

The scale factor K, if desired, may be applied to COEFA[0] and COEFC[0]:

$$\text{COEFA[0]} \leftarrow K$$

and

$$\text{COEFC[0]} \leftarrow K d_0$$

It is assumed that the latest input datum has been stored in the location DATA[0]. The routine DIRECT2 begins by specifying the address register to be used. Next, the address register is loaded with the ending address of the data array. The next instruction, RPTZ, clears the multiplier-accumulator pipeline and causes the next instruction to be executed (in this case) 11 times, with no looping overhead, except for the initial RPTZ instruction. The MAC instruction causes the coefficients and data to be multiply-accumulated, with the data index, AR0, decremented after each MAC instruction. The final product is accumulated using the APAC instruction, and the output of the filter is stored in the accumulator buffer by the SACB instruction. The next step is to execute a loop to compute the feedback term and to update the DATA array in anticipation of the storage of a new input datum. This loop works as before, except the MACD instruction is used instead of the MAC instruction, causing the DATA array to be shifted as the array is traversed. Since the results of the multiply-accumulate sequence are to be stored at the beginning of the DATA array, AR0 must be adjusted so that it points to DATA[0] (after the loop, it points to DATA[-1]). The ADRK instruction is used to add 1 to AR0. Notice that the ADRK instruction is placed *before* the APAC instruction, which is followed by a NOP. When an instruction modifies an AR* on the C50, the results are not available for two cycles. After the NOP, the SACH (store accumulator high) applies a right arithmetic shift of the most significant 16 bits of the accumulator (which is 32 bits) by NORMF bits and stores the result in DATA[0]. The normalization factor, NORMF, depends on the precision of the input data and coefficient data.

Example 14.2: Direct-II Architecture

Problem Statement. Convert the eighth-order Chebyshev-II filter described below and having the transfer function

$$H(z) = 0.0888 \frac{(z^8 + 4.43z^7 + 10.76z^6 + 17.46z^5 + 20.48z^4 + 17.46z^3 + 10.76z^2 + 4.43z + 1)}{(z^8 + 1.10z^7 + 1.97z^6 + 1.55z^5 + 1.22z^4 + 0.61z^3 + 0.24z^2 + 0.061z + 0.008)}$$

into a direct-II form.

Analysis and Conclusions. The transfer function can be directly converted into a state-variable model using the presented direct-II mapping rules. The resulting architecture is shown in Fig. 14.5. The filter is seen to consist of a chain of eight shift registers and a set of feedforward plus feedback paths. The states of the system reside in the shift registers, which are indexed from right to left. The elements of the state-variable four-tuple $S = (\mathbf{A}, \mathbf{b}, \mathbf{c}, d_0)$ can be produced using a general-purpose digital computer as shown in the Computer Study.

Computer Study. The designed Chebyshev-II filter is saved under the file name "scaled.IIR" with the engineering report saved as "scaled.RPT," and a direct-II architecture file as "scaled.arc." Sections of the engineering report are reproduced below. The developed filter is defined in terms of the set of specifications given in the report file (coefficient indices reversed).

```
DESIGN DEFINITION
File                          : scaled.IIR {derived from scale.IIR}
Source Module                 : Infinite Impulse Response Filter
Filter Type                   : Chebyshev Type II
Order                         : 8
Frequency Response            : Lowpass
Desired Passband Attenuation  : 0.1000000000000000 dB
Desired Stopband Attenuation  : 40.00000000000000 dB
Actual Stopband Attenuation   : 40.03955716783712 dB
Sampling Frequency            : 44100.00000000000 Hz
Passband Edge                 : 10000.00000000000 Hz
Stopband Edge                 : 12500.00000000000 Hz

TRANSFER FUNCTION
Scale Factor = 0.08883285197457290
Numerator Coefficients              Denominator Coefficients
b 8: 1.000000000000000              a 8: 1.000000000000000
b 7: 4.437919605564610              a 7: 1.104614299236229
b 6: 10.76921619855683              a 6: 1.978668209561079
b 5: 17.46970831199284              a 5: 1.556364236494628
b 4: 20.40285887953456              a 4: 1.226408547493966
b 3: 17.46970831199284              a 3: 0.6160515120514172
b 2: 10.76921619855682              a 2: 0.2446494077658335
b 1: 4.437919605564609              a 1: 0.06099774584323624
b 0: 0.9999999999999999             a 0: 0.007910400932499942
```

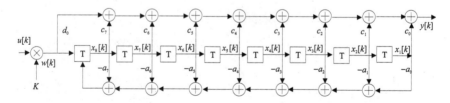

Figure 14.5 High-order digital filter reduced to a direct-II architecture showing state assignments and DSP coefficients. Note that the states $x_i[k]$ found in the text are carried as states $x_{i-1}[k]$ in the software.

The state-variable model presented below was produced using the direct-II architecture option found in the Architecture menu.

```
File : scaleb.arc
Source Module : Architecture

DIRECT-II STATE VARIABLE FILTER DESCRIPTION
Scale Factor = 0.08883285197457290

A Matrix
Row 1
0.000000000000000          1.000000000000000
0.000000000000000          0.000000000000000
0.000000000000000          0.000000000000000
0.000000000000000          0.000000000000000
Row 2
0.000000000000000          0.000000000000000
1.000000000000000          0.000000000000000
0.000000000000000          0.000000000000000
0.000000000000000          0.000000000000000
Row 3
0.000000000000000          0.000000000000000
0.000000000000000          1.000000000000000
0.000000000000000          0.000000000000000
0.000000000000000          0.000000000000000
Row 4
0.000000000000000          0.000000000000000
0.000000000000000          0.000000000000000
1.000000000000000          0.000000000000000
0.000000000000000          0.000000000000000
Row 5
0.000000000000000          0.000000000000000
0.000000000000000          0.000000000000000
0.000000000000000          1.000000000000000
0.000000000000000          0.000000000000000
Row 6
0.000000000000000          0.000000000000000
0.000000000000000          0.000000000000000
0.000000000000000          0.000000000000000
1.000000000000000          0.000000000000000
Row 7
0.000000000000000          0.000000000000000
0.000000000000000          0.000000000000000
0.000000000000000          0.000000000000000
0.000000000000000          1.000000000000000
Row 8
-0.007910400932499942      -0.06099774584323624
-0.2446494077658335        -0.6160515120514172
-1.226408547493966         -1.556364236494628
-1.978668209561079         -1.104614299236229


B Vector                C' Vector               D Scalar
0.000000000000000       0.9920895990674999      1.000000000000
0.000000000000000       4.376921859721373
```

```
0.000000000000000        10.52456679079099
0.000000000000000        16.85365679994142
0.000000000000000        19.17645033204060
0.000000000000000        15.91334407549821
0.000000000000000        8.790547988995748
1.000000000000000        3.333305306328382
```

The DSP coefficients are

```
MICROPROCESSOR COEFFICIENTS
Scale factor = 0.08888285197457290
a0 : -0.007910400932499942
a1 : -0.06099774584323624
a2 : -0.2446494077658335
a3 : -0.6160515120514172
a4 : -1.226408547493966
a5 : -1.556364236494628
a6 : -1.978668209561079
a7 : -1.104614299236229
b7 : 1.000000000000000
c0 : 0.9920895990674999
c1 : 4.376921859721373
c2 : 10.52456679079099
c3 : 16.85365679994142
c4 : 19.17645033204060
c5 : 15.91334407549821
c6 : 8.790547988995748
c7 : 3.333305306328382
d0 : 1.000000000000000
```

The DSP coefficients are the actual parameters that would be downloaded to a DSP microprocessor for execution. The total DSP coefficient count is $16 < 2N + 1$ (less input scale factor). The largest coefficient is $c_5 = 19.1764$, which is bounded by 2^5. ❖

14.4 Change of Architecture

The architecture defined by a state-determined system can be extended in terms of similarity transforms. Using a linear transform, a system characterized by $\boldsymbol{S} = (\mathbf{A}, \mathbf{b}, \mathbf{c}, d_0)$ can be mapped into a new system, say $\hat{\boldsymbol{S}} = (\hat{\mathbf{A}}, \hat{\mathbf{b}}, \hat{\mathbf{c}}, d_0)$. The original system, having a state vector $\mathbf{x}[k]$, is defined with respect to a given architecture (e.g., direct-II). The state vector of the transformed system, $\hat{\mathbf{x}}[k]$, corresponds to another architecture. The transformation of the architecture defined by $\boldsymbol{S} = (\mathbf{A}, \mathbf{b}, \mathbf{c}, d_0)$ into another characterized by $\hat{\boldsymbol{S}} = (\hat{\mathbf{A}}, \hat{\mathbf{b}}, \hat{\mathbf{c}}, d_0)$ is mathematically defined by

$$\hat{\mathbf{x}}[k] = \mathbf{T}\mathbf{x}[k] \tag{14.20}$$

where $\mathbf{T}$ is a nonsingular $N \times N$ matrix. Applying this transform to a state-determined system defined by Eq. (14.4) will result in

$$\hat{\mathbf{x}}[k+1] = \mathbf{T}\mathbf{x}[k+1]$$

$$= \mathbf{T}\mathbf{A}\mathbf{x}[k] + \mathbf{T}\mathbf{b}u[k]$$

$$= \mathbf{T}\mathbf{A}\mathbf{T}^{-1}\hat{\mathbf{x}}[k] + \mathbf{T}\mathbf{b}u[k]$$

$$= \hat{\mathbf{A}}\hat{\mathbf{x}}[k] + \hat{\mathbf{b}}u[k] \qquad \text{(state equation)} \qquad (14.21)$$

and

$$y[k] = \mathbf{c}^T\mathbf{x}[k] + du[k]$$

$$= \mathbf{c}^T\mathbf{T}^{-1}\hat{\mathbf{x}}[k] + du[k]$$

$$= \hat{\mathbf{c}}^T\hat{\mathbf{x}}[k] + \hat{d}u[k] \qquad \text{\{output equation\}} \qquad (14.22)$$

where

$$\hat{\mathbf{A}} = \mathbf{T}\mathbf{A}\mathbf{T}^{-1}; \quad \mathbf{A} = \mathbf{T}^{-1}\hat{\mathbf{A}}\mathbf{T}$$

$$\hat{\mathbf{b}} = \mathbf{T}\mathbf{b}; \quad \mathbf{b} = \mathbf{T}^{-1}\hat{\mathbf{b}}$$

$$\hat{\mathbf{c}} = \mathbf{c}^T\mathbf{T}^{-1}; \quad \mathbf{c}^T = \hat{\mathbf{c}}^T\mathbf{T}$$

$$\hat{d} = d \qquad (14.23)$$

Therefore, the nonsingular transform $\mathbf{T}$ defines the new architecture, or wiring diagram, given by the four-tuple $\hat{S} = (\hat{\mathbf{A}}, \hat{\mathbf{b}}, \hat{\mathbf{c}}, d_0)$.

The ability to map one architecture into another can be very useful in developing and defining systems having special attributes such as

- Minimal dynamic range requirements
- Minimal noise gain
- Minimal coefficient roundoff errors

plus others. In this section, a number of architectures will be presented, each having potentially different attributes. A detailed examination of these special properties will be deferred to Chap. 15. In this section, architecture will simply be presented as an algorithmic statement without any attempt to justify or motivate an individual structure.

14.5 Cascade Architecture

The most popular IIR filter architecture found in common use is called the *cascade architecture*. A cascade architecture is shown in Fig. 14.6 and implements the transfer function having the factored form

$$H(z) = K \prod_{i=1}^{Q} H_i(z) \tag{14.24}$$

where $H_i(z)$ is a first- or second-order subsystem defined with respect to real coefficients. Each subsystem is represented by a state-determined model $S_i = (A_i, b_i, c_i, d_i)$ and

$$\sum_{i=1}^{Q} \text{order}[H_i(z)] = N \tag{14.25}$$

A cascade architecture, as the name implies, links the output of one subsystem to the input of its successor. Specifically, if $S_i = (A_i, b_i, c_i, d_i)$ and $S_{i+1} = (A_{i+1}, b_{i+1}, c_{i+1}, d_{i+1})$ are chained together by mapping of the output of S_i, namely $y_i[k]$, to the input of S_{i+1}, namely $u_{i+1}[k]$, then they form a cascaded pair. Each individual first- or second-order subfilter S_i is implemented as a biquad, or direct-II filter. Following this procedure, the state-variable model for a cascade system is given by $S = (A, b, c, d_0)$, where

$$A = \begin{bmatrix} A_1 & 0 & 0 & \cdots & 0 \\ b_2 c_1^T & A_2 & 0 & \cdots & 0 \\ b_3 d_2 c_1^T & b_3 c_1^T & A_3 & \cdots & 0 \\ \vdots & \vdots & \vdots & \ddots & \vdots \\ b_Q(d_{Q-1} d_{Q-2} \cdots d_2) c_1^T & b_Q(d_{Q-1} d_{Q-2} \cdots d_3) c_2^T & b_Q(d_{Q-1} d_{Q-2} \cdots d_4) c_3^T & \cdots & A_Q \end{bmatrix} \tag{14.26}$$

$$b = \begin{bmatrix} b_1 \\ d_1 b_2 \\ \vdots \\ (d_{Q-1} \cdots d_1) b_Q \end{bmatrix} \tag{14.27}$$

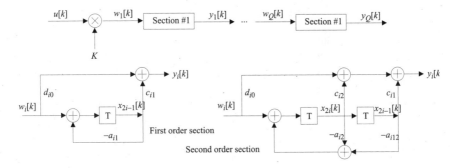

Figure 14.6 Cascade architecture.

$$\mathbf{c} = \begin{bmatrix} (d_Q d_{Q-1} \ldots d_2) \mathbf{c}_1 \\ (d_Q d_{Q-1} \ldots d_3) \mathbf{c}_2 \\ \vdots \\ \mathbf{c}_Q \end{bmatrix} \tag{14.28}$$

$$d_0 = d_Q d_{Q-1} \ldots d_2 d_1 \tag{14.29}$$

The elements of $\mathbf{A}$ having indices a_{ij} correspond to the gain of the path coupling of information from S_i into S_k where $k \leq i$. Information is seen to move forward or remain within a subsystem, but it never flows backward. It can also be seen that the construction rules for a cascade design are also straightforward. A cascade implementation of an Nth-order system requires, at most, N multiplies from $\mathbf{A}$ (since the terms a_{ij} correspond to a mapping of S_i into S_k for $k > i$ are not physical multiplications). Furthermore, there are N multiplies from $\mathbf{b}$ and $\mathbf{c}$, and one from d_0, for a total complexity measure of $M_{\text{multiplier}} \leq 3N + 1$. This is technically larger than that computed for a direct-II filter. In practice, however, many cascade coefficients are unity, which will often reduce the complexity of a cascade architecture to be similar to that of a direct-II.

The following TI TMS320C50 code is typical of a cascade implementation:

```
.mmregs              ; Symbols for memory mapped registers.
STATEV .set 02000h   ; Location of state variables in data memory.
COEFC .set 02040h    ; Location of coefs in data memory space.
COEFA .set 02080h    ; Location of coefs in data memory space.
NSECT .set 6         ; Number of second-order sections.
NORMF .set 1         ; Normalization factor.
```

```
CASCADE              ; Initialize address registers.
    LAR AR0,#COEFC
    LAR AR1,#STATEV
    LAR AR2,#STATEV+2
    LAR AR3,#COEFA
    SPLK #2,INDX         ; Initialize index.
    MAR *,AR1            ; Initialize ARP<=1.
    SPLK #NSECT-1        ; Set number of sections to loop through.
    RPTB CLOOP-1         ; Repeat for NSECTs.
    LACL #0              ; Clear accumulator.
    LT *+,AR0            ; T<=x[n].
    MPY *+,AR1           ; P<=d_0*x[n]
    LT *+,AR0            ; T<=x[n-1]
    MPYA *+,AR1          ; Acc<=Acc+P, P<=c_2*x[n-1]
    LT *+,AR0            ; T<=x[n-2]
    MPYA *+,AR1          ; Acc<=Acc+P, P<=c_1*x[n-2]
    APAC                 ; Acc<=Acc+P
    SACH *,NORMF,AR2     ; Next section x[n]<=Acc
    LACL #0              ; Clear accumulator.
    LT *-,AR3            ; T<=x[n-2]
    MPY *+,AR2           ; P<=a_1*x[n-2]
    LTD *-,AR3           ; Acc<=Acc+P, T<=x[n-1], x[n-2]<=x[n-1]
    MPY *+,AR2           ; P<=a_2*x[n-1]
    APAC                 ; Acc<=Acc+P
    ADD *+               ; Acc<=Acc+x[n]
CLOOP SACH *0+,NORMF,AR1 ; x[n-1]<=Acc
RET                      ; Return from routine.
```

The above code segment implements a function that will compute a filter step with a specified number of cascaded second-order sections. First-order sections are not handled, since a first-order section may be implemented using a second-order section. Also, scaling is not handled in this code. The implementation shown above is based on the second-order section shown in Fig. 14.6. Comparing the above implementation to that of the direct-II implementation, the absence of a MAC instruction is notable. Due to the dynamic memory addressing requirements of this code, the MAC instruction (and its variants) cannot be used—at least not without writing self-modifying code. It is possible to produce a more efficient implementation of the cascade filter architecture on the TMS320C50 than that shown above by using a slightly modified definition of the second-order section shown in Fig. 14.6, namely the direct-II architecture shown in Fig. 9.4.

Example 14.3: Cascade Architecture

Problem Statement. Implement the eight-order digital filter studied in Example 14.2 as a cascade filter. The cascade factorization is shown in the Computer Study section.

Analysis and Conclusions. The transfer function can be directly mapped into a state-variable model. The filter is seen to consist of a chain of four second-order subfilters. Each second-order section is implemented as a direct-II filter. The states of the system reside in the shift registers, which are indexed from right to left in block (paired) form as shown in Fig. 14.7. The state-variable model shown in Fig. 14.7, namely $S = (\mathbf{A}, \mathbf{b}, \mathbf{c}, d_0)$, can be implemented using a general-purpose digital computer.

Computer Study. The designed Chebyshev-II filter is saved under the file name "scalec.IIR" along with an engineering report saved as "scalec.RPT" and a cascade architecture file saved as "scalec.arc." Sections of the engineering report are reproduced below. The developed filter is defined in terms of the following specifications (see report file). The resulting cascade architecture shown in Fig. 14.7 can also be viewed by loading "scalec.arc" into the Schematic menu.

```
SECOND-ORDER CASCADE FILTER SECTIONS
    Scale Factor = 0.08883285197457290
    Section 1
    b2 = 1.000000000000000      a2 = 1.000000000000000
    b1 = 0.4541774847813810      a1 = 0.5174788167607780
    b0 = 1.000000000000000      a0 = 0.09123912551991054
    Section 2
    b2 = 1.000000000000000      a2 = 1.000000000000000
    b1 = 0.7535171024450185      a1 = 0.3741770017630547
```

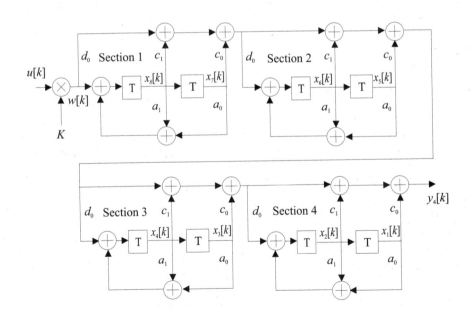

Figure 14.7 High-order digital filter reduced to a cascade architecture showing state assignments and DSP coefficients. Note that the states $x_i[k]$ in the text are carried as states $x_{i-1}[k]$ in the software.

```
b0 = 1.000000000000000          a0 = 0.2311930335002332
Section 3
b2 = 1.000000000000000          a2 = 1.000000000000000
b1 = 1.327487768897543          a1 = 0.1778396572194128
b0 = 0.9999999999999999         a0 = 0.4720529007738288
Section 4
b2 = 1.000000000000000          a2 = 1.000000000000000
b1 = 1.902737249440667          a1 = 0.03511882349298313
b0 = 1.000000000000000          a0 = 0.7944232896725211
```

```
ARCHITECTURE REPORT
File                    : scalec.arc
Source Module           : Architecture
Order                   : 8
CASCADE STATE VARIABLE FILTER DESCRIPTION
```

State variable description for each second-order section is shown below.

```
Section 1
Scale Factor = 0.08883285197457290
A Matrix
0.000000000000000        1.000000000000000
-0.09123912551991054    -0.5174788167607780
B Vector                C' Vector               D Scalar
0.000000000000000        0.9087608744800895     1.000000000000000
1.000000000000000       -0.06330133197939697
Section 2
Scale Factor = 1.000000000000000
A Matrix
0.000000000000000              1.000000000000000
-0.2311930335002332           -0.3741770017630547
B Vector                C' Vector               D Scalar
0.000000000000000        0.7688069664997668     1.000000000000000
1.000000000000000        0.3793401006819637
Section 3
Scale Factor = 1.000000000000000
A Matrix
0.000000000000000              1.000000000000000
-0.4720529007738288           -0.1778396572194128
B Vector                C' Vector               D Scalar
0.000000000000000        0.5279470992261710     1.000000000000000
1.000000000000000        1.149648111678130
Section 4
Scale Factor = 1.000000000000000
A Matrix
0.000000000000000              1.000000000000000
-0.7944232896725211           -0.03511882349298313
B Vector                C' Vector               D Scalar
0.000000000000000        0.2055767103274789     1.000000000000000
1.000000000000000        1.867618425947684
```

The total state-variable description is given by

```
Scale Factor = 0.08883285197457290
A Matrix
Row 1
 0.000000000000000              1.000000000000000
 0.000000000000000              0.000000000000000
 0.000000000000000              0.000000000000000
 0.000000000000000              0.000000000000000
Row 2
-0.09123912551991054           -0.5174788167607780
 0.000000000000000              0.000000000000000
 0.000000000000000              0.000000000000000
 0.000000000000000              0.000000000000000
Row 3
 0.000000000000000             -0.000000000000000
 0.000000000000000              1.000000000000000
 0.000000000000000              0.000000000000000
 0.000000000000000              0.000000000000000
Row 4
 0.9087608744800895            -0.06330133197939697
-0.2311930335002332            -0.3741770017630547
 0.000000000000000              0.000000000000000
 0.000000000000000              0.000000000000000
Row 5
 0.000000000000000             -0.000000000000000
 0.000000000000000              0.000000000000000
 0.000000000000000              1.000000000000000
 0.000000000000000              0.000000000000000
Row 6
 0.9087608744800895            -0.06330133197939697
 0.7688069664997668             0.3793401006819637
-0.4720529007738288           -0.1778396572194128
 0.000000000000000              0.000000000000000
Row 7
 0.000000000000000             -0.000000000000000
 0.000000000000000              0.000000000000000
 0.000000000000000              0.000000000000000
 0.000000000000000              1.000000000000000
Row 8
 0.9087608744800895            -0.06330133197939697
 0.7688069664997668             0.3793401006819637
 0.5279470992261710             1.149648111678130
-0.7944232896725211           -0.03511882349298313
B Vector            C' Vector                    D Scalar
 0.000000000000000   0.9087608744800895          1.000000000000000
 1.000000000000000  -0.06330133197939697
 0.000000000000000   0.7688069664997668
 1.000000000000000   0.3793401006819637
 0.000000000000000   0.5279470992261710
 1.000000000000000   1.149648111678130
 0.000000000000000   0.2055767103274789
 1.000000000000000   1.867618425947684
```

The DSP coefficients are

```
MICROPROCESSOR COEFFICIENTS
    Scale factor = 0.08883285197457290
    Cascade Sections
    Section 1
    a0 : -0.09123912551991054
    a1 : -0.5174788167607780
    b1 :  1.000000000000000
    c0 :  0.9087608744800895
    c1 : -0.06330133197939697
    d0 :  1.000000000000000
    Section 2
    a0 : -0.2311930335002332
    a1 : -0.3741770017630547
    b1 :  1.000000000000000
    c0 :  0.7688069664997668
    c1 :  0.3793401006819637
    d0 :  1.000000000000000
    Section 3
    a0 : -0.4720529007738288
    a1 : -0.1778396572194128
    b1 :  1.000000000000000
    c0 :  0.5279470992261710
    c1 :  1.149648111678130
    d0 :  1.000000000000000
    Section 4
    a0 : -0.7944232896725211
    a1 : -0.03511882349298313
    b1 :  1.000000000000000
    c0 :  0.2055767103274789
    c1 :  1.867618425947684
    d0 :  1.000000000000000
```

The DSP coefficient count is $16 < 3N + 1$. The largest coefficient in this case is $c_1 = 1.8676 < 2^1$ from Section 4.   ❖

14.6 Parallel Architecture

A parallel IIR filter architecture is shown in Fig. 14.8 that implements the transfer function

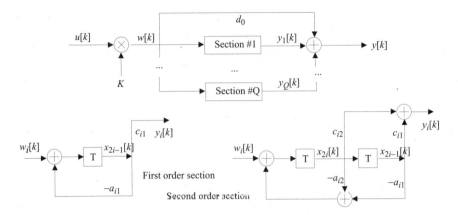

Figure 14.8 Parallel Architecture.

$$H(z) = K\left(d_0 + \sum_{i=1}^{Q} H_i(z)\right) \qquad (14.30)$$

where $H_i(z)$ is a first- or second-order subsystem defined with respect to real coefficients. Each subsystem is represented by a state-determined model $\mathbf{S}_i = (\mathbf{A}_i, \mathbf{b}_i, \mathbf{c}_i, d_i)$ and

$$\sum_{i=1}^{Q} \text{order}(H_i(z)) = N \qquad (14.31)$$

A parallel architecture, as the name implies, consists of Q subfilters operating in parallel with their outputs sent to a common adder. Specifically, a possibly scaled common input $u[k]$ is presented to all subfilters $\mathbf{S}_i = (\mathbf{A}_i, \mathbf{b}_i, \mathbf{c}_i, d_i)$, producing independent output responses $y_i[k]$ that are added to form $y[k]$. Following this procedure, the state-variable model for a parallel system, is given by $\mathbf{S} = (\mathbf{A}, \mathbf{b}, \mathbf{c}, d_0)$ where

$$\mathbf{A} = \begin{bmatrix} \mathbf{A}_1 & 0 & 0 & \cdots & 0 \\ 0 & \mathbf{A}_2 & 0 & \cdots & 0 \\ 0 & 0 & \mathbf{A}_3 & \cdots & 0 \\ \vdots & \vdots & \vdots & \ddots & \vdots \\ 0 & 0 & 0 & \cdots & \mathbf{A}_Q \end{bmatrix} \qquad (14.32)$$

$$\mathbf{b} = \begin{bmatrix} \mathbf{b}_1 \\ \mathbf{b}_2 \\ \vdots \\ \mathbf{b}_Q \end{bmatrix} \tag{14.33}$$

$$\mathbf{c} = \begin{bmatrix} \mathbf{c}_1 \\ \mathbf{c}_2 \\ \vdots \\ \mathbf{c}_Q \end{bmatrix} \tag{14.34}$$

$$d_0 = \sum_{i=1}^{Q} d_i \tag{14.35}$$

The construction rules for a parallel design are likewise straightforward. Each subfilter found along the diagonal of $\mathbf{A}$ is implemented as a direct-II filter. A parallel implementation of an Nth-order system can also be seen to require, at most, N multiplies from $\mathbf{A}$, N from $\mathbf{b}$ and $\mathbf{c}$, and one from d_0 for a total complexity measure of $M_{\text{multiplier}} \leq 3N + 1$, the same metric found in the cascade case. Again, this is generally considered to be a conservative complexity measure.

The implementation of a parallel IIR filter on a modern DSP microprocessor is a straightforward process. The implementation of a parallel architecture filter using second-order sections as in the previous section would require only some minor modifications to the code previously presented for cascade implementation. Therefore, it will not be given here. One strong advantage of the cascade and parallel architectures is the ease with which they may be used for parallel processing. In both cases, the subfilters may be distributed among more than one processor resource with relatively minor effort.

Example 14.4: Parallel Architecture

Problem Statement. The eight-order digital filter studied in Example 14.2 is to be converted into parallel form.

Analysis and Conclusions. The transfer function can be directly mapped into a state-variable model using the parallel mapping rule. The filter is seen to consist of a collection of four second-order biquad subfilters connected in parallel. The states of the system reside in the shift registers which are indexed as shown in Fig. 14.9. The state-variable model as the four-tuple $\boldsymbol{S} = (\mathbf{A}, \mathbf{b}, \mathbf{c}, d_0)$ can be implemented using a general-purpose digital computer as shown in the Computer Study.

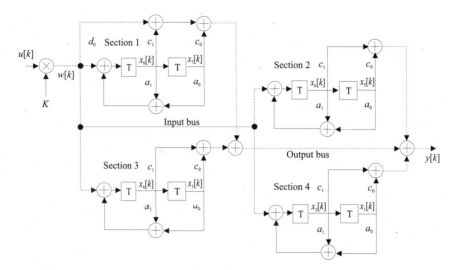

Figure 14.9 High-order digital filter reduced to a parallel architecture showing state assignments and DSP coefficients. Note that the states $x_i[k]$ in the text are carried as states $x_{i-1}[k]$ in the software.

Computer Study. The designed Chebyshev-II filter is saved under the file name "scalep.IIR" along with an engineering report saved as "scalep.RPT" and a parallel architecture file saved as "scalep.arc." Sections of the engineering report are reproduced below. The developed filter is defined in terms of the following specifications (see report file).

State variable description for each second-order biquad section.

```
PARALLEL STATE VARIABLE FILTER DESCRIPTION
Section 1
Scale Factor = 0.08883285197457290
A Matrix
 0.000000000000000                   1.000000000000000
-0.09123912551991054                -0.5174788167607780
B Vector            C' Vector         D Scalar
 0.000000000000000   9.071936917931088   1.000000000000000
 1.000000000000000   23.93838601217097
Section 2
Scale Factor = 0.08883285197457290
A Matrix
 0.000000000000000                   1.000000000000000
-0.2311930335002332                 -0.3741770017630547
B Vector            C' Vector         D Scalar
 0.000000000000000   7.519023765565351   0.000000000000000
 1.000000000000000  -8.262697588676973
Section 3
Scale Factor = 0.08883285197457290
```

```
A Matrix
  0.000000000000000                    1.000000000000000
 -0.4720529007738288                  -0.1778396572194128
B Vector           C' Vector          D Scalar
  0.000000000000000  - 1.434379819318134  0.000000000000000
  1.000000000000000  -11.16703262387511
Section 4
Scale Factor = 0.08883285197457290
A Matrix
  0.000000000000000                    1.000000000000000
 -0.7944232896725211                  -0.03511882349298313
B Vector           C' Vector          D Scalar
  0.000000000000000  -2.779371502242076  0.000000000000000
  1.000000000000000  -1.175350493290505
```

The total state-variable description is given by

```
Scale Factor = 0.08883285197457290
A Matrix
Row 1
  0.000000000000000                    1.000000000000000
  0.000000000000000                    0.000000000000000
  0.000000000000000                    0.000000000000000
  0.000000000000000                    0.000000000000000
Row 2
 -0.09123912551991054                 -0.5174788167607780
  0.000000000000000                    0.000000000000000
  0.000000000000000                    0.000000000000000
  0.000000000000000                    0.000000000000000
Row 3
  0.000000000000000                    0.000000000000000
  0.000000000000000                    1.000000000000000
  0.000000000000000                    0.000000000000000
  0.000000000000000                    0.000000000000000
Row 4
  0.000000000000000                    0.000000000000000
 -0.2311930335002332                  -0.3741770017630547
  0.000000000000000                    0.000000000000000
  0.000000000000000                    0.000000000000000
Row 5
  0.000000000000000                    0.000000000000000
  0.000000000000000                    0.000000000000000
  0.000000000000000                    1.000000000000000
  0.000000000000000                    0.000000000000000
Row 6
  0.000000000000000                    0.000000000000000
  0.000000000000000                    0.000000000000000
 -0.4720529007738288                  -0.1778396572194128
  0.000000000000000                    0.000000000000000
Row 7
  0.000000000000000                    0.000000000000000
  0.000000000000000                    0.000000000000000
  0.000000000000000                    0.000000000000000
  0.000000000000000                    1.000000000000000
```

```
Row 8
0.000000000000000              0.000000000000000
0.000000000000000              0.000000000000000
0.000000000000000              0.000000000000000
-0.7944232896725211           -0.03511882349298313
B Vector          C' Vector            D Scalar
0.000000000000000    9.071936917931088   1.000000000000000
1.000000000000000   23.93838601217097
0.000000000000000    7.519023765565351
1.000000000000000  - 8.262697588676973
0.000000000000000  - 1.434379819318134
1.000000000000000  -11.16703262387511
0.000000000000000  - 2.779371502242076
1.000000000000000  - 1.175350493290505
```

The DSP coefficients are:

```
MICROPROCESSOR COEFFICIENTS
Scale factor = 0.08883285197457290
Parallel Sections
Section 1
a0 : -0.09123912551991054
a1 : -0.5174788167607780
b1 :  1.000000000000000
c0 :  9.071936917931088
c1 : 23.93838601217097
d0 :  1.000000000000000
Section 2
a0 : -0.2311930335002332
a1 : -0.3741770017630547
b1 :  1.000000000000000
c0 :  7.519023765565351
c1 : -8.262697588676973
d0 :  0.000000000000000
Section 3
a0 : - 0.4720529007738288
a1 : - 0.1778396572194128
b1 :   1.000000000000000
c0 : - 1.434379819318134
c1 : -11.16703262387511
d0 :   0.000000000000000
Section 4
a0 : -0.7944232896725211
a1 : -0.03511882349298313
b1 : 1.000000000000000
c0 : -2.779371502242076
c1 : -1.175350493290505
d0 : 0.000000000000000
```

The total DSP coefficient count is $16 < 3N + 1$. Note also that only d_0 of the first subfilter has a non-zero value. The largest parallel coefficient is $c_1 = 23.938 < 2^5$ from Section 1.   ❖

14.7 Normal Architecture

First- and second-order normal subfilters are shown in Fig. 14.10. They are seen to be more complicated than the three previous structures but justify the added complexity by reducing errors due to coefficient roundoff. A normal filter satisfies the commutative property $AA^T = A^T A$ and can be defined in terms of the eigenvectors of A. The contemporary design of a normal filter has been reduced to a system of algebraic equations. The process begins with a transfer function having the familiar form

$$H(z) = K\left(d_0 + \sum_{i=1}^{Q} H_i(z)\right) \tag{14.36}$$

where Q is the number of first- and second-order pole pairs in the factored form of $H(z)$. The basic structure of a first- and second-order normal IIR filter is shown in Fig. 14.10.

Suppose λ_i is a real eigenvalue of A, then the transfer function $H_i(z)$ of the normal filter found in Fig. 14.10a satisfies

$$H_i(z) = \frac{n_{i1}}{z - \lambda_i} = \frac{n_{i1}z^{-1}}{1 - \lambda_i z^{-1}} \tag{14.37}$$

If λ_i is complex, then the transfer function $H_i(z)$ of the normal filter found in Fig. 14.10b satisfies

$$H_i(z) = \frac{n_{i1}z^{-1} + n_{i2}z^{-2}}{1 - p_{i1}z^{-1} - p_{i2}z^{-2}} = \frac{\gamma_i z^{-1}}{(1 - \lambda_i z^{-1})} + \frac{\gamma_i^* z^{-1}}{(1 - \lambda_i^* z^{-1})} \tag{14.38}$$

where

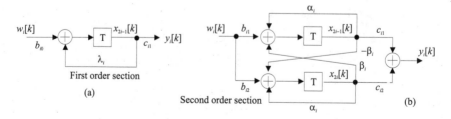

First order section

(a)

Second order section

(b)

Figure 14.10 Normal architecture showing first- and second-order sections.

$$\lambda_i = (\alpha_i + j\beta_i) = r_i e^{j\phi_i}$$

$$\lambda_i^* = (\alpha_i - j\beta_i) = r_i e^{-j\phi_i} \qquad (14.39)$$

and

$$\gamma_i = \left. \frac{(1 - \lambda_i^* z^{-1})}{z^{-1}} H_i(z) \right|_{z = \lambda_i}$$

$$\gamma_i^* = \mathrm{Re}(\gamma_i) - j\,\mathrm{Im}(\gamma_i) \qquad (14.40)$$

Assume that $H_i(z)$ is the ith second-order subfilter of $H(z)$. A 2×2 matrix $\mathbf{A}_i$, having the required commutative property, can be defined to be

$$\mathbf{A}_i = r_i \begin{bmatrix} \cos(\phi_i) & \sin(\phi_i) \\ -\sin(\phi_i) & \cos(\phi_i) \end{bmatrix} \qquad (14.41)$$

such that

$$\mathbf{A}_i \mathbf{A}_i^T = \mathbf{A}_i^T \mathbf{A}_i = r_i^2 \mathbf{I} \qquad (14.42)$$

The elements of the $\mathbf{b}_i$ vector satisfy

$$b_{i1}^2 + b_{i2}^2 = 2(1 - r_i^2) \qquad (14.43)$$

and

$$\frac{b_{i2}}{b_{i1}} = \frac{r_i^2 \sin(2\phi_i)\sqrt{1 + r_i^4 - 2r_i^2 \cos(2\phi_i)}}{1 - r_i^2 \cos(2\phi_i)} \qquad (14.44)$$

where b_{i1} and b_{i2} are nonunique. The elements of the vector c_i satisfy

$$c_i = \begin{cases} c_{i1} = \dfrac{b_{i1}(\alpha_i + \alpha_i^*) + jb_{i2}(\alpha_i - \alpha_i^*)}{2(1 - r_i^2)} \\[4mm] c_{i2} = \dfrac{b_{i1}j(\alpha_i^* - \alpha_i) + b_{i2}(\alpha_i + \alpha_i^*)}{2(1 - r_i^2)} \end{cases} \qquad (14.45)$$

and the scalar d_0 has a value given by Eq. (14.36).

In some cases, it is desired to have symmetric path gains through a second-order filter. In such cases, an additional constraint can be placed on the design in the form of (see Sec. 17.5)

$$b_1 c_1 = b_2 c_2 \tag{14.46}$$

Example 14.5: Normal Section Architecture

Problem Statement. Reduce the second-order section given by

$$H(z) = \frac{\left(z - \frac{1}{2}\right)}{\left(z^2 - z + \frac{1}{2}\right)}$$

to a normal second-order architecture.

Analysis and Conclusions. The poles of $H(z)$ are located at

$$\lambda = \frac{1}{2}(1 + j) = \frac{e^{j\pi/4}}{\sqrt{2}}$$

$$\lambda^* = \frac{1}{2}(1 - j) = \frac{e^{-j\pi/4}}{\sqrt{2}}$$

which defines $r = 0.707$ and $\phi = \pi/4$. Applying Eqs. (16.41) through (16.45), one obtains

$$\mathbf{A} = r \begin{bmatrix} \cos(\phi) & \sin(\phi) \\ -\sin(\phi) & \cos(\phi) \end{bmatrix} = \frac{1}{\sqrt{2}} \begin{bmatrix} \cos(\pi/4) & \sin(\pi/4) \\ -\sin(\pi/4) & \cos(\pi/4) \end{bmatrix}$$

$$b_1^2 + b_2^2 = 2(1 - r^2) = 2\left(\frac{1}{2}\right) = 1$$

and

$$\frac{b_2}{b_1} = \frac{r^2 \sin(2\phi) + \sqrt{1 + r^4 - 2r^2 \cos(2\phi)}}{1 - r^2 \cos(2\phi)} = \frac{1 + \sqrt{5}}{2}$$

Solving for the elements of $\mathbf{b}$, let $b_1 = 0.526$ and $b_2 = 0.851$ (nonunique). The vector $\mathbf{c}$ then is assigned

$$\mathbf{c} = \begin{bmatrix} c_1 = b_1 = 0.526 \\ c_2 = b_2 = 0.851 \end{bmatrix}$$

This reduces to the state representation $S_i = (A_i, b_i, c_i, d_i)$, where

$$A = \begin{bmatrix} 0.5 & 0.5 \\ -0.5 & 0.5 \end{bmatrix}; \quad b = \begin{bmatrix} 0.526 \\ 0.851 \end{bmatrix} = c^T; \quad d = 0$$

and is graphically interpreted in Fig. 14.11. The information contained in Fig. 14.11 can be examined using Mason's gain formula to verify that the architecture does, in fact, agree with $H(z)$. An analysis of the system shown in Fig. 14.11 would yield:

Feedforward path	Gain	Δ
1	$M_1 = b_1 c_1 z^{-1}$	$\Delta_1 = 1 - a_{22} z^{-1}$
2	$M_2 = b_2 c_2 z^{-1}$	$\Delta_2 = 1 - a_{11} z^{-1}$
3	$M_3 = b_1 a_{12} c_2 z^{-1}$	$\Delta_3 = 1$
4	$M_4 = b_1 a_{22} c_1 z^{-1}$	$\Delta_4 = 1$

The system characteristic equation is defined by the feedback path gains and is given by (note the presence of two non-touching loops)

$$\Delta = 1 - (a_{11} z^{-1} + a_{22} z^{-1} + a_{21} a_{12} z^{-2}) + a_{11} a_{22} z^{-2}$$

and the transfer function is given by

$$H(z) = N(z)/D(z) = (M_1 \Delta_1 + M_2 \Delta_2 + M_3 \Delta_3 + M_4 \Delta_4)/\Delta$$

Substituting the computed values into the above equations results in

$$\begin{aligned} N(z) &= (M_1 \Delta_1 + M_2 \Delta_2 + M_3 \Delta_3 + M_4 \Delta_4) \\ &= (0.576^2 + 0.81^2 + (0.576 \times 0.5 \times 0.851) - (0.576 \times 0.5 \times 0.851)) z^{-1} \\ &\quad - (0.576^2 \times 0.5 + 0.851^2 \times 0.5) z^{-2} \\ &\sim z^{-1} - 0.5 z^{-2} \end{aligned}$$

and

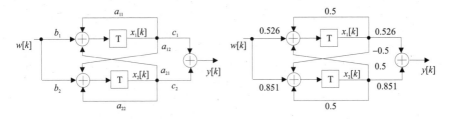

Figure 14.11   Second-order normal filter section.

$$D(z) = \Delta = 1 \times (0.5 + 0.5)z^{-1} + (0.25 + 0.25)z^{-2} = 1 - z^{-1} + 0.5z^{-2}$$

which agrees with the known original transfer function.   ❖

It will be shown in Chap. 17 that a normal filter can, to some degree, suppress the effects of coefficient roundoff error. At this point, however, only normal first- and second-order sections are defined. The low-order sections can be used to create high normal systems having either a cascade or parallel form.

Recall that a cascade architecture implements the transfer function having a factored form

$$H(z) = K \prod_{i=1}^{Q} H_i(z) \tag{14.47}$$

where $H_i(z)$ is a first- or second-order subsystem defined with respect to real coefficients. Each subsystem is represented by a state-determined model $S_i = (A_i, b_i, c_i, d_i)$, each having a normal form. Similarly, a parallel architecture implements the transfer function having a factored form

$$H(z) = K \left(d_0 \sum_{i=1}^{Q} H_i(z) \right) \tag{14.48}$$

where, once again, $H_i(z)$ is a first- or second-order subsystem defined with respect to real coefficients. Each subsystem again is represented by a state-determined model $S_i = (A_i, b_i, c_i, d_i)$ having the normal form.

Example 14.6: High-Order Normal IIR filter

Problem Statement. The eight-order digital filter studied in Example 14.2 is to be converted into both cascade-normal and parallel-normal forms.

Analysis and Conclusions. The transfer function can be directly mapped into a state-variable model using the cascade-normal or parallel-normal mapping rules. The state-variable model four-tuple $S = (A, b, c, d_0)$ can be implemented using a general-purpose digital computer as shown in the Computer Study.

Computer Study. The designed Chebyshev-II filter is saved under file name "scalen.IIR" along with an engineering report saved as "scalen.RPT" and a cascade normal architecture file as "scalen.arc." Sections of the engineering report are reproduced below. The developed filter is defined in terms of the following specifications listed in the report file. The resulting normal architecture can be constructed directly from the state-variable description.

CASCADE NORMAL STATE VARIABLE FILTER DESCRIPTION

State variable description for each second-order section
    Section 1
    Scale Factor = 0.08883285197457290
    A Matrix
    -0.2587394083803891   0.1558622599312509
    -0.1558622599312509   -0.2587394083803891
    B Vector   C' Vector   D Scalar
     0.9924982076722217   2.945119942463238   1.000000000000000
     0.9123974225783447   -3.273055712734053
    Section 2
    Scale Factor = 1.000000000000000
    A Matrix
    -0.1870885008815274   0.4429344492564740
    -0.4429344492564740   -0.1870885008815274
    B Vector   C' Vector   D Scalar
     0.9367163255006228   1.063619143630472   1.000000000000000
     0.8125124359295338   -0.7593352273628449
    Section 3
    Scale Factor = 1.000000000000000
    A Matrix
    -0.08891982860970646   0.6812827348860745
    -0.6812827348860745   -0.08891982860970646
    B Vector   C' Vector   D Scalar
     0.7561223714610351   1.235050590535010   1.000000000000000
     0.6958255225474866   0.3101333930643790
    Section 4
    Scale Factor = 1.000000000000000
    A Matrix
    -0.01755941174649154   0.8911312791792454
    -0.8911312791792454   -0.01755941174649154
    B Vector   C' Vector   D Scalar
     0.4573431313195357   2.289373298313999   1.000000000000000
     0.4494337335912824   1.825829286067065

The Total State Variable Description

    Scale Factor = 0.08883285197457290
    A Matrix
    Row 1
    -0.2587394083803891   0.1558622599312509
     0.000000000000000   0.000000000000000
     0.000000000000000   0.000000000000000
     0.000000000000000   0.000000000000000
    Row 2
    -0.1558622599312509   -0.2587394083803891
     0.000000000000000   0.000000000000000
     0.000000000000000   0.000000000000000
     0.000000000000000   0.000000000000000
    Row 3
     2.758741930662770   -3.065924720391064

```
-0.1870885008815274    0.4429344492564740
 0.000000000000000      0.000000000000000
 0.000000000000000      0.000000000000000
Row 4
 2.392946578555454     -2.659398470086622
-0.4429344492564740    -0.1870885008815274
 0.000000000000000      0.000000000000000
 0.000000000000000      0.000000000000000
Row 5
 2.226871075132490     -2.474830647436561
 0.8042262292132281    -0.5741503528474985
-0.08891982860970646    0.6812827348860745
 0.000000000000000      0.000000000000000
Row 6
 2.049289622929506     -2.277475701640209
 0.7400933464081836    -0.5283648313684661
-0.6812827348860745    -0.08891982860970646
 0.000000000000000      0.000000000000000
Row 7
 1.346930376597748     -1.496909548645087
 0.4864389096793632    -0.3472767506033551
 0.5648419044133230     0.1418373771108155
-0.01755941174649154    0.8911312791792454
Row 8
 1.323636251615396     -1.471021649226341
 0.4780263228410055    -0.3412708662810686
 0.5550733980782675     0.1393844087562566
-0.8911312791792454    -0.01755941174649154
B Vector               C' Vector                D Scalar
 0.9924982076722217     2.945119942463238       1.000000000000000
 0.9123974225783447    -3.273055712734053
 0.9367163255006228     1.063619143630472
 0.8125124359295338    -0.7593352273628449
 0.7561223714610351     1.235050590535010
 0.6958255225474866     0.3101333930643790
 0.4573431313195357     2.289373298313999
 0.4494337335912824     1.825829286067065
```

The DSP microprocessor coefficients are given by

```
MICROPROCESSOR COEFFICIENTS
Scale factor = 0.08883285197457290
Cascade Sections
Section 1
a0,0: -0.2587394083803891
a0,1:  0.1558622599312509
a1,0: -0.1558622599312509
a1,1: -0.2587394083803891
b0  :  0.9924982076722217
b1  :  0.9123974225783447
c0  :  2.945119942463238
c1  : -3.273055712734053
d0  :  1.000000000000000
```

```
Section 2
a0,0: -0.1870885008815274
a0,1:  0.4429344492564740
a1,0: -0.4429344492564740
a1,1: -0.1870885008815274
b0 :   0.9367163255006228
b1 :   0.8125124359295338
c0 :   1.063619143630472
c1 :  -0.7593352273628449
d0 :   1.000000000000000
Section 3
a0,0: -0.08891982860970646
a0,1:  0.6812827348860745
a1,0: -0.6812827348860745
a1,1: -0.08891982860970646
b0 :   0.7561223714610351
b1 :   0.6958255225474866
c0 :   1.235050590535010
c1 :   0.3101333930643790
d0 :   1.000000000000000
Section 4
a0,0: -0.01755941174649154
a0,1:  0.8911312791792454
a1,0: -0.8911312791792454
a1,1: -0.01755941174649154
b0 :   0.4573431313195357
b1 :   0.4494337335912824
c0 :   2.289373298313999
c1 :   1.825829286067065
d0 :   1.000000000000000
```

The total number of DSP coefficients is 32, which is twice the count for the previously studied architectures. The largest coefficient is $c_1 = -3.2730 > -2^2$ from Section 1.

For the parallel normal filter (saved as "scalenp.*")

```
PARALLEL NORMAL STATE VARIABLE FILTER DESCRIPTION
State variable description for each second-order section
Section 1
Scale Factor = 0.08883285197457290
A Matrix
-0.2587394083803891     0.1558622599312509
-0.1558622599312509    -0.2587394083803891
B Vector              C' Vector            D Scalar
 0.9924982076722217   22.34197330863682    1.000000000000000
 0.9123974225783447    1.933387254101800
Section 2
Scale Factor = 0.08883285197457290
A Matrix
-0.1870885008815274     0.4429344492564740
-0.4429344492564740    -0.1870885008815274
```

```
B Vector                 C' Vector                D Scalar
  0.9367163255006228      5.780825470174547       0.000000000000000
  0.8125124359295338    -16.83382379964643
Section 3
Scale Factor = 0.08883285197457290
A Matrix
-0.08891982860970646      0.6812827348860745
-0.6812827348860745     -0.08891982860970646
B Vector                 C' Vector                D Scalar
  0.7561223714610351     -8.423642063404602       0.000000000000000
  0.6958255225474866     -6.895016430253565
Section 4
Scale Factor = 0.08883285197457290
A Matrix
-0.01755941174649154      0.8911312791792454
-0.8911312791792454     -0.01755941174649154
B Vector                 C' Vector                D Scalar
  0.4573431313195357     -4.691386737118036       0.000000000000000
  0.4494337335912824      2.158767655336534
```

The Total State Variable Description

```
Scale Factor = 0.08883285197457290
A Matrix
Row 1
-0.2587394083803891      0.1558622599312509
 0.000000000000000       0.000000000000000
 0.000000000000000       0.000000000000000
 0.000000000000000       0.000000000000000
Row 2
-0.1558622599312509     -0.2587394083803891
 0.000000000000000       0.000000000000000
 0.000000000000000       0.000000000000000
 0.000000000000000       0.000000000000000
Row 3
 0.000000000000000       0.000000000000000
-0.1870885008815274      0.4429344492564740
 0.000000000000000       0.000000000000000
 0.000000000000000       0.000000000000000
Row 4
 0.000000000000000       0.000000000000000
-0.4429344492564740     -0.1870885008815274
 0.000000000000000       0.000000000000000
 0.000000000000000       0.000000000000000
Row 5
 0.000000000000000       0.000000000000000
 0.000000000000000       0.000000000000000
-0.08891982860970646     0.6812827348860745
 0.000000000000000       0.000000000000000
Row 6
 0.000000000000000       0.000000000000000
 0.000000000000000       0.000000000000000
```

```
-0.6812827348860745    -0.08891982860970646
 0.000000000000000      0.000000000000000
Row 7
 0.000000000000000      0.000000000000000
 0.000000000000000      0.000000000000000
 0.000000000000000      0.000000000000000
-0.01755941174649154    0.8911312791792454
Row 8
 0.000000000000000      0.000000000000000
 0.000000000000000      0.000000000000000
 0.000000000000000      0.000000000000000
-0.8911312791792454    -0.01755941174649154
B Vector              C' Vector             D Scalar
 0.9924982076722217   22.34197330863682     1.000000000000000
 0.9123974225783447    1.933387254101800
 0.9367163255006228    5.780825470174547
 0.8125124359295338  -16.83382379964643
 0.7561223714610351   -8.423642063404602
 0.6958255225474866   -6.895016430253565
 0.4573431313195357   -4.691386737118036
 0.4494337335912824    2.158767655336534
```

The DSP microprocessor coefficients are given by

```
MICROPROCESSOR COEFFICIENTS
Scale factor = 0.08883285197457290
Parallel Sections
Section 1
a0,0: -0.2587394083803891
a0,1:  0.1558622599312509
a1,0: -0.1558622599312509
a1,1: -0.2587394083803891
b0  :  0.9924982076722217
b1  :  0.9123974225783447
c0  : 22.34197330863682
c1  :  1.933387254101800
d0  :  1.000000000000000
Section 2
a0,0: -0.1870885008815274
a0,1:  0.4429344492564740
a1,0: -0.4429344492564740
a1,1: -0.1870885008815274
b0  :  0.9367163255006228
b1  :  0.8125124359295338
c0  :  5.780825470174547
c1  : -16.83382379964643
d0  :  0.000000000000000
Section 3
a0,0: -0.08891982860970646
a0,1:  0.6812827348860745
a1,0: -0.6812827348860745
a1,1: -0.08891982860970646
```

```
b0 :    0.7561223714610351
b1 :    0.6958255225474866
c0 :   -8.423642063404602
c1 :   -6.895016430253565
d0 :    0.000000000000000
Section 4
a0,0: -0.01755941174649154
a0,1:  0.8911312791792454
a1,0: -0.8911312791792454
a1,1: -0.01755941174649154
b0 :    0.4573431313195357
b1 :    0.4494337335912824
c0 :   -4.691386737118036
c1 :    2.158767655336534
d0 :    0.000000000000000
```

The total number of DSP coefficients is again 32, which means that 32 multiplies will be required per filter cycle. The largest coefficient is $c_0 = 22.3419 < 2^5$ from Section 1.   ❖

14.8 Ladder/Lattice Architecture

Ladder/lattice filters will be shown to have a low sensitivity to coefficient roundoff errors. A ladder architecture also enjoys several other unique properties that are attributable to its structure but will not be developed at this point. A ladder architecture can be motivated by expanding a transfer function $H(z)$ using long division as follows:

$$H(z) = \frac{\displaystyle\sum_{i=0}^{N} b_i z^{-i}}{\displaystyle\sum_{i=0}^{N} a_i z^{-i}} = A_0 + \cfrac{1}{B_1 z^{-1} + \cfrac{1}{A_1 + \cfrac{1}{B_2 z^{-1} + \cfrac{1}{A_2 + \dots}}}} \qquad (14.49)$$

which is interpreted in Fig. 14.12. It should be clear why the architecture is called ladder (if the figure is rotated 90°) or lattice (if left on its side).

The most popular implementation of this formula is called the *Gray-Markel* filter. The Gray-Markel form is shown in Fig. 14.13. Notice that each of the N first-order sections has two inputs and two outputs. The ith section is shown to have inputs $v_i[k]$ and $x_i[k]$ where $x_i[k]$ is the ith state variable and is loaded into the section's shift register during each filter cycle. The outputs are $v_{i+1}[k]$ and $x_{i+1}[k]$, respectively. The model for the first-order ith subsystem, denoted $G_i(z)$ in Fig. 14.13, is given by

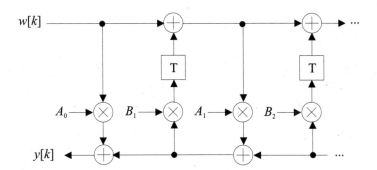

Figure 14.12   Basic ladder/lattice architecture.

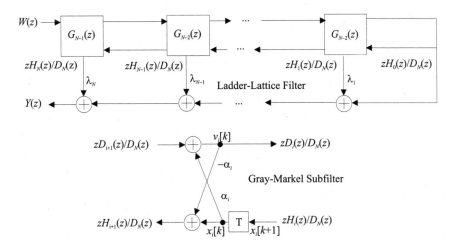

Figure 14.13   Gray-Markel ladder/lattice filter.

$$\begin{bmatrix} D_{i+1}(z) \\ H_{i+1}(z) \end{bmatrix} = \begin{bmatrix} 1 & \alpha_i \\ \alpha_i z^{-1} & z^{-1} \end{bmatrix} \begin{bmatrix} D_i(z) \\ H_i(z) \end{bmatrix} \tag{14.50}$$

or

$$\begin{bmatrix} D_i(z) \\ H_i(z) \end{bmatrix} = \frac{1}{(1 - \alpha_i^2)} \begin{bmatrix} 1 & -\alpha_i \\ -\alpha_i z & z \end{bmatrix} \begin{bmatrix} D_{i+1}(z) \\ H_{i+1}(z) \end{bmatrix} \tag{14.51}$$

The production rule for coefficients of the subfilters is given by

$$zH_i(z) = D_i(z^{-1})z^{-1} \tag{14.52}$$

$$\alpha_{i-1} = a_{i,i} \text{ (filter coefficients)} \tag{14.53}$$

$$D_{i-1}(z) = \frac{D_i(z) - \alpha_{i-1}zH(z)}{1 - \alpha_{i-1}^2} = \sum_{j=0}^{i-1} a_{i-1,j} z^{-j} \tag{14.54}$$

$$\lambda_i = n_{i,i} \text{ (output coefficients)} \tag{14.55}$$

$$N_{i-1}(z) = N_i(z) - zH_i(z)\lambda_i = \sum_{j=0}^{i-1} n_{i-1,j} z^{-j} \tag{14.56}$$

and

$$N_N(z) = \sum_{i=0}^{N} \lambda_i zH_i(z) \tag{14.57}$$

which collectively define

$$H(z) = \frac{\displaystyle\sum_{i=0}^{N} \lambda_i zH_i(z)}{D_N(z)} \tag{14.58}$$

It can be noted that some equations are expressed as increasing powers of z, while others use decreasing powers. These are often referred to as *forward-* and *reverse-order* equations.

Example 14.7: Gray-Markel Ladder/Lattice Architecture

Problem Statement. Implement the given third-order transfer function as a Gray-Markel filter.

$$H(z) = \frac{N_3(z)}{D_3(z)} = \frac{1 + 2z^{-1} + 3z^{-2} + 4z^{-3}}{1 + \frac{21}{32}z^{-1} + \frac{21}{64}z^{-2} + \frac{1}{8}z^{-3}}$$

Analysis and Conclusions. The coefficients translate to the following parameter set

$$n_{30} = 1; \, n_{31} = 2; \, n_{32} = 3; \, n_{33} = 4$$

$$a_{30} = 1; \ a_{31} = \frac{21}{32}; \ a_{32} = \frac{21}{64}; \ a_{33} = \frac{1}{8}$$

For $i = 3$, it follows that

$$\alpha_2 = a_{33} = \frac{1}{8}, \quad \lambda_3 = n_{33} = 4$$

$$zH_3(z) = D_3(z^{-1})z^{-3} = \frac{1}{8} + \frac{21}{64}z^{-1} + \frac{21}{32}z^{-2} + z^{-3}$$

$$N_2(z) = N_3(z) - zH_3(z)\lambda_3 = \frac{1}{2} + \frac{11}{16}z^{-1} + \frac{3}{8}z^{-2}$$

and

$$D_2(z) = \frac{D_3(z) - \alpha_2 zH_3(z)}{1 - \alpha_2^2} = 1 + \frac{5}{8}z^{-1} + \frac{1}{4}z^{-2}$$

which produces the intermediate coefficient set

$$n_{20} = \frac{1}{2}, \quad n_{21} = \frac{11}{16}, \quad n_{22} = \frac{3}{8}$$

and

$$a_{20} = 1, \quad a_{21} = \frac{5}{8}, \quad a_{22} = \frac{1}{4}$$

For $i = 2$, it follows that

$$\alpha_1 = a_{22} = \frac{1}{4}, \quad \lambda_2 = n_{22} = \frac{3}{8}$$

$$zH_2(z) = D_2(z^{-1})z^{-2} = \frac{1}{4} + \frac{5}{8}z^{-1} + z^{-2}$$

$$N_1(z) = N_2(z) - zH_2(z)\lambda_2 = \frac{13}{32} + \frac{29}{64}z^{-1}$$

and

$$D_1(z) = \frac{D_2(z) - \alpha_1 zH_2(z)}{1 - \alpha_1^2} = 1 + \frac{1}{2}z^{-1}$$

which produces the intermediate coefficient set

$$n_{10} = \frac{13}{32}, \quad n_{11} = \frac{29}{64}$$

and

$$a_{10} = 1, \quad a_{11} = \frac{1}{2}$$

For $i = 1$, it follows that

$$\alpha_0 = a_{11} = \frac{1}{2}, \quad \lambda_1 = n_{11} = \frac{29}{64}$$

$$zH_1(z) = D_1(z^{-1})z^{-1} = \frac{1}{2} + z^{-1}$$

$$N_0(z) = N_1(z) - zH_1(z)\lambda_1 = \frac{23}{128}$$

and

$$D_0(z) = \frac{D_1(z) - \alpha_0 z H_1(z)}{1 - \alpha_0^2} = 1.0$$

which produces the intermediate coefficient

$$\lambda_0 = n_{00} = \frac{23}{128}$$

and as a check,

$$zH_0(z) = 1.0$$

The resulting third-order ladder filter is shown in Fig. 14.14. To partially verify that the architecture shown in Fig. 14.13 produces the desired transfer function,

$$H(z) = \frac{\displaystyle\sum_{i=0}^{N} \lambda_i z H_i(z)}{D_N(z)} = \frac{1 + 2z^{-1} + 3z^{-2} + 4z^{-3}}{1 + \frac{21}{32}z^{-1} + \frac{21}{64}z^{-2} + \frac{1}{8}z^{-3}}$$

The information contained in the diagram in Fig. 14.14 can be examined using Mason's gain formula to verify that the architecture does, in fact, agree with $H(z)$. An analysis of the system shown in Fig. 14.14 is summarized in Table 14.1. The system characteristic equation $\Delta = D(z)$ satisfies $\Delta = 1 - (-\alpha_1\alpha_2 z^{-1} - \alpha_0\alpha_1 z^{-1} - \alpha_0 z^{-1} - \alpha_0\alpha_2 z^{-2} - \alpha_1 z^{-2} - \alpha_2 z^{-3})$ {single paths} $+ (\alpha_0\alpha_1\alpha_2 z^{-2})$ {dual paths}, which numerically simplifies to $\Delta = 1 + (\alpha_1\alpha_2 + \alpha_0\alpha_1 + \alpha_0)z^{-1} + (\alpha_0\alpha_2\alpha_1 + \alpha_0\alpha_1\alpha_2)z^{-2} + \alpha_2 z^{-3}$. Substituting the known values of $\alpha_2 = 1/8$, $\alpha_1 = 1/4$, and $\alpha_0 = 1/2$, one obtains

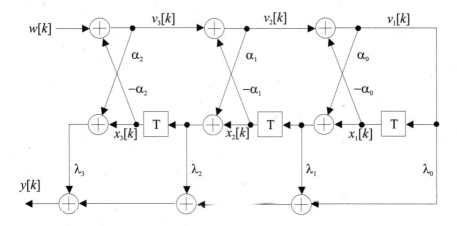

Figure 14.14 Third-order Gray-Markel ladder/lattice filter.

Table 14.1   Mason's Gain Formula Analysis of the System Shown in Fig. 14.14

Feedforward path	Gain	D
1	$M_1 = \alpha_2 \lambda_3$	$\Delta_1 = 1 - (-\alpha_0 \alpha_1 z^{-1} - \alpha_0 z^{-1} - \alpha_1 z^{-2})$
2	$M_2 = \alpha_1 \lambda_3 z^{-1}$	$\Delta_2 = 1 - (-\alpha_0 z^{-1})$
3	$M_3 = \alpha_0 \lambda_3 z^{-2}$	$\Delta_3 = 1$
4	$M_4 = \lambda_3 z^{-3}$	$\Delta_4 = 1$
5	$M_5 = \alpha_1 \lambda_2$	$\Delta_5 = 1 - (-\alpha_0 z^{-1})$
6	$M_6 = \alpha_0 \lambda_2 z^{-1}$	$\Delta_6 = 1$
7	$M_7 = \lambda_2 z^{-2}$	$\Delta_7 = 1$
8	$M_8 = \alpha_0 \lambda_1$	$\Delta_8 = 1$
9	$M_9 = \lambda_1 z^{-1}$	$\Delta_9 = 1$
10	$M_{10} = \lambda_0$	$\Delta_{10} = 1$

$$\Delta = 1 + \left(\frac{1}{32} + \frac{1}{8} + \frac{1}{2}\right)z^{-1} + \left(\frac{1}{16} + \frac{1}{4} + \frac{1}{64}\right)z^{-2} + \frac{1}{8}z^{-3}$$

$$= 1 + \frac{21}{32}z^{-1} + \frac{21}{64}z^{-2} + \frac{1}{8}z^{-3}$$

which agrees with the known $D(z)$. The numerator, $N(z)$, of the original transfer function can be reconstructed by forming

$$N(z) = \sum_{m=1}^{10} M_i(z)\Delta_i(z) = 1 + 2z^{-2} + 3z^{-2} + 4z^{-3}$$ ❖

Example 14.8: Gray-Markel Ladder/Lattice Architecture

Problem Statement. Implement the given third-order transfer Chebyshev-I function as a Gray-Markel filter.

$$H(z) = \frac{N_3(z)}{D_3(z)} = \frac{0.0154 + 0.0462z^{-1} + 0.0462z^{-2} + 0.0154z^{-3}}{1.0 - 2.0z^{-1} + 1.572z^{-2} - 0.4583z^{-3}}$$

Analysis and Conclusions. The coefficient translate to the following parameter set

$$n_{30} = n_{33} = 0.0154; \quad n_{31} = n_{32} = 0.0462$$

and

$$a_{30} = 1, a_{31} = -2.0; \quad a_{32} = 1.572, a_{33} = -0.4583$$

For $i = 3$, it follows that

$$\alpha_2 = a_{33} = -0.4583, \lambda_3 = n_{33} = 0.0154$$

$$zH_3(z) = D_3(z^{-1})z^{-3} = -0.4583 + 1.572z^{-1} - 2.0z^{-2} + z^{-3}$$

$$N_2(z) = N_3(z) - zH_3(z)\lambda_3 = 0.0224578 + 0.0219912z^{-1} + 0.076886z^{-2}$$

and

$$D_2(z) = \frac{D_3(z) - \alpha_2 zH_3(z)}{1 - \alpha_2^2} = 1.0 - 1.607107z^{-1} + 0.835463z^{-2}$$

which produces the intermediate coefficient set

$$n_{20} = 0.022457, \quad n_{21} = 0.0219912, \quad n_{22} = 0.076886$$

and

$$a_{20} = 1.0, \quad a_{21} = -1.607107, \quad a_{22} = 0.835463$$

For $i = 2$, it follows that

$$\alpha_1 = a_{22} = 0.835463, \lambda_2 = n_{22} = 0.076886$$

$$zH_2(z) = D_2(z^{-1})z^{-2} = 0.83546264 - 1.6071075z^{-1} + z^{-2}$$

$$N_1(z) = N_2(z) - zH_2(z)\lambda_2 = -0.04174415 + 0.01454909z^{-1}$$

and

$$D_1(z) = \frac{D_2(z) - \alpha_1 zH_2(z)}{1 - \alpha_1^2} = 1.0 - 0.87558719z^{-1}$$

which produces the intermediate coefficient set

$$n_{10} = -0.04174415, \quad n_{11} = 0.01454909$$

and

$$a_{10} = 1.0, \quad a_{11} = -0.87558719$$

For $i = 1$, it follows that

$$\alpha_0 = a_{11} = 0.87558719, \lambda_1 = n_{11} = 0.01454909$$

$$zH_1(z) = D_1(z^{-1})z^{-1} = 0.87558713 + z^{-1}$$

$$N_0(z) = N_1(z) - zH_1(z)\lambda_1 = 0.08564588$$

and

$$D_0(z) = \frac{D_1(z) - \alpha_0 zH_1(z)}{1 - \alpha_0^2} = 1.0$$

which produces the intermediate coefficient set

$$\lambda_0 = n_{00} = 0.08564588$$

and

$$zH_0(z) = 1.0$$

The resulting filter is also shown in Fig. 14.14 for the coefficients defined above. ❖

A Gray-Markel architecture possesses the state-determined model $S = (\mathbf{A}_x, \mathbf{A}_v, \mathbf{C}, \mathbf{b}, \mathbf{c}_x, \mathbf{c}_u, \mathbf{d})$. Notice that there are additional matrices and vector present that are not found in the standard four-tuple model $S = (\mathbf{A}, \mathbf{b}, \mathbf{c}, d_0)$. The new terms are required to model the introduced variable $v_i[k]$. The state equations are now expanded to read as

$$\mathbf{v}[k] = \mathbf{A}_v \mathbf{x}[k] + \mathbf{1} w[k] \tag{14.59}$$

where $\mathbf{A}_v$ is an $N \times N$ matrix and $\mathbf{1}$ is an N-vector given by

$$\mathbf{A}_v = \begin{bmatrix} -\alpha_0 & -\alpha_1 & -\alpha_2 & \cdots & -\alpha_{N-1} \\ 0 & -\alpha_1 & -\alpha_2 & \cdots & -\alpha_{N-1} \\ 0 & 0 & -\alpha_2 & \cdots & -\alpha_{N-1} \\ \vdots & \vdots & \vdots & \ddots & \vdots \\ 0 & 0 & 0 & \cdots & -\alpha_{N-1} \end{bmatrix} \tag{14.60}$$

and

$$\mathbf{1} = \begin{bmatrix} 1 \\ 1 \\ \vdots \\ 1 \end{bmatrix} \tag{14.61}$$

and, as before

$$\mathbf{x}[k+1] = \mathbf{A}_x \mathbf{x}[k] + \mathbf{b} w[k] \tag{14.62}$$

where

$$\mathbf{A}_x = \begin{bmatrix} -\alpha_0 & -\alpha_1 & -\alpha_2 & \cdots & -\alpha_{N-2} & -\alpha N - 1 \\ 0 & -\alpha_0 & -\alpha_2 & \cdots & -\alpha_0\alpha_{N-2} & -\alpha_0\alpha_{N-1} \\ 0 & 0 & -\alpha_2 & \cdots & -\alpha_1\alpha_{N-2} & -\alpha_1\alpha_{N-2} \\ \vdots & \vdots & \vdots & \ddots & & \vdots \\ 0 & 0 & 0 & \cdots & 1-\alpha_{N-2}^2 & -\alpha_{N-2}\alpha_{N-1} \end{bmatrix} \tag{14.63}$$

$$\mathbf{b} = \begin{bmatrix} 1 \\ \alpha_0 \\ \vdots \\ \alpha_{N-3} \\ \alpha_{N-2} \end{bmatrix}$$

Lastly, for $\mathbf{c}_x$ being an N-vector, $\mathbf{C}$ an $N \times N$ matrix, and d_0 a scalar, define

$$y[k] = \mathbf{c}_x^T(\mathbf{x}[k] + \mathbf{C}\mathbf{v}[k]) + d_0 u_0[k] \tag{14.64}$$

where

$$\mathbf{c}_x = \begin{bmatrix} \lambda_1 \\ \lambda_2 \\ \cdots \\ \lambda_{N-1} \end{bmatrix} \tag{14.65}$$

$$\mathbf{C} = \begin{bmatrix} \alpha_0 & 0 & \vdots & 0 \\ 0 & \alpha_1 & \vdots & 0 \\ \vdots & \vdots & \ddots & \vdots \\ 0 & 0 & \cdots & \alpha_{N-1} \end{bmatrix} \tag{14.66}$$

and

$$d_0 = \lambda_0 \tag{14.67}$$

Finally $\mathbf{c}_u$ is the N-vector defining

$$u_0[k] = -\mathbf{c}_u^T \mathbf{x}[k] + w[k] \tag{14.68}$$

where

$$\mathbf{c}_u = \begin{bmatrix} \alpha_0 \\ \alpha_1 \\ \vdots \\ \alpha_{N-1} \end{bmatrix} \tag{14.69}$$

The complicated structure is due to multiplicity of data paths found in a ladder/lattice filter. In addition, it can be seen that the filter has essentially $2N$ states when $v[k]$ and $x[k]$ are jointly considered. For notational simplicity, the standard state four-tuple $\mathcal{S} = (\mathbf{A}, \mathbf{b}, \mathbf{c}, d_0)$ is often the only set of parameters published since the others, namely $\mathbf{A}_v$, $\mathbf{C}$, and $\mathbf{c}_u$, may be derived from the standard state four-tuple.

The actual number of coefficients needed to implement a Gray-Markel filter can be determined as follows. There are three multiplications required per state (i.e., $+\alpha_i$, $-\alpha_i$, and λ_i). Therefore, there are on the order of $3N$ multiplies required per filter cycle, which is higher than the others considered.

Example 14.9: Gray-Markel Ladder Architecture

Problem Statement. Convert the eighth-order digital filter studied in Example 14.2 into a Gray-Markel ladder filter.

Analysis and Conclusions. The transfer function can be directly mapped into a state-variable model using the cascade or parallel normal mapping rules. The state-variable model as the four-tuple $S = (\mathbf{A}, \mathbf{b}, \mathbf{c}, d_0)$ can be implemented using a general-purpose digital computer as shown in the Computer Study section.

Computer Study. The designed Chebyshev-II filter is saved under file name "scale1.IIR" along with an engineering report saved as "scale1.RPT" and a Gray-Markel architecture file as "scale1.arc." Sections of the engineering report are reproduced below. The developed filter is defined in terms of the following specifications listed in the report file. The resulting Gray-Markel architecture can be constructed directly from the state-variable description.

```
LADDER-2-MULTIPLIER STATE VARIABLE FILTER DESCRIPTION
Scale Factor = 0.08883285197457290
A Matrix
Row 1
-0.1303196838616324    -0.9017030078897827
-0.4232598905961886    -0.5495184223233259
-0.3232406881092252    -0.1717719296330049
-0.05226307419325777   -0.007910400932499942
Row 2
 0.9830167799982041    -0.1175096509252795
-0.05515909513380443   -0.07161306707331885
-0.04212462428561076   -0.02238526356607577
-0.006810907306502393  -0.001030880948742155
Row 3
 0.000000000000000      0.1869316855625184
-0.3816547164696836    -0.4955024142997909
-0.2914671007404515    -0.1548872656211126
-0.04712577120162741   -0.007132832314449339
Row 4
 0.000000000000000      0.000000000000000
 0.8208510650125025    -0.2325891073131611
-0.1368148182853474    -0.07270416814396187
-0.02212086306525877   -0.003348155433261914
Row 5
 0.000000000000000      0.000000000000000
 0.000000000000000      0.6980295035272828
```

```
-0.1776267129604877    -0.09439183977136222
-0.02871952207644594   -0.004346911040372334
Row 6
 0.000000000000000      0.000000000000000
 0.000000000000000      0.000000000000000
 0.8955154575506745    -0.05552367673242193
-0.01689355206493213   -0.002556963440641138
Row 7
 0.000000000000000      0.000000000000000
 0.000000000000000      0.000000000000000
 0.000000000000000      0.9704944041901540
-0.008977329102728788 -0.001358784832346236
Row 8
 0.000000000000000      0.000000000000000
 0.000000000000000      0.000000000000000
 0.000000000000000      0.000000000000000
 0.9972685710758701    -0.0004134218708336599
B Vector               C' Vector              D Scalar
 1.000000000000000     -0.2015773006278755     1.000000000000000
 0.1303196838616324    -0.6621659811434043
 0.9017030078897827    -1.133545966135784
 0.4232598905961886     0.8566094690634870
 0.5495184223233259     3.746732915315314
 0.3232406881092252     4.969420379155345
 0.1717719296330049     3.281455654005956
 0.05226307419325777    0.9920895990674999
```

The DSP microprocessor coefficients are given below.

```
MICROPROCESSOR COEFFICIENTS
Scale factor = 0.08883285197457290
a0 :   -0.1303196838616324         α0
a1 :   -0.9017030078897827         α1
a2 :   -0.4232598905961886         α2
a3 :   -0.5495184223233259         α3
a4 :   -0.3232406881092252         α4
a5 :   -0.1717719296330049         α5
a6 :   -0.05226307419325777        α6
a7 :   -0.007910400932499942       α7
v0 :    0.4972088264885172         λ0
v1 :   -0.1391443221823015         λ1
v2 :   -1.231370095361671          λ2
v3 :   -1.706437077145716          λ3
v4 :    0.1616332763953049         λ4
v5 :    3.727384318627080          λ5
v6 :    5.109909345221221          λ6
v7 :    3.333305306328381          λ7
v8 :    0.9999999999999999         λ8
```

It can be seen that there are $2N$ DSP coefficients, but $3N$ coefficient multiplies are required per filter cycle. The largest coefficient is $v_6 = \lambda_6 = 5.1099 < 2^3$. ❖

14.9 Summary

In Chap. 14, the concept of IIR filter architecture was developed. The basic filter forms called direct-II, cascade, parallel, normal, and lattice/ladder were developed in the context of a state-variable model. In all cases, mapping methods that can convert a transfer function $H(z)$ into a state-determined architecture were shown to be implementable with a general-purpose digital computer. Furthermore, the state-variable model can be directly mapped into code for DSP microprocessors. The true value of the state-variable model will be shown to be its ability to isolate and analyze individual states or collections of states in terms of their dynamic range requirements and noise sensitivity. This thesis will be developed in the next chapter.

14.10 Self-Study Problems

1. Pole-Zero Pairing   A second-order IIR filter has a transfer function given by

$$H(z) = H_1(z)H_2(z)$$

$$H_1(z) = (z - z_1)/(z - p_1)$$

and

$$H_2(z) = (z - z_2)/(z - p_2)$$

where $z_1 = 1.0$, $z_2 = -1.0$, $p_1 = 0.5$, and $p_2 = -0.5$. Evaluate the two possible pairings and, in each case, determine what the maximum gain of $H_1(z)$ and $H_2(z)$ would be. What is the better pairing strategy?

2. Direct-II IIR Filter   A third-order IIR filter is given by

$$H(z) = (1 + z^{-1})^3/(1 - 1.759z^{-1} + 1.181z^{-2} - 0.278z^{-3})$$

Implement the IIR filter as a direct-II IIR filter. Verify the transfer function by performing a computer-aided Mason's gain formula analysis of the third-order design.

3. Cascade IIR Filter   Consider the third-order IIR filter studied in Exercise 14.2. Implement as a cascade IIR filter. Verify the transfer function by performing a computer-aided Mason's gain formula analysis of the third-order design.

4. Parallel IIR Filter   Consider the third-order IIR filter studied in Exercise 14.2. Implement as a parallel IIR filter. Verify the transfer function by performing a computer-aided Mason's gain formula analysis of the third-order design.

5. Normal IIR Filter   Consider the third-order IIR filter studied in Exercise 14.2. Implement as a cascade and parallel normal IIR filter. Verify the transfer function by performing a computer-aided Mason's gain formula analysis of the third-order design.

6. Gray-Markel IIR Filter   Consider the third-order IIR filter studied in Exercise 14.2. Implement as a Gray-Markel IIR filter. Verify the transfer function by performing a computer-aided Mason's gain formula analysis of the third-order design.

15

Finite Wordlength Effects

15.1 Introduction

In Chap. 13, IIR filters were introduced. In Chap. 14, the companion study of architecture was presented. It was suggested that the dynamic range and precision parameters are strongly influenced by the choice of architecture. It is assumed that in a floating-point environment, dynamic range is essentially infinite, and filters operate free of overflow. Furthermore, floating-point arithmetic is often assumed to be error-free, which is not always a safe assumption. Fixed-point implementations of digital filters have limited dynamic range and precision and can exhibit a variety of pathological behaviors due to these limitations. The errors associated with a fixed-point FIR filter or IIR filter implementation are called *finite wordlength effects* and refer to a class of errors that occur whenever finite-precision arithmetic is performed. While finite wordlength effects are normally associated with fixed-point designs, they are technically present in floating-point systems as well. However, in the floating-point systems these errors are negligible in all but the most extreme cases.

15.2 Fixed-Point Systems

It has been established that an $[N{:}F]$ formatted digital word can be thought of as an N-bit binary array consisting of a sign-bit, I integer bits, and F fractional bits where $N = I + F + 1$. The binary point indicates the boundary between the integer and fractional data fields. The dynamic range of an $[N{:}F]$ system is on the order of 2^I, and significance of the least significant bit is 2^{-F}. The error variance associated with rounding or truncating a real number to one with an $[N{:}F]$ format was given in Chap. 1 as $\sigma^2 = Q^2/12$, where Q is the quantization step size, which is related to F. A fixed-point system possesses both a limited dynamic range and precision that can compromise system performance. As a general rule, finite wordlength effects appear in the following forms:

- Overflow saturation (see Chap. 16)
- Arithmetic rounding (see Chap. 17)
- Coefficient rounding (see Chap. 15)
- Data scaling (see Chap. 15)
- Zero-input limit cycling (see Chap. 15)

Any of these problems can compromise a digital filter to the point that it is no longer viable. Fortunately, corrective actions are available and can take the form of

- Scale the input or coefficients.
- Increase the wordlength.
- Select an alternative architecture.

This process is illustrated in Fig. 15.1, beginning with the specification of a filter's transfer function or coefficients. The ideal filter, defined over a set of real or complex coefficients, would then be mapped into a fixed-point design with a given architecture carrying coefficients of finite precision. The performance of the candidate fixed-point filter would be analyzed mathematically and/or experimentally to determine if it still meets the design specifications. If the candidate design fails this test, then the design is often altered in terms of redefining the data format [N:F], or architecture, and the test is repeated.

15.3 Saturation Effects

Of the referenced finite wordlength effects, run-time *register overflow* is potentially the most disastrous. Register overflow can introduce major nonlinear distortions into a system's output, which can often ren-

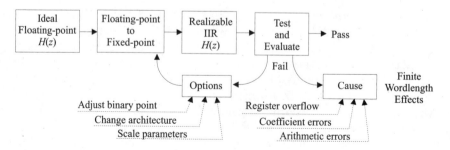

Figure 15.1   Fixed-point filter design procedure.

der a filter useless. Therefore, it is incumbent on the filter designer to eliminate, or at least mitigate, the effects of overflow. Unless the elimination of register overflow can be guaranteed, several safeguards actions should be taken. First, if fixed-point format is used, it should be two's complement. Recall that a string of two's complement numbers can be added (accumulated) and, as long as the final result is a valid two's complement number, then run-time accumulator overflow can be ignored. If this case cannot be mathematically guaranteed, then saturating arithmetic should be used. Saturating arithmetic will, upon detecting an overflow, clamp the accumulator output to the most positive or negative (i.e., two's complement extremal value). That is, the output of an N-bit two's complement saturating accumulator is defined to be

$$
ACC_{\text{output}} = \begin{bmatrix} (2^{N-1} - 1) \text{ if } ACC \geq 2^{N-1} \\ ACC \qquad \text{if } -2^{N-1} < ACC < 2^{N-1} - 1 \\ -2^{N-1} \qquad \text{if } ACC \leq -2^{N-1} \end{bmatrix} \qquad (15.1)
$$

Register overflow can be studied experimentally by continually scaling the input downward until overflows cease to occur. Unfortunately, many other possible untested inputs probably exist that can overflow an experimentally scaled system. One could take the position that no overflow will occur by aggressively scaling the input with a large scale factor (i.e., increase the number of integer bits, I). While ensuring that the overflow will not occur, such a policy can reduce the precision of the resulting filter to a point where it is of no practical value (i.e., excessively decreasing the number of fractional bits, F). In some implementation environments, such as standard cell, it is possible to increase N to allow I to increase without decreasing F; however, this increases the cost of the implementation. Since the total number of bits available, N, is either fixed (e.g., in the case of a DSP microprocessor) or it is undesirable to increase N due to cost or speed constraints, the number of integer and fractional bits are traded to arrive at an acceptable solution. Since dynamic range and precision metrics are inversely related in fixed-point systems, it is important to mathematically define the dynamic range bound requirements on all filter states so that intelligent overflow scaling can take place. This question is mathematically addressed in Chap. 16 and therein is called the *binary-point assignment problem*.

15.4 Arithmetic Errors

The next most severe type of error is attributed to finite precision arithmetic effects. Consider the diagram of a multiply-accumulate

(MAC) unit shown in Fig. 15.2, where the fixed-point MAC operation is interpreted in block diagram form. It is assumed that a and x are N-bit words which are presented to a full-precision multiplier producing a $2N$-bit product. The $2N$-bit full precision product may be reduced to an S-bit word, $S \leq 2N$, before being presented to a T-bit accumulator. The T-bit accumulator is finally reduced to an M-bit final result where $M \leq T$. It is generally assumed that the error associated with the case $M \ll 2N$ is given by

$$e = \{y[k] - Q_M(y[k])\} \tag{15.2}$$

where $Q_M(q)$ denotes the quantization of q to a digital word of M-bit precision. The error variance is given by

$$\sigma^2 = \frac{Q^2}{12} \tag{15.3}$$

where Q is the weight of the least significant bit of the M-bit accumulator output. The problem is not necessarily the error itself but, rather, what happens to the error after it is produced. Errors generated internally to an IIR filter can recirculate and remain within the system indefinitely. The analysis of this type of error will be performed in Chap. 17.

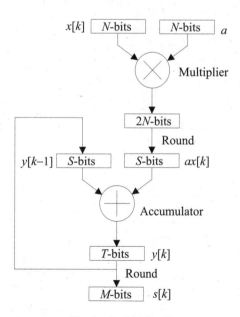

Figure 15.2 Fixed-point MAC unit.

Example 15.1: Finite Wordlength Effects

Problem Statement. Consider an eighth-order lowpass Chebyshev II filter (see Computer Study) having the transfer function

$$H(z) = 0.0888 \frac{(z^8 + 4.43z^7 + 10.76z^6 + 17.46z^5 + 20.48z^4 + 17.46z^3 + 10.76z^2 + 4.43z + 1)}{(z^8 + 1.10z^7 + 1.97z^6 + 1.55z^5 + 1.22z^4 + 0.61z^3 + 0.24z^2 + 0.061z + 0.008)}$$

Implement as a [16:15] fixed-point direct-II and cascade normal IIR filters. Test the system for finite wordlength effects using a unit step function.

Analysis and Conclusions. The simulated step-responses (see Computer Study) of the lowpass direct-II and normal IIR filters are shown in Fig. 15.3. The data format for coefficients and arithmetic is assumed to be 16 bits wide with 15 fractional bits (i.e., [16:15]). Also shown in Fig. 15.3 is the ideal and a highly distorted direct-II response due to 93 register overflows occurring in the first 101 samples. The overflow introduce grossly nonlinear behavior in the discrete-time domain. The normal filter's response is far less distorted and suffers only 19 register overflows during the same time period. However, the cascade normal filter requires 25 multiply-accumulates per filter cycle, while the direct-II needs 18. Therefore, the maximum real-time bandwidth of the direct-II can be 40 percent higher than the cascade. As a result, architecture can, as is often the case, define a trade-off between bandwidth and precision.

Computer Study. The designed filter is defined in terms of the following specifications:

```
DESIGN DEFINITION

    File                         : scale.IIR
    Source Module                : Infinite Impulse Response Filter
    Filter Type                  : Chebyshev Type II
    Order                        : 8
    Frequency Response           : Lowpass
    Desired Passband Attenuation : 0.1000000000000000 dB
```

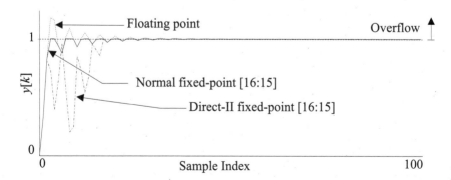

Figure 15.3 Direct-II and normal eighth-order IIR filter architecture and responses.

```
Desired Stopband Attenuation : 40.00000000000000 dB
Actual Stopband Attenuation  : 40.03955716783712 dB
Sampling Frequency           : 44100.00000000000 Hz
Passband Edge                : 10000.00000000000 Hz
Stopband Edge                : 12500.00000000000 Hz
```

TRANSFER FUNCTION

```
Scale Factor = 0.08883285197457290
Numerator Coefficients Denominator Coefficients
b 8: 1.000000000000000     a 8: 1.000000000000000
b 7: 4.437919605564610     a 7: 1.104614299236229
b 6: 10.76921619855683     a 6: 1.978668209561079
b 5: 17.46970831199284     a 5: 1.556364236494628
b 4: 20.40285887953456     a 4: 1.226408547493966
b 3: 17.46970831199284     a 3: 0.6160515120514172
b 2: 10.76921619855682     a 2: 0.2446494077658335
b 1: 4.437919605564609     a 1: 0.06099774584323624
b 0: 0.9999999999999999    a 0: 0.007910400932499942
```

ZEROS AND POLES OF TRANSFER FUNCTION

```
Zeros
-0.2270887423906905        +j  0.9738740694152477
-0.2270887423906905        +j -0.9738740694152477
-0.3767585512225092        +j  0.9263114994863855
-0.3767585512225092        +j -0.9263114994863855
-0.6637438844487714        +j  0.7479599293122967
-0.6637438844487714        +j -0.7479599293122967
-0.9513686247203337        +j  0.3080547676919496
-0.9513686247203337        +j -0.3080547676919496
Poles
-0.2587394083803890        +j  0.1558622599312509
-0.2587394083803890        +j -0.1558622599312509
-0.1870885008815274        +j  0.4429344492564740
-0.1870885008815274        +j -0.4429344492564740
-0.08891982860970640       +j  0.6812827348860745
-0.08891982860970640       +j -0.6812827348860745
-0.01755941174649156       +j  0.8911312791792454
-0.01755941174649156       +j -0.8911312791792454
```

The implemented direct-II filter was saved as architecture file "**scaled.arc.**"
The implemented cascade normal IIR filter was saved as architecture file
"**scalen.arc.**" The simulated step response shown in Fig. 15.3 was produced
using the following sequence of Siglab commands:

```
>hd=rarc("chap15\scaled.arc",1)   # direct-II
>hn=rarc("chap15\scalen.arc",1)   # cascaded-normal
>x=ones(101)                      # input step
>y=filt(hd,x)                     # ideal filter
>y15d=fxpfilt(hd,x,16,15)         # [16:15] direct-II
# 93 overflows detected - select Cancel
```

```
>y15n=fxpfilt(hn,x,16,15)          # [16:15] cascade-normal
# 19 overflows detected - select Cancel
>ograph(y,y15d,y15n)               # overlay graph
```

Refer to Fig. 15.3 and note that only the transient response was contaminated by finite wordlength effect errors. This is due to the nature of the test conducted, namely a study of the IIR filter's step-response. Since the step-response transient period lasts for only a few samples, there would be little noticeable difference between the step-responses of the three filters. However, if the transient response is important, as in a motor control application, the filter responses would significantly differ.   ❖

15.5 Coefficient Sensitivity

A third source of finite wordlength error is attributed to coefficient rounding. A general model for a monic IIR filter is

$$H(z) = \frac{\sum\limits_{m=0}^{M} b_m z^{-m}}{1 + \sum\limits_{m=1}^{N} a_m z^{-m}} = d_0 + \frac{\sum\limits_{m=1}^{N} c_m z^{-m}}{1 + \sum\limits_{m=1}^{N} a_m z^{-m}} \qquad (15.4)$$

whose coefficients are defined over the real field. Suppose that the filter is implemented using a direct-II architecture having quantized and rounded coefficients. Then

$$H'(z) = \frac{\sum\limits_{m=0}^{M} b'_m z^{-m}}{1 + \sum\limits_{m=1}^{N} a'_m z^{-m}} = d'_0 + \frac{\sum\limits_{m=1}^{N} c'_m z^{-m}}{1 + \sum\limits_{m=1}^{N} a'_m z^{-m}} \qquad (15.5)$$

where

$$a'_m = Q_V(a_m) = a_m + \Delta a_m, \, m \in \{1, 2, 3, ..., N\}$$

$$b'_m = Q_V(b_m) = b_m + \Delta b_m, \, m \in \{1, 2, 3, ..., M\}$$

$$c'_m = Q_V(c_m) = c_m + \Delta c_m, \, m \in \{1, 2, 3, ..., N\}$$

$$d'_0 = Q_V(d_0) = d_0 + \Delta d_0$$

$$(15.6)$$

after quantizing the coefficients to V significant bits. Comparing $H(z)$ with $H'(z)$ is difficult, in general, due to the nonlinear nature of the problem. Small changes in a denominator coefficient, for example, can radically alter the transfer function's pole location.

The poles of the IIR filter are denoted p_m and defined in Eq. (15.4). Specifically,

$$D(z) = 1 + \sum_{m=1}^{N} a_m z^{-m} = \prod_{m=1}^{N} (1 - p_m z^{-1}) \qquad (15.7)$$

If the coefficients a_m are quantized to values $a'_m = Q_V(a_m) = a_m + \Delta a_m$, then it follows that

$$D'(z) = 1 + \sum_{m=1}^{N} a'_m z^{-m} = \prod_{m=1}^{N} (1 - p'_m z^{-1}) \qquad (15.8)$$

where

$$p'_m = p_m + \Delta p_m, \, m \in \{1, 2, 3, ..., N\} \qquad (15.9)$$

The roots of $D'(z)$ define the pole locations for fixed-point filter and are assumed to have moved from their ideal location $z = p_m$ by an amount Δp_m. The incremental change in pole location Δp_m, as a function of coefficient quantization, can be approximated using a partial derivative

$$\Delta p_m = \sum_{k=1}^{N} \frac{\partial p_m}{\partial a_k} \Delta a_k \qquad (15.10)$$

It should be apparent that this line of analysis can be very tedious and limiting. Furthermore, partial derivative or *sensitivity analysis* studies are only valid for small incremental changes in parameters. This makes its value questionable in the study of a practical fixed-point filter. In lieu of a formal mathematical test, simulation can be used to qualify the effects of coefficient rounding.

If the original filter $H(z)$ is a highly frequency-selective filter (i.e., high-Q), then some of the filter poles are close to the periphery of the unit circle. Therefore, it is quite possible that the poles of the quantized filter could migrate to locations that significantly alter the realized frequency response. In severe cases, the realized poles could move outside the unit circle, causing the filter to become unstable.

Example 15.2: Coefficient Sensitivity

Problem Statement. Consider again the eighth-order Chebyshev II filter studied in Example 15.1. Isolate and analyze the effect of coefficient rounding if 12-bit data registers are assumed.

Analysis and Conclusions. The largest direct-II coefficient for the target Chebyshev II filter can be determined to be $19.1765 < 2^5$, which implies that at least a [12:6] format is needed to ensure that five integer bits are available for coefficient encoding. Therefore, the maximum rounding error is $2^{-6}/2 = 0.0078$. Using simulation, the difference between the ideal and finite precision magnitude frequency response is shown in Fig. 15.4.

Computer Study. The following sequence of Siglab commands was used to create the data shown in Fig. 15.4:

```
>h=rarc("chap15\scaled.arc",1)    # Read Direct-II state model.
>h                                # Direct-II state model.
      0        1        0        0        0        0        0        0
      0        0        1        0        0        0        0        0
      0        0        0        1        0        0        0        0
      0        0        0        0        1        0        0        0
      0        0        0        0        0        1        0        0
      0        0        0        0        0        0        1        0
      0        0        0        0        0        0        0        1
 -8e-03   -0.061   -0.245   -0.616   -1.226   -1.556   -1.979   -1.105
      0        0        0        0        0        0        0        1
  0.992    4.377    10.52    16.85    19.18    15.91    8.791    3.333
      1        0        0        0        0        0        0        0
  0.089        0        0        0        0        0        0        0
>max(mag(h))
19.1765
>wordlen(12,6,1)                        # Set fixed-point format.
>hf=fxpt(h); mag(max(hf-h))             # Maximum quantization error
0.011                                   #<0.014
# Compute floating-point and fixed-point magnitude freq response.
>y=filt(h,impulse(1024));yf=filt(hf,impulse(1024))
>ograph(mag(pfft(y)),mag(pfft(yf)))
```

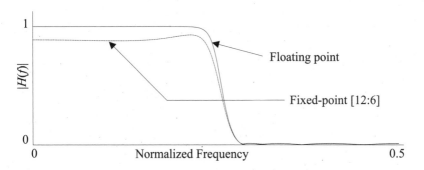

Figure 15.4 Comparison of an eighth-order Chebyshev-II IIR filter using a direct-II implementation with floating-point and [12:6] fixed-point coefficients.

The response of the simulated floating-point and fixed-point filters can be compared using a soundboard. However, according to the data found Fig. 15.4, only a minor amplitude difference exists between the floating-point and fixed-point magnitude frequency responses. Therefore, it can be assumed that the audio responses of the two filters will be very similar. The following sequence of Siglab commands may be used to test this hypothesis using a 16-bit sound-board sampled at 11,025 Hz.

```
>x=mkcos(0.8*2^15,1000/11025,0,10e3)  # long 1kHz sinusoid
>y=filt(h,x);yf=filt(hf,x)            # h and hf previously defined
>wavout(y); wavout(yf)                                              ❖
```

The results of a sensitivity analysis, whether mathematical or simulation, must be repeated for each architecture. This is based on the fact that each distinct architecture consists of a unique set of filter coefficients. It is often considered to be a good design practice to select the architecture whose largest coefficient, in absolute value, is closest to unity and all the coefficients being on the same order of magnitude. In this case, the fractional precision is equally used across the coefficient field. If, however, a few coefficients are significantly larger than the others, then too many integer bits may be required to cover these anomalies, leaving too few fractional bits of precision to preserve the significance of the smaller coefficients.

15.6 Second-Order Section

It was noted that rounding or truncating the feedback coefficients can significantly affect the pole locations of the realized filter. In fact, it is possible to have a stable filter become unstable after coefficient quantization. In an ideal situation (i.e., infinite precision), the poles of an IIR filter are invariant with respect to the choice of architecture. Again, the poles of a transfer function,

$$H(z) = K\frac{N(z)}{D(z)} = K\frac{\displaystyle\sum_{m=0}^{M} b_m z^{-m}}{\displaystyle\sum_{m=0}^{N} a_m z^{-m}} \tag{15.11}$$

are defined to be the roots of $D(z) = 0$, which can be assumed to be

$$D(z) = \sum_{m=0}^{N} a_m z^{-m} = \prod_{m=0}^{N} (1 - p_m z^{-1}) \tag{15.12}$$

where the poles p_m are, in general, complex. If, for example, a direct-II architecture is chosen and the coefficients are exact, then the characteristic equation, say $D^{II}(z)$, is given by Eq. (15.12), and the poles would be located at $z = p_m$. Specifically,

$$D^{II}(z) = D(z) = \sum_{m=0}^{N} a_m^{II} z^{-m} = \prod_{m=0}^{N} (1 - p_m z^{-1}) \qquad (15.13)$$

Rounding or truncating the coefficients a_m to V significant bits, however, would result in a new characteristic polynomial

$$D_R^{II}(z) = \sum_{m=0}^{N} Q_V(a_m^{II}) z^{-m} = \prod_{m=0}^{N} (1 - p_m^{II} z^{-1}) \qquad (15.14)$$

where $Q_V(v)$ again is the quantized value of v, with

$$Q_V(a_m^{II}) = a_m + \Delta a_m$$

and

$$p_m^{II} = p_m + \Delta p_m^{II}$$

The movement of the poles is seen to be the collective action of all $(N + 1)$ feedback coefficients.

Suppose that the IIR filter is implemented with a cascade architecture consisting of a collection of second-order filters. Then the characteristic equation, say $D^C(z)$, is given by

$$D(z) = D^C(z) = \prod_{m=1}^{N/2} (1 + a_{1:m}^C z^{-1} + a_{2:m}^C z^{-2}) = \prod_{m=0}^{N} (1 - p_m z^{-1}) \quad (15.15)$$

and the pole locations remain unchanged. If the coefficients of the cascade architecture are truncated or rounded, then the poles of the fixed-point filter's cascaded characteristic equation is given by

$$D_R^C(z) = \sum_{m=1}^{N/2} (1 + Q_V(a_{1:m}^C) z^{-1} + Q_V(a_{2:m}^C) z^{-2}) = \prod_{m=0}^{N} (1 - p_m^C z^{-1}) \quad (15.16)$$

where $Q_V(a_{i:m}^C) = a_{i:m} + \Delta a_{i:m}$ and where, in general, $\Delta p_m^{II} \neq \Delta p_m^C$. The movement of the poles for the cascade architecture are seen to be defined by the rounding or truncation of pairs of feedback coefficients which are distinctly different than those of a direct-II filter. It should therefore be apparent that the pole sensitivity is architecture dependent.

Since second-order sections are considered to be digital filter building blocks, they have been the focus of special study. If the characteristic equation for the second-order section is

$$D(z) = 1 - 2r\cos(\phi)z^{-1} + r^2 z^{-2} \tag{15.17}$$

then the poles are located at $z = re^{\pm j\phi}$. A second-order direct-II filter section would possess a state matrix given by

$$\mathbf{A} = \begin{bmatrix} 0 & 1 \\ -r^2 & 2r\cos(\phi) \end{bmatrix} \tag{15.18}$$

and that of a quantized second direct-II section would be

$$Q_V(\mathbf{A}) = \begin{bmatrix} 0 & 1 \\ Q_V(-r^2) & Q_V(2r\cos(\phi)) \end{bmatrix} \tag{15.19}$$

where the eigenvalues (poles) of $Q_V(\mathbf{A})$ are given by

$$\det(z\mathbf{I} - Q_V(\mathbf{A})) = \det \begin{bmatrix} z & -1 \\ -Q_V(-r^2) & z - Q_V(2r\cos(\phi)) \end{bmatrix}$$

$$= z^2 - Q_V(2r\cos(\phi))z - Q_V(-r^2) \tag{15.20}$$

and displayed, in the first quadrant, in Fig. 15.5 for all possible four fractional bit coefficients. It can be seen that the roots of Eq. (15.20) provided a nonuniform coverage in the z-plane. The poles lie on a circular loci at various radii from the origin. Notice also that the quantized poles are more dense nearer the periphery of the unit circle and sparse in the interior.

If the second-order section is implemented as a normal filter, it would have a state matrix of the form

$$\mathbf{A} = \begin{bmatrix} r\cos(\phi) & -r\sin(\phi) \\ r\sin(\phi) & r\cos(\phi) \end{bmatrix} \tag{15.21}$$

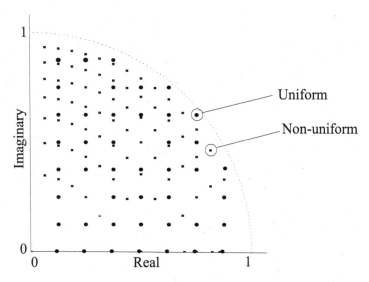

Figure 15.5 Admissible locations of the quantized poles for a four fractional bit direct-II IIR filter (uniform pole spacing) and normal IIR filter (nonuniform pole spacing).

and that of a quantized second-order normal section satisfies

$$Q_V(\mathbf{A}) = \begin{bmatrix} Q_V[r\cos(\phi)] & Q_V[-r\sin(\phi)] \\ Q_V[r\sin(\phi)] & Q_V[r\cos(\phi)] \end{bmatrix} \tag{15.22}$$

The eigenvalues (poles) of $Q_V(\mathbf{A})$ are given by the roots of

$$\det(z\mathbf{I} - Q_V(\mathbf{A})) = \det \begin{bmatrix} z - Q_V[r\cos(\phi)] & -Q_V[-r\sin(\phi)] \\ -Q_V[r\sin(\phi)] & z - Q_V[r\cos(\phi)] \end{bmatrix}$$

$$= z^2 - 2Q_V[r\cos(\phi)]z + \{Q_v[r\cos(\phi)]\}^2$$

$$-\{Q_V[-r\sin(\phi)]\}^2 \tag{15.23}$$

which are displayed in this first quadrant in Fig. 15.5 for all possible four fractional bit coefficients. It can be seen that the quantized pole coverage is now uniformly distributed over the z-plane. Because of this feature, the normal architecture is considered to have low coefficient roundoff sensitivity. However, this is true only in a global sense. If the poles are known to be close to the unit circle, the quantization pattern of the direct-II can result in a design having a lower coefficient round-off error budget.

Example 15.3: Pole Sensitivity

Problem Statement. Analyze how quantization affects the location of the poles of a stable second-order direct-II and normal IIR filter. The largest coefficient of the direct-II second-order section is bounded by two, while that of a normal is bounded by unity. Therefore, the minimum data format for a direct-II is $[F + 2{:}F]$ and, for a normal filter, $[F + 1{:}F]$. The results, in the first quadrant of the unit circle are shown in Fig. 15.5.

Computer Study. The following sequence of Siglab commands was used to produce the data shown in Fig. 15.5. The s-file "`poles.s`" defines the function "`poles(n)`" where n is the number of fractional bits of coefficient precision.

```
>include "chap15\poles.s"
>poles(4) # Set x-axis to 1.33 in MinMax menu to reproduce figure. ❖
```

15.7 Scaling

Register overflow can be inhibited by reducing the dynamic range of the input. Scaling the input by a positive constant, $K < 1$, reduces the chance of run-time overflow. Unfortunately, scaling also reduces the output precision. In particular, the precision of a scaled data word will be reduced by k bits, where

$$k = \log_2(K) \qquad (15.24)$$

Equivalently, the output precision will be reduced by a like amount. Therefore, the error variance, compared to the unscaled system, can be expressed as

$$\sigma_K^2 = \frac{(KQ)^2}{12} = K^2\sigma^2 \qquad (15.25)$$

As a result, scaling (if required) should only be used minimally to the point where overflow is controlled.

Extending the wordlength of a system upward will increase the dynamic range of the system. This is done, however, at an additional system cost and possible reduction in the maximum real-time bandwidth. In the context of a modern DSP microprocessor, the logical choices are 16 bits or 24 bits. In the context of a direct ASIC implementation, any wordlength may be chosen; however, every additional bit of precision adds to the expense of the implementation and will tend to reduce the peak execution speed of the system.

15.8 Limit Cycling

Another finite wordlength effect is called *zero input limit cycling* or simply *limit cycling*. Limit cycling causes a digital filter to produce small amplitude changes to appear at the system's output during periods when the input is zero. In voice communication applications, limit cycling can manifest itself as undesirable audible "clicking" sounds that are audible during quiet (unvoiced) periods. The first-generation DSP microprocessors were imprecise (e.g., eight bits) and, as a result, limit cycling was an annoying problem that could be reduced only with substantial engineering effort. With the advent of 16-bit, 24-bit and floating-point DSP microprocessors, limit cycling has become a secondary problem in DSP microprocessor-based implementations. However, in direct ASIC implementations constructed with small wordlength arithmetic operations, limit cycling may still be a concern.

Limit cycling is caused when the response of an unforced stable LTI system does not successfully decay to zero in time due to finite wordlength effects. Consider the simple first-order system

$$y[k] = ay[k-1] + x[k] \qquad (15.26)$$

If $|a| < 1$, then ideally $y[k] \to 0$ when $x[k] = 0$. If the filter is implemented in fixed-point, then

$$y[k] = Q_V[ay[k-1]] + x[k] \qquad (15.27)$$

where $Q_V[q[k]]$ denotes the quantized value of $q[k]$ to V significant bits. Refer to Fig. 15.6. Suppose that at sample instance k, the first-order system has a value $y[k]$ and an input $x[k] = 0$ for all $k > K$. If the decay rate of the quantized system is too slow to allow the output to decay by an amount more than half a quantization interval (i.e., $\Delta[k] = (y[k]-y[k-1]) < Q/2$), then $y[k]$ would be rounded back to its previous value $y[k-1]$. As such, the output would never be able to decay to zero and would have some constant offset (possibly oscillating), which is called limit cycling.

Example 15.4: Limit Cycling

Problem Statement. Suppose a system $y[k] = ay[k-1] + x[k]$ has an [N:4] format and $a \in \{1/2, -1/2, 3/4, -3/4\}$. Investigate the zero input limit cycling properties of the system.

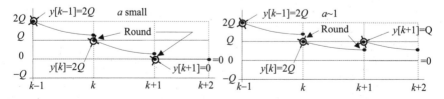

Figure 15.6 Limit cycling interpretation for $y[k+1] = ay[k] + x[k]$. Responses show the unforced first-order system $y[k] = ay[k-1] + x[k]$, $x[k] = 0$ for two different values of a.

Analysis and Conclusions. The implementation of these four filters would yield the results shown below. The data is interpreted in Table 15.1 for a two's complement implementation using as an initial condition $x[0] = 1 \rightarrow 0_\Diamond 1111$ where $\Diamond$ denotes the binary point.

The results are graphically interpreted in Fig. 15.6 for the four-bit case. The hand-calculated example considers a rounded four-bit fractional case system, whereas the simulator assumed four-bit and eight-bit rounded integer arithmetic. It can be seen that limit cycling can take place under certain circumstances and that the severity of limit cycling is a function of decay rate of the response (i.e., "a") and the number of bits of arithmetic precision maintained after multiplication. For contemporary 16-bit or 24-bit designs, especially those using full-precision multipliers and extended-precision accumulators, the effect of limit cycling is generally very small and negligible.

Computer Study. The s-file "limitcyc.s" defines the function "limit-cyc(a,n,m)" where a is the filter coefficient, n is the number of bits of integer precision used in computing the rounded impulse response, and m is the length of the time-series to be saved. The following sequence of Siglab commands demonstrates the effects of limit cycling with the results shown in Fig. 15.7.

```
>include "chap15\limitcyc.s"
>n=4;m=16;a=1/2;b=-a;c=3/4;d=-c
```

Table 15.1   First-Order System Limit Cycling Example

k	$a = 1/2 = 0_\Diamond 1000$	$a = -1/2 = 1_\Diamond 1000$	$a = 3/4 = 0_\Diamond 1100$	$a = 3/4 = 1_\Diamond 1100$
0	$0_\Diamond 1111$ (15/16)	$0_\Diamond 1111$ (15/16)	$0_\Diamond 1111$ (15/16)	$1_\Diamond 1111$ (15/16)
1	$0_\Diamond 1000$ (8/16)	$1_\Diamond 1000$ (−8/16)	$0_\Diamond 1011$ (11/16)	$1_\Diamond 0101$ (−11/16)
2	$0_\Diamond 0100$ (4/16)	$0_\Diamond 0100$ (4/16)	$0_\Diamond 0110$ (6/16)	$0_\Diamond 0110$ (6/16)
3	$0_\Diamond 0010$ (2/16)	$1_\Diamond 1110$ (−2/16)	$0_\Diamond 0101$ (5/16)	$0_\Diamond 1100$ (−4/16)
4	$0_\Diamond 0001$ (1/16)	$0_\Diamond 0001$ (1/16)	$0_\Diamond 0100$ (4/16)	$0_\Diamond 0011$ (3/16)
5	$0_\Diamond 0001$ (1/16)	$1_\Diamond 1111$ (−1/16)	$0_\Diamond 0011$ (3/16)	$1_\Diamond 1101$ (−3/16)
6	$0_\Diamond 0001$ (1/16)	$0_\Diamond 0000$ (0/16)	$0_\Diamond 0010$ (2/16)	$0_\Diamond 0010$ (2/16)
7	$0_\Diamond 0001$ (1/16)	$0_\Diamond 0000$ (0/16)	$0_\Diamond 0010$ (2/16)	$1_\Diamond 1111$ (−1/16)
8	$0_\Diamond 0001$ (1/16)	$0_\Diamond 0000$ (0/16)	$0_\Diamond 0010$ (2/16)	$0_\Diamond 0001$ (1/16)
9	$0_\Diamond 0001$ (1/16)	$0_\Diamond 0000$ (0/16)	$0_\Diamond 0010$ (2/16)	$1_\Diamond 1111$ (−1/16)
k	limit cycling at $y[k] = 1/16$	no limit cycling	limit cycling at $y[k] = 2/16$	limit cycling at $y[k] = (-1)^k(1/16)$

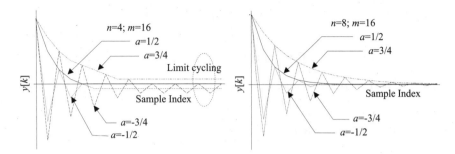

Figure 15.7 Limit cycling for the unforced system $y[k] = ay[k{-}1] + x[k]$, $x[k] = 0$, using a four bit and eight bit rounded integer arithmetic.

```
>ograph(limitcyc(a,n,m),limitcyc(b,n,m),limitcyc(c,n,m),limitcyc(d,n,m))
>n=8;m=16;a=1/2;b=-a;c=3/4;d=-c
>ograph(limitcyc(a,n,m),limitcyc(b,n,m),limitcyc(c,n,m),limitcyc(d,n,m))
```
❖

15.9 Summary

In Chap. 15, the concept of finite wordlength effect was introduced. It was shown that implementing digital filters using fixed-point arithmetic can result in a number of errors ranging from catastrophic register overflow to benign zero-input limit cycling. Other error sources, such as arithmetic and coefficient rounding, can also degrade the performance of a fixed-point design. Scaling can be used to control these errors in some cases. In the next two chapters, overflow and arithmetic errors are mathematically analyzed. Because fixed-point solutions generally are less costly and can operate at higher bandwidths than floating-point designs, engineers generally are willing to pursue a variety of fixed-point options before finally discarding a design in favor of a floating-point solution.

15.10 Self-Study Problems

1. Butterworth Bilinear z-Transform Direct-II Design A lowpass Butterworth IIR filter is implemented that meets or exceeds the following specifications:

Desired passband attenuation = 3.0 dB

Desired stopband attenuation = 20.0 dB

Sampling frequency = 10,000.0 Hz

Passband edge = 2000.0 Hz

Stopband edge = 2500.0 Hz

using a direct-II architecture and saved as "prob15_1.(IIR,ARC)." Compare the floating-point magnitude frequency response to that of a direct-II implementation using [16:15], [16:12], [16:8], and [16:4] coefficient formats (i.e., coefficient roundoff error). Identify the cause of any major differences.

2. Cascade IIR Filter Repeat problem 15.1 for a cascade architecture (saved as "prob15_2.(IIR,ARC)").

3. Parallel IIR Filter Repeat problem 15.1 for a parallel architecture (saved as "prob15_3.(IIR,ARC)").

4. Normal IIR Filter Repeat problem 15.1 for a normal architecture (saved as "prob15_4.(IIR,ARC)").

5. Zero-Input Limit Cycling. Reconsider the system studied in Example 15.4. Derive and verify a relationship between the coefficient a and the number of fractional bits that will ensure that no zero input limit cycling can take place.

16

Overflow Prevention

16.1 Introduction

In Chap. 15, finite wordlength effects were introduced. Potentially the most grievous finite wordlength effect is called *run-time register overflow*, also known simply as *register overflow, large signal limit cycling,* or *saturation*. Register overflow can occur whenever the dynamic range requirements of system's variables exceed the dynamic range capabilities of the system's storage registers. If run-time register overflow occurs, gross nonlinearities can be introduced into the output. In the most severe case, the saturated output may suddenly go from a large positive value to a large negative value. This obviously can create havoc in audio, servomechanical, biomedical, or other applications. Therefore, register overflow is a problem that must be controlled. Unfortunately, providing this control can be a very challenging problem and often requires detailed mathematical analysis and/or computer simulations. In Chap. 16, the mechanics of managing register overflow are presented.

16.2 Finite Wordlength Effects

Register overflow can introduce major nonlinear distortions into a system's output, which can render a filter useless. Register overflow can often be eliminated by properly scaling selected system parameters so that the signals, found both within and without the filter, reside in a highly restricted range so as to not exceed any specified dynamic range limits. Managing register overflow in a fixed-point filter is called solving the *binary-point assignment problem*. Unfortunately, this is generally a nontrivial problem. Before this question can be understood and resolved, a mathematical framework for the binary-point assignment problem must first be developed.

16.3 Norm Bounds

Recall from Chap. 9 that a single-input single-output Nth-order proper IIR digital filter can be reduced to a state variable model of the form

$$\mathbf{x}[k+1] = \mathbf{A}\mathbf{x}[k] + \mathbf{b}v[k] \quad \text{(state equation)} \tag{16.1}$$

and

$$y[k] = \mathbf{c}^T\mathbf{x}[k] + d_0 v[k] \quad \text{(output equation)} \tag{16.2}$$

The input to the system is defined to be $v[k]$ which, according to Fig. 16.1, is related to the system input $u[k]$ through a proportionality constant k (i.e., $v[k] = ku[k]$). It will be shown that the general purpose of the constant k is to ensure that the filter meets prespecified absolute passband gain specifications. The constant k can also be used simply to amplify or attenuate the input to increase filter precision or reduce the possibility of run-time register overflow. It will also be shown that an advantage of a state variable model is found in the fact that states are directly correlated to information stored at the shift register level. Therefore, measuring a filter's register dynamic range requirements is equivalent to quantifying the amplitude bounds on the system's state variables. This is the key to the binary-assignment problem.

Fortunately, classical methods have been developed that can be used to compute a state bound in an ℓ_p sense where the ℓ_p norm of a causal time-series $s[k]$, denoted $\|s\|_p$, is given by

$$\|s\|_p = \left(\sum_{k=0}^{\infty} |s[k]|^p \right)^{1/p} \tag{16.3}$$

where $p > 0$. For $p \to \infty$, it follows that $\|s\|_\infty = \max(|s[k]|)$ for all $k \geq 0$.

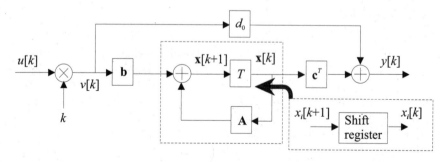

Figure 16.1 State variable model of an IIR showing the states residing at the shift register level.

Example 16.1: Norms

Problem Statement. Consider the time-series $s[k]$, having sample values $s[k]$ = $4 \times (u[k] - u[k - 1024])$ = 4 for $k \in \{0,1,2,...,1023\}$ and $s[k] = 0$ elsewhere. What is the ℓ_p norm of $s[k]$, namely $\|s\|_p$, for $p \in \{1/2,1,2,\infty\}$?

Analysis and Conclusion. Direct computation yields:

$$\|s\|_{1/2} = \left(\sum_{k=0}^{\infty} |s[k]|^{1/2} \right)^2 = (4^{1/2} \times 2^{10})^2 = 2^{22}$$

$$\|s\|_1 = \left(\sum_{k=0}^{\infty} |s[k]| \right) = (4 \times 2^{10})^2 = 2^{12}$$

$$\|s\|_2 = \sqrt{\sum_{k=0}^{\infty} |s[k]|^2} = \sqrt{4^2 \times 2^{10}} = \sqrt{2^{14}} = 2^7$$

$$\|s\|_{\infty} = \max(s[k]) = 2^2$$

It can therefore be seen that the value of the norm of a common time-series is dependent on the choice of norm. ❖

Example 16.2: Norms of Common Signals

Problem Statement. Compute the ℓ_p norms of a Kronecker impulse $\delta[k]$ and a decaying exponential $y[k] = \{1,a,a^2,a^3,a^4,...\}$, $a \in (0,1)$, for $p \in \{1,2,\infty\}$.

Analysis and Conclusions. Direct computation yields:

$$(i) \begin{cases} \|\delta[k]\|_1 = \sum_{k=0}^{\infty} |\delta[k]| = 1 \\[2mm] \|\delta[k]\|_2 = \sqrt{\sum_{k=0}^{\infty} \delta[k]^2} = 1 \\[2mm] \|\delta[k]\|_{\infty} = \max(\delta[k]) = 1 \end{cases}$$

$$(ii) \begin{cases} \|y[k]\|_1 = \sum_{k=0}^{\infty} |a^k| = \dfrac{1}{1-|a|} \\[3mm] \|y[k]\|_2 = \sqrt{\sum_{k=0}^{\infty} |a^{2k}|} = \sqrt{\dfrac{1}{1-a^2}} \\[3mm] \|y[k]\|_{\infty} = \max(a^k u[k]) = 1 \end{cases}$$

The norms are known to be hierarchically related by the relationship $\|y[k]\|_1 \geq \|y[k]\|_2 \geq \|y[k]\|_\infty$. For example, for $a = 1/2$, it follows that $\|y[k]\|_1 = 2 \geq \|y[k]\|_2 = 1.154 \geq \|y[k]\|_\infty = 1$. Notice also that as $a \to 1$, $\|y\|_1 \to \infty$, and $\|y\|_2 \to \infty$. As a result, virtually any interesting infinite duration signal, such as $s[k] = \cos(\omega k)u[k]$, has ℓ_1 and ℓ_2 norms which tend toward infinity.   ❖

16.4 Overflow Scaling

Developing a state bound begins with computing the impulse response measured at a shift-register location. If the impulse response, measured at the ith state is denoted $x_i[k]$, then the forced response to an input $v[k]$, also measured at the ith state location, is given by the linear convolution sum $y_i[k] = x_i[k]*v[k]$. From Holder's inequality,[*] it follows that

$$\left|y_i[k]\right| = \sum_{m=0}^{\infty} \left|x_i[k-m]v[m]\right| \leq \left\|x_i\right\|_p \left\|v\right\|_q \tag{16.4}$$

where $\dfrac{1}{p} + \dfrac{1}{q} = 1$

Observe that if $\|v\|_q \leq 1$, then the state bound in register i will be bounded by $\|x_i\|_p$. This can serve as a measure of the dynamic range requirement of the ith state. If $p = q = 2$ (i.e., ℓ_2 norms), then $y_i[k]$ will be bounded if both $x_i[k]$ and $v[k]$ have finite ℓ_2 bounds. It may be recalled that a process having a finite ℓ_2 norm is also of finite energy. Classifying signals on the basis of finite energy (not to be confused with power) is a natural way to partition the signal space. Methods of processing signals of finite energy have been extensively studied and cataloged by engineers, leaving a rich mathematical legacy. Another advantage of working with ℓ_2 norms is that this study is well developed in the context of mathematical optimization of linear systems (e.g., least-squares optimization).

16.5 ℓ_2 Scaling

Norms, and the ℓ_2 norm in particular, can be used to quantifying the dynamic range requirements of each state variable. The ℓ_2 bound, in a

---

[*] Holder inequality: $\Sigma\,|x_iy_i| \geq |x|_p\,|y|_q$ if $1/p + 1/q = 1$.

few cases, can be computed in closed form. For example, consider a first-order system having an impulse response

$$h[k] = a^k u[k] \tag{16.5}$$

where $|a| < 1$. The ℓ_2 norm squared given by

$$\|h[k]\|_2^2 = \sum_{k=0}^{\infty} (a^k)^2 = \sum_{k=0}^{\infty} (a^{2k}) = \frac{1}{1-a^2} \tag{16.6}$$

Notice that as $a \to 1$, the ℓ_2 norm (squared) rapidly approaches infinity.

In some cases, the ℓ_2 norm of a second-order system can be computed in closed form. Consider, for example, a causal second-order system having an impulse response given by

$$h[k] = \frac{r^k \sin(\omega[k+1]T_s)}{\sin(\omega T_s)} \tag{16.7}$$

The ℓ_2 norm squared bound can be shown to be given by

$$\|h[k]\|_2^2 = \frac{(1+r^2)}{(1-r^2)(r^4+1-2r\cos(2\omega T_s))} \tag{16.8}$$

It is unrealistic to develop such formula for every possible second-order case. Furthermore, for higher-order systems, a more general computing method is required.

The forced, or inhomogeneous, response of a state-determined LTI system to an arbitrary input is given by the state determined convolution sum

$$\mathbf{x}[k] = \sum_{n=0}^{\infty} \mathbf{A}^k \mathbf{b} v[k+1-n] \tag{16.9}$$

Suppose that the input signal process $v[k]$ is a zero mean wide sense stationary random process. The whiteness of $v[k]$ implies that the signals autocorrelation function satisfies

$$E(v[k+1-n]v[k+1-m]) = \delta[n-m] \tag{16.10}$$

This leads to the definition of a non-negative definite $N \times N$ matrix $\mathbf{K}$ given by

$$K = E(\mathbf{x}[k]\mathbf{x}^T[k]) = \sum_{n=0}^{\infty} \sum_{m=0}^{\infty} \mathbf{A}^n \mathbf{b} E(v[k+1-n]v[k+1-m])\mathbf{b}^T(\mathbf{A}^T)^m$$

$$= \sum_{n=0}^{\infty} \mathbf{A}^n \mathbf{b}\mathbf{b}^T(\mathbf{A}^T)^n \qquad (16.11)$$

Recall that the impulse response, measured at the ith state or shift register location, is defined by $x_i[k]$. It then follows that

$$K_{ii} = E(x_i^2[k]) = \sum_{k=0}^{\infty} x_i^2[k] = \|x_i\|_2^2 \qquad (16.12)$$

That is, the ith on-diagonal element of $\mathbf{K}$ is the ℓ_2 norm squared value of $x_i[k]$. Thus, Eq. (16.12) can provide a gateway to computing the state norms provided that the elements of the matrix $\mathbf{K}$ can be efficiently computed. Fortunately such methods exist.

Recall that the z-transform of the Nth-order state determined system given in Eq. (16.1) is

$$\mathbf{X}(z) = (z\mathbf{I} - \mathbf{A})^{-1}\mathbf{b}V(z) \qquad (16.13)$$

where $v[k]$ is shown in Fig. 16.1. This defines the z-transform of $x_i[k]$ to be

$$X_i(z) = (z\mathbf{I} - \mathbf{A})^{-1}\mathbf{b}V(z) \}_i \qquad (16.14)$$

where $\}_i$ denotes the ith element of a column vector. The inverse z-transform of $X_i(z)$ would therefore provide a computing formula of $x_i[k]$ found in Eq. (16.12). This in turn would have to be evaluated after Eq. (16.12) to produce the needed state bound, which is by itself a challenging problem. However, alternative methods exists which can produce the ℓ_2 norm squared value of $x_i[k]$, namely K_{ii}, directly.

Suppose again that the input process is a zero mean wide sense stationary random process. Then,

$$E(\mathbf{x}[k+1]\mathbf{x}^T[k+1]) = \mathbf{A}E(\mathbf{x}[k]\mathbf{x}^T[k])\mathbf{A}^T + \mathbf{b}E(v[k]^2)\mathbf{b}^T$$

$$+ \mathbf{A}E(\mathbf{x}[k]v[k])\mathbf{b}^T + \mathbf{b}E(v[k]\mathbf{x}^T[k])\mathbf{A}^T \qquad (16.15)$$

with $E(v[k]) = 0$, $E(\mathbf{x}[k]v[k]) = \mathbf{0}$, and $E(v[k]^2) = 1$. Based on the wide sense stationarity assumption, the statistics will not vary in time. From this it is known that

$$E(\mathbf{x}[k+1]\mathbf{x}^T[k+1]) \;=\; E(\mathbf{x}[k]\mathbf{x}^T[k]) \stackrel{\wedge}{=} \mathbf{K} \qquad (16.16)$$

Upon substituting these expected values into Eq. (16.15), it follows that

$$E(\mathbf{x}[k+1]\mathbf{x}^T[k+1]) \;=\; \mathbf{K} \;=\; \mathbf{AKA}^T + \mathbf{bb}^T \qquad (16.17)$$

The matrix $\mathbf{K}$ is often referred to as a *Lyapunov stability matrix*, is extensively studied in linear systems, and was previously introduced in Eq. (9.53). Assuming $\mathbf{K}$ can be readily computed, the on-diagonal terms of the matrix $\mathbf{K}$ can be used to establish ℓ_2 norm squared state bounds, namely $K_{ii} = \|x_i\|^2$. In addition, the matrix equation $\mathbf{K} = \mathbf{AKA}^T + \mathbf{bb}^T$ has a solution if all the eigenvalues of $\mathbf{A}$ are interior to the unit circle [i.e., $S = (\mathbf{A},\mathbf{b})$ is BIBO stable].

If the system is stable, then the elements of $\mathbf{K}$ are bounded and satisfy

$$\mathbf{K} \;=\; \sum_{i=0}^{\infty} \mathbf{A}^i \mathbf{bb}^T (\mathbf{A}^T)^i \;=\; \mathbf{AKA}^T + \mathbf{bb}^T \qquad (16.18)$$

The value of $\mathbf{K}$ can be recursively solved by using the following procedure

$$\mathbf{K}_0 = \mathbf{bb}^T$$
$$\mathbf{Q}_0 = \mathbf{A}$$
$$\text{do } \{$$
$$\mathbf{K}_{i+1} = \mathbf{Q}_i \mathbf{K}_i \mathbf{Q}_i^T$$
$$\mathbf{Q}_{i+1} = \mathbf{Q}_i^2$$
$$\} \text{ while } \mathbf{Q}_{i+1} > \varepsilon$$

$$(16.19)$$

where ε is a small stopping threshold. Alternatively, $\mathbf{K}$ can be computed using the algorithm developed below.

Suppose that the characteristic polynomial of a state-determined discrete time system, namely $\Delta(z) = \det(z\mathbf{I} - \mathbf{A})$, satisfies

$$\Delta(z) = z^N + p_1 z^{N-1} + \ldots + p_{N-1} z + p_N \tag{16.20}$$

Define a coefficient vector $\mathbf{r} = [r_0\ r_1\ \ldots\ r_{N-1}]^\mathrm{T}$ as

$$r_i + \sum_{j=1}^{N} p_j r_{|i-j|} = \begin{cases} 1 \text{ if } i = 0 \\ 0 \text{ if } i \neq 0 \end{cases} \tag{16.21}$$

The values of $\mathbf{r} = [r_0\ r_1\ \ldots\ r_{N-1}]^\mathrm{T}$ are then used to define a Toeplitz matrix $\mathbf{R}$, which is defined to be

$$\mathbf{R} = \begin{bmatrix} r_0 & r_1 & r_2 & \cdots & r_{N-1} \\ r_1 & r_0 & r_1 & \cdots & r_{N-2} \\ r_2 & r_1 & r_0 & \cdots & r_{N-3} \\ \vdots & \vdots & \vdots & \ddots & \vdots \\ r_{N-1} & r_{N-2} & r_{N-3} & \cdots & r_0 \end{bmatrix} \tag{16.22}$$

Next define

$$\mathbf{x}[1] = \mathbf{b} \tag{16.23}$$

and

$$\mathbf{x}[k+1] = \mathbf{A}\mathbf{x}[k] + p_k \mathbf{b} \tag{16.24}$$

and collect the terms to form another $N \times N$ matrix

$$\mathbf{X} = [\mathbf{x}[N], \ldots, \mathbf{x}[2], \mathbf{x}[1]] \tag{16.25}$$

The matrices $\mathbf{X}$ and $\mathbf{R}$ can then be combined to define

$$\mathbf{K} = \mathbf{X}\mathbf{R}\mathbf{X}^T \tag{16.26}$$

Regardless of the method used to compute $\mathbf{K}$, the on-diagonal terms provide a direct measure of the ℓ_2 norm squared bounds on the state vectors.

Once the analysis is performed for a given system $\mathbf{S} = (\mathbf{A}, \mathbf{b})$, the maximum value of K_{ii} can be determined, $K_{max} = \max(K_{ii})$. To protect the states from encountering run-time dynamic range overflow in an ℓ_2

sense, the data format [N:F] must be carefully chosen. In the context of K_{max}, the size of the integer field, in bits (i.e., $I = N\text{-}F\text{--}1$ bits) needs to satisfy

$$I = \left\lceil \log_2 \sqrt{K_{max}} \right\rceil \tag{16.27}$$

That is, at least I integer bits need to be allocated within an N bit data field to protect against run-time overflow in an ℓ_2 sense.

Example 16.3: ℓ_2 State Bounds

Problem Statement. In Example 14.5, a second-order system having a transfer function given by

$$H(z) = \frac{z - 0.5}{z^2 - z + 0.5}$$

has a direct-II model given by $S = (\mathbf{A}, \mathbf{b})$ where

$$\mathbf{A} = \begin{pmatrix} 0 & 1 \\ -0.5 & 1 \end{pmatrix}, \quad \mathbf{b} = \begin{pmatrix} 0 \\ 1 \end{pmatrix}$$

What are the ℓ_2 bounds on the state vectors?

Analysis and Conclusions. Computing

$$\Delta(z) = \det(z\mathbf{I} - \mathbf{A}) = z^2 - z + 0.5$$

defines $p_1 = -1$ and $p_2 = 0.5$ in Eq. (16.20). Based on knowledge of p_1 and p_2, the solution to Eq. (16.21) is given in terms of

$$r_0 + p_1 r_1 + p_2 r_2 = 1$$

$$r_1 + p_1 r_0 + p_2 r_1 = 0$$

$$r_2 + p_1 r_1 + p_2 r_0 = 0$$

which can be expressed in matrix-vector form as

$$\mathbf{P}\mathbf{r} = \begin{pmatrix} 1 & -1 & 0.5 \\ -1 & 1.5 & 0 \\ 0.5 & -1 & 1 \end{pmatrix} \begin{pmatrix} r_0 \\ r_1 \\ r_2 \end{pmatrix} = \begin{pmatrix} 1 \\ 0 \\ 0 \end{pmatrix} = \mathbf{s}$$

or, solving the equation $\mathbf{P}^{-1}\mathbf{s} = \mathbf{r}$, yields

$$\begin{pmatrix} r_0 \\ r_1 \\ r_2 \end{pmatrix} = \begin{pmatrix} \dfrac{12}{5} \\ \dfrac{8}{5} \\ \dfrac{2}{5} \end{pmatrix}$$

The 2×2 matrix $\mathbf{R}$ can then be defined to be

$$\mathbf{R} = \begin{pmatrix} \dfrac{12}{5} & \dfrac{8}{5} \\ \dfrac{8}{5} & \dfrac{12}{5} \end{pmatrix}$$

Finally,

$$\mathbf{x}[1] = \begin{pmatrix} 0 \\ 1 \end{pmatrix}; \quad \mathbf{x}[2] = \mathbf{Ax}[1] + p_1\mathbf{b} = \begin{pmatrix} 1 \\ 0 \end{pmatrix}$$

which implies that $\mathbf{X} = \mathbf{I}$. Computing $\mathbf{K} = \mathbf{XRX}^T$ yields

$$\mathbf{K} = \begin{pmatrix} \dfrac{12}{5} & \dfrac{8}{5} \\ \dfrac{8}{5} & \dfrac{12}{5} \end{pmatrix}$$

Therefore, $K_{11} = K_{22} = 12/5$, which is the same result obtained using Eq. (16.7) where $r^2 = 1/2$ and $\omega = \pi/4$. Notice also that $\|x_1[k]\|_2 = \|x_2[k]\|_2$ as expected, since the two states are tied together through a shift register.

Computer Study. Using a stand-alone Lyapunov computational routine, the value $\mathbf{K}$ matrix can be numerically computed using the function "k(A,b)" defined in the s-file "k.s." The function implements the computing procedure developed in Eqs. (16.20) through (16.26). The following sequence of Siglab commands demonstrates the use of the function to determine the $\mathbf{K}$ matrix for the given system.

```
>include "chap16\k.s"          # Get function definition.
>A = {[0,1],[-0.5,1]};A        # Define A matrix.
      0       1
   -0.5       1
>b = {0,1};b                   # Define b matrix.
      0
      1
>K = k(A,b)                    # Compute K matrix.
>K                             # Echo results.
```

```
2.4    1.6
1.6    2.4
```

The ℓ_2 norm squared bound can be experimentally verified by driving the system with an unit ℓ_2 norm impulse and measuring the ℓ_2 norm squared state bound. Since the second-order system under study has a direct-II architecture, computing the state bound on $x_1[k]$ is the same as $x_2[k]$. The behavior of state $x_1[k]$ can be isolated by assigning the output $y[k] = x_1[k]$. This can be accomplished in general using the s-file "**state.s**," which makes the output assignment $y[k] = x_1[k]$ by setting $\mathbf{c}^T = (1,0)$ and $d_0 = 0$. This mapping can also be implemented manually. In either case, this action will mathematically tie state $x_1[k]$ directly to the output.

If the input $v[k]$ is chosen to have a unit ℓ_2 bound, then the resulting output should achieve a value $\|x_1[k]\|_2^2 = \|y[k]\|_2^2 = 12/5 - 2.4$ in an ℓ_2 norm squared sense. For $v[k] = \delta[k]$, it follows that $\|v[k]\|_2 = 1$. The following sequence of Siglab commands was used to compute the ℓ_2 norm squared state bound on $x_1[k]$ or K_{11} for a unit norm input. The mapping $y[k] = x_1[k]$ is manually implemented in this case. The results are presented in Fig. 16.2.

```
# define state 4-tuple S = (A,b,c,d) and k:
# A = {[0,1],[-0.5,1]}
# b = {0 = b₁, 1 = b₂}
# cᵀ = {1 = c₁,0 = c₂}
# d = {0 = d₀}
# k = {1 = k}
>h1 = {[0,1],[-0.5,1],[0,1],[1,0],[[0,0],[1,0]]}
>v = impulse(51)      # Define test input signal.
>y = filt(h1,v)       # Linear convolution.
>ograph(y,intg(y^2))  # Impulse response and running l₂ norm squared.
```

Observe that the ℓ_2 norm squared state bound increases rapidly and then tops out at a steady-state value of as predicted. ❖

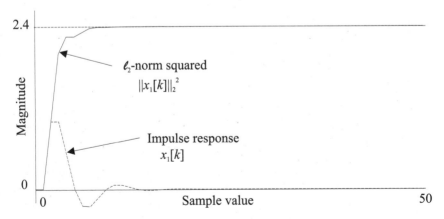

Figure 16.2 The ℓ_2 norm squared and impulse response of a simple LTI filter.

The ℓ_2 norm squared state bound information is contained in the Lyapunov matrix $\mathbf{K}$. Specifically, $\mathbf{K}$ is defined by the two-tuple $(\mathbf{A},\mathbf{b})$ from $\mathbf{S} = (\mathbf{A},\mathbf{b},\mathbf{c},d_0)$. However, this model ignores the possible presence of an input or front-end scale factor k shown in Fig. 16.1. The presence of k results in a general transfer function given by

$$H(z) = k\frac{N(z)}{D(z)} = k\frac{(z_N + \alpha_{N-1}z^{N-1} + ... + \alpha_1 z^{-1} + \alpha_0)}{D(z)} \qquad (16.28)$$

It should be appreciated that the shape of the frequency response of the filter defined by $H(z)$ will not change due to the presence of k. In fact, the magnitude frequency response of $H(z)$ will have the same relative passband to stopband gain ratios regardless of the value of k chosen (i.e., relative gains remain invariant to k). The only purpose that k has is to adjust the absolute passband gain to a value of unity, or 0 dB. Therefore, there is often a high degree of freedom associated with the choice of k. It should be remembered, however, that the value of k described in Eq. (16.28), is not explicitly incorporated into the state four-tuple $\mathbf{S} = (\mathbf{A},\mathbf{b},\mathbf{c},d_0)$ as specified in Eq. (16.1). As a result, the actual ℓ_2 norm for the system described in Eq. (16.28) will differ from the computed ℓ_2 norm by a factor k^2. That is, the actual ℓ_2-squared dynamic range growth measured at the ith state is $k^2 K_{ii}$.

The state equation for the system modeled by Eq. (16.29) would have the effect of embedding k into the state four-tuple $\mathbf{S} = (\mathbf{A},\mathbf{b},\mathbf{c},d_0)$ and therefore requires no correction.

$$H(z) = \frac{N(z)}{D(z)} = \frac{(kz^N + k\alpha_{N-1}z^{N-1} + ... + k\alpha_1 z^{-1} + k\alpha_0)}{D(z)}$$

$$= \frac{(\eta_N z^N + \eta_{N-1}z^{N-1} + ... + \eta_1 z^{-1} + \eta_0)}{D(z)} \qquad (16.29)$$

Determining the overflow dynamic range requirements of the state or shift registers is not the only concern. Other registers exist within an IIR that also need to be protected. In particular, coefficient registers must have a sufficiently large integer field to encode the largest filter coefficient magnitude, say $coef_{max}$, without introducing overflow error. Since the coefficients are constants, their absolute values also equal their ℓ_2 norms. That is,

$$|coef_{max}| = \|coef_{max}\|_2 \qquad (16.30)$$

Lastly, the output $y[k]$ is assumed to be stored in a register. For an $[N{:}F]$ format, this technically requires that

$$\|y[k]\|_2 \le 2^I \qquad (16.31)$$

where $F = N - I - 1$. However, in practice this condition is often not invoked because:

- The output register may be defined by a host external to the DSP processor and made to be sufficiently large to hold $y[k]$, or
- The DSP processor may possess an extended precision accumulator.

The following example explores these cases.

Example 16.4: ℓ_2 State Bounds on High-Order Systems

Problem Statement. The eighth-order Chebyshev II filter was studied in Example 14.2 and has a transfer function given by

$$H(z) = 0.0888 \frac{(z^8 + 4.43z^7 + 10.76z^6 + 17.46z^5 + 20.48z^4 + 17.46z^3 + 10.76z^2 + 4.43z + 1)}{(z^8 + 1.10z^7 + 1.97z^6 + 1.55z^5 + 1.22z^4 + 0.61z^3 + 0.24z^2 + 0.061z + 0.008)}$$

where $k = 0.0888$. The state determined system $S = (\mathbf{A}, \mathbf{b})$ has been detailed in Example 14.2 for a number of selected architectures. What are the ℓ_2 bounds on the states for each architecture? Discuss and test the performance of each architecture.

Analysis and Conclusions. The computing formula for the ℓ_2 state bounds is given in Eq. (16.18) for the input scale factor assumed to be $k = 0.0888$ and $k = 1$. The output is assumed to be saved in an extended precision output register.

Computer Study. An eighth-order Chebyshev II filter was created and saved as file "scale.iir." After generation, it was implemented using cascade, direct-II, parallel, cascade-normal, parallel-normal, and ladder architectures using the options found in the architecture menu. The architecture and report files are saved under the following file names in the directory "chap16."

- `scalec.(ARC,RPT)`   —Cascade
- `scalep.(ARC,RPT)`   —Parallel
- `scaled.(ARC,RPT)`   —Direct-II
- `scalel.(ARC,RPT)`   —Ladder
- `scalen.(ARC,RPT)`   —Normal (cascade)
- `scalenp.(ARC,RPT)`   —Normal (parallel)

The calculations are based only on state two-tuple $S = (\mathbf{A}, \mathbf{b})$ and the Lyapunov computing formula built into the software.

Cascade—K Matrix Trace.

$K_{11}=1.300947889519952$; $K_{22}=1.300947889519952$
$K_{33}=2.512067965072942$; $K_{44}=2.512067965072942$
$K_{55}=2.778069019155927$; $K_{66}=2.778069019155927$
$K_{77}=7.455869902166960$; $K_{88}=7.455869902166949$

max $K_{ii}=7.455869902166949$

Direct-II—K Matrix Trace.

$K_{11}=10.95889897287866$; $K_{22}=10.95889897287866$
$K_{33}=10.95889897287866$; $K_{44}=10.95889897287866$
$K_{55}=10.95889897287866$; $K_{66}=10.95889897287866$
$K_{77}=10.95889897287866$; $K_{88}=10.95889897287866$

max $K_{ii}=10.95889897287866$

Parallel—K Matrix Trace.

$K_{11}=1.300947889519952$; $K_{22}=1.300947889519952$
$K_{33}=1.163977916723221$; $K_{44}=1.163977916723221$
$K_{55}=1.305784491175111$; $K_{66}=1.305784491175110$
$K_{77}=2.711861898037760$; $K_{88}=2.711861898037761$

max $K_{ii}=2.711861898037761$

Cascade Normal—K Matrix Trace.

$K_{11}=0.9999999999999918$; $K_{22}=0.9999999999999926$
$K_{33}=1.327270385794750$; $K_{44}=2.512241143038886$
$K_{44}=1.827256173177680$ $K_{66}=2.358884073470594$
$K_{66}=4.066636791512576$; $K_{88}=1.482972740293089$

max $K_{ii}=4.066636791512576$

(Notice: The report file shows that the K for each individual second-order subfilter is unity.)

Parallel Normal—K Matrix Trace.

$K_{11}=0.9999999999999918$; $K_{22}=0.9999999999999926$
$K_{33}=0.9999999999999840$; $K_{44}=0.9999999999999987$
$K_{55}=0.9999999999999916$; $K_{66}=1.000000000000001$
$K_{77}=0.9999999999999968$; $K_{88}=1.000000000000003$

max $K_{ii}=1.000000000000001$

Ladder—K Matrix Trace.

$K_{11}=1.300947889519952$; $K_{22}=1.300947889519952$
$K_{33}=2.512067965072942$; $K_{44}=2.512067965072942$
$K_{55}=2.778069019155927$; $K_{66}=2.778069019155927$
$K_{77}=7.455869902166960$; $K_{88}=7.455869902166949$

max $K_{ii}=7.455869902166960$

Comparing these filter architectures, the number of integer bits that must be allocated to eliminate run-time register overflow can be computed and is reported in Table 16.1.

It can be seen that the parallel normal architecture has a definite advantage in dynamic range management while the others have one to two fewer bits of precision. If the input is scaled by $k = 0.0888$ rather than $k = 1$, the data shown in Table 16.2 would apply. The coefficient register requirements also need to be quantified. The maximum coefficient values per architecture are reported in Table 16.3.

Table 16.1 ℓ_2 **Dynamic Range Requirements for** $k = 1$

Architecture	$K_{\max}$	$\sqrt{K_{\max}}$	$I = \left\lceil \log_2(\sqrt{K_{\max}}) \right\rceil$
Cascade	7.455869902166949	2.73	2
Direct-II	10.95889897287866	3.31	2
Parallel	2.711861898037761	1.64	1
Cascade normal	4.066636791512576	2.01	~1
Parallel normal	1.000000000000001	1.00	~0
Ladder	7.455809902166960	2.73	2

Table 16.2 ℓ_2 **Dynamic Range Requirements for** $k = 0.0888$

Architecture	$k^2 \times K_{\max}$	$\sqrt{K_{\max}}$	$I = \left\lceil \log_2(\sqrt{\text{maximum}}) \right\rceil$
Cascade	0.05879	0.2424	−3
Direct-II	0.08641	0.2934	−2
Parallel	0.02138	0.1462	−3
Cascade normal	0.03206	0.1790	−3
Parallel normal	0.00788	0.0888	−4
Ladder	0.05879	0.2425	−3

Table 16.3 **Coefficient Dynamic Range Requirements**

Architecture	Mag. maximum DSP coefficient	$I = \left\lceil \log_2(\sqrt{\text{maximum}}) \right\rceil$
Cascade	1.86762	1
Direct-II	19.1764	5
Parallel	22.9383	5
Cascade normal	3.2730	2
Parallel normal	22.3419	5
Ladder	5.1099	3

It can be seen that, in each case, the minimum integer field size is established by the maximum coefficient. In fact, based on what is given, choosing k = 0.0888 would be counterproductive unless maintaining absolute passband and stopband gains is necessary. In particular, one to five bits of integer precision will be needed to cover the coefficients. This overspecifies the scaled state dynamic range requirements, which need only –3 to –4 integer bits of precision. However, as shall be seen, the ℓ_2 norm is not strong enough to serve as a reliable method of eliminating run-time register overflow. A more precise method of assigning the binary point will be developed later in this section.

To further test the eighth-order Chebyshev-II filter, audio soundboard trials can be conducted. The tests include creating a sinusoid located in the passband of the filter, and then filtering it with the given IIR filter architectures for various fixed-point formats. Assume a 16-bit solution with 15 or 12 fractional bits of accuracy. The 15 fractional bit solution will technically cause some of the filter coefficients to be improperly encoded, as suggested in Table 16.3. The effect of this coefficient roundoff error can be seen in the filter's magnitude frequency response. However, this does not necessarily mean that the differences can be discerned by comparing their audio responses. Listeners with an acute sense of hearing will be able to differentiate between the responses while others will detect no differences. To conduct such an experiment, consider the following sequence of Siglab commands.

```
>hc=rarc("chap16\scalec.arc",1)      # cascade
>hd=rarc("chap16\scaled.arc",1)      # direct-II
>hl=rarc("chap16\scalel.arc",1)      # ladder
>hn=rarc("chap16\scalen.arc",1)      # normal - cascade
>hc[11,0]                            # scale factor k
0.0888329
>max(mag(hc))                        # maximum coefficient
1.86762
>max(mag(hd))
19.1765
>max(mag(hl))
4.96942
>max(mag(hn))
3.27306
>x = mkcos(1,1/16,0,10000)          # test signal
>wavout(0.8*2^15*filt(hc,x))        # floating point response
>wavout(0.8*2^15*filt(hd,x))
>wavout(0.8*2^15*filt(hl,x))
>wavout(0.8*2^15*filt(hn,x))
>wavout(0.8*2^15*fxpfilt(hc,x,16,15)) # Fixed-point, 1 overflow
>wavout(0.8*2^15*fxpfilt(hd,x,16,15)) # Fixed-point, 20 overflows
>wavout(0.8*2^15*fxpfilt(hl,x,16,15)) # Fixed-point, 1 overflow
>wavout(0.8*2^15*fxpfilt(hn,x,16,15)) # Fixed-point, 1 overflow
>wavout(0.8*2^15*fxpfilt(hc,x,16,12)) # Fixed-point, 0 overflows
>wavout(0.8*2^15*fxpfilt(hd,x,16,12)) # Fixed-point, 0 overflows
>wavout(0.8*2^15*fxpfilt(hl,x,16,12)) # Fixed-point, 0 overflows
>wavout(0.8*2^15*fxpfilt(hn,x,16,12)) # Fixed-point, 0 overflows
```

It can be noted that all audio files essentially sound the same.   ❖

16.6 Change of Architecture

The study of Lyapunov matrices can be seen to be specific to the state two tuple $\mathbf{S} = (\mathbf{A,b})$. Suppose a system having a state vector $\mathbf{x}[k]$ with a given architecture is to be rewired into a new architecture having a state vector $\hat{\mathbf{x}}[k]$. The transformed architecture can be mathematically represented by

$$\hat{\mathbf{x}}[k] = \mathbf{T}\mathbf{x}[k] \tag{16.32}$$

Applying this transform to the state determined system defined by Eq. (16.1) {assuming the input is $v[k]$}, results in

$$\hat{\mathbf{x}}[k+1] = \mathbf{T}\mathbf{x}[k+1] = \mathbf{T}\mathbf{A}\mathbf{x}[k] + \mathbf{T}\mathbf{b}v[k]$$

$$= \mathbf{T}\mathbf{A}\mathbf{T}^{-1}\hat{\mathbf{x}}[k] + \mathbf{T}\mathbf{b}v[k] = \hat{\mathbf{A}}\hat{\mathbf{x}}[k] + \hat{\mathbf{b}}v[k] \quad \text{(state equation)} \tag{16.33}$$

yielding

$$\hat{\mathbf{A}} = \mathbf{T}\mathbf{A}\mathbf{T}^{-1}; \quad \mathbf{A} = \mathbf{T}^{-1}\hat{\mathbf{A}}\mathbf{T}$$

$$\hat{\mathbf{b}} = \mathbf{T}\mathbf{b}; \quad \mathbf{b} = \mathbf{T}^{-1}\hat{\mathbf{b}} \tag{16.34}$$

Substituting this result into Eq. (16.18) results in:

$$\mathbf{K} = \mathbf{T}^{-1}\hat{\mathbf{A}}\mathbf{T}\mathbf{K}\mathbf{T}^{T}\hat{\mathbf{A}}^{T}(\mathbf{T}^{T})^{-1} + \mathbf{T}^{-1}\hat{\mathbf{b}}\hat{\mathbf{b}}^{T}(\mathbf{T}^{T})^{-1} \tag{16.35}$$

Therefore, the Lyapunov matrix for the arbitrary system $\mathbf{S} = (\hat{\mathbf{A}}, \hat{\mathbf{b}})$ satisfies

$$\hat{K} = \mathbf{T}\mathbf{K}\mathbf{T}^{T} \tag{16.36}$$

Consider a special case where $\mathbf{T}$ is chosen to be a diagonal matrix having the form

$$\mathbf{T} = s\begin{pmatrix} \dfrac{1}{\sqrt{K}_{11}} & 0 & \cdots & 0 \\ 0 & \dfrac{1}{\sqrt{K}_{22}} & \cdots & 0 \\ \vdots & \vdots & \ddots & \vdots \\ 0 & 0 & \cdots & \dfrac{1}{\sqrt{K}_{NN}} \end{pmatrix} \tag{16.37}$$

where K_{ii} is the ith on-diagonal element of $\mathbf{K}$ and s is the desired maximal ℓ_2 norm value of the states. For a given nonsingular linear transform $\mathbf{T}$, the resulting subsystem $\hat{S} = (\hat{\mathbf{A}}, \hat{\mathbf{b}})$ possesses a Lyapunov matrix

$$\hat{\mathbf{K}} = \mathbf{TKT} = s^2\mathbf{I} \tag{16.38}$$

Therefore, $\left\|\hat{x}_i[k]\right\|_2^2 = s^2$ for all i. That is, the state dynamic range requirements are now the same at each shift-register location, which results in a balanced system. The system $\hat{S} = (\hat{\mathbf{A}}, \hat{\mathbf{b}})$ is therefore called a *scaled system*. The question remains of how to assign a value to s. For $s = 1$, the ℓ_2 norm of all the states is normalized to unity. This is a popular choice, since normalization simplifies the analysis of a filter at a system level. Another logical choice is to assign s to be the largest ℓ_2 norm for a specific design (i.e., state, coefficient, and possibly output register requirements). Consider again Example 16.4, which established the largest ℓ_2 state and coefficient norm for each architecture and reduce the information to tables. Using the largest norm as an ℓ_2 scaling criteria, the corresponding values of s are summarized in Table 16.4. The choice of s guarantees that the data format will not be overspecified in the context of the state or coefficient range.

Example 16.5: Linear Transformation

Problem Statement: Consider again Example 16.3, which was characterized by a direct-II architecture having state model $S = (\mathbf{A}, \mathbf{b})$ given by

$$\mathbf{A} = \begin{pmatrix} 0 & 1 \\ -0.5 & 1 \end{pmatrix}, \quad \mathbf{b} = \begin{pmatrix} 0 \\ 1 \end{pmatrix}$$

Table 16.4 Possible Assignment of Scale Factor s

Architecture	s
Cascade	1.8676
Direct-II	19.1764
Parallel	22.9383
Cascade normal	3.2730
Parallel normal	22.3419
Ladder	5.1099

with $K_{11} = K_{22} = 12/5$, and the maximum coefficient being unity. Scale the system to one having unit register norms.

Analysis and Conclusions: From Eq. (16.36), let

$$
\mathbf{T} = \begin{pmatrix} \sqrt{\dfrac{5}{12}} & 0 \\ 0 & \sqrt{\dfrac{5}{12}} \end{pmatrix}
$$

Then

$$
\hat{\mathbf{A}} = \mathbf{T}\mathbf{A}\mathbf{T}^{-1} = \sqrt{\frac{5}{12}}\sqrt{\frac{12}{5}}\mathbf{I}\mathbf{A}\mathbf{I} = \mathbf{A}
$$

$$
\hat{\mathbf{b}} = \mathbf{T}\mathbf{b} = \sqrt{\frac{5}{12}}\mathbf{I}\mathbf{b} = \sqrt{\frac{5}{12}}\mathbf{b}
$$

It can be seen that the production rule for the Lyapunov matrix $\hat{\mathbf{K}}$ is defined in terms of the two matrices $\hat{\mathbf{R}}$ and $\hat{\mathbf{X}}$ from Eq. (16.26). Since the matrix $\hat{\mathbf{R}}$ is defined in terms of the eigenvalues of $\hat{\mathbf{A}}$, and $\mathbf{A} = \hat{\mathbf{A}}$, it follows that $\hat{\mathbf{R}} = \mathbf{R}$. However, the matrix $\mathbf{X}$ is a different case and is given in terms of

$$
\hat{\mathbf{x}}[1] = \hat{\mathbf{b}} = \sqrt{\frac{5}{12}}\mathbf{b} = \sqrt{\frac{5}{12}}\begin{pmatrix} 0 \\ 1 \end{pmatrix}
$$

$$
\hat{\mathbf{x}}[2] = \hat{\mathbf{A}}\hat{\mathbf{x}}[1] + p_1\hat{\mathbf{b}} = \sqrt{\frac{5}{12}}(\mathbf{A}\mathbf{x}[1] + p_1\mathbf{b}) = \sqrt{\frac{5}{12}}\begin{pmatrix} 1 \\ 0 \end{pmatrix}
$$

Therefore

$$
\hat{\mathbf{X}} = \sqrt{\frac{5}{12}}\mathbf{I}
$$

which results in a Lyapunov matrix for the scaled system given by

$$
\hat{\mathbf{K}} = \hat{\mathbf{X}}\hat{\mathbf{R}}\hat{\mathbf{X}}^T = \sqrt{\frac{5}{12}}\mathbf{I}\hat{\mathbf{R}}\mathbf{I}\sqrt{\frac{5}{12}} = \begin{pmatrix} 1 & \dfrac{2}{3} \\ \dfrac{2}{3} & 1 \end{pmatrix}
$$

The on-diagonal terms are now seen to be unity, as expected.   ❖

The ℓ_2 state bound is often referred to as a *conservative bound*. This can create difficulties when designing a high-order IIR, since the integer field (i.e., I bits) can be under-assigned, leading to possible dynamic range overflow during run-time. The basic problem with the ℓ_2 norm is actually found in the ℓ_2 assumption itself. From Eq. (16.4), it follows if $p = q$, then $p = q = 2$. The input signal, therefore, must also be ℓ_2 bounded (assume $\|v\|_2 \leq 1$). This is generally an unrealistic requirement or assumption as demonstrated in Example 16.2. Here it was noted that most interesting input test signals are unbounded in an ℓ_2 sense. The value of the ℓ_2 state norm is, in fact, one of providing a quick comparison of architectural choices. The ℓ_2 norm, nevertheless, needs to be replaced with a more precise scheme.

16.7 ℓ_f State Bound

The ℓ_f norm technique is related to the $\mathcal{L}_\infty$ norm in the frequency domain, which is often used to scale continuous-time signals and systems. Specifically, the $\mathcal{L}_p$ norm of a continuous-time function $x(t)$ is given by

$$X = \|x(t)\|_p = \left(\int_{-\infty}^{\infty} |x(s)|^p ds \right)^{\frac{1}{p}} \tag{16.39}$$

The use of this norm can be motivated as follows. Assume that the discrete-time impulse response measured from a signal injection point to the output is $h[k]$. Let the DTFT of $h[k]$ be $H(e^{j\phi})$ where $-\pi \leq \phi \leq \pi$. The $\mathcal{L}_p$ norm of $H(e^{j\phi})$ is therefore given by

$$\left\| H(e^{j\phi}) \right\|_p = \left| \int_{\phi=-\pi}^{\pi} \left| H(e^{j\phi}) \right|^p d\phi \right|^{\frac{1}{p}} \tag{16.40}$$

For $p = 2$, Eq. (16.40) is part of Parseval's energy theorem. For $p = \infty$,

$$\left\| H(e^{j\phi}) \right\|_\infty = \max_{0 \leq \phi \leq \pi} \left| H(e^{j\phi}) \right| \tag{16.41}$$

Based on Holder inequality, consider an input signal $v[k]$ with an $\mathcal{L}_1$ frequency domain norm $|V(e^{j\phi})|_1 = V$. Upon passing the signal $v[k]$ through a linear filter having an $\mathcal{L}_\infty$ norm of $|H(e^{j\phi})|_\infty = H_{max}$, the output will be bounded by VH_{max}. Suppose that the input to the filter is a sinusoid having a unit amplitude envelope represented by

$$V(e^{j\phi}) = \frac{1}{2}\delta(\phi - \phi_0) \pm \frac{1}{2}\delta(\phi + \phi_0) \tag{16.42}$$

It then follows that

$$\left|V(e^{j\phi})\right|_1 = \frac{1}{2}\int_{-\pi}^{\pi}\left|\delta(\phi - \phi_0) + \delta(\phi + \phi_0)\right|d\phi = \frac{1}{2}(1 + 1) = 1 \qquad (16.43)$$

Based on this assumption, the output is a sinusoid whose amplitude is bounded by

$$\left|Y(e^{j\phi})\right| \leq \left\|V(e^{j\phi})\right\|_1\left\|H(e^{j\phi})\right\|_\infty = \left\|H(e^{j\phi})\right\|_\infty = H_{max} \qquad (16.44)$$

This implies that for a unit envelope sinusoidal input, the maximal steady-state output is bounded by H_{max}. This provides motivation for the use of an ℓ_f norm.

The ℓ_f state bound assumes that the worst case input is a sinusoid of the form $v[k] = \cos(\omega_0 kT_s + \phi)$. This implicitly assumes that the linear system will be evaluated at steady state, that any overflows that may have occurred took place during the transient period, and that their effects have since decayed to zero. If the discrete-time LTI system, defined by $S = (\mathbf{A}, \mathbf{b})$, has a transfer function vector satisfying

$$\mathbf{H}(z) = (z\mathbf{I} - \mathbf{A})^{-1}\mathbf{b} \qquad (16.45)$$

where $\mathbf{H}(z)$ is an N vector, then the frequency response measured at each state location is given by the N vector

$$\mathbf{X}(e^{j\omega T_s}) = \mathbf{H}(e^{j\omega T_s})V(e^{j\omega T_s}) \qquad (16.46)$$

Based on the assumption that the worst-case input is a mono-frequency sinusoid $v[k] = \cos(\omega_0 kT_s + \phi)$, and knowledge that a linear system:

- cannot create new frequencies, but
- can alter the amplitude of the input, and
- can alter the phase of the input

Then it follows that the ith state will be bounded by:

$$\left|x_i[k]\right| \leq H_i^{max}\left|v[k]\right| \qquad (16.47)$$

where

$$H_i^{max} = \max\left|H_i(e^{j\omega T_s})\right| \qquad (16.48)$$

for all $\omega \in [-\pi, \pi]$, and $H_i(e^{j\omega T})$ is the ith element of $\mathbf{H}(e^{j\omega T_s})$. Based on such an analysis, it can be seen that the worst case steady-state output measured at the ith shift register location is given by

$$x_i^{\max}[k] = H_i^{\max} \cos(\omega_i k T_s + \phi_i) \tag{16.49}$$

where ω_i is called the maximizing frequency. Equivalently,

$$\|x_i[k]\|_f \le H_i^{\max} \tag{16.50}$$

which is called the ℓ_f state bound of state $x_i[k]$. Therefore if

$$H^{\max} = \max\left\{ H_i^{\max}\Big|_{\forall i} \right\} \tag{16.51}$$

then the integer field should be I bits wide where

$$I = \left\lceil \log_2(H^{\max}) \right\rceil \tag{16.52}$$

to prevent register overflow in an ℓ_f sense.

Example 16.6: ℓ_f State Bound

Problem Statement. A first-order system is assumed to have an impulse response

$$h[k] = a^k u[k]$$

where $|a| < 1$. What is $\|h[k]\|_f$?

Analysis and Conclusions. The z-transform of $h[k]$ is given by

$$H(z) = \frac{z}{z-a}$$

The maximal ℓ_f state bound is given by

$$H^{\max} = \max\left(|H_i(e^{j\omega T_s})|_{\forall \omega}\right) = \max\left(\left|\frac{e^{j\omega T_s}}{e^{j\omega T_s} - a}\right|_{\forall \omega}\right)$$

which, for $0 \le a < 1$, is maximized when $e^{j\omega T_s} = 1$ (i.e., $\omega = 0$), which results in $H^{\max} = 1/(1-a)$.

If $-1 < a < 0$, the maximum occurs when $e^{j\omega T_s} = -1$ (i.e., $\omega = \pi/T_s$) which results in $H^{\max} = 1/(1+a)$. Combining the two cases, it follows that

$$H^{\max} = \frac{1}{1 - |a|}$$

Notice that as $a \to 1$, the ℓ_f norm rapidly approaches infinity.   ❖

The question, of course, is how to effectively compute the ℓ_f state bound for a high-order system having an arbitrary and asymptotically stable architecture $S = (\mathbf{A}, \mathbf{b})$. The answer to the question is motivated in Example 16.3. The output $y[k]$ can be assigned to be $x_i[k]$, namely $y[k] = x_i[k]$, by defining $\mathbf{c}^T = \mathbf{e}_i$, where $\mathbf{e}_i$ is called the ith natural basis vector satisfying $\mathbf{e}_i = (0,\dots,1(i\text{th location}),\dots,0)$, and $d_0 = 0$. This mapping can be performed using software (e.g. program "state.s" in Example 16.3) or manually. Once $x_i[k]$ is produced and allowed to decay to essentially zero (guaranteed by the asymptotically stable assumption), then the DFT of $x_i[k]$ is taken, producing $X_i(n) = \text{DFT}(x_i[k])$. The maximal magnitude value of $|X_i(n)|$ is then determined and set equal to $H_i^{\max}$. Care must be taken to ensure that $x_i[k]$ has properly converged to a neighborhood of zero and that the DFT contains a sufficiently large number of sample values to produce a high-resolution spectrum.

Example 16.7: ℓ_f Norm

Problem Statement. The eighth-order Chebyshev II filter studied in Example 16.4 is implemented using the cascade, direct-II, ladder/lattice, and normal architectures. Determine the ℓ_f state bounds for each architecture.

Analysis and Conclusions. To study the ℓ_f state bounds, the state four-tuple $S = (\mathbf{A}, \mathbf{b}, \mathbf{c}, d_0)$ is modified to $S' = (\mathbf{A}', \mathbf{b}', \mathbf{c}', d_0')$ where:

Original	Modified
k	0.08883
$\mathbf{A}$	$\mathbf{A}' = \mathbf{A}$
$\mathbf{b}$	$\mathbf{b}' = \mathbf{b}$
$\mathbf{c}$	$\mathbf{c}' = [0,0,\dots,0,1,0,\dots,0] = \mathbf{e}_i^T$
d_0	$d_0' = 0$

The decision on whether to maintain k at $k = 0.08883$ or unity is a design choice. Maintaining $k = 0.08883$ will result in a unity passband gain. If $k = 1$, the passband gain increases to 11.257 (i.e., 1/0.08883). Once k is chosen, the DFT of the impulse response is measured at the output of each shift register (state location), and the maximum value is assigned to $H^{\max}$. A useful observation is to note that only one register has to be tested for a direct-II architecture, since all registers are chained together. For cascade architectures consisting of a collection second-order direct-II sections, however, only one of the two registers per second-order section need be tested.

Technically, the ℓ_f bound of the output $y[k]$ also needs to be computed. If an absolute unity passband is to be retained, then the local passband gain for an elliptic or Chebyshev-II IIR filter can reach a value of $\|y[k]\|_f = 1 + \varepsilon$, $\varepsilon > 0$. If so, then possible an additional integer bit may be required to keep the output register from overflowing during run time. This problem can be managed by replacing the input scale factor k by $k/(1 + \varepsilon)$. This would bound the output by unity in an ℓ_f-sense and leave the passband gain essentially unaffected. Finally, to complete the analysis, the magnitude bound on the coefficients will need to be determined to protect the coefficients from overflowing their registers. For $k = 0.08883$, the ℓ_f data and coefficient register bounds arc reported in Table 16.5.

The production of the state norms are shown in Fig. 16.3. The data indicates that the integer bit field (i.e., I) is established by the filter coefficient dynamic range requirements. Setting aside the coefficient problem, it is noted that the next largest dynamic range requirements is established by $\|y[k]\|_f = 1$, which corresponds to $k = 0.08883$. The value of k, however, can be scaled by s to produce a new scaling factor $k' = sk$. The cascade filter, for example, requires that the integer field be set to one bit to cover the maximum coefficient (i.e., 1.86). Therefore, s would be chosen so that

$$\max_s \{ s \|x_i[k]\|_f, s \|y[k]_f\| \} \le 2^1 = 2$$

In this case, $s = 2$ and $k' = 0.17766$. As a result, the coefficients, states, and output would fully use the resulting $[N{:}N - 2]$ register format.

Computer Study. The following sequence of Siglab commands is used to compute the ℓ_f state bounds.

```
>include "chap16\state.s"
>include "chap16\lfnorm.s"
>h=rarc("chap16\scalec.arc",1); lfnorm(h)
0.46103
```

Table 16.5 Summary of the Derived ℓ_f Data and Coefficient Bounds

Architecture	H_{max}	@ state	$\|y[k]\|_f$	Max. coefficient	Max. $\le 2^I$	I
Cascade	0.461	[7,8]	1	1.86	1.86	1
Direct-II	0.992	[1,...,8]	1	19.17	19.17	5
Ladder/lattice	0.992	[1,2]	1	5.11	5.11	3
Normal (cascade)	0.297	[7,8]	1	3.2	3.2	2

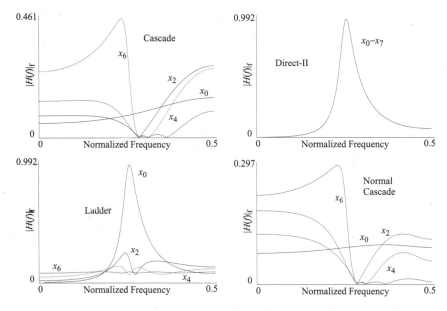

Figure 16.3 ℓ_f state norms for cascade, direct-II, lattice/ladder, and normal architecture filters.

```
>h=rarc("chap16\scaled.arc",1); lfnorm(h)
0.922066
>h=rarc("chap16\scalel.arc",1); lfnorm(h)
0.922066
>h=rarc("chap16\scalen.arc",1); lfnorm(h)
0.296677                                                    ❖
```

16.8 ℓ_1 State Bound

In the study of ℓ_2 state bounds, it was noted that its practical use was compromised by the fact that many important inputs to a discrete-time LTI system do not have a finite ℓ_2 bound. The ℓ_f state bound was somewhat more practical but suffers from the fact that it was based on a steady-state sinusoidal assumption. This leaves the system vulnerable to potential run-time register overflows during the transient period. Ideally, what is desired is to produce a state bound on a sample-by-sample basis for all possible inputs meeting the physically meaningful requirement that $\|v\|_\infty = \max(|v[k]|) \leq 1$. This means that it is required that the input be bounded by unity on a sample-by-sample basis. From Holder's inequality, it follows that if $p = \infty$, then $q = 1$. Thus the state impulse response $x_i[k]$ must be ℓ_1 bounded. Whereas the ℓ_1 norm of some simple state impulse responses are computable in closed form, this is generally not the case.

Example 16.8: ℓ_1 State Bound

Problem Statement. Consider a first-order system having an impulse response

$$h[k] = a^k u[k]$$

where $|a| < 1$. What is $\|h[k]\|_1$?

Analysis and Conclusions. Direct computation of the ℓ_1 norm of $h[k]$ is given by

$$\|h[k]\|_1 = \sum_{k=0}^{\infty} |a^k| = \frac{1}{1 - |a|}$$

Notice that as $a \to 1$, the ℓ_1 norm (squared) rapidly approaches infinity. Refer to Example 16.2, where the ℓ_2 norm of $h[k]$ is given by

$$\|h[k]\|_2 = \sqrt{\sum_{k=0}^{\infty} |a^k|^2} = \sqrt{\frac{1}{1 - a^2}}$$

From Example 16.6, the ℓ_f norm of $h[k]$ was shown to be given by

$$\|h[k]\|_f = \frac{1}{1 - |a|}$$

For $a = 0.9$, for example, the norms are seen to be ordered as

$$\|h[k]\|_2 = \sqrt{5.263} = 2.29 < \|h[k]\|_f = 10 = \|h[k]\|_1 \qquad \diamondsuit$$

The impulse response of an Nth-order LTI IIR filter, measured at the ith shift register location, is given by the inverse z-transform of

$$x_i[k] = \sum_{j=1}^{N} \alpha_{i:j} k^{[g(j)-1]} \lambda_j^k \overset{z}{\leftrightarrow} (e_i(z\mathbf{I} - \mathbf{A})^{-1} \mathbf{b}) \qquad (16.53)$$

where λ_j is the jth eigenvalue of $\mathbf{A}$, $g[j] \geq 1$ corresponds to the multiplicity of λ_j, and $\mathbf{e}_i = (0, 0, \ldots, 1(i\text{th location}), \ldots, 0)$. In general, λ_j is complex and may appear with a multiplicity of up to N. If the system is stable, then the system's impulse response will decay to zero. That is, for a stable filter, $x_i[k] \to 0$ as $k \to \infty$. It follows that the ℓ_1 bound on $x_i[k]$ is given by

$$\|x_i\|_1 = \sum_{k=0}^{\infty} \left| \sum_{j=1}^{N} \alpha_{j} k^{[g(j:i)-1]} \lambda_{j:i}^k \right| < M \qquad (16.54)$$

which, unfortunately, does not usually have a simple closed-form solution. Unlike steady-state analysis, where only the dominant eigenvalue trajectory is analyzed, the entire infinite sum needs to be included in the production of the ℓ_1 norm. A careful approximation, however, can be made. An estimate of the ℓ_1 state bound can be obtained by using a finite sum to approximate the infinite sum,

$$\|\hat{x}_i\|_1 = \sum_{k=0}^{N_1} |x_i[k]| \le \sum_{k=0}^{\infty} |x_i[k]| = \|x_i\|_1 \tag{16.55}$$

The problem, of course, is to estimate the indexing parameter N_1 so that the difference between the computed (estimated) and actual ℓ_1 norm is less than some acceptable value $\varepsilon > 0$. A selection rule for N_1 can be based on the easily computed ℓ_2 norm of a filter state. For a stable IIR filter, there is an integer N_2 such that

$$\sum_{k=0}^{N_2} \|x_i[k]\|_2^2 - K_{ii} < \varepsilon \tag{16.56}$$

where K_{ii} is the ℓ_2 norm squared of $x_i[k]$ (i.e., $\|x_i\|_2^2$) given by Eq. (16.18), and ε is an acceptable error threshold. The value of N_1 in Eq. (16.55) can be approximated by assuming that, near convergence, the solution is defined only by a stable system's dominant eigenvalue (pole), assumed to be λ_{max} (i.e,. closest to the periphery of the unit circle), where

$$\lambda_{max} = \left\{ \lambda_i = a_i e^{j\omega_i} \middle| |\lambda_i| < 1.0, i \in \{1, 2, ..., n\} \right\} \tag{16.57}$$

Under this assumption, the system's impulse response due to this eigenvalue is convergent and has the form

$$x_i[k] \propto k^{s-1} a_i^k \cos(k\omega_1 + \phi) \tag{16.58}$$

where s is the multiplicity of the eigenvalue. If the maximal eigenvalue has low multiplicity, as is generally the case,

$$\sum_{k=0}^{\infty} |x_i[k]|^p \le (1 - |a_{max}|^p)^{-s} \tag{16.59}$$

where equality holds for $s = 1$. From this it follows that the relationship between N_1 and N_2 can be expressed as

$$r = \frac{N_1}{N_2} = \left(\frac{1 - |a_{max}|^2}{1 - |a_{max}|}\right)^s \qquad (16.60)$$

Notice that, for $|\lambda_{max}| \to 1.0$, the correction term r becomes unbounded. For a value $|\lambda_{max}|$ less than unity, a stopping rule for Eq. (16.60) can be defined that will result in an approximate ℓ_1 state bound.

The advantage of estimating an ℓ_1 bound can be appreciated when the worst case is considered. The worst-case ℓ_∞ input, having a unit bound, measured at state location i, is given by

$$v_i[k] = \text{sign}(x_i[k - m]) \qquad (16.61)$$

Convoluting the input defined by Eq. (16.61) with the filter's impulse response measured at state location i, namely $x_i[k]$, results in

$$\sum_{m=0}^{\infty} x_i[k - m]v[k] = \sum_{m=0}^{\infty} x_i[k - m]\text{sign}(x_i[k - m])$$

$$= z_i[k] \le \|x_i\|_1 \|v\|_\infty = \|x_i\|_1 \qquad (16.62)$$

which is the actual ℓ_1 state norm that is assumed to be sufficiently well approximated by Eq. (16.55). Once the output has a computed estimate of the ℓ_1 bound, the filter's binary point can be defined.

Example 16.9: ℓ_1 Bound Estimate

Problem Statement. A sixth-order elliptic lowpass filter having the following specifications was designed and analyzed.

- −0.5 dB passband over $f \in [0,5]$ kHz
- −60 dB stopband over $f \in [7, 22.05]$ kHz
- Sample frequency = 44.1 kHz

The magnitude of the dominant eigenvalue of the system is $\lambda_{max} = 0.964$. What is the approximate ℓ_1 worst-case state gain?

Analysis and Results. For the sake of providing a common analysis forum, the input scale factor $k = 0.0031998736$ is set to unity. The magnitude frequency response of the filter, measured at states two, four, and six, is shown in Fig. 16.4. From this, the ℓ_f norm can be estimated to be 52 dB (~8.6 bits) from states $x_5[k]$ and $x_6[k]$. Therefore, $\|x_6[k]\|_f = 398.1$, which requires an integer field size of $I = 9$ bits to protect against register overflow in an ℓ_f sense.

The largest theoretical determined value of K_{ii} was for states x_5 and x_6, and in particular $\|x_6\|_2^2 = \|x_6\|_2^2 > 16 \times 10^3$ whose square root value is $\|x_6\|_2 \approx 256$

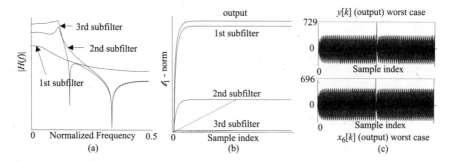

Figure 16.4 (a) Magnitude frequency response measured at the output, states two, four, and six, corresponding to second-order subsystems one, two, and three; (b) running computation of the ℓ_1 state norm out to steady state for each subsystem; and (c) worst-case state and output responses to an input with a unit ℓ_1 bound.

which, therefore, requires $I = 8$ integer bits to protect against register overflow. Using a value of ε in Eq. (16.56) of 1 (≈ 0.006 percent of K_{66}), the value of N_2 was experimentally found to be $N_2 = 112$. Using Eq. (16.60) and the posted value of λ_{max} defines $N_1 \sim 1.96\, N_2 \sim 220$, which corresponds to $\|x_6\|_1 \sim 696$. The experimentally derived ℓ_1 norms at states two, four, and six are shown in Fig. 16.4 and support the previous calculations. Based on the estimated ℓ_1 norm, the output would need to be bounded by $696 < 1024 = 2^{10}$, or ten integer bits would be needed at the register level to ensure that no run-time overflow occurs. The testing of this hypothesis with the known worst-case input [see Eq. (16.62)] results in the data shown in Fig. 16.4, where the maximal output achieves the computed ℓ_1 bound on the order of 696. Finally, notice that $\|x\|_2 \leq \|x\|_f \leq \|x\|_1$. This will be explored further in the next example.

Observe that the ℓ_1 norm measured at the output is shown in Fig. 16.4 and is seen to be about $730 < 2^{10}$, which exceeds $\|x_6\|_1$. If the output of the state registers are sent to an extended-precision register, then this need not be a problem. If the output accumulator is a single-precision device, then output register overflow is a potential problem. The worst-case output in this case is seen, however, to be bounded by ten bits, just as $\|x_6\|_1$. Therefore, ten bits of integer precision is sufficient to protect against state, coefficient, and output register overflow.

The analysis can be scaled to optimize the use of an ascribed data format by identifying the worst-case ℓ_1 requirement; here it is the output $\|y[k]\|_1 = 730 < 2^{10}$. By using an input scale factor $s = 1024/730 = 1.4$, the output and state variables will have an ℓ_1 norm that will come close to exceeding a ten-bit integer range but will not exceed it.

This analysis can be extended to the original filter definition, which included the input scale factor $k = 0.0031998736$. If this scale factor is reinstated, the ℓ_1 norm will be decreased by k. The maximum value of k and the ℓ_1 norms of states and output is $k \times \|y[k]\|_1 = 2.336 < 2^2$. The maximum coefficient in the filter architecture is measured to be $coef_{max} = 2.6023 < 2^2$. Therefore, two bits will be needed to protect the system from overflow. Rescaling the system, to maximize the use of the two integer bit dynamic range, would require

that k be adjusted to $k' = (4/2.6023)k = 0.004918$. This, however, would change
the unscaled passband gain from 1.0 to $(4/2.6023) = 1.54$.

In comparison, if one assumes that the worst-case input for a $k = 1$ lowpass
filter is a unit step, the maximum step response value measured at state $x_6[k]$
is approximately 248. This corresponds to an integer bit assignment of eight
bits. However, this is not the worst-case bound (which is ten bits), so setting
the word format at $[N:N - 9]$ could result in run-time overflows.

Computer Study. The data used in the example is produced using the following sequence of Siglab commands contained in the script "llnorms.s."

```
>include "chap16\llnorms.s" # Everything else comes from the script.
showinc on
h=rarc("chap16\llnorm.arc",1)
include "chap16\state.s"
h[9,0]=1 # kill k
yf=mag(pfft(filt(h,impulse(1024))))
xf=zeros(3,512) # storage for 3 second-order sections
for (i=0:2)
xf[i,]=mag(pfft(filt(state(h,2*i),impulse(1024))))
end
ograph(20*log(xf[0,]),20*log(xf[1,]),20*log(xf[2,]))
max(xf)
351.755
# compute l1 norm
x1=zeros(3,1024) # storage for l1 norm
for (i=0:2)
x1[i,]=filt(state(h,2*i),impulse(1024))
end
y=filt(h,impulse(1024))
x1m=mag(x1)
ograph(intg(mag(y)),intg(x1m[0]),intg(x1m[1]),intg(x1m[2]))
ywc=sign(y); x6wc = sign(x1[2,])
graph(filt(h,rev(ywc)&rev(ywc)),filt(state(h,5),(rev(x6wc)&rev(x6wc))))
```

For the stopping rule set to analyze the largest ℓ_2 state (i.e., x_6) at $\varepsilon = 1$ (≈ 0.006
percent of K_{66}), the value of N_2 is found to be 112, and at $N_1 = 220$, the ℓ_1 norm
at state six is found to be 696. These results may be computed using the following sequence of Siglab commands.

```
>include "chap16\state.s"
>h=rarc("chap16\llnorm.arc",1)        # Read architecture file.
>h[9,0]=1                             # Set k=1.
>K66=16537
>y=filt(state(h,5),impulse(1024))     # Output y[k]=x6[k].
>yi=intg(y^2)
>for (i=0:1023)
>>if (mag(yi[i]-K66)>1); N=i; end
>>end
>N
```

```
112
>y11=intg(mag(y))
>y11[int(1.9*N)]                         # After Eq. (16.58).
696.417
```

To test the previously defined eighth-order Chebyshev-II filter using a sound-board, again create a sinusoid located in the passband of the filter. Assume an eight-bit solution with five fractional bits of accuracy. This will provide two bits of integer precision. According to the data in Table 16.6 and the computation of the maximum coefficients (shown below), the direct-II and ladder filters will suffer some coefficient distortion. The effect of using the [8:5] format is studied in the context of their audio quality using the sequence of Siglab commands.

```
>hc=rarc("chap16\scalec.arc",1)     # cascade
>hd=rarc("chap16\scaled.arc",1)     # direct-II
>hl=rarc("chap16\scalel.arc",1)     # ladder
>hn=rarc("chap16\scalen.arc",1)     # normal - cascade
>hc[11,0]                            # scale factor k
0.0888329
>max(mag(hc))                        # Maximum coefficients.
1.86762
>max(mag(hd))
19.1765
>max(mag(hl))
4.96942
>max(mag(hn))
3.27306
>x=mkcos(1,1/16,0,10000)            # Generate test signal.
>wavout(0.8*2^7*filt(hc,x))         # Floating point simulation.
>wavout(0.8*2^7*fxpfilt(hc,x,8,5))  # Fixed-point simulations.
>wavout(0.8*2^7*fxpfilt(hd,x,8,5))
>wavout(0.8*2^7*fxpfilt(hl,x,8,5))
>wavout(0.8*2^7*fxpfilt(hn,x,8,5))
```

It can be noted that the fixed-point cascade plays the best. The direct-II is hopelessly distorted. If the format is selected to eliminate coefficient and register overflow, then the formats would have been selected to be:

Cascade = [8:5]

Direct-II = [8:2]

Ladder = [8:4]

Normal = [8:5]

Upon repeating the experiment for the new formats, the cascade design is heard to play the best. It should be noted that the direct-II has no output due to the fact that some critical coefficients are rounded to a value of zero, which nulls the output. ❖

The ℓ_1 norm is more pessimistic than the ℓ_2 or ℓ_f bound in that $\|x\|_2 \leq \|x\|_f \leq \|x\|_1$. The ℓ_1 norm is the most conservative of the set and therefore the safest. However, at times, all of these norms are considered to be too conservative. In Example 16.9, for the $k = 1$ (unscaled) case and a 16-bit implementation, a [16:5] format was required to ensure that no run-time overflow would occur. A [16:7] format would save two bits of precision and may perform without ever encountering register overflow. However, it may. If conservative bounds are to be relaxed, then it is critically important that saturating arithmetic be used to ensure that if overflow does occur, it will be kept under control.

Example 16.10: Reconciliation

Problem Statement. Complete an ℓ_1 state norm analysis of the eighth-order Chebyshev II filter studied in Example 16.4. Show that $\|x\|_2\|x\|_f\|x\|_1$.

Analysis and Conclusions. The ℓ_2 state norm was computed in terms of the on-diagonal terms of the matrix. Furthermore it is assumed that the input scale factor k equaled unity. For the ℓ_1 and ℓ_f state norm calculations, $k = 0.08883$. Therefore, the ℓ_2 state norm needs to be corrected by $K_{ii} = k^2 K_{ii}$ if it is to be compared to the ℓ_1 and ℓ_f state norm calculation. The ℓ_1 state norms are estimated using a worst-case input signal and are summarized in Table 16.6 and in Fig. 16.5, from which it is apparent that $\|x\|_2 \leq \|x\|_f \leq \|x\|_1$.

Computer Study. The following sequence of Siglab commands was used to estimate the ℓ_1 state norms.

```
>include "chap16\state.s"
>include "chap16\l1_8.S"
>h=rarc("chap16\scalec.arc",1)
>l1norm(h)
0.929623
>h=rarc("chap16\scaled.arc",1)
>l1norm(h)
```

Table 16.6   Results of State Norm Analysis for Cascade, Direct-II, Ladder/Lattice, and Normal Cascade Architecture Filters

Architecture	$\max(k\sqrt{K_{ii}})$	ℓ_f state norm	ℓ_1 state norm
Cascade	0.243	0.461	0.929
Direct	0.294	0.992	1.158
Ladder/lattice	0.243	0.992	1.158
Normal cascade	0.179	0.297	0.606

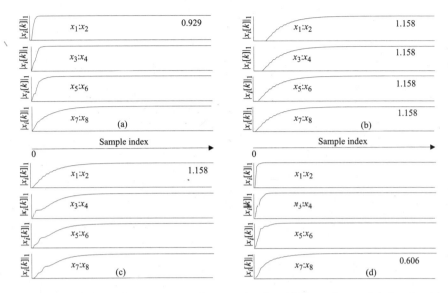

Figure 16.5 Estimated ℓ_1 state norms by second-order subfilters for an eighth-order (a) cascade, (b) direct-II, (c) ladder/lattice, and (d) normal cascade filter. The largest second-order section gain is shown numerically.

```
1.15858
>h=rarc("chap16\scale1.arc",1)
>l1norm(h)
1.15858
>h=rarc("chap16\scalen.arc",1)
>l1norm(h)
0.606557                                                    ❖
```

16.9 Summary

In Chap. 16, the binary point assignment problem was mathematically explored. It was found that the easily computed ℓ_2 state norm has limited value and, while it is generally used to compare architectures, it is not used to scale them. The ℓ_f state norm was also found to be easy to compute and has some practical value if it is assumed that the input to a system is dominated by a sinusoid. The most effective state norm is the ℓ_1 norm, since it requires only limited knowledge of the input signal space. Mathematically, the ℓ_1 state norm is difficult to compute and therefore often needs to be approximated using a computer simulation. Regardless, the methods presented can be used to scale a filter so that it can be protected from run-time overflow in the chosen norm sense.

16.10 Self-Study Problems

1. Butterworth Bilinear z-Transform Direct-II Design Design a lowpass Butterworth IIR filter that meets or exceeds the following specifications:

Desired passband attenuation = 3.0 dB
Desired stopband attenuation = 20.0 dB
Sampling frequency = 10,000.0 Hz
Passband edge = 2000.0 Hz
Stopband edge = 2500.0 Hz

The filter is implemented using a direct-II architecture and saved as "prob15_1.(IIR,ARC)." Compare the floating-point magnitude frequency response of a direct-II implementation using the [16:15], [16:12], [16:8], and [16:4] coefficient formats. Identify the cause of any major differences (i.e., coefficient roundoff error).

Compute the maximum ℓ_2, ℓ_f, and ℓ_1, state, output, and coefficient norms. Repeat the reconciliation study reported in Example 16.9. Recommend a register format $[N{:}F]$ for your system if the input is assumed to be a uniformly distributed random noise over the interval $[-1,1]$.

2. Cascaded IIR Filter Repeat problem 1 for a cascade architecture (saved as "prob15_2.(IIR,ARC)").

3. Parallel IIR Filter Repeat problem 1 for a parallel architecture (saved as "prob15_3.(IIR,ARC)").

4. Normal IIR Filter Repeat problem 1 for a normal architecture (saved as "prob15_4.(IIR,ARC)").

17

Noise Gain

17.1 Introduction

In Chap. 16, methods were presented that can protect a system having the form

$$\frac{Y(z)}{X(z)} = H(z) = \frac{N(z)}{D(z)} = k\frac{\displaystyle\sum_{i=0}^{M} b_i z^i}{1 + \displaystyle\sum_{i=1}^{N} a_i z^i} \qquad (17.1)$$

from encountering run-time register overflow. In particular, the mathematical techniques developed in Chap. 16 were oriented toward solving the binary point assignment problem. It was noted that the solution to this problem was a function of filter type and architecture. By properly selecting the location of the binary point, overflow problems can be either eliminated or mitigated. It should be remembered, however, that meeting posted precision constraint is often also part of the filter design process. In Chap. 17 the question of quantifying the expected precision of a filter will be addressed.

The difference between the ideal and computed outputs using digital computational elements having finite precision may be computed as $e[k] = y_{ideal}[k] - y_{finite}[k]$. Impreciseness of this type was previously studied in terms of quantization errors. Quantization errors, such as those found at the output of an analog-to-digital converter (ADC), were produced on a sample-by-sample basis. The quantization error was reported to be a uniformly distributed random process with zero mean and variance $\sigma^2 = Q^2/12$, where Q is the quantization step size. An error injected at a point internal to a filter shown in Fig. 17.1, how-

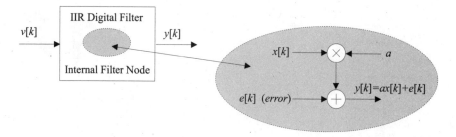

Figure 17.1 Physical error model where $e[k]$ is a white uniformly distributed random process.

ever, could potentially circulate within that structure—possibly forever. In the process of recalculating, the effect of the error could, in fact, render the filter output useless. This is but one possible manifestation of a class of errors called *finite wordlength effects*.

17.2 Coefficient Roundoff Errors

Coefficient roundoff errors are the result of converting real or complex coefficients into digital words of finite precision. The error is defined to be the difference between the ideal and that encoded coefficient c_i and is given by $e = c_{ideal} - c_{finite}$. Again, the roundoff error is assumed to be a uniform random process having a zero mean and variance $\sigma^2 = Q^2/12$, where Q is again the quantization step size. If the largest coefficient to be encode is bounded in absolute value by $|c_i| \leq V \leq 2^I$, then the data format must be at least $[N{:}F]$ having $F = N - I - 1$ and I fractional and integer bits respectively. In this case, $Q = 2^{-F}$.

The effect of coefficient roundoff on the performance of an IIR filter can be evaluated analytically or experimentally. Coefficient sensitivity was introduced in Chap. 15 based on migration of pole locations due to coefficient rounding. A more general analysis can be provided by using classical sensitivity analysis schemes that assume that coefficient roundoff errors are sufficiently small that their effect can be modeled using partial derivatives. The effect of rounding the coefficients of a transfer function $H(z,c)$, where z is the z-transform operator and c is a filter coefficient, is modeled by

$$H_{finite}(z, c) = H_{ideal}(z, c) + \sum_{i=1}^{P} \frac{\partial H(z, c)}{\partial c} \Delta c \qquad (17.2)$$

Sensitivity techniques are, unfortunately, mathematically cumbersome and at times very complicated. In addition, it is only valid for small roundoff errors ($\Delta c \sim 0$). The need for this analysis method is somewhat limited in the era of the 16- or 24-bit fixed-point and floating-point DSP microprocessors (floating-point arithmetic is generally assumed to be essentially free of roundoff error). In such cases, a well designed filter will contain a sufficiently large number of fractional bits so as to reduce the effects of coefficient roundoff to an insignificant level. This was not the case when dealing with first-generation eight-bit DSP microprocessors, where coefficient errors were often devastating. The most serious problem generally encountered in such cases is the rounding of the input scale factor k [see Eq. (17.1)]. Since the scaling coefficient k often has a small value (e.g., $k < 2^{-10}$), it can be easily quantized to zero, which will null the input and render a filter useless.

Example 17.1: FIR Filter Coefficient Roundoff

Problem Statement. A 51st-order equiripple FIR filter, having a passband defined over $f \in [15,35]$ kHz, stopbands ranging over $f \in [0,10]$ kHz and $f \in [40,50]$ kHz, with $f_s = 100$kHz, is to be analyzed. What is the coefficient roundoff error?

Analysis and Conclusions. The FIR filter has a transfer function given by

$$H(z, c) = \sum_{i=0}^{L-1} c_i z^{-i}$$

From Eq. (17.2), assuming the coefficient roundoff errors are small (i.e., $\Delta c_i = \Delta \sim 0$), it follows that

$$H_{finite}(z, c) = H_{ideal}(z, c) + \sum_{i=0}^{L-1} \Delta c_i z^{-i}$$

The incremental change in $H(z,c)$, due to coefficient rounding, becomes

$$\Delta H(z, c) = H_{finite}(z, c) - H_{ideal}(z, c) = \sum_{i=0}^{L-1} \Delta c_i z^{-i}$$

which implies that $\Delta H(z,c)$ is defined by a small-amplitude moving average filter. The validity of this error model is based on the smallness of the incremental error Δc_i.

Computer Study. The validity of the error model can be tested using computer simulation. The following sequence of Siglab commands can be used to compare the predicted value of $\Delta H(z,c)$ to the simulated change in the transfer function if a $[N:8]$ format is used ($N > 8$). That is, $|\Delta c_i| = \leq 2^{-8}$.

```
> h=rf("chap17\symmetri.imp")    # 51st order FIR.
> wordlen(16,8,1)                 # Set fixed-point format [16:8].
> h8=fxpt(h)                      # Quantize to [16:8].
```

```
> deltah=h-h8                    # Compute error.
> ograph(2^(-8)*ones(len(h)),mag(deltah)) # Absolute error and bound.
```

Refer to Fig. 17.2 and notice that the individual coefficient errors are bound, as the model predicts. It appears that the coefficient roundoff error is *not* uniformly distributed as suggested by the model. Examining the filter's unquantized coefficients, it is apparent that 38 of the 51 coefficients are near zero; in fact, the largest of the near-zero coefficients is 1.35×10^{-5}.   ❖

The previous example demonstrates that the effects of coefficient roundoff errors can be simulated. Considering the limitations associated with analytic roundoff error estimation methods, simulation offers a great many advantages. First, simulation can work in both the discrete-time and frequency domains. Second, simulation can produce both a statistical roundoff error model as well as quantify the maximum error. Third, simulation can be used to investigate the effect of one, some, or all coefficients being rounded. Fourth, simulation is highly synergistic with a state variable modeling paradigm that provides a mechanism that permits roundoff errors to be studied at each shift register location and not just at the output.

Example 17.2: Simulation

Problem Statement. The 51st-order equiripple FIR filter studied in Example 17.1 is again used as the filter model. Using simulation, investigate coefficient roundoff errors.

Analysis and Conclusions. The coefficient roundoff study of the FIR filter is to be conducted using simulation.

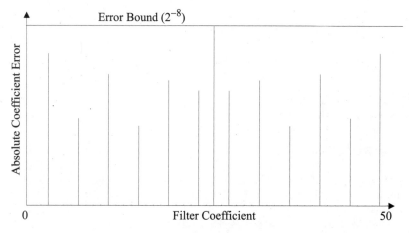

Figure 17.2 Graph of the magnitude frequency response of the error model and actual error.

Computer Study. The following sequence of Siglab commands computes floating point, and fixed-point [16,12], [16,8], and [16,4] filter magnitude frequency responses.

```
> h=rf("chap17\symmetri.imp")   # 51st order FIR impulse response.
> wordlen(16,12,0);h12=fxpt(h)   # Truncate to [16:12].
> wordlen(16,8,0);h8=fxpt(h)     # Truncate to [16:8].
> wordlen(16,4,0);h4=fxpt(h)     # Truncate to [16:4].
> hf=mag(pfft(h&zeros(512)))     # Compute magnitude frequency responses.
> hf12=mag(pfft(h12&zeros(512)))
> hf8=mag(pfft(h8&zeros(512)))
> hf4=mag(pfft(h4&zeros(512)))
> ograph(h,h12,h8,h4)            # Display impulse responses.
> ograph(hf,hf12,hf8,hf4)        # Display magnitude frequency responses
```

Refer to Fig. 17.3 and observe that coefficient rounding does not have a measurable effect on system performance until only four fractional bits of precision remain. However, it should be appreciated that most seriously effected filter coefficients are those with intrinsically small values. These coefficients provide the fine detail to the filter's magnitude frequency response. Destroying them often leaves a fixed-point high-order FIR filter with a magnitude frequency response very similar to that of a low-order floating point FIR filter. The magnitude filter response is seen to remain reasonably precise if the coefficients maintain at least eight bits of fractional precision.   ❖

An IIR filter's sensitivity to coefficient rounding is strongly influenced by the choice of architecture and the design specifications (i.e., transition bandwidth and so on). It is known that the dynamic range requirements of two architectures can vary widely. It may be recalled that a normal architecture was claimed to have a low roundoff error sensitivity (see Sec. 14.7). The veracity of this claim can be substantiated by studying how the location of the realized poles are effected by coefficient rounding.

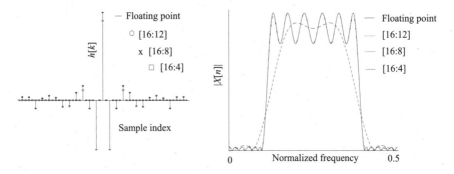

Figure 17.3 Discrete-time and frequency response of a typical FIR filter for ideal (floating-point) and fixed-point [16:12], [16:8], and [16:4] data formats. The magnitude frequency response of the 12 and 8 fractional bit simulated FIR filter overlaying the ideal.

In Sec. 15.6, the coefficient roundoff sensitivity of a second-order section was studied. The question of managing roundoff errors in a second-order section will be addressed again from a design perspective. Recall that a stable second-order IIR filter has a transfer function of the form

$$H(z) = \frac{b_0 + b_1 z^{-1} + b_2 z^{-2}}{1 + a_1 z^{-1} + a_2 z^{-2}} \tag{17.3}$$

Assuming that the poles are complex, they may be located at

$$p = -\frac{a_1}{2} \pm j\frac{\sqrt{a_1^2 - 4a_2}}{2} = \sigma + j\omega \tag{17.4}$$

and reside anywhere inside the unit circle on the z-plane. As a point of comparison, the direct-II implementation of the system defined by Eq. (17.3) can be modeled by the partial state determined system $\mathbf{S} = (\mathbf{A}_D)$, where

$$\mathbf{A}_D = \begin{bmatrix} 0 & 1 \\ -a_2 & -a_1 \end{bmatrix} \tag{17.5}$$

Upon rounding the coefficients in Eq. (17.5), using an $[N{:}F]$ format, a new state matrix results and is given by

$$\hat{\mathbf{A}}_D = \begin{bmatrix} 0 & 1 \\ \langle -a_2 \rangle & \langle -a_1 \rangle_R \end{bmatrix} \tag{17.6}$$

where $<.>_R$ denotes rounding to R fractional bits of accuracy. The poles of the rounded system are given by the roots of

$$\det(z\mathbf{I} - \hat{\mathbf{A}}_D) = z^2 + \langle a_1 \rangle_R z + \langle a_2 \rangle_R \tag{17.7}$$

which results in a permissible pole distribution as shown in Fig. 17.4. The rounded or finite precision poles reside on the inscribed circles. It can be seen that the rounding error is nonuniformly distributed within the unit circle.

A normal filter would implement the feedback structure modeled by the partial state determined system $\mathbf{S} = (\mathbf{A}_N)$, where

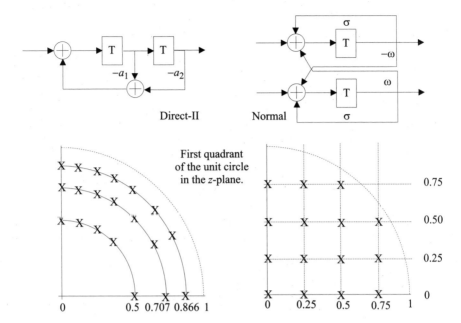

Figure 17.4 Admissible quantization locations of direct-II and normal IIR filters having two fractional bits of precision.

$$\mathbf{A}_N = \begin{bmatrix} \sigma & \omega \\ -\omega & \sigma \end{bmatrix} \tag{17.8}$$

Upon rounding the coefficients using an $[N{:}F]$ format, a new state matrix results and is given by

$$\hat{\mathbf{A}}_N = \begin{bmatrix} \langle\sigma\rangle_R & \langle\omega\rangle_R \\ \langle-\omega\rangle_R & \langle\sigma\rangle_R \end{bmatrix} \tag{17.9}$$

The poles of the rounded system are given by the roots of

$$\det(z\mathbf{I} - \hat{\mathbf{A}}_N) = (z - \langle\sigma\rangle_R)^2 + (\langle\omega\rangle_R)^2 \tag{17.10}$$

which results in pole distributions as shown in Fig. 17.4. In the case of the normal architecture, the rounded values reside on the intersection of the vertical and horizontal grid (i.e., a Cartesian grid). It can be seen that the rounded pole locations are uniformly distributed within the unit circle. In contrast, the rounded poles for the direct-II architecture are nonuniformly distributed within the unit circle, and it tends to place more possible pole locations nearer to the edge of the unit circle than to the center. The normal filter is considered to have superior roundoff error sensitivity in a general sense. However, in certain regions of the z-plane, such as near the center of the unit circle, the direct-II architecture can be superior.

Ideally, a good filter would consist of a set of coefficients that are essentially all of the same order of magnitude. This will allow all the coefficients to be identically encoded with the same number of fractional bits with minimum waste of precision. The antithesis is a case where there is a mix of large and small critical coefficients. Here, a large integer field may be required to code the coefficient set leaving to few fractional bits to accurately represent the smaller coefficients. Again, the effects of coefficient rounding can be readily evaluated using simulation.

Example 17.3: IIR Filter Coefficient Roundoff Error

Problem Statement. A fourth-order Chebyshev II filter having the specifications shown in the Computer Study is implemented as a direct-II and parallel normal architecture. Experimentally examine the effects of coefficient roundoff error.

Analysis and Conclusions. The experimental study is reported below and in Fig. 17.5. Observe that the largest coefficient, in absolute value, is 3.75 for the direct-II filter and 70.48 for the parallel normal design. Notice also that the direct-II filter requires a format $[N:N-3]$ if the coefficients are to be encoded without overflow. The parallel filter requires a format $[N:N-8]$ if the coefficients are to be encoded without overflow. For the sake of commonality, assume that $N = 16$, and each filter is given eight fractional bits (i.e., [16:8]).

Referring to Fig. 17.5, it can be seen that the direct-II implementation suffers more from coefficient roundoff error than the parallel normal form when both designs are required to use an eight-bit fractional word. However, the di-

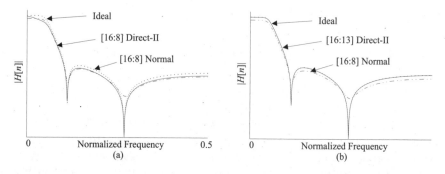

Figure 17.5 (a) Comparison of a direct-II and parallel normal fourth-order IIR filter using a [16:8] coefficient format and, (b) comparison of a direct-II [16:13] and parallel normal [16:8] coefficient format.

rect-II can operate with five more additional fractional bits of precision, since it requires five fewer integer bits than a parallel normal filter to protect the against coefficient register overflow. When the five bits are given back to the direct-II implementation, the direct-II design produces a magnitude frequency response that is superior to that of the parallel normal design. While other factors, such as arithmetic register overflow requirements, will typically effect the final choice of the data format (i.e., $[N:F]$), it can be seen that the study of coefficient roundoff errors is more subtle than simply comparing architectures using a common data format.

Computer Study. Chebyshev II filter specifications.

File	: cheby2d.IIR
Source Module	: Infinite Impulse Response Filter
Filter Type	: Chebyshev Type II
Order	: 4
Frequency Response	: Lowpass
Desired Passband Attenuation	: 1.000000000000000 dB
Desired Stopband Attenuation	: 40.00000000000000 dB
Sampling Frequency	: 1000.000000000000 Hz
Passband Edge	: 35.00000000000000 Hz
Stopband Edge	: 100.0000000000000 Hz

TRANSFER FUNCTION
Scale Factor = 0.01000000000000000

Numerator Coefficients	Denominator Coefficients
b 4: 1.000000000000000	a 4: 1.000000000000000
b 3: 0.000000000000000	a 3: 3.977269408560995
b 2: 69.29403176688086	a 2: 7.916265377314252
b 1: 0.000000000000000	a 1: 9.311816620479314
b 0: 600.2078548136868	a 0: 6.002078548136869

The analysis of the filter proceeds using the following sequence of Siglab commands

```
> hd=rarc("chap17\cheby2d.arc",1)   # Direct-II filter.
> hp=rarc("chap17\cheby2p.arc",1)   # Normal filter.
> max(mag(hd))
3.75233                             # max coef modulus for direct-II <2²
> max(mag(hp))
48.1553                             # max coef modulus <2⁶
> wordlen(16,8,1);hd8=fxpt(hd)      # Set [N:F] and quantize.
> yd8=filt(hd8,impulse(1024));yfd8=20*log(mag(pfft(yd8)))
> wordlen(16,8,1);hp8=fxpt(hp)      # Set [N:F] and quantize.
> yp8=filt(hp8,impulse(1024));yfp8=20*log(mag(pfft(yp8)))
> y=filt(hd,impulse(1024));yf=20*log(mag(pfft(y)))
> ograph(yf,yfd8,yfp8)              # Plot ideal and fixed-point spectra.
```

Simulating the Direct-II with three additional fractional bits results in the following (see Fig. 17.5).

```
> wordlen(16,13,1);hd13=fxpt(hd)    # Set [N:F] and quantize
> yd13=filt(hd13,impulse(1024));yfd13=20*log(mag(pfft(yd13)))
> ograph(yf,yfd13,yfp8)
```

❖

17.3 Arithmetic Roundoff Errors

The effects of arithmetic roundoff, in many cases, can be far more severe than those introduced by coefficient rounding. Consider, initially,
a simple direct implementation of an FIR filter having a transfer function given by

$$H(z) = \sum_{i=0}^{L-1} h_i z^{-i} \tag{17.11}$$

and an architecture shown in Fig. 17.6. The inhomogeneous response
shall be assumed to be given by the convolution sum

$$y[k] = \sum_{m=0}^{L-1} h_m x[k-m] \tag{17.12}$$

It can be seen that the effects of individual arithmetic roundoff errors,
denoted $e_i[k]$ in Fig. 17.6, are additive. Specifically, the arithmetic error at sample k is given by

$$e[k] = y[k] - \langle y[k] \rangle_R = \sum_{m=0}^{L-1} h_m x[k-m] - \sum_{m=0}^{L-1} \langle h_m x[k-m] \rangle_R = \sum_{m=0}^{L-1} e_m[k]$$

$$= \sum_{m=0}^{L-1} h_m x[k-m] - \langle h_m \times [k-m] \rangle_R = \sum_{m=0}^{L-1} e_m[k]$$

$$\tag{17.13}$$

Suppose that the tap weight multiplication is performed with finite
precision, and the least significant bit has a weight of $Q = 2^{-F}$. The er-

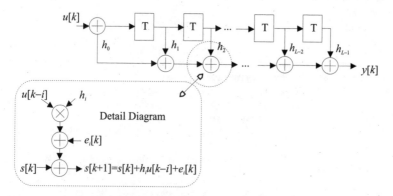

Figure 17.6 Basic FIR filter showing noise being injected into a multiplier
output due to rounding of results received by the output accumulator.

ror variance measured at the output of an individual multiplier, due to rounding to the F most significant fractional bits, is given by

$$\sigma_{h_i}^2 = \frac{Q^2}{12} \tag{17.14}$$

The tap-weight multiplication imprecision, in bits, is given by

$$B = \log_2(\sigma) = \log_2\left(\frac{Q}{\sqrt{12}}\right) = \log_2(Q) - 1.79248 \tag{17.15}$$

The errors introduced by each tap weight multiplication are assumed to be statistically independent from the others. Therefore, the total error variance due to L noise independent sources is given by

$$\sigma^2 = \sum_{i=0}^{L-1} \sigma_{h_i}^2 = L\frac{Q^2}{12} \tag{17.16}$$

Thus, the arithmetic error in an FIR filter grows linearly with filter order like the coefficient roundoff error model. The value of σ^2 is called the *noise power gain* or simply *noise gain*. Obviously, as the noise gain increases, the output precision decreases.

Example 17.4: FIR Filter Noise Gain

Problem Statement. A 16-bit, 64th-order ($L = 64$) FIR filter, whose coefficients are all bounded by unity, is found to have a worst-case gain of $G_{max} = 1.8$. What is the noise model?

Analysis and Conclusions. Based on the given data, an overflow-free FIR filter can be implemented using a [16:14] format. Therefore, $Q = 2^{-14}$ and

$$\sigma^2 = L\frac{Q^2}{12} = 2^6\frac{(2^{-14})^2}{12} = \frac{2^{-22}}{12}$$

The maximal output precision, in bits, therefore is given by

$$B = \log_2(\sigma) = \log_2\left(\frac{2^{-11}}{\sqrt{12}}\right) = \log_2(2^{-11}) - 1.79248 = -12.79248$$

which means that the output contains almost 13 bits of fractional precision. ❖

The study of IIR filters is substantially more difficult than that of FIR filters. The fundamental difference is that noise introduced by arithmetic rounding in an FIR filter can belong only to a feedforward path, whereas the noise injected into an IIR filter can belong to a feed-back path. As such, IIR filter rounding errors can recirculate and build up over time. The graph of the error variance versus the fractional wordlength can exhibit radical localized changes as suggested in Fig. 17.7. The three regions illustrated in this figure correspond to the following possible modes of operation:

1. *Nonlinear (rounding),* resulting in gross run-time errors introduced due to insufficient coefficient and data precision (i.e., F too small).

2. *Linear,* resulting in $\log(\sigma)$ precision versus the fractional wordlength F following the relationship that satisfies $\log(\sigma) \propto -\alpha F$.

3. *Nonlinear (saturation),* resulting in gross run-time saturation overflows occur due to insufficient integer word length $I = N - F - 1$.

Because of the complexity associated with the study of nonlinear systems, the error performance envelope of a system is often determined experimentally. Here, nonlinear arithmetic errors for a given IIR filter are computed (after Fig. 17.7) over a collection of architectural choices. A best design is selected in the context of minimizing the overall filter noise gain while simultaneously eliminating or minimizing run-time register overflow.

Example 17.5: IIR Filter Noise Gain

Problem Statement. The eighth-order IIR filter studied in Example 14.2 has the following specifications:

```
Source Module                   : Infinite Impulse Response Filter
Filter Type                     : Chebyshev Type II
Order                           : 8
Frequency Response              : Lowpass
Desired Passband Attenuation    : 0.1000000000000000 dB
```

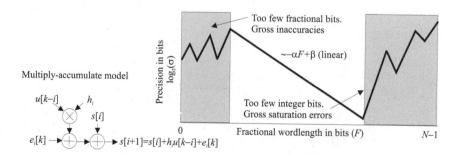

Figure 17.7 Error variance vs. fractional wordlength for a typical IIR filter.

```
Desired Stopband Attenuation   : 40.00000000000000 dB
Sampling Frequency             : 44100.00000000000 Hz
Passband Edge                  : 10000.00000000000 Hz
Stopband Edge                  : 12500.00000000000 Hz
```

Implement the IIR filter as a direct-II and cascade normal filter. Experimentally compare their noise gains for a [16:13] design.

Analysis and Conclusions. The problem specifies an experimental study. The computer simulation, shown below, demonstrates that the performance of a fixed-point filter is strongly influenced by the choice of architecture.

Computer Study. The magnitude frequency responses of the direct-II and cascade normal filter implementations for a [16:13] fixed-point format is simulated using the sequence of Siglab commands shown below.

```
> hd=rarc("chap17\scaled.arc",1)  # Direct-II.
> hn=rarc("chap17\scalen.arc",1)  # Cascade normal.
> i=impulse(2^12)                  # Create test impulse.
> y=filt(hd,i)                     # Create ideal, Direct-II and CN
> yd=fxpfilt(hd,i,16,13)           # impulse responses.
> yn=fxpfilt(hn,i,16,13)
> ograph(mag(pfft(y)),mag(pfft(yd)),mag(pfft(yn)))# Graph results.
> ograph(mag(pfft(y)),mag(pfft(yd)))
> ograph(mag(pfft(y)),mag(pfft(yn)))
```

The results are shown in Fig. 17.8. The effect of architecture is easily seen.

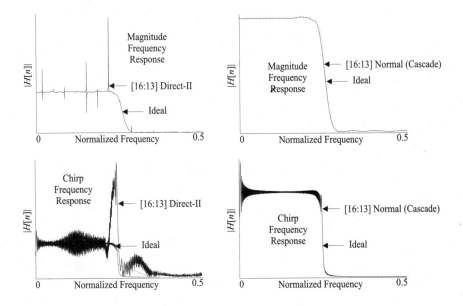

Figure 17.8 Magnitude frequency responses of direct-II and cascade normal [16:13] implementations and chirp magnitude frequency responses of direct-II and cascade normal [16:13] implementations.

Finite wordlength effects can also be studied in an audio format. The filters studied above can be investigated in terms of their ability to filter a chirp (linearly swept FM signal) over their passbands. The following program generates a chirp and plays it through an assumed 16-bit soundboard using a sampling rate of $f = 11,025$ Hz.

```
> include "chap17\chirp.s"          # Chirp generator s-file.
> x=chirp(2^14,0,0,0.25)            # Long baseband chirp.
> yd=fxpfilt(hd,i,16,13)            # takes time
> yn=fxpfilt(hn,i,16,13)            # takes time
> ograph(mag(pfft(y)),mag(pfft(yd)))  # Graph results.
> waveout(2^15*x)                   # Play chirp.
> waveout(2^15*yd)                  # Direct-II filtered chirp.
> waveout(2^15*yn)                  # Normal filtered chirp.
```

The sound produced by the direct-II implementation is terribly distorted due to rounding and overflow contamination. The normal filter, however, plays reasonably well. The chirp spectra shown in Fig. 17.8 support this conclusion.   ❖

17.4 State-Determined Errors

As stated, the study of errors in an IIR filter can be challenging. Consider first posing a simple question regarding what happens to noise injected into the ith shift register of an Nth-order IIR filter. Assume that the impulse response, as measured between the ith shift register and output, is $g_i[k]$. It should be fully appreciated that $g_i[k]$ includes the effects of all the feedback loops that pass through the ith shift register. Thus noise injected into the ith shift register will circulate within the filter before passing to the output. Therefore, the problem becomes one of measuring the effects of the injected noise over time.

Statistically, the error variance, measured at the output due to the locally injected noise, is given by

$$\sigma_i^2 = k_i \frac{Q^2}{12} \sum_{k=0}^{\infty} g_i[k] g_i^*[k] = k_i \frac{Q^2}{12} \|g_i\|_2^2 \tag{17.17}$$

where k_i reflects the strength of the injected noise relative to a value of $Q^2/12$. If each full-precision multiplication is rounded before it is summed, then k_i equals the number of multiplications sent to the adder. If a collection of full-precision multipliers are sent to a full or extended precision adder (as is the case for some DSP microprocessors) before rounding, then $k_i = 1$. The variance of the injected noise is scaled or amplified by the ℓ_2 norm squared of the path response between the ith register and output. Since there are N shift registers in an Nth-order canonic IIR filter architecture, the noise variance measured at the output due to each register being stimulated by statistically independent uniform noise sources is given by

$$\sigma^2 = \sum_{i=1}^{N} \sigma_i^2 = \frac{Q^2}{12} \sum_{i=1}^{N} k_i \|g_i\|_2^2 = \frac{Q^2}{12} G^2 \tag{17.18}$$

The value σ^2 is the roundoff error variance measured at the output and G^2 is again referred to as the filter's *noise power gain*. The rms value of G^2, namely G, is called the *noise gain*. In addition, the form assumed by Eq. (17.18) is reminiscent of the computing formulas development for use with ℓ_2 scaling policies in a state-determined IIR filter. This connection can be expanded by formally casting the noise gain problem in a formal state variable setting later in this section.

Example 17.6: Arithmetic Roundoff Errors

Problem Statement. A simple first-order IIR filter is given by

$$y[k] = ay[k-1] + v[k]$$

for $|a| < 1$. What are noise power and noise gain?

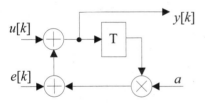

Analysis and Conclusions. The only noise input found in Fig. 17.9 is located at the output of the multiplier and is denoted $e[k] \in U[-Q/2, Q/2]$. The resulting noise model is given by

$$y[k] = ay[k-1] + \varepsilon[k]$$

Figure 17.9 Noise gain model of a first-order system.

The impulse response measured between the noise source and output is defined to be

$$g[k] = a^k u[k]$$

where $u[k]$ is the unit step function. Continuing,

$$\|g[k]\|_2^2 = \frac{1}{1 - |a|^2} = G^2$$

Substituting this result in Eq. (17.18), and recognizing that $k_1 = 1$ (i.e., one multiplier, noise source is attached to shift register), the noise power gain is given by G^2, and the output roundoff noise variance is defined by

$$\sigma^2 = \frac{Q^2}{12} \frac{1}{1 - |a|^2}$$

The noise measured at the output, in bits, is therefore

$$B = \log_2(\sigma) = -\log_2(\sqrt{1 - |a|^2}) + \log_2(Q) - 1.79248$$

The effective decrease in the output precision over that of a single isolated multiply, is seen to be $\log_2(\sqrt{1 - |a|^2})$ bits. This value can be observed to increase rapidly as $a \to 1.0$. ❖

Consider again an N dimensional state variable model of an Nth order discrete-time LTI system given by

$$\mathbf{x}[k + 1] = \mathbf{A}\mathbf{x}[k] + \mathbf{b}u[k] \tag{17.19}$$

and

$$y[k] = \mathbf{c}^T\mathbf{x}[k] + d_0 u[k] \tag{17.20}$$

Suppose that the only noise gain associated with the feedback structure is to be quantified. Then, Eq. (17.20) will need to be modified by setting $d_0 = 0$. Furthermore, to measure the noise gain, noise must be independently injected into each shift register. In fact, noise should be the only system driving force. This can be accomplished by replacing $\mathbf{b}u[k]$ in Eq. (17.19) with

$$\mathbf{b}u[k] \leftarrow \mathbf{e}[k] = \begin{bmatrix} e_1[k] \\ e_2[k] \\ \vdots \\ e_N[k] \end{bmatrix} \tag{17.21}$$

where $e_i[k]$ is sent to the ith shift register with strength $k_i Q^2/12$. The resulting LTI is seen to be a multiple-input, single-output (MISO) system. Combining these modifications, the following modified state determined system results:

$$\mathbf{x}[k + 1] = \mathbf{A}\mathbf{x}[k] + \mathbf{e}[k] \tag{17.22}$$

and

$$y[k] = \mathbf{c}^T\mathbf{x}[k] + 0\mathbf{e}[k] \tag{17.23}$$

Note that the input is now decoupled from the output. The z-transform of Eq. (17.23) results in a noise gain transfer function vector that satisfies

$$\mathbf{G}(z) = \mathbf{c}^T(z\mathbf{I} - \mathbf{A})^{-1} \tag{17.24}$$

Referring to the state transition matrix associated with Eq. (9.41), it can be seen that

$$(z\mathbf{I} - \mathbf{A})^{-1} \leftrightarrow \Phi[k-1] = \mathbf{A}^{k-1} \tag{17.25}$$

Therefore, Eq. (17.24) implies that the vector $\mathbf{g}[k]$ satisfies

$$\mathbf{g}[k] = \mathbf{c}^T\Phi[k-1] = \mathbf{c}^T\mathbf{A}^{k-1} \tag{17.26}$$

Define a matrix $\mathbf{W}[k]$ to be the dyadic product

$$\mathbf{W}[k] = \mathbf{g}[k]\mathbf{g}^T[k] = (\mathbf{A}^T)^{k-1}\mathbf{cc}^T(\mathbf{A})^{k-1} \tag{17.27}$$

The elements of $\mathbf{W}[k]$ are seen to relate to $\|g_i[k]\|_2^2$, which is used in Eq. (17.17) to define the noise power gain. The question, therefore, is *how can $\|g_i[k]\|_2^2$ be efficiently computed?* The answer, fortunately, is found in Sec. 16.5, where ℓ_2 scaling techniques were introduced. More specifically, define the matrix $\mathbf{W}$ to be

$$\mathbf{W} = \sum_{k=1}^{\infty} \mathbf{W}[k] = \sum_{k=1}^{\infty} \mathbf{g}[k]\mathbf{g}^T[k] = \sum_{k=0}^{\infty} (\mathbf{A}^T)^k\mathbf{cc}^T(\mathbf{A}^T)^k \tag{17.28}$$

It can be seen that $\mathbf{W}$ is a Lyapunov matrix that satisfies

$$\mathbf{W} = \mathbf{A}^T\mathbf{WA} + \mathbf{cc}^T \tag{17.29}$$

where the on-diagonal term, W_{ii}, is defined to be

$$W_{ii} = \|g_i[k]\|_2^2 \tag{17.30}$$

Therefore, from Eq. (17.18), it can be seen that a state-determined system $\mathbf{S} = (\mathbf{A},\mathbf{b},\mathbf{c},d_0)$ exhibits an output roundoff error variance, or noise power gain measure, given by

$$\sigma^2 = \frac{Q^2}{12}G = \frac{Q^2}{12}\sum_{i=1}^{N} k_i W_{ii} \tag{17.31}$$

It can be seen that everything required to compute Eq. (17.31) is specified in terms of the elements of the state coefficients found in $\mathbf{S} =$

$(\mathbf{A},\mathbf{b},\mathbf{c},d_0)$, except of k_i. The value of k_i can be directly equated to the number of noise sources in a feedback path having variance $Q^2/12$. The determination of k_i can be automated by searching the ith row of the $\mathbf{A}$ matrix and counting the number of non-zero, nontrivial ($\neq \pm 1$ or 0) coefficients. Recall that, if M multipliers are attached to the input of the ith shift register and products are rounded before being summed, then $k_i = M$. If, however, all multiplies are maintained as full-precision products and are accumulated using an extended precision adder and then rounded, then $k_i = 1$. Implicit in this model is the assumption that $(\mathbf{c},d_0)$, elements of the state four-tuple, do not appreciably add to the noise power gain.

The computational rule for $\mathbf{W}$ can be derived from the technique used to compute the ℓ_2 Lyapunov matrix $\mathbf{K}$. Specifically, $\mathbf{W}$ can be computed using a modification of the procedure developed to compute the Lyapunov matrix $\mathbf{K}$ (see Sec. 16.5). First determine, from the state-determined discrete-time system, the characteristic equation $\Delta(z) = \det(z\mathbf{I} - \mathbf{A})$, satisfying

$$\Delta(z) = z^N + p_1 z^{N-1} + \dots + p_{N-1}z + p_N \tag{17.32}$$

Define a coefficient vector to be $r = [r_0\ r_1\ \dots\ r_{N-1}]^T$ where

$$r_i + \sum_{j=1}^{N} p_j r_{|i-j|} = \begin{cases} 1 & \text{if } i = 0 \\ 0 & \text{if } i \neq 0 \end{cases} \tag{17.33}$$

which are used to form a Toeplitz matrix $\mathbf{R}$ given by

$$\mathbf{R} = \begin{bmatrix} r_0 & r_1 & r_2 & \cdots & r_{N-1} \\ r_1 & r_0 & r_1 & \cdots & r_{N-2} \\ r_2 & r_1 & r_0 & \cdots & r_{N-3} \\ \vdots & \vdots & \vdots & \ddots & \vdots \\ r_{N-1} & r_{N-2} & r_{N-3} & \cdots & r_0 \end{bmatrix} \tag{17.34}$$

Next define

$$\mathbf{x}[1] = \mathbf{c} \tag{17.35}$$

and

$$\mathbf{x}[k+1] = \mathbf{A}^T \mathbf{x}[k] + p_k \mathbf{c} \tag{17.36}$$

and collect the terms to form an $N \times N$ matrix,

$$\mathbf{X} = [\mathbf{x}[N]|...|\mathbf{x}[2]|\mathbf{x}[1]]$$ (17.37)

The matrices $\mathbf{X}$ and $\mathbf{R}$ can then be combined to create

$$\mathbf{W} = \mathbf{X}^T \mathbf{R} \mathbf{X}$$ (17.38)

Example 17.7: Lyapunov Solution

Problem Statement. A second-order system has a transfer function given by

$$H(z) = \frac{z - 0.5}{z^2 - z + 0.5}$$

For a Direct-II implement, has a state determined model given by $\mathbf{S} = (\mathbf{A}, \mathbf{b}, \mathbf{c}, d_0)$, having the following structure:

$$\mathbf{A} = \begin{bmatrix} 0 & 1 \\ -0.5 & 1 \end{bmatrix}, \quad \mathbf{b} = \begin{bmatrix} 0 \\ 1 \end{bmatrix}, \quad \mathbf{c} = \begin{bmatrix} -0.5 \\ 1 \end{bmatrix}, \quad d_0 = 0$$

What is the noise power gain?

Analysis and Conclusions. Solving the system determinant,

$$\Delta(z) = \det(z\mathbf{I} - \mathbf{A}) = z^2 - z + 0.5 = 0$$

defines $p_1 = -1$ and $p_2 = 0.5$. Based on this result, it follows that

$$r_0 + p_1 r_1 + p_2 r_2 = 1$$

$$r_1 + p_1 r_0 + p_2 r_1 = 0$$

$$r_2 + p_1 r_1 + p_2 r_0 = 0$$

which can be expressed in matrix-vector form as

$$\mathbf{P}\mathbf{r} = \begin{bmatrix} 1 & -1 & 0.5 \\ -1 & 1.5 & 0 \\ 0.5 & -1 & 1 \end{bmatrix} \begin{bmatrix} r_0 \\ r_1 \\ r_2 \end{bmatrix} = \begin{bmatrix} 1 \\ 0 \\ 0 \end{bmatrix} = \mathbf{s}$$

Solving the equation $\mathbf{P}^{-1}\mathbf{s} = \mathbf{r}$ yields

$$\begin{bmatrix} r_0 \\ r_1 \\ r_2 \end{bmatrix} = \begin{bmatrix} \dfrac{12}{5} \\ \dfrac{8}{5} \\ \dfrac{2}{5} \end{bmatrix}$$

The value of $\mathbf{R}$ then can be determined, and it is

$$\mathbf{R} = \begin{bmatrix} \dfrac{12}{5} & \dfrac{8}{5} \\ \dfrac{8}{5} & \dfrac{12}{5} \end{bmatrix}$$

Finally,

$$\mathbf{x}[1] = \begin{bmatrix} -0.5 \\ 1 \end{bmatrix}; \quad \mathbf{x}[2] = \mathbf{A}^T \mathbf{x}[1] + p_1 \mathbf{c} = \begin{bmatrix} 0 \\ -0.5 \end{bmatrix}$$

which implies that

$$\mathbf{X} = \begin{bmatrix} 0 & -0.5 \\ -0.5 & 1 \end{bmatrix}$$

Computing $\mathbf{W} = \mathbf{X}^T \mathbf{R} \mathbf{X}$ yields

$$\mathbf{W} = \begin{bmatrix} \dfrac{3}{5} & \dfrac{-4}{5} \\ \dfrac{-4}{5} & \dfrac{7}{5} \end{bmatrix}$$

from which it follows that $W_{11} = 0.6$ and $W_{22} = 1.4$. Notice that there is only one nontrivial multiplication in the feedback structure (see the $\mathbf{A}$ matrix), and it is defined by the feedback coefficient $a_{21} = -0.5$ attached to the $x_2[k]$ shift register. Therefore $k_1 = 0$ and $k_2 = 1$ and the noise power gain is given by $G^2 = W_{22} = 1.4$. The noise variance measured at the output is therefore given by

$$\sigma^2 = 1.4 \frac{Q^2}{12}$$

Computer Study. The filter can be analyzed using a general-purpose digital computer. The filter $H(z)$ was configured using a direct-II architecture, and the results are reported below and saved under the file name "wave.RPT".

```
DIGITAL FILTER ARCHITECTURE
File                              : wave.arc
Source Module                    : Transfer Function
Order                            : 2
TRANSFER FUNCTION
Scale Factor: 1.000000000000000
Numerator Coefficients           Denominator Coefficients
b2 : 0.000000000000000           a2 : 1.000000000000000
b1 : 1.000000000000000           a1 : -1.000000000000000
b0 : -0.5000000000000000         a0 : 0.5000000000000000

DIRECT-II STATE VARIABLE FILTER DESCRIPTION
Scale Factor = 1.000000000000000
A Matrix
      0.000000000000000            1.000000000000000
     -0.5000000000000000           1.000000000000000
B Vector              C' Vector              D Scalar
      0.000000000000000   -0.5000000000000000   0.000000000000000
      1.000000000000000    1.000000000000000
K Matrix Trace
      2.400000000000000            2.400000000000000
W Matrix Trace
      0.6000000000000001           1.400000000000000
Unscaled System                   Scaled System
State X(i)        Noise Sources    State X(i)        Noise Sources
    1                  0               1                  0
    2                  1               2                  2
sigma^2(y) = 1.4000   * sigma^2(x) sigma^2(y) = 6.7200   * sigma^2(x)
```

Observe that the machine computed results are $W_{11} = 0.6$ and $W_{22} = 1.4$ as expected. The value of k_1 and k_2 are found under the "Unscaled System" label found in the text file "wave.RPT" and are seen to be zero and one, respectively. Therefore,

$$\sigma^2 = 1.4 \frac{Q^2}{12}$$

as predicted. The entries listed under "Scaled System" will be developed later in this section.

The Lyapunov matrix $\mathbf{W}$ can be computed using the function "W(A,c)" defined in the s-file "w.s" using the following sequence of Siglab commands:

```
> include "chap17\w.s"        # Get function definition.
> A = {[0,1],[-0.5,1]};A       # Define state matrix A, vector c.
        0            1
      -0.5           1
> c = {-0.5,1};c
      -0.5
        1
> W = W(A,c)                    # Compute Lyapunov matrix.
> W
        0.6          -0.8
       -0.8           1.4
```

The results computed here correspond to those computed manually above for $\mathbf{W}$.❖

It has been previously established that a system possessing a state variable model $S = (\mathbf{A}, \mathbf{b}, \mathbf{c}, d_0)$ can be transformed into another, say $\hat{S} = (\hat{\mathbf{A}}, \hat{\mathbf{b}}, \hat{\mathbf{c}}, \hat{d}_0)$, using a linear transform $\mathbf{T}$. Suppose again that S and $\hat{S}$ are mathematically connected by

$$\mathbf{z}[k] = \mathbf{T}\mathbf{x}[k] \tag{17.39}$$

It has been previously established that [see Eq. (14.23)]:

$$\hat{\mathbf{A}} = \mathbf{T}\mathbf{A}\mathbf{T}^{-1}; \quad \mathbf{A} = \mathbf{T}^{-1}\hat{\mathbf{A}}\mathbf{T}$$

$$\hat{\mathbf{b}} = \mathbf{T}\mathbf{b}; \quad \mathbf{b} = \mathbf{T}^{-1}\hat{\mathbf{b}}$$

$$\hat{\mathbf{c}}^T = \mathbf{c}^T\mathbf{T}^{-1}; \quad \mathbf{c}^T = \hat{\mathbf{c}}^T\mathbf{T}$$

$$\hat{d}_0 = d_0 \tag{17.40}$$

Substituting these results in Eq. (17.29), the noise power gain matrix for the transformed system becomes:

$$\mathbf{W} = \mathbf{A}^T\mathbf{W}\mathbf{A} + \mathbf{c}\mathbf{c}^T = \mathbf{T}^T\hat{\mathbf{A}}(\mathbf{T}^T)^{-1}\mathbf{W}\mathbf{T}^{-1}\hat{\mathbf{A}}\mathbf{T} + \mathbf{T}^T\hat{\mathbf{c}}\hat{\mathbf{c}}^T\mathbf{T} \tag{17.41}$$

which can be simplified to read

$$\hat{\mathbf{W}} = (\mathbf{T}^T)^{-1}\mathbf{W}\mathbf{T}^{-1} \tag{17.42}$$

which is the noise power gain matrix for the scaled system $\hat{S}$. Suppose that the linear transform $\mathbf{T}$ is chosen to be a diagonal matrix

$$\mathbf{T} = \begin{bmatrix} T_{11} & 0 & \cdots & 0 \\ 0 & T_{22} & \cdots & 0 \\ \vdots & \vdots & \ddots & \vdots \\ 0 & 0 & \cdots & T_{NN} \end{bmatrix} \tag{17.43}$$

where

$$T_{ii} = \frac{1}{\sqrt{K_{ii}}} \tag{17.44}$$

Then the new system, it may be recalled, produces the ℓ_2 state norm of unity for all i. The result is a Lyapunov matrix for a scaled system in which $\hat{\mathbf{K}} = \mathbf{I}$. The noise power gain of the scaled system will, in general, differ from the unscaled system.

Example 17.8: Scaled System

Problem Statement. In Example 17.7, the second-order system

$$H(z) = \frac{z - 0.5}{z^2 - z + 0.5}$$

was implemented as a direct-II filter and had a state-determined model given by $S = (\mathbf{A}, \mathbf{b}, \mathbf{c}, d_0)$ having the structure

$$\mathbf{A} = \begin{bmatrix} 0 & 1 \\ -0.5 & 1 \end{bmatrix}, \quad \mathbf{b} = \begin{bmatrix} 0 \\ 1 \end{bmatrix}, \quad \mathbf{c} = \begin{bmatrix} -0.5 \\ 1 \end{bmatrix}, \quad d_0 = 0$$

What is the noise power gain for the scaled system?

Analysis and Conclusions. From Example 17.7, the elements of the linear transform $\mathbf{T}$ can be shown to be given by

$$\mathbf{T} = \begin{bmatrix} \dfrac{1}{\sqrt{K_{11}}} & 0 \\ 0 & \dfrac{1}{\sqrt{K_{22}}} \end{bmatrix} = \sqrt{\frac{5}{12}}\mathbf{I}$$

which results in a scaled system given by $\hat{S} = (\hat{\mathbf{A}}, \hat{\mathbf{b}}, \hat{\mathbf{c}}, \hat{d}_0)$ where

$$\hat{\mathbf{A}} = \mathbf{A}, \quad \hat{\mathbf{b}} = \sqrt{\frac{5}{12}}\mathbf{b}, \quad \hat{\mathbf{c}}^T = \sqrt{\frac{12}{5}}\mathbf{c}^T, \quad \hat{d}_0 = d_0$$

It then follows that

$$\hat{\mathbf{W}} = (\mathbf{T}^T)^{-1}\mathbf{W}\mathbf{T}^{-1} = \sqrt{\frac{12}{5}}\mathbf{I}\mathbf{W}\mathbf{I}\sqrt{\frac{12}{5}} = \frac{12}{5}\mathbf{W}$$

where

$$\hat{\mathbf{W}} = \begin{bmatrix} \dfrac{3}{5} & \dfrac{-4}{5} \\ \dfrac{-4}{5} & \dfrac{7}{5} \end{bmatrix}$$

Notice that the scaled noise power gain matrix satisfies the relationship

$$\hat{W}_{ii} = K_{ii}W_{ii} \qquad ❖$$

If $K_{ii} > 1$, then scaling the input will further reduce the output precision and add to the effective noise power as measured at the output.

The resulting state norms of the scaled system are all unity and noise power gain is given by

$$G^2 = \sum_{i=1}^{N} k_i K_{ii} W_{ii} = \sum_{i=1}^{N} k_i \hat{W}_{ii} \qquad (17.45)$$

and the noise variance follows as

$$\sigma^2 = \frac{Q^2}{12} G^2 = \frac{Q^2}{12} \sum_{i=1}^{N} k_i \hat{W}_{ii} = \frac{Q^2}{12} \sum_{i=1}^{N} k_i K_{ii} W_{ii} \qquad (17.46)$$

To appreciate how the significance of scaling, consider the following simple example.

Example 17.9: Scaled Noise Power Gain

Problem Statement. A second-order lowpass Butterworth IIR filter is assumed to be given by

$$H(z) = K \frac{1 + 2z^{-1} + z^{-2}}{1 + 0.2230z^{-1} + 0.1804z^{-2}}$$

What is the scaled and unscaled noise power grain (ignore the effect of the scale factor K)?

Analysis and Conclusions. The noise power calculation is summarized in the Computer Study. Fig. 17.10 shows that the computed error variance closely follows the predicted noise power gain curve ($k_2 W_{22}$ in bits) for fractional wordlengths ranging from 5 to 15 bits with a cascade implementation. For smaller fractional wordlengths, there is too little precision left in the data to

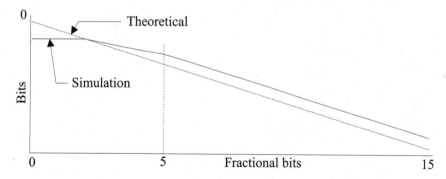

Figure 17.10 Roundoff error response as a function of fractional wordlength for a 16-bit solution and its comparison to noise power gain calculation. It can be seen that the mathematically predicted and simulated data are in good agreement.

maintain agreement between the computed error variance and the theoretical prediction, which is based on a linear state model.

Computer Study. The second-order Butterworth, saved under file name "second.IIR" is architected as a cascade (equivalent to a direct-II and parallel second-order section), as well as a normal and ladder filter.

```
File                          : second.IIR
Source Module                 : Infinite Impulse Response Filter
Filter Type                   : Butterworth
Order                         : 2
Frequency Response            : Lowpass
Desired Passband Attenuation  : 1.000000000000000 dB
Desired Stopband Attenuation  : 20.00000000000000 dB
Sampling Frequency            : 44100.00000000000 Hz
Passband Edge                 : 10000.00000000000 Hz
Stopband Edge                 : 20000.00000000000 Hz
TRANSFER FUNCTION
Scale Factor = 0.3508520186818723
Numerator Coefficients        Denominator Coefficients
b 2: 1.000000000000000        a 2: 1.000000000000000
b 1: 2.000000000000000        a 1: 0.2230283267256145
b 0: 1.000000000000000        a 0: 0.1803797480018745


DIRECT-II STATE VARIABLE FILTER DESCRIPTION
Scale Factor = 0.3508520186818723
A Matrix
   0.000000000000000            1.000000000000000
  -0.1803797480018745          -0.2230283267256145
B Vector C' Vector D Scalar
   0.000000000000000            0.8196202519981255     1.000000000000000
   1.000000000000000            1.776971673274385
K Matrix Trace
   1.071898619073553            1.071898619073553
W Matrix Trace
   0.7861373928047978           3.514784715037589
Unscaled System                             Scaled System
   State X(i)        Noise Sources          State X(i)        Noise Sources
      1                  0                      1                  0
      2                  2                      2                  3
sigma^2(y) = 7.0296     * sigma^2(x)    sigma^2(y) = 11.302  * sigma^2(x)


CASCADE STATE VARIABLE FILTER DESCRIPTION
Scale Factor = 0.3508520186818723
A Matrix
   0.000000000000000            1.000000000000000
  -0.1803797480018745          -0.2230283267256145
B Vector C' Vector D Scalar
   0.000000000000000            0.8196202519981255     1.000000000000000
   1.000000000000000            1.776971673274385
K Matrix Trace
   1.071898619073553            1.071898619073553
```

```
W Matrix Trace
   0.7861373928047978              3.514784715037589
Unscaled System Scaled System
     State X(i)           Noise Sources          State X(i)        Noise Sources
        1                     0                     1                   0
        2                     2                     2                   3
sigma^2(y) = 7.0296      * sigma^2(x)    sigma^2(y) = 11.302  * sigma^2(x)
```

CASCADE NORMAL STATE VARIABLE FILTER DESCRIPTION
Scale Factor = 0.3508520186818723
A Matrix
```
  -0.1115141633628073              0.4098101259990627
  -0.4098101259990627             -0.1115141633628073
```
B Vector   C' Vector   D Scalar
```
  0.9403453053324387       1.823177841365533       1.000000000000000
  0.8689023021810298       0.07199307586498294
```
K Matrix Trace
```
  1.000000000000000        1.000000000000000
```
W Matrix Trace
```
  3.466991180445419        0.5948416453086210
```
Unscaled System   Scaled System
```
     State X(i)           Noise Sources          State X(i)        Noise Sources
        1                     3                     1                   3
        2                     3                     2                   3
sigma^2(y) = 12.185      * sigma^2(x)    sigma^2(y) = 12.185  * sigma^2(x)
```

LADDER-2-MULTIPLIER STATE VARIABLE FILTER DESCRIPTION
Scale Factor = 0.3508520186818723
A Matrix
```
  -0.1889462498006703             -0.1803797480018745
   0.9642993146862627             -0.03408207692494414
```
B Vector   C' Vector   D Scalar
```
  1.000000000000000        1.622107500398659       1.000000000000000
  0.1889462498006703       0.8196202519981255
```
K Matrix Trace
```
  1.071898619073553        1.033631103785778
```
W Matrix Trace
```
  3.031309777701076        0.7861373928047978
```
Unscaled System   Scaled System
```
     State X(i)           Noise Sources          State X(i)        Noise Sources
        1                     2                     1                   3
        2                     3                     2                   3
sigma^2(y) = 8.4210      * sigma^2(x)    sigma^2(y) = 12.185  * sigma^2(x)
```

The results of analyzing this second-order system are summarized in Table 17.1. The reported unscaled and scaled values multiply the noise variance constant $Q^2/12$. This model assumes that multiplier outputs are rounded prior to be added and sent to a shift register. Based on this analysis it can be seen that all architectures perform about the same in terms of roundoff error sensitivity. More specifically, comparing the architectures in terms of the number of bits contaminated by noise (i.e., $\log_2(\sigma)$), the data shown in Table 17.2 result for $Q = 2^{-F}$.

Table 17.1   Dynamic Range and Noise Gain Parameters

Type	K_{11}	K_{22}	W_{11}	W_{22}	Unscaled	Scaled
Direct/cascade/parallel	1.0718	1.0718	0.7861	3.5147	7.0296	11.302
Normal	1.0	1.0	3.4669	0.5948	12.185	12.185
Ladder	1.071	1.0336	3.0310	0.7861	8.2410	12.185

Table 17.2   Noise Gain, in Bits, of a Scaled-State System

Type	Scaled σ^2	Scaled σ	$b = \log_2(\sigma)$
Direct/cascade/parallel	11.302	3.362	$1.749 < 2$
Normal	12.185	3.490	$1.803 < 2$
Ladder	12.185	3.490	$1.803 < 2$

Note that each architecture presented indicates about a two-bit loss of fractional precision within the IIR filter. While appearing to integrate the effects of register overflow scaling into roundoff error analysis, it does not take into account the binary point assignment requirements imposed by the coefficient dynamic range. In the study of the second-order system, the maximal coefficient and ℓ_2 state norm (for an unscaled filter), are summarized in Table 17.3. Observe that, in all cases, one less fractional bit will available after accounting for the coefficient requirements (i.e., $I = 1$). This, in effect, will increase the estimated output error variance by $2^2 = 4$, standard deviation by 2, or, equivalently, add one additional bit to the error budget. The change in fractional bit precision is shown in Table 17.3.

Again each architecture produces a similar loss in precision. Specifically, when the systems are overflow scaled, it can be seen that about three bits of precision is lost inside the IIR filter.

It should be strongly pointed out that each filter type and architecture can bring about different relationships within a list of design parameters. Each

Table 17.3   Noise Gain for Overflow-Free Operation

Type	$\max \sqrt{K_{ii}}$	$\max$ |coef|	$\max (\sqrt{K_{ii}}, \|\text{coef}\|)$	I (bits)	Scaled σ $(2 \times \sigma)$	$\log_2(\sigma)$ $b = b + 1$
Direct/cascade/parallel	~1.0	1.7769	1.7769	1	6.724	$2.749 < 3$
Normal	1.0	1.8231	1.8231	1	6.981	$2.803 < 3$
Ladder	~1.0	1.6221	1.6221	1	6.981	$2.803 < 3$

situation requires that the design be completely analyzed in terms of roundoff error and its relationship to overflow scaling and coefficient range requirements. The majority of IIR filter designs found in contemporary practice are cascade. The cascade architecture is generally considered to have a good blend dynamic range requirements and roundoff error variance. Consider the second order IIR filter once again architected as a cascade filter. The following sequence of Siglab commands uses the files "second.ARC", "K.s", "W.s", and "var_2.s" to simulate the fixed-point response of the second order IIR filter for $F \in \{0,1,2,...,15\}$. It should also be noted that K_{ii} and W_{ii} correspond to $K_{i-1,i-1}$ and $W_{i-1,i-1}$ in the software.

```
> include "chap17\k.s"              # Get functions.
> include "chap17\w.s"
> include "chap17\state.s"
> include "chap17\var_2.s"
> h = rarc("chap17\second.arc",1)   # Get cascade architecture filter.
> h                                 # Print state information.
            0              1
         -0.18         -0.223
            0              1
          0.82          1.777
            1              0
          0.351            0
> A = h[0:1,0:1];A                  # Extract state information.
            0              1
         -0.18         -0.223
> b = h[2,]';b
            0
            1
> c = h[3,]';c
          0.82
          1.777
> K = K(A,b);K                      # Compute and display K matrix.
          1.072         -0.203
         -0.203          1.072
> W = W(A,c);W                      # Compute and display W matrix.
          0.786          1.354
          1.354          3.515
> roundoff(h,0,16,2*W[1,1])         # Plot tested and predicted roundoff.
```

For multiply-add-round structures, $k_1 = 0$ and $k_2 = 1$. The latter is simulated and reported in Fig. 17.10. This noise power gain is displayed on a semi-log scale. On the semi-log axis is the theoretical noise gain $G \propto \alpha^2$, which reduces to $\log_2(G) \propto 2\log_2(\alpha)$ in bits.   ❖

Example 17.9: High-Order Noise Power Gain

Problem Statement. An eighth-order Chebyshev II filter, having the transfer function

$$H(z) = 0.0888 \frac{(z^8 + 4.43z^7 + 10.76z^6 + 17.46z^5 + 20.48z^4 + 17.46z^3 + 10.76z^2 + 4.43z + 1)}{(z^8 + 1.10z^7 + 1.97z^6 + 1.55z^5 + 1.22z^4 + 0.61z^3 + 0.24z^2 + 0.061z + 0.008)}$$

was studied in Example 15.1 as a direct-II, cascade, parallel, normal, and ladder/lattice filter. Each architecture has a unique state-variable model and, therefore, noise power gain. Compute the noise power gain for a scaled and unscaled direct-II, cascade, and lattice/ladder system. Interpret the statistical precision of the resulting output. Assume that all the products are rounded before being summed and sent to a shift register.

Analysis and Conclusions. The elements of the state variable four-tuple $S = (\mathbf{A}, \mathbf{b}, \mathbf{c}, d_0)$ can be used to compute the noise power gain for scaled and unscaled systems. A summary of the computed values of W_{ii} for each state for each architecture is shown below. (See Example 16.4 for a summary of K_{ii}.) The state model assumes that the input scale factor is $k = 1$. Also reported are the number of non-trivial coefficient multiplies attached to each shift-register for each architecture (i.e., k_i).

```
File                         : scale.arc
Source Module                : Architecture
Order                        : 8
CASCADE STATE VARIABLE FILTER DESCRIPTION
W Matrix Trace
  59.85366413164084   37.35309535368246
  40.69069144302898   48.67773851790900
  11.68652793957771   41.47945348895648
   6.058506840236071   9.532824166912910   Wmax = W11 = 59.85366413164084
Unscaled System Scaled System
```

State X(i)	Noise Sources	State X(i)	Noise Sources
1	0	1	0
2	2	2	3
3	0	3	0
4	4	4	5
5	0	5	0
6	6	6	7
7	0	7	0
8	8	8	9

sigma^2(y) = 594.56   * sigma^2(x)   sigma^2(y) = 2203.5   * sigma^2(x)

```
DIRECT-II STATE VARIABLE FILTER DESCRIPTION
W Matrix Trace
   0.9881505888415242   20.01021036520583
 125.9786591582922     369.0375798288538
 572.1751052036090     517.1590627374259
 209.1314073652742      62.46665701954116   Wmax=W66=517.1590627374259
Unscaled System Scaled System
```

State X(i)	Noise Sources	State X(i)	Noise Sources
1	0	1	0
2	0	2	0
3	0	3	0
4	0	4	0
5	0	5	0
6	0	6	0
7	0	7	0
8	8	8	9

sigma^2(y) = 499.73   * sigma^2(x)   sigma^2(y) = 6161.1   * sigma^2(x)

```
LADDER-2-MULTIPLIER STATE VARIABLE FILTER DESCRIPTION
W Matrix Trace
    2.336093421163991  2.482187597220112
   12.83952769455967   18.92320774231405
   35.58254381830981   33.34891475755572
   11.61158723281273    0.9881505888415248   Wmax = W55 = 35.58254381830981
Unscaled System                             Scaled System
    State X(i)         Noise Sources          State X(i)      Noise Sources
        1                  8                      1               9
        2                  9                      2               9
        3                  8                      3               8
        4                  7                      4               7
        5                  6                      5               6
        6                  5                      6               5
        7                  4                      7               4
        8                  3                      8               3
  sigma^2(y) = 705.86    * sigma^2(x)    sigma^2(y) = 1365.1  * sigma^2(x)
```

It may be recalled that the ladder/lattice filter generally requires more coefficient multiplies per filter cycle than the others. The exact count is defined by the rounding rule implemented (i.e., round before or after accumulation) and the number of noise sources as specified by scanning the state four-tuple $S = (\mathbf{A}, \mathbf{b}, \mathbf{c}, d_0)$. Assuming that full-precision products are rounded before being accumulated, the data shown in Table 17.5 results.

The unscaled noise power gains would indicate that all the architectures are essentially the same, operating at about 4.5 bits loss of precision. The scaled noise power gains, however, can be ranked ordered as ladder/lattice, cascade, and lastly direct-II (remember that the ladder/lattice power gain estimate may be underestimated). In either case, a cascade architecture is generally a good compromise. The actual statistical performance of a given architecture is a function of the complex interaction of arithmetic and coefficient roundoff errors, which can at various times behave linearly or nonlinearly. Therefore, the scaled or unscaled noise models provide only a guideline to how an IIR filter may behave during run-time.

The noise gain, in bits, can be experimentally studied by comparing the fixed-point system impulse response to the floating-point response, which is considered to be error free. The results of such an experiment are shown in Fig. 17.11a for the cascade, direct-II, and lattice/ladder cases. The 16-bit architectures were studied over all possible fractional wordlengths $F \in \{0, 1, 2, \dots, 15\}$.

Table 17.5   Eighth-Order IIR Filter Example

Architecture	max (|coef|)	Unscaled G^2	Scaled G^2
Cascade	1.8676	594.56 ~ 4.61 bits	2203.5 ~ 5.555 bits
Direct-II	19.1764	499.73 ~ 4.48 bits	6161.1 ~ 6.29 bits
Ladder/lattice	5.1099	705.86 ~ 4.73 bits	1365.1 ~ 5.20 bits

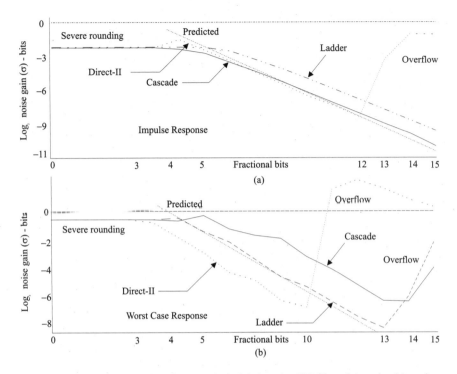

Figure 17.11 Noise gain in bits for a typical eighth-order IIR filter driven by (a) an impulse and (b) a unit-bound worst-case input.

It can be seen that, initially ($F \sim 0$), all of the architectures suffer from severe coefficient quantization. An insufficient fractional data field will result in small, but critical, coefficients being rounded to zero. Here, for example, the input scale factor $k = 0.0888$ will be mapped to zero when $F \leq 3$ bits. For larger values of F, the system enters a normal mode of operation (no severe quantization or overflow errors). Here, the measured roundoff errors of the direct-II and cascade are nearly identical. The lattice/ladder response is similar, although somewhat higher. It should be remembered that the lattice/ladder was promoted as a low coefficient roundoff error sensitivity filter, which is not the same as claiming it to be immune to other roundoff error types. At 12 bits, the direct-II filter begins to suffer from saturation errors, whereas the cascade and lattice/ladder filters' performance continues to improve out to the maximum value of $F = 15$. The results are summarized in Table 17.6.

A problem associated with experimentally determining the effects of roundoff noise is choosing a valid forcing function. This is particularly true when attempting to verify the scaled error variance predications. This is due to an acknowledged weakness of the ℓ_2 scaling paradigm in that it is known to underestimate the overflow prevention scaling policy. As a result, unexpected overflows can occur whenever the input to the system is something other than a simple impulse. Suppose that the input $u[k]$ is bounded by unity but other-

Table 17.6   Noise Gain Performance of a Typical Eighth-Order IIR Filter Using an Impulse

Type	Severe rounding range	Normal operation	Overflow saturation
Cascade	$F \in \{0,\ldots,5\}$	$F \in \{6,\ldots,15\}$	Not applicable
Direct-II	$F \in \{0,\ldots,4\}$	$F \in \{5,\ldots,12\}$	$F \in \{11,\ldots,15\}$
Ladder/lattice	$F \in \{0,\ldots,6\}$	$F \in \{7,\ldots,15\}$	Not applicable

wise arbitrary (i.e., $|u[k]| \leq 1$). A popular choice is simply a uniformly distributed random noise $u[k] \in U[-1,1]$. The most aggressive input is given by the worst case input formula presented Sec. 16.7. An example of this choice of forcing function is shown in Fig. 17.11b and summarized in Table 17.7. It can be seen that, in normal operation, the slope of each error trajectory is similar to that of the direct-II. Normal operation for the direct-II is over the range $F \in \{4,5,6,\ldots,10\}$ and $F \in \{11,12,13\}$ for the lattice/ladder. The lattice/ladder filter has an internal error of about five bits over F at $F = 13$. It is interesting to note that the optimal operating point of the lattice/ladder filter is at $F = 13$ or $I = 2$, a point where the maximum lattice/ladder coefficient will be rounded from 5.1099 to 4. It can also be noted that the lattice/ladder superiority is present only over a subrange of F. For the simulation study, it would appear that the lattice/ladder, when set to $F = 13$, would provide the best statistical performance with the direct-II ($F = 10$) and cascade ($F = 13$) behaving about the same. The maximum run-time bandwidth of the lattice/ladder would normally be inferior to the direct-II or cascade, which require fewer multiply-accumulations per filter cycle.

Computer Study. The following program was used to produce the data shown in Fig. 17.11a and Table 17.6. The cascade, direct-II, and lattice/ladder are loaded as architecture files "scalec.arc", "scaled.arc", and "scalel.arc". The analysis is provided by the function "roundoff(h,n)" which is defined in the s-file "varfun.s" and takes an impulse forcing function, presents it to the filter, and runs a 16-bit fixed-point simulation over $F \in \{0,1,2,\ldots,15\}$.

Table 17.7   Noise Gain Performance of a Typical Eighth-Order IIR Filter Using a Worst-Case Input

Type	Severe rounding range	Normal operation	Overflow saturation
Cascade	$F \in \{0,\ldots,6\}$	$F \in \{7,\ldots,13\}$	$F \in \{14,15\}$
Direct-II	$F \in \{0,\ldots,3\}$	$F \in \{4,\ldots,10\}$	$F \in \{11,\ldots,15\}$
Ladder/lattice	$F \in \{0,\ldots,5\}$	$F \in \{6,\ldots,13\}$	$F \in \{14,15\}$

```
> include "chap17\state.s"        # Define state function.
> include "chap17\varfun.s"       # Define roundoff function.
> hc=rarc("chap17\scalec.arc",1)  # Cascade.
> hd=rarc("chap17\scaled.arc",1)  # Direct-II.
> hl=rarc("chap17\scalel.arc",1)  # Lattice/ladder.
> vc=roundoff(hc,16)              # cascade simulation--takes time
> vd=roundoff(hd,16)              # Direct-II simulation--takes time
> vl=roundoff(hl,16)              # lattice/ladder simulation--takes time
> ograph(vd,vc,,vl,zeros(len(vd))) # Plot results.
```

The following sequence of Siglab commands was used to produce the data shown in Fig. 17.11(b) and Table 17.7. The cascade, direct-II, and lattice/ladder are loaded into the program as architecture files "scalec.arc", "scaled.arc", and "scalel.arc". The analysis is provided by the function "roundoff(h,n)", which is defined in the s-file "varfun.s" and computes the input-output worst-case forcing fucntion, presents it to the filter, and runs a 16-bit fixed-point simulation over $F \in \{0,1,2,...,15\}$.

```
> include "chap17\state.s"        # Define state function.
> include "chap17\worstvar.s"     # Define roundoff function.
> hc=rarc("chap17\scalec.arc",1)  # Cascade.
> hd=rarc("chap17\scaled.arc",1)  # Direct-II.
> hl=rarc("chap17\scalel.arc",1)  # Lattice/ladder.
> vc=roundoff(hc,16)              # Cascade simulation--takes time
> vd=roundoff(hd,16)              # Direct-II simulation--takes time
> vl=roundoff(hl,16)              # lattice/ladder simulation--takes time
> ograph(vd,vc,,vl,zeros(len(vd))) # Plot results.                    ❖
```

17.5 Section Optimal Filter

A useful extension of the normal filter studied in Sec. 14.7 is called a *section optimal* or *section filter*. Section filters have a cascade system architecture defined in terms of first- and second-order sections. Recall that the noise power gain of a scaled state-determined system is given by

$$\sigma^2 = \frac{Q^2}{12} \sum_{i=1}^{N} k_i K_{ii} W_{ii} \qquad (17.47)$$

Assume, at this point, that $k_i = 1$ for all i. The section filter is developed to minimize the σ^2 and thereby minimize the output roundoff error variance of the scaled system. Assume that K and W are positive definite symmetric matrices and that the eigenvalues of KW are $\lambda_i^2 > 0$, $i \in \{1,2,3,...,N\}$. It then follows that

$$\sigma^2 = \frac{Q^2}{12} \left(\sum_{i=1}^{N} K_{ii} W_{ii} \right) \geq \frac{Q^2}{12} \left[\frac{1}{N} \left(\sum_{i=1}^{N} \lambda_i \right)^2 \right] \qquad (17.48)$$

where equality holds if and only it $\mathbf{K} = \mathbf{DWD}$ for some diagonal matrix $\mathbf{D}$ whose on-diagonal elements are

$$D_{ii} = K_{ii}W_{ii} = K_{jj}W_{jj} \tag{17.49}$$

for $i,j \in \{12,3,...,N\}$. In such cases, the minimum roundoff error variance is given by

$$\sigma^2_{min} = \frac{Q^2}{12N}\left(\sum_{i=1}^{N}\lambda_i\right)^2 \tag{17.50}$$

It will be shown that the key to developing a section filter is creating balanced second-order sections having a state-determined model $\mathcal{S} = (\mathbf{A},\mathbf{b},\mathbf{c},d_0)$ that satisfies

$$a_{11} = a_{22}$$

$$b_1c_1 = b_2c_2 \tag{17.51}$$

The design of a section filter begins with a normal filter. Assume that the second-order subfilters of a normal filter are defined by $\mathcal{S}_N = (\mathbf{A}_N,\mathbf{b}_N,\mathbf{c}_N,d_N)$, where

$$\mathbf{A}_N = \begin{bmatrix} \alpha & \beta \\ -\beta & \alpha \end{bmatrix}, \quad \mathbf{b}_N = \begin{bmatrix} b_{N1} \\ b_{N2} \end{bmatrix}, \quad \mathbf{c}_N = \begin{bmatrix} c_{N1} \\ c_{N2} \end{bmatrix}, \quad d_N = d_0 \tag{17.52}$$

The normal filter, as given in Eq. (17.52), automatically satisfies the condition $a_{11} = a_{22}$. The second condition given in Eq. (17.51), namely, $b_1c_1 = b_2c_2$, will be satisfied using a rotational matrix transform. The required rotation is demonstrated in Fig. 17.12. In Fig. 17.12, b_N and c_N are considered to be a two-dimensional vector having the polar representation shown in Eq. (17.53).

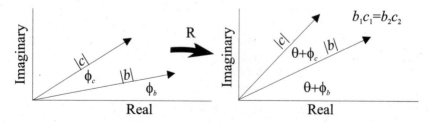

Figure 17.12   Rotational matrix mapping of a vector pair.

$$b_N = r_b e^{j\phi_b}, \quad c_N = r_c e^{j\phi_c} \tag{17.53}$$

The condition $b_1 c_1 = b_2 c_2$ requires that the vectors b and c be rotated by an amount θ in order that

$$r_b r_c \cos(\phi_b + \theta) \cos(\phi_c + \theta) = r_b r_c \sin(\phi_b + \theta) \sin(\phi_c + \theta) \tag{17.54}$$

or

$$\cos(\phi_b + \theta) \cos(\phi_c + \theta) - \sin(\phi_b + \theta) \sin(\phi_c + \theta) = \cos(\phi_b + \phi_c + 2\theta) = 0 \tag{17.55}$$

which implies that

$$(\phi_b + \phi_c + 2\theta) = \left\{ \frac{\pi}{2}, \frac{3\pi}{2} \right\} \tag{17.56}$$

or

$$\theta = \frac{\left\{ \dfrac{\pi}{2}, \dfrac{3\pi}{2} \right\} - \phi_b - \phi_c}{2} \tag{17.57}$$

Let $\mathbf{R}(\phi)$ be a 2×2 rotational matrix that will rotate a vector by an angle θ and leave the vector length unchanged. Specifically, let $\mathbf{R}(\theta)$ be given by

$$\mathbf{R}(\theta) = \begin{bmatrix} \cos(\theta) & -\sin(\theta) \\ \sin(\theta) & \cos(\theta) \end{bmatrix} \tag{17.58}$$

An interesting property of $\mathbf{R}(\theta)$ is that

$$[R(\theta)]^{-1} = \begin{bmatrix} \cos(\theta) & -\sin(\theta) \\ \sin(\theta) & \cos(\theta) \end{bmatrix}^{-1}$$

$$= \begin{bmatrix} \cos(\theta) & \sin(\theta) \\ -\sin(\theta) & \cos(\theta) \end{bmatrix} = \begin{bmatrix} \cos(-\theta) & -\sin(-\theta) \\ \sin(-\theta) & \cos(-\theta) \end{bmatrix} = \mathbf{R}(-\theta) = \mathbf{R}(\theta)^T \tag{17.59}$$

which implies that if $\mathbf{R}(\theta)$ rotates a vector through an angle θ, then $\mathbf{R}(\theta)^{-1}$ rotates the same vector through an angle $-\theta$. Applying $\mathbf{R}(\theta)$ to the state-elements of $\mathcal{S}_N$, a section filter $\mathcal{S}_S$ results and is given by

$$\mathbf{A}_s = \mathbf{R}(\theta)\mathbf{A}_N\mathbf{R}(-\theta) = \mathbf{A}_N$$

$$\mathbf{b}_s = \mathbf{R}(\theta)\mathbf{b}_N$$

$$\mathbf{c}_S^T = \mathbf{c}_N^T\mathbf{R}(-\theta) \text{ or } \mathbf{c}_S = \mathbf{R}^T(-\theta)\mathbf{c}_N = \mathbf{R}(\theta)\mathbf{c}_N$$

$$d_s = d_N$$

$$\text{(17.60)}$$

Once the filter $\mathcal{S}_S$ is designed, it can be renormalized using a scaling transform $\mathbf{T}$ where

$$\mathbf{T} = \begin{bmatrix} \dfrac{1}{\sqrt{K_{11}}} & 0 \\ 0 & \dfrac{1}{\sqrt{K_{22}}} \end{bmatrix} \qquad \text{(17.61)}$$

and K_{ii} is obtained from the on-diagonal terms of the Lyapunov matrix $\mathbf{K}$ for $\mathcal{S}_S$.

Example 17.10: Section Optimal Filter

Problem Statement. The second-order system studied in Example 17.8 is to be transferred into a section optimal filter.

Analysis and Conclusions. The section filter is reduced to the computer program shown in the Computer Study. The result of comparing the second-order normal and section filter system is summarized in Table 17.8. In both cases, it is assumed that $k_1 = k_2 = 1$, which may or may not reflect reality.

The noise power gain for the section filter is seen to be smaller that of the normal filter by 25 percent. The noise gain advantage is somewhat smaller. The advantage is case dependent.

Table 17.8   Section Optimal and Normal IIR Filter Comparison

Type	K_{11}	K_{22}	W_{11}	W_{22}	Unscaled	Scaled
Normal	1.0	1.0	3.4	0.5	4.0	4.0
Section (normalized)	1.0	1.0	0.9	1.9	2.9	2.9

Computer Study. The following sequence of Siglab commands converts a normal filter to a section optimal filter. The following sequence of Siglab commands uses files "secondn.ARC", "k.s", "w.s", and "sectop.s".

```
> showinc on
> include "chap17\sectopt.s"       # Script, the rest follows.
h=rarc("chap17\secondn.arc",1)     # Normal filter
h
   -0.112              0.41
   -0.41              -0.112
    0.94               0.869
    1.823              0.072
        1              0
    0.351              0
A=h[0:1,0:1]; b=h[2,]';c=h[3,]'    # strip [A,b,c]
bmag=mag(<b[0],b[1]>);bmag         # analyze b vector
1.28033
phib=atan(b[1]/b[0]);phib
0.745931
cmag=mag(<c[0],c[1]>);cmag         # analyze c vector
1.8246
phic=atan(c[1]/c[0]);phic
0.0394672
theta=((pi/2)-phib-phic)/2;theta   # define rotational angle
0.392699
R={[cos(theta),-sin(theta)],[sin(theta),cos(theta)]} # rotational matrix
# section optimal filter
AS=R*A*inv(R)                      #AS=A
AS
   -0.112              0.41
   -0.41              -0.112
bS=R*b;bS
    0.536
    1.163
cS=R*c;cS
    1.657
    0.764
bS[0]*cS[0]                        # test to see if b0c0=b1c1
0.888486
bS[1]*cS[1]
0.888486
# analyze section optimal filter
include "chap17\K.S"
K=K(AS,bS);K                       # K for normal filter
      0.5                0.5
      0.5                1.5
include "chap17\W.S"
W=X'*R*X
# echo "Lyapunov matrix W\n";W
end                                # end function
W=W(AS,cS)
W
    0.683              0.161
    0.161              3.056
```

```
K[0,0]*W[0,0] + K[1,1]*W[1,1]
4.92518
# optionally re-normalize filter using diagonal scaling matrix
T={[1/sqrt(K[0,0]),0],[0,1/sqrt(K[1,1])]}
# normalized section optimal filter
A=T*AS*inv(T);A
  -0.112            0.71
  -0.237           -0.112
B=T*bS;B
   0.758
   0.949
C=inv(T)'*cS;C
   1.172
   0.936
B[0]*C[0]                          # test to see if b0c0=b1c1
0.888486
B[1]*C[1]
0.888486
# analyze normalized section optimal filter
K=K(A,B); K
     1             0.577
   0.577             1
W=W(A,C); W
   0.982           0.62
    0.62          1.966
K[0,0]*W[0,0] + K[1,1]*W[1,1]
2.94839                                                        ❖
```

17.6 Extended-Precision Register Arithmetic

Virtually every generation of DSP microprocessor has employed extended-precision multiply-accumulators (MACs). A typical extended-
precision MAC accepts two N-bit operands and produces a full-precision $2N$-bit product, which is sent to a $2N + M$ bit accumulator. The
additional M bits of register precision at the accumulator level means
that 2^M worst-case products can be sequentially added without the
possibility of encountering an accumulator overflow. Since testing for
overflow after each MAC cycle is time consuming, the additional register space allows the testing to be deferred if the dynamic range of the
internal variables is known to be a valid $2N + M$-bit word. Extended-
precision accumulators have been shown to be a very important attribute in many signal-processing applications—DFTs in particular.
The consequence of this in a simulation study is subtle and involves a
three-step process, namely:

- Specify the data format (i.e, [N:F])
- Round the DSP coefficients to an [N:F] format
- Perform a simulation using a [$M + N$:F] format

17.7 Summary

In Chap. 17, a study of how finite wordlength effects affect the output of a filter was presented. The analysis of these errors is mathematically tractable if the filter is modeled in state-variable form. Simulation can also be used to quantify the finite wordlength behavior of a filter. It was found that the extent of the roundoff error is architecture dependent. In addition, overflow scaling can further reduce the precision of the filter, although it too is architecture dependent. As such, careful attention should be given to choosing an architecture that provides the best overall design trade-offs between speed and accuracy. It was also found that architectures can be developed that have a low noise gain but at the cost of added complexity. The value of the potential increase in precision must be weighed against the added complexity of low noise gain filter design. A successful design involves making these design trade-offs while requiring that the final solution meets user imposed design objectives in terms of cost, frequency response, precision, and bandwidth.

In the absence of a definitive study or simulation, a cascade architecture is generally assumed to provide a good design compromise. However, even taking this pedestrian viewpoint will not relieve the designer from properly scaling the design to suppress overflows. These questions are, of course, generally avoided if the floating-point path is taken. Design cost and power consumption considerations now become important limitations.

17.8 Self-Study Problems

1. **Chcbyshev Bilinear z-Transform Direct-II Design** A fourth-order low-pass Chebyshev-II IIR filter is designed that meets or exceeds the following specifications:

Desired passband attenuation = 3.0 dB

Desired stopband attenuation = 20.0 dB

Sampling frequency = 10,000.0 Hz

Passband edge = 2000.0 Hz

Stopband edge = 2500.0 Hz

The filter is implemented using a direct-II architecture and saved as "p17_1.(IIR,ARC,RPT)." Recommend a data format $[N{:}F]$ for your system if the input is assumed to be a uniformly distributed random noise (i.e., $u[k] = 2 \times$ rand(n), $-1 \le u[k] \le 1$). Measure the input and output noise power and compare.

2. Cascaded IIR   Repeat problem 17.1 for a cascade architecture (saved as "p17_2.(IIR,ARC,RPT)").

3. Parallel IIR   Repeat Problem 17.1 for a parallel architecture (saved as "p17_3.(IIR,ARC,RPT)").

4. Normal IIR   Repeat Problem 17.1 for a normal architecture (saved as "p17_4.(IIR,ARC,RPT)").

18

Multirate Signal Processing

18.1 Introduction

In previous chapters, the design of digital filters was considered from an algorithmic and architectural standpoint. In general, the design process is a set of steps that can be expedited with the aid of a digital computer aided engineering (CAE) tools. There are times, however, when a design attempt results in a poor or questionable outcome. In particular, digital filters are considered to be intrinsically difficult to design when any of the following conditions exist.

- The filter order is very high.
- The filter has a narrow passband relative to the sampling frequency.
- The filter has a narrow transition band relative to the sampling frequency.

To achieve a satisfactory design, a way must be found to relax the design requirements in some acceptable sense. The methods developed in this section can often achieve that goal.

18.2 Multirate Systems

Digital signal processing systems generally accept a time series input and produce a time series output. In between, a signal can be modified in terms of its time and/or frequency domain attributes. One of the important functions that a digital signal processing system can serve is that of *sample rate converter*. As the name implies, a sample rate converter changes a system's sample rate from a value of f_{in} samples per second to a rate of f_{out} samples per second. Systems that contain multiple sample rates are called *multirate* systems.

Multirate systems are routinely found in audio signal processing applications where various subsystems having different sample rates (e.g., 48 kHz and 44.1 kHz) need to be interconnected. At other times, multirate systems are used to reduce the computational requirements of a system. Suppose, for example, an algorithm requires K operations to be completed per cycle. By reducing the sample rate of a signal or system by a factor M, the arithmetic bandwidth requirements are reduced from Kf_s operations per second to Kf_s/M (i.e., an M-fold decrease in computational bandwidth requirements). In other applications, resampling a signal at a lower rate will allow it to pass through a channel of limited bandwidth.

If a time series $x[k]$ is accepted at a sample rate f_{in} and exported at a rate f_{out}, such that $f_{in} > f_{out}$, then the signal is said to be *decimated*[*] by M, where

$$M = \frac{f_{out}}{f_{in}} \qquad (18.1)$$

If M is an integer, then the decimated time series is given by $x_d[k] = x[Mk]$, where every Mth sample of the original time series is retained. Therefore, the effective sample rate is reduced from f_s to $f_{dec} = f_s/M$ samples per second, as shown in Fig. 18.1 for $M = 2$.

18.3 Decimation

Decimation is the act of reducing a system's sample rate. The spectral properties of a decimated signal can be examined by exploring a decimated signal in the transform domain. A signal decimated by M is modeled as shown in Eq. (18.2).

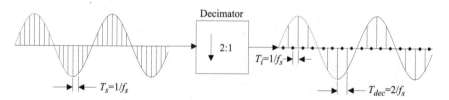

Figure 18.1 Decimation by $M = 2$.

---

[*] *Decimation* originally referred to a disciplinary method employed by the Romans in dealing with mutinous soldiers. The mutineers would be force to select balls from an urn containing ten times as many white balls as black balls. The holders of black balls would be put to the sword. Therefore, every tenth soldier would be slain or decimated.

$$x_d[k] = x[kM] = \sum_{m=-\infty}^{\infty} x[m]\delta[m-kM] \tag{18.2}$$

with the z-transform given by

$$X_d(z) = \sum_{k=-\infty}^{\infty} x_d[k]z^{-k}$$

$$= \sum_{k=-\infty}^{\infty}\left(\sum_{m=-\infty}^{\infty} x[m]\delta[m-km]^{-k}\right) = \sum_{k=-\infty}^{\infty} x[kM](z^{-Mk}) \tag{18.3}$$

or

$$X_d(z) = X(z^M) \tag{18.4}$$

The frequency signature of the decimated signal, relative to the undecimated parent signal, is given by

$$X_d(e^{j\omega}) = X(e^{jM\omega}) \tag{18.5}$$

which is seen to be a frequency-scaled version of the original signal spectrum repeated on $2\pi/M$ centers when plotted on a frequency axis defined with respect to the original sampling frequency (see Fig. 18.2).

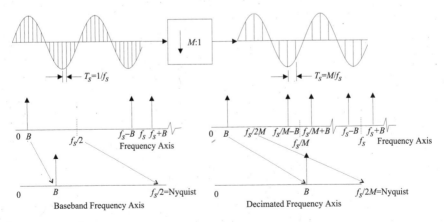

Figure 18.2 Spectrum of a decimated by M signal plotted against the original and decimated sampling frequency.

Example 18.1: Decimation

Problem Statement. An analog signal given by $x(t) = \cos(2\pi f_s t/16)$ is sampled at an f_s to form a time series $x[k]$. The DFT spectrum is given by $X(n)$. The time series and spectrum are shown in Fig. 18.3. What is the spectrum of the decimated by four version of $x[k]$?

Analysis and Conclusions. Equation (18.5) defines the shape of the resulting decimated spectrum, which is given by $X_d(n) = X(4n)$. The decimated time series and spectrum are also displayed in Fig. 18.3.

Computer Study. The s-file "dec1.s" contains a script that is used to produce the data shown in Fig. 18.3. The following Siglab command executes that script.

```
> include "chap18\dec1.s"
```

It can be seen that the spectrum is originally located at $f_s/16$ relative to f_s and is scaled proportionally when related to $f_s/4$ along the decimated frequency axis with respect to the decimated sample frequency as shown in Eq. (18.5).

The effects of decimation can also be studied using a soundboard. Assuming that a 16-bit soundboard is available having sample rates $f_1 = 11,025$ and $f_2 = 4 \times f_1 = 44,100$ Hz, the test shown below can be conducted. A long, pure tone $x[k]$ is created at the frequency $f = f_2/32$ and decimated by four to produce $y[k]$. Playing $x[k]$ at f_2 and $y[k]$ at f_1, both signals sound to be identical.

```
> x=mkcos(0.9*2^15,1/32,0,40000)   # ~ 16-bit amplitude.
> y=dec(x,4); z = y&y&y&y           # 40000 sample decimated signal.
> wavout(x)                          # Play at 44100 Hz.
> wavout(y)                          # Play at 11025 Hz.          ❖
```

One should be aware that Shannon's sampling theorem also applies to decimation. Suppose the highest frequency found in $x[k]$ is B Hz. Aliasing can be avoided if the decimating sampling rate exceeds $f_d = 2B$ Hz. This means that there is a practical upper bound to the deci-

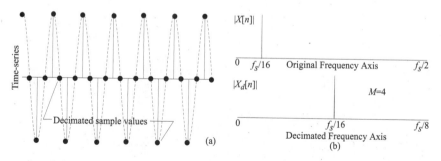

Figure 18.3   (a) The parent time series $x[k]$ and decimated by $M = 4$ time series $x_d[k]$, and (b) magnitude frequency responses $X[n]$ sampled at f_s and $X_d[n]$ sampled at $f_s/4$.

mation rate. Referring to Fig. 18.4, it can be seen that, for unaliased decimation to take place,

$$\frac{f_s}{M} - B > B \tag{18.6}$$

or

$$M < \frac{f_s}{2B} \tag{18.7}$$

Increasing the decimation rate beyond this value will introduce aliasing errors into the decimated time series as also shown in Fig. 18.4. In practice, the maximal decimation rate is rarely used. Instead, a more conservative value is generally employed which will allow for a well defined guardband to be established as shown in Fig. 18.4.

Example 18.2: Decimation Rate

Problem Statement. The highest frequency found in the signal $x(t) = \cos(2\pi \times 10^3 t)$ is $B = 10^3$ Hz. Suppose $x(t)$ is highly *oversampled* at a rate $f_s = 10^5$ Hz. What is the maximum decimation rate such that no aliasing will occur? What would happen if the signal is decimated beyond that maximum rate?

Analysis and Conclusions. The minimum lower bound on the sampling rate (i.e., Nyquist frequency) is 2×10^3 Hz. Therefore, the maximum decimation rate is bounded by $M < 10^5/2 \times 10^3 = 50$. The spectrum of the undecimated and decimated by 16 signals are reported in Fig. 18.5a over the common frequency range $f \in [0, 10^5/2)$. It can be seen that the decimated spectrum contains copies of the baseband signal on $f_s/16$ Hz centers. The resulting time series produced with a decimation by $M = 16 < 50$ operation is shown in Fig. 18.5b. The general shape of the original sinusoidal envelope is seen to persist in the decimated signal.

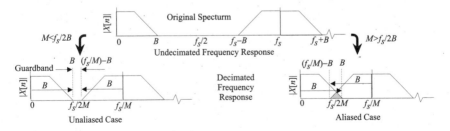

Figure 18.4 Unaliased and aliased decimation cases.

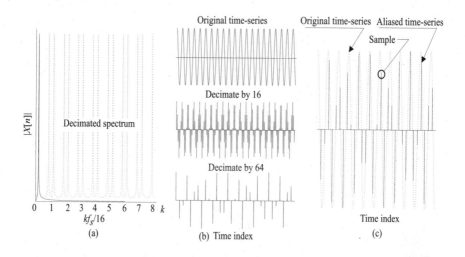

Figure 18.5 (a) Magnitude frequency responses (line spectrum) of $x[k]$ and the decimated by 16 signal, (b) original $x[k]$ and decimated by 16 and 64 time series, and (c) overlay of $x[k]$, the decimated by 64 time series, and aliased image having a normalized frequency of $1/64 - 1/100$, showing agreement at the sample instances.

To demonstrate what happens if the decimation rate is too high, consider the case where $M = 64 > 50$. The resulting time series is also shown in Fig. 18.5(b) and is seen to impersonate (alias) a signal having a frequency equal to $(1/64 - 1/100) \times 10^3$ Hz, which is superimposed over the decimated time series in Fig. 18.5(c).

Computer Study. The decimation time series is simulated using s-file "dec2.s" as shown below. To be able to compare signal over a common time and frequency interval, the decimated signals are displayed with respect to the original sample rate.

```
> include "chap18\dec2.s"
```

To this point, it has been assumed that the signal being decimated is a baseband signal. This need not be the case. Consider the band-limited signal shown in Fig. 18.6. Decimation can be used to rearrange the spectrum so as to place copies of the signals spectrum at baseband. In particular, consider

$$\frac{mf_s}{2M} \le f_s \le \frac{(m+1)f_s}{2M}$$

where M is the decimation rate, and m is a positive integer. The resulting decimated spectrum appears at baseband as shown in Fig.

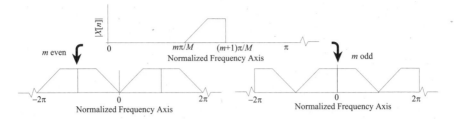

Figure 18.6 Decimation of a band-limited signal for *m* even and odd.

18.6. If *m* is odd, the spectrum is a reflection of the original spectrum, which can be compensated for (i.e., reversed) by multiplying the decimate time series $x_d[k]$ by $(-1)^k$.

Example 18.3: Decimating Bandlimited Signals

Problem Statement. Experimentally verify the predicted spectral behavior of a decimated bandlimited signal for both odd and even values of *m*. Verify that reflected spectrum can be corrected using a $(-1)^k$ modulation signal.

Analysis and Conclusions. Use a time series $x[k]$ having a shaped spectrum with edges defined at $f_{LO} = mf_s/M$ and $f_{HI} = (m + 1)f_s/M$. Decimate by *M* for *m* = 1(odd) and *m* = 2(even). The results are shown in Fig. 18.7. It can be seen that for *m* odd (*m* = 1) and a decimation factor of *M* = 8, the baseband spectrum is reflected about $f = 0$ (dc). The reflection can be corrected by modulating the decimated signal by $(-1)^k$. For *m* even (*m* = 2), the decimated spectrum is not reflected.

Computer Study. The s-file "`band.s`" defines the function "`band(k,m)`" which generates a 1024 sample signal that, in the frequency domain, is divided into *m* bands, and places an obviously oriented spectral signature in the *k*th band. The following sequence of Siglab commands creates two signals, one of which has an oriented spectral signature in band two of four (even band), while the other has an oriented spectral signature in band one of four (odd band). Next, the two signals are decimated by four, and their spectra are computed. Both signals are aliased to the baseband (see Fig. 18.7). The odd band signal is seen to have a reversed spectrum. Finally, the reversed spectrum is corrected using the binary modulation signal, which is produced using $(-1)^k = \cos(\pi k)$.

```
> include "chap18\band.s"        # Define the band function.
> xe=band(2,4)                   # Create even band (2 of 4) signal.
> xo=band(1,4)                   # Create odd band (1 of 4) signal.
> xa=mkcos(1,0.5,0,512)          # Create {1,-1,1,-1,..} signal.
> xef=mag(fft(xe))               # Compute spectrum of even signal.
> xof=mag(fft(xo))               # Compute spectrum of odd signal.
> xed4f=mag(fft(dec(xe,4)))  # Spectrum of even signal decimated by 4.
> xod4f=mag(fft(dec(xo,4)))  # Spectrum of odd signal decimated by 4.
> xod4fc=mag(fft(dec(xo,4).*xa)) # Corrected reversed odd spectrum.
> graph(xef,xed4f,xof,xod4f,xod4fc)      # Plot results.           ❖
```

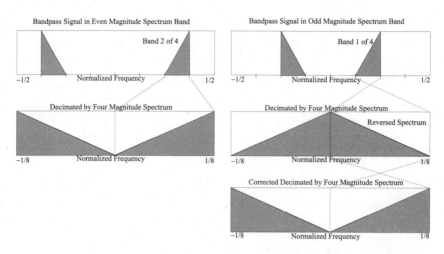

Figure 18.7 Even (left) and odd (right) bandpass spectra decimated into the baseband.

18.4 Interpolation

The antithesis of decimation is called *interpolation*. The use of the term *interpolation* is somewhat unfortunate, since it has previously been used to define methods of reconstructing a facsimile of $x[k]$ from a sparse set of samples contained in $x_d[k]$. In the context of decimation and interpolation, interpolation simply refers to a mechanism that increases the effective sample rate of a signal. Formally, suppose a signal $x_d[k]$ is interpolating by a factor N to create $x_i[k]$, then

$$x_i[k] = \begin{cases} x_d[k/N] & \text{if } k = 0 \bmod N \\ 0 & \text{otherwise} \end{cases} \tag{18.8}$$

or as it sometimes appears,

$$x_i[k] = \sum_{m=-\infty}^{\infty} x_d[k]\delta[k-mN] \tag{18.9}$$

That is, the act of interpolation inserts $N - 1$ zeros in between the samples of the original time series. This action is sometimes referred to as *zero-padding*. The result is that a time series sampled at a rate f_{in} is interpolated into a new time series sampled at rate $f_{out} = Nf_{in}$, as shown in Fig. 18.8.

Interpolation is often directly linked to decimation. Suppose $x_d[k]$ is a decimated by M version of a time series $x[k]$, which was sampled at a

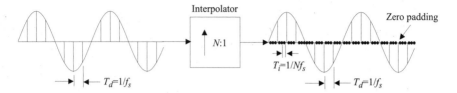

Figure 18.8 Interpolation example.

rate f_s. Then, $x_d[k]$ contains only every Mth sample of $x[k]$ and is defined with respect to a decimated sample rate $f_d = f_s/M$. Interpolating $x_d[k]$ by N would result in a time series $x_i[k]$, where $x_i[Nk] = x_d[k]$ and zero otherwise. The sample rate of the interpolated signal would be increased from f_d to $f_i = Nf_d = Nf_s/M$. If $N = M$, it can be seen that the output sample rate would be restored to f_s, although the original signal would *not* be restored.

Relative to the decimated $x_d[k]$, the frequency-domain signature of an interpolated by N signal, $x_i[k]$, can be defined in terms of the z-transform of Eq. (18.9),

$$X_i(z) = \sum_{m=-\infty}^{\infty} x_d[m]z^{-mN} = X_d(z^N) \tag{18.10}$$

which defines the frequency signature of $x_i[k]$ to be

$$X_i(e^{j\omega}) = X_d(e^{jN\omega}) \tag{18.11}$$

Therefore, the frequency-domain representation of the zero-padded interpolated signal is that of the original signal scaled by N as shown in Fig. 18.9. As expected, the complex exponential gives rise to periodic replications of the spectrum on centers that are integer multiples of the original sample rate.

Example 18.4 Interpolation

Problem Statement. Interpolation can be explained in terms of applying a gating function to a predecimated signal as suggested in Fig. 18.10. For a given N, compute the frequency response of the interpolated signal and compare to that of the original time series.

Analysis and Conclusions. The result of interpolating a low-frequency multitone process sampled at a rate of f_s/N by an interpolation factor $N = 16$ is displayed in Fig. 18.10. Observe that the resulting interpolated spectrum contains copies of the baseband spectrum located on f_s/N ($N = 16$) centers.

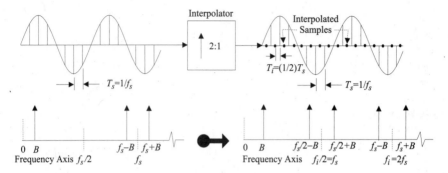

Figure 18.9 Frequency response of a zero-padded interpolated signal for $N = 2$.

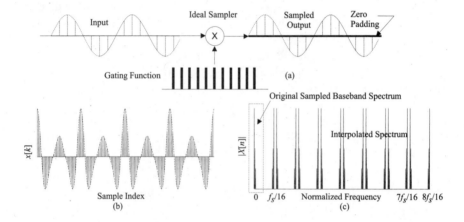

Figure 18.10 (a) Gating function model, (b) interpolated signal and signal envelope, and (c) interpolated spectrum and spectrum of the envelope.

Computer Study. The s-file "dec4.s" defines a function that produces an interpolated time series and spectrum. The function first decimates the multitone time series by N and then interpolates the decimated time series by N. The format of the function is given by "interp(N,f)" where N is the decimation/interpolation factor and f is a seed frequency for the generation of a multitone sinusoidal process. The following sequence of Siglab commands produces the data shown in Fig. 18.10.

```
> include "chap18\dec4.s"
> interp(16,1/256)                                        ❖
```

The signals found at the output of the interpolators shown in Figs. 18.9 and 18.10 are seen to contain multiple copies of the baseband spectrum. The unwanted copies generally need to be removed before the interpolated signal can be made useful. If the interpolated signal is first converted into an analog domain using a DAC, a simple RC lowpass circuit can sometimes be used to eliminate the unwanted extraneous spectral components. If the elimination of the unwanted copies of the baseband spectrum from the interpolated spectrum is to be performed digitally, an *ideal baseband filter* is optimal (see Fig. 18.11). Recall that an ideal baseband filter has a magnitude frequency response given by $|H(\omega)| = 1$ if $\omega \in [-\pi f_s/N, \pi f_s/N]$ and zero elsewhere. For an arbitrary interpolation integer N, the time-domain impulse response of the ideal baseband filter is given by

$$h\lfloor k \rfloor = \sum_{i=-\infty}^{\infty} \frac{\sin[\pi(k-i)/M]}{\pi(k-i)} \tag{18.12}$$

The ideal lowpass filter's impulse response has a classic $\sin(x)/x$ shape and is non-causal [i.e., persists for all time $k \in (-\infty, \infty)$] and is therefore not physically realizable. As a result, the ideal Shannon interpolation filter is generally only approximated by a practical and realizable interpolator, such as a zero or first order hold circuit. Another approximate interpolation technique is a simple lowpass filter having a passband covering the range filter $f \in [-f_s/2N, f_s/2N]$.

18.5 Polyphase Representation

Interpolation and decimation systems can be studied in a piecemeal fashion. However, a more robust and convenient means of mathematical modeling multirate systems is called the *polyphase decomposition*. Polyphase models are defined in the z-domain and are used to represent an arbitrary time series $x[k]$, sampled at a rate of f_s samples per second. The defining discrete-time signal model is given by

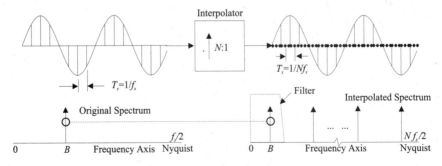

Figure 18.11 Spectra of decimated signal, zero-filled interpolated signal, and filtered baseband signal. Also shown is an overlay of an ideal anti-aliasing filter.

$$X(z) = \sum_{k=-\infty}^{\infty} x[k]z^{-k} \tag{18.13}$$

Suppose the time series $x[k]$ is partitioned into M sequences of data, as shown in Fig. 18.12, where the ith data sequence is given by

$$x_i[k] = \{..., x[i], x[i+M], x[i+2M]...\} \tag{18.14}$$

The ith block can be seen to be the original time series delayed by "i" samples and then decimated by M (i.e., retain every Mth sample beginning with $x[i]$). The z-transform of the decomposed time series can be expressed as

$$X(z) = \begin{bmatrix} ... \ (x[0] + z^{-M}x[M] + z^{-2M}x[2M] + ...) \\ ... \ (x[1]z^{-1} + z^{-(M+1)}x[M+1] + z^{-(2M+1)}x[2M+1] + ...) \\ ... \qquad ... \qquad ... \qquad ... \qquad ... \\ ... \ (x[M-1]z^{-(M-1)} + z^{-(2M-1)}x[2M-1] + z^{-(3M-1)}x[3M-1] + ...) \end{bmatrix} \tag{18.15}$$

Alternatively, the above expression can be regrouped and factored to read

$$X(z) = \begin{bmatrix} ... \ (x[0] + z^{-M}x[M] + z^{-2M}x[2M] + ...) \\ ... \ z^{-1}(x[1] + z^{-M}x[M+1] + z^{-2M}x[2M+1] + ...) \\ ... \qquad ... \qquad ... \qquad ... \qquad ... \\ ... \ z^{-(M-1)}(x[M-1] + z^{-M}x[2M-1] + z^{-2M}x[3M-1] + ...) \end{bmatrix} \tag{18.16}$$

The ith row of Eq. (18.16) can now be defined in terms of the ith *polyphase function* given by

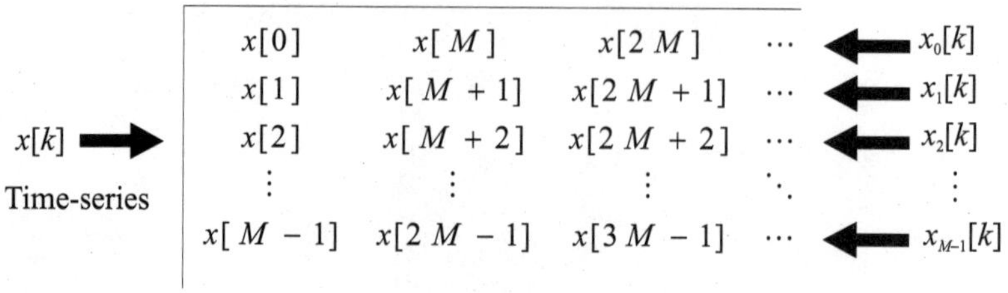

Figure 18.12 Block decomposition of the signal space.

$$P_i(z) = \sum_{k=-\infty}^{\infty} x(kM - i)z^{-k} \tag{18.17}$$

and as a result, Eq. (18.16) can be used to synthesize $X(z)$ as

$$X(z) = \sum_{i=0}^{M-1} z^{-i} P_i(z^M) \tag{18.18}$$

The resulting compact representation is called an M-component polyphase decomposition of a time series $x[k]$.

Example 18.5: Polyphase Decomposition

Problem Statement. Consider the time series

$$x[k] = \{...,0,1,2,3,4,3,2,1,0,1,2,3,4,3,2,1,0,1,2,3,4,3,2,1,...\}$$

where $x[0] = 0$, and $M = 4$. What is its polyphase representation?

Analysis and Conclusions. From Eq. (18.17) it follows that $P_0(z) = \{... + 0z^0 + 4z^{-1} + 0z^{-2} + 4z^{-3} + ...\}$, $P_1(z) = \{... + 1z^0 + 3z^{-1} + 1z^{-2} + 3z^{-3} + ...\}$, $P_2(z) = \{... + 2z^0 + 2z^{-1} + 2z^{-2} + 2z^{-3} + ...\}$, and $P_3(z) = \{... + 3z^0 + 1z^{-1} + 3z^{-2} + 1z^{-3} + ...\}$. Therefore, $X(z) = P_0(z^4)z^0 + P_1(z^4)z^{-1} + P_2(z^4)z^{-2} + P_3(z^4)z^{-3}$, which is the correct result. Notice that $P_0(z)$ represents the z-transform of a decimated by four version of $x[k]$ containing $x[0]$. ❖

Polyphase representation methods will be used to model and analyze decimation systems. Interpolation can be described in terms of the *transposed polyphase function,* denoted $Q_i(z)$, which is related to the polyphase function $P_i(z)$ through

$$Q_i(z) = P_{M-1-i}(z) \tag{18.19}$$

The results expressed in Eq. (18.18) can therefore be defined in transpose polyphase form as

$$X(z) = \sum_{i=0}^{M-1} z^{-(M-1-i)} Q_i(z^M) \tag{18.20}$$

The mechanics of the polyphase and transpose polyphase decomposition are summarized in Fig. 18.13. Notice that, for the case $M = 4$, the solution consists of four parallel channels along with $M - 1 = 3$ shift registers to properly phase the signals. Each channel consists of a 4:1 decimator and a path from input to output which is being clocked at $f_s/4$. The bandwidth requirement of each individual channel, therefore,

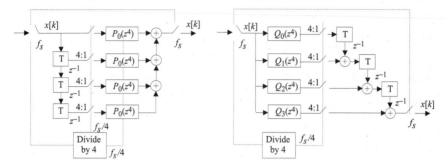

Figure 18.13   Example $M = 4$ polyphase multirate system.

is $1/M = 1/4$th that of a direct input/output path. The interleaved polyphase components can then be recombined using Eq. (18.18) to reconstruct the original signal at the original sample rate. This points out a possible utility of multirate systems, which is channel bandwidth compression.

A polyphase representation also can be used to explain the act of decimation. Consider a time series $x[k]$, which is decimated by a factor M to produce a new time series $x_d[k]$ given by

$$x_d[k] = x[Mk] \tag{18.21}$$

Suppose that the z-transform of $x[k]$ is known, and is given by

$$X(z) = \sum_{k=-\infty}^{\infty} x[k]z^{-k} \tag{18.22}$$

with respect to a sample frequency f_s and a sample delay of $T_s = 1/f_s$. The z-transform of the decimated time series can then be expressed as

$$X_d(z) = \sum_{k=-\infty}^{\infty} x_d[k]z^{-k} = P_0(z) \tag{18.23}$$

With some algebraic manipulation, Eq. (18.23) can be expressed as

$$X_d(z) = P_0(z) = \frac{1}{M}\sum_{k=0}^{M-1} X(W_M^k z^{1/M}) \tag{18.24}$$

where $W_M = e^{-j2\pi/M}$. The term z^{-1} refers to a single clock delay relative to the decimated sample rate $f_d = f_s/M$ or, equivalently, a delay of $T_d = M/f_s$. Therefore, $z^{-1/M}$ explicitly corresponds to a delay $T_d/M = T_s = 1/f_s$. This is illustrated in the following example.

Example 18.6: Polyphase Spectrum

Problem Statement . The time series $x[k] = a^k u[k]$, where $|a| < 1$, represents a decaying exponential signal being sampled at a rate of f_s samples per second. Assume that the decimation rate is $M = 2$ and $M = 4$. What is the polyphase representation of $x[k]$ and its z-transform?

Analysis and Conclusions. From Table 2.1, it follows that $X(z) = 1/(1 - az^{-1})$ where z^{-1} corresponds to a delay of $1/f_s$ seconds. The decimated signal is given by $x_d[k] = x[2k] = \{1, a^2, a^4, ...\}$. The z-transform of the decimated signal satisfies $X_d(z) = 1/(1 - a^2 z^{-1})$, where z^{-1} noncorresponds to the delay of the decimated signal, which is given by $T_d = 2/f_s$ seconds.

For $M = 2$, Eq. (18.24) requires $W_2^0 = 1$, $W_2^1 = -1$. It then follows that

$$X_d(z) = \frac{1}{2}[X(W_2^0 z^{1/2}) + X(W_2^1 z^{1/2})]$$

where, at the decimated sample rate of $f_d = f_s/2$, the undecimated component terms are

$$X(W_2^0 z^{1/2}) = X(z^{1/2}) = \sum_{k=0}^{\infty} a^k z^{-k/2} = \frac{1}{1 - az^{-1/2}} \leftrightarrow \{1, a, a^2, a^3, ...\}$$

and

$$X(W_2^1 z^{1/2}) = X(-z^{1/2}) = \sum_{k=0}^{\infty} (-1)^{-k/2} a^k z^{-k/2} = \frac{1}{1 + az^{-1/2}} \leftrightarrow \{1, -a, a^2, -a^3, ...\}$$

It can immediately be seen that combining these terms, according to Eq. (18.24), results in $X_d(z)$ or $x_d[k]$.

For $M = 4$, it follows that $x_d[k] = x[4k] = \{1, a^4, a^8, ...\}$ and

$$X_d(z) = \frac{1}{4}[X(W_4^0 z^{1/4}) + X(W_4^1 z^{1/4}) + X(W_4^2 z^{1/4}) + X(W_4^3 z^{1/4})]$$

where, at the sample rate of $f_d = f_s/4$, the component terms are defined with respect to f_s and satisfy

$$X(W_4^0 z^{1/4}) = X(z^{1/4}) = \sum_{k=0}^{\infty} a^k z^{-k/4} = \frac{1}{1 - az^{-1/4}} \leftrightarrow \{1, a, a^2, a^3, a^4, ...\}$$

$$X(W_4^1 z^{1/4}) = X(-jz^{1/4}) = \sum_{k=0}^{\infty} (-j)^{-k/4} a^k z^{-k/4} = \frac{1}{1 + jaz^{-1/4}} \leftrightarrow \{1, -ja, -a^2, ja^3, a^4, ...\}$$

$$X(W_4^0 z^{1/4}) = X(-z^{1/4}) = \sum_{k=0}^{\infty} (-a)^k z^{-k/4} = \frac{1}{1 + az^{-1/4}} \leftrightarrow \{1, -a, a^2, -a^3, a^4, ...\}$$

and

$$X(W_4^3 z^{1/4}) = X(jz^{1/4}) = \sum_{k=0}^{\infty} (j)^{-k/4} a^k z^{-k/4} = \frac{1}{1-jaz^{-1/4}} \leftrightarrow \{1, -ja, -a^2, -ja^3, a^4, \ldots\}$$

It can immediately be seen that combining these terms, in either the transform or time domain, will produce $X_d(z)$ or $x_d[k] = \{1, a^4, a^8, \ldots\}$. ❖

A logical question to pose relates to the location of the decimator in a signal processing stream. The two systems shown in Fig. 18.14 are functionally equivalent. The topmost path of Fig. 18.14 consists of a decimator and a filter. The bottom path consists of a filter that is identical to that found in the top loop, except its clock is running at M times the rate of that found in the top loop. Both designs have the same number of coefficients and therefore the same arithmetic complexity. The major difference between the circuits is found in the rate at which the coefficient multiplication must be performed. The topmost filter has an internal data rate $1/M$th that of the bottom filter. Therefore, the top architecture is generally preferred due to its lower real-time computational requirement.

Example 18.7: Polyphase Filter Description

Problem Statement. Consider a filter $H(z) = 2 + 3z^{-1} + 3z^{-2} + 2z^{-3}$ and $M = 2$. Implement $H(z)$ using the polyphase architectures shown in Fig. 18.15.

Analysis and Conclusions. Observe that

$$H(z) = P_0(z^2) + z^{-1}P_1(z^2) = [2 + 3z^{-1}] + z^{-1}[3 + 2z^{-1}]$$

It then follows that the circuits shown in Fig. 18.15 are equivalent. The data being filtered by circuit A arrives at the input at a rate half of that seen by circuit B. Therefore, circuit A would have the lowest demands placed on its arithmetic unit, which is obviously desirable. ❖

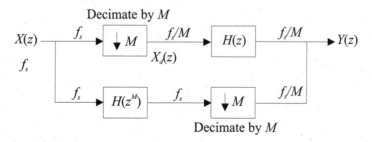

Figure 18.14 Equivalent decimated systems.

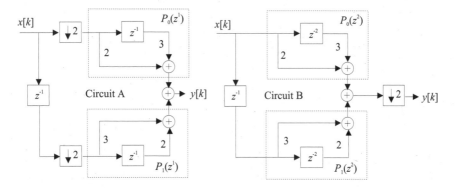

Figure 18.15 Equivalent decimating filters.

Closely aligned to the process of decimation and interpolation is the concept of *oversampling*. Oversampling corresponds to the intentional increase in sample rate to achieve a desirable effect. Two cases where oversampling is commonly encountered are:

- Replacing expensive, low-frequency, high-precision ADCs/DACs with inexpensive, mid-range, low-precision ADCs/DACs

- Relaxing the design requirements on the analog conditioning systems interfaced to a DSP system

An example of the latter follows.

Example 18.8: Audio CD-ROM

Problem Statement. Assume that an audio signal is bandlimited to 20 kHz. The class of signal can be digitized above the Nyquist frequency using an industry standard 44.1 kHz ADC or at a higher rate. Compare the analog anti-aliasing filter requirements of each design strategy.

Analysis and Conclusions. Based on a 44.1 kHz sampling rate, the audio spectrum is copied on 44.1 kHz centers. Between the spectrum centered at 0 Hz and 44.1 kHz is a frequency band void of spectral energy of width $\Delta f = f_s - 2 \times 20 \times 10^3 = 4.1$ kHz.

To ensure that no aliasing errors occur, an analog anti-aliasing filter is used. The lowpass anti-aliasing analog filter would have a 20 kHz passband, a stopband beginning at 24.1 kHz, and a transition band of Δf. This is generally considered to be a steep-skirt analog filter. A relaxed design requirement would, for example, increase the width of the anti-aliasing filter's transition band. This immediately suggests that a higher sample rate be used.

Suppose four-times over-sampling is used (i.e., $f_s = 4 \times 44.1 \times 10^3 = 176.4 \times 10^3$), then the unoccupied frequency range existing between the baseband audio spectrum and that centered about f_s is given by $\Delta f = 176.4 \times 10^3 - 2 \times 20 \times 10^3 = 136.4$ kHz. While the anti-aliasing filter would again have a passband of

20 kHz. The transition band, however, is now 136.4 kHz, which is easily realized with a low-order analog filter.

To demonstrate the design differences, refer to Fig. 18.16 where the ±20 kHz baseband spectrum is displayed along with the shape of the analog filter for f_s = 44.1 kHz and 136.4 kHz. For f_s = 44.1 kHz, a difficult to design steep skirt filter results. For f_s = 136.4 kHz, a simple low-order filter results.   ❖

Finally, a commonly encountered problem is interfacing two systems having dissimilar rates—say f_{in} and f_{out}, respectively. Such a system was earlier called a *sample rate converter*. If the ratio of f_{in} to f_{out}, or vice versa, is a rational fraction, then direct decimation of interpolation can be used. Suppose that the sample rates are related by the rational number $k = N/M$, where $f_{in} = kf_{out}$. The system described in Fig. 18.17, called a non-integer sample-rate converter, will perform a rational conversion. The filters, whether separate or combined, either "clean up" an interpolated spectrum or provide anti-aliasing services to the decimator.

In some cases, the rates to be converted are related to one another by a rational k where the numerator and denominator are not small integers. For example, consider the problem of converting compact disc digital audio, which is sampled at 44.1 kHz, to the 48 kHz sampling rate used by digital audio tape. Here, k = 48,000/44,100 = 160/ 147. Since this fraction cannot be reduced any further, it is necessary

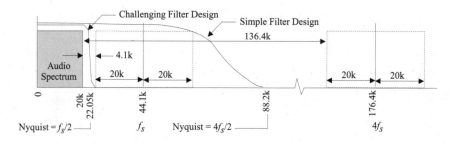

Figure 18.16   Oversample experiment.

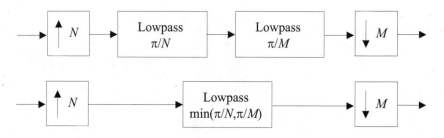

Figure 18.17   Non-integer decimated system.

to interpolate the 44.1 kHz CD audio by a factor of 160 to 7.056 MHz. While, this is unattractive, it is not impossible. Now, before decimating the signal, it is necessary to anti-alias filter the 7.056 MHz interpolated data stream using a lowpass filter with a passband of 1/160th of the Nyquist frequency, (1/160) × 3.528 MHz = 22,050 Hz. This is an *extremely* steep-skirt lowpass filter and is not likely to be practical. Current industry practice is to use a DAC operated at the input sample rate connected to an ADC operated at the output sample rate. In fact, for digital audio applications there are currently single-chip devices that use this process.

18.6 Subband Filters

Multirate systems often appear as a bank of filters where each filter maps the input into a *subband*. Refer to Fig. 18.18, which separates the filters of a multirate system into analysis and synthesis sections. The analysis filters, denoted $H_i(z)$, convolve the input into M subbands. The synthesis filters, denoted $F_i(z)$, put the signal back together from the subband components. This structure is common to many bandwidth compression and signal enhancement applications. The subband filters can be designed so that the data rate requirements of each channel shown in Fig. 18.17 are 1/Mth that of the input. Thus, low data rate channels can be used to communicate analysis data to the synthesizer for reconstruction.

18.7 DFT Filter Bank

Subband architectures are used in many applications. Some have unique properties. An interesting application of a subband architecture is called a *uniform DFT filter bank*. The ith filter in a DFT filter bank is denoted $H_i(z)$ in Fig. 18.18 and is defined in terms of H_0 as

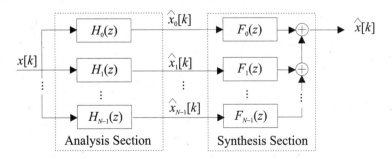

Figure 18.18 Typical subband decomposition system showing the analysis (left) and synthesis (right) filters.

$$H_i(z) = H_0(W_M^i z) \qquad (18.25)$$

for $i \in \{0,1,2,...,M-1\}$. The frequency response of the ith filter is given by

$$H_i(e^{j\omega}) = H_0(e^{j(\omega - 2i\pi/M)}) \qquad (18.26)$$

It can be seen that the frequency response envelope of $H_0(z)$ filter is copied to new center frequencies located at $f_i = if_s/M$ as shown in Fig. 18.19.

Consider the case where $H_0(z)$ is defined in terms of the polyphase representation

$$H_0(z) = \sum_{k=0}^{M-1} z^{-k} P_{0k}(z^M) \qquad (18.27)$$

From Eq. (18.26), this implies that

$$H_i(z) = \sum_{k=0}^{M-1} W_M^{-ik} z^{-k} P_{0k}(z^M) \qquad (18.28)$$

which has the structure of a DFT (i.e, $X[n] = (W^{nk} x[k])$). That is,

$$h_i(k) = \sum_{j=0}^{M-1} W_M^{-ik} y_j[k] \qquad (18.29)$$

where $y_j[k]$ is the output of the jth polyphase filter shown in Fig. 18.20. Equivalently, it follows that $Y_k(z) = z^{-k} P_{0k}(z^M)$. According to Eq. (18.28) or (18.29), the polynomial outputs are combined using an

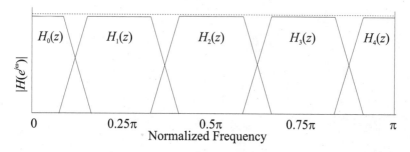

Figure 18.19   Uniform DFT filter bank magnitude frequency response for $M = 8$.

M-sample DFT as shown in Fig. 18.20a to synthesize the frequency-selective filters $h_i[k]$. The DFT filter bank can, therefore, leverage the reduced arithmetic complexity of a well designed DFT. To motivate this claim, consider that each filter $H_i(z)$ is on the order N. A filter bank would therefore require MN multiplies to complete a filter cycle. Each polyphase filter, however, is of order N/M, which reduces the complexity of each filter cycle to $MN/M = N$ multiplies, that is, in N multiplies, all the intermediate values of $y_i[k]$. (See Fig. 18.20a.) Adding decimators as shown in Fig. 18.20b, the bandwidth requirement of each of the polyphase filters $H_i(z)$ can be reduced. The multiply count of an M sample DFT would have to be added to this count to complete the multiply audit for a filter cycle. The multiply count of a DFT is dependent on M and can, for some choices, be extremely small.

It can be noted that the data moving through the polyphase filters is slowed by a factor of M through decimation. By placing decimator in the filter bank at the locations suggested in Fig. 18.20b, a further complexity reduction can be realized.

Example 18.9: DFT Filter Bank

Problem Statement. The filter studied in Example 18.7 is to be interpreted in the context of a DFT filter bank. What is the impulse response of the system?

Analysis and Conclusions. The generating filter bank is given by

$$H_0(z) = 2 + 3z^{-1} + 3z^{-2} + 2z^{-3}$$

which has a polyphase representation given by

$$H_0(z) = P_{00}(z^2) + z^{-1} P_{01}(z^2)$$

where

$$P_{00}(z) = 2 + 3z^{-1} \quad \text{and} \quad P_{01}(z) = 3 + 2z^{-1}$$

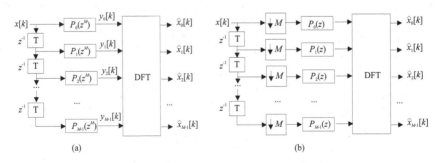

Figure 18.20   (a) DFT filter bank and (b) DFT filter bank with decimators.

Then, from Eqs. (18.27) and (18.28), it follows that

$$H_0(z) = \sum_{i=0}^{1} W_2^0 z^{-i} P_{0i}(z^2) = P_{00}(z^2) + z^{-1} P_{01}(z^2) = 2 + 3z^{-1} + 3z^{-2} + 2z^{-3}$$

$$H_1(z) = \sum_{i=0}^{1} W_2^{-i} z^{-i} P_{0i}(z^2) = P_{00}(z^2) - z^{-1} P_{01} z^2 = 2 - 3z^{-1} + 3z^{-2} - 2z^{-3}$$

The impulse response is graphically interpreted in Fig. 18.21. The impulse response measured along the top path is seen to be equal to $\{2,3,3,2\}$, which corresponds to $h_0[k]$. The impulse response measured along the bottom path is seen to be equal to $\{2,-3,3,-2\}$, which corresponds to $h_1[k]$.   ❖

18.8 Quadrature Mirror Filter

Multirate systems are often used to reduce the sample rate to a value that can be passed through a bandlimited communication channel. *Quadrature mirror filters* (QMFs) are a popular means of performing a subband signal decomposition into channels having reduced bandwidths. The basic architecture of a two-channel QMF system is shown in Fig. 18.22. The two-channel QMF system establishes two input-out-

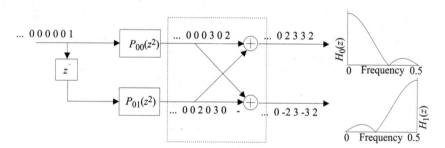

Figure 18.21   DFT filter bank and impulse response for M = 2.

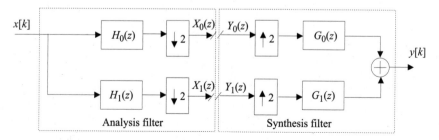

Figure 18.22   QMF filter architecture.

put paths, each having a bandwidth requirement that is half the original bandwidth requirements. Using this technique, a 2N-channel QMF can be designed using an N-level binary tree architecture. The top path shown in Fig. 18.22 contains a lowpass filter, and the bottom path contains a highpass filter. The signals at the top and bottom, after decimation, can be expressed in terms of Eq. (18.24) to read

$$X_0(z) = \frac{1}{2}\left\{ X(z^{1/2})H_0(z^{1/2}) + X(-z^{1/2})H_0(-z^{1/2}) \right\}$$

$$X_1(z) = \frac{1}{2}\left\{ X(z^{1/2})H_1(z^{1/2}) + X(-z^{1/2})H_1(z^{1/2}) \right\}$$

$$\text{(18.30)}$$

Noting that $Y_0(z)$ and $Y_1(z)$ are $X_0(z)$ and $X_1(z)$ upsampled by a factor of two, it follows that $Y(z)$ can be expressed as

$$Y(z) = G_0(z)Y_0(z^2) + G_1(z)Y_1(z^2)$$

$$= \frac{1}{2}\{[H_0(z)G_0(z) + H_1(z)G_1(z)]X(z)\}\{\text{unaliased term}\}$$

$$+ \frac{1}{2}\{[H_0(-z)G_0(z) + H_1(-z)G_1(z)]X(-z)\}\{\text{aliased term}\}$$

$$\text{(18.31)}$$

which consists of both an alias-free and an aliased term. The production of the aliased terms is graphically interpreted in Fig. 18.23.

The first term (unaliased) in Eq. (18.29) will reconstruct $X(z)$, provided that the second term (aliased) equals zero. Specifically, it is required that

$$\{[H_0(-z)G_0(z) + H_1(-z)G_1(z)]X(-z)\} = 0 \qquad \text{(18.32)}$$

which is satisfied if $G_0(z) = H_1(-z)$ and $G_1(z) = -H_0(-z)$. A special case assumes that $H_0(z)$ and $H_1(z)$ are subband filters satisfying the mirror filter relationship

$$H_1(z) = H_0(-z) \qquad \text{(18.33)}$$

In the z-domain, the assumption results in the QMF condition

$$Y(z) = k[H_0^2(z) - H_0^2(-z)]X(z) = T(z)X(z) \qquad \text{(18.34)}$$

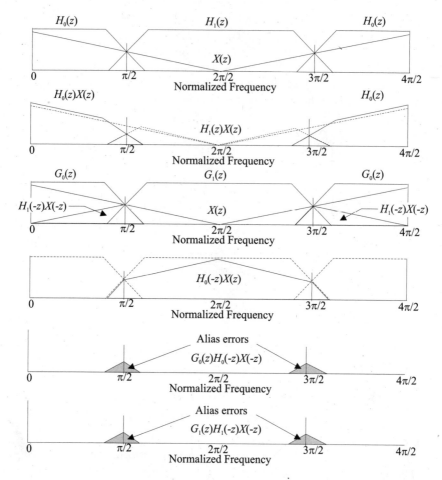

Figure 18.23 Graphic interpretation of aliasing in a QMF.

where k is a real scale factor introduced by decimation. Recall that Eq. (18.33) defines $H_1(z)$ to be the mirror image of $H_0(z)$ in that

$$H_1(e^{j\omega}) = H_0(e^{j(\omega - \pi)}) \tag{18.35}$$

From an implementation standpoint, the conversion of a filter having a z-transform

$$H(z) = \sum_{i=-L}^{L} h_i z^{-i} \tag{18.36}$$

to its mirror image is given by

$$H_{\text{mirror}}(z) = H(-z) = \sum_{i=-L}^{L} h_i(-z)^{-i} = \sum_{i=-L}^{L} (-1)^i h_i z^{-i} \qquad (18.37)$$

Equation (18.37) demonstrates that the coefficients of the parent filter are modified by $(-1)^i$ (i.e., sign change) in the mirrored version.

Example 18.10: Mirror Filter

Problem Statement. A 63rd-order lowpass FIR filter, having a transition band that is symmetric about the quarter sampling frequency (i.e., $f_{\text{transition}} = f_s/4 = f_{\text{Nyquist}}/2$) is called a *halfband* filter (see Example 11.2). A known property of a halfband filter is that, except for the center tap coefficient, every other filter coefficient is zero (i.c., the even numbered). That is, the impulse response of a typical FIR filter is $h[k] = \{...,h_{-5},h_{-4},h_{-3},h_{-2},h_{-1},h_0,h_1,h_2,h_3,h_4,h_5,...\}$, whereas a halfband FIR filter has an impulse response given by $h_{\text{halfband}}[k] = \{...,h_{-5},0,h_{-3},0,h_{-1},h_0,h_1,0,h_3,0,h_5,...\}$. As a result, compared to an arbitrary FIR filter, a halfband FIR filter needs approximately half the number of multiply coefficients per filter cycle. Use Eq. (18.33) to produce the mirror image of a 63rd-order lowpass FIR filter.

Analysis and Conclusions. Upon multiplying the coefficients of the original filter by $(-1)^i$, a FIR filter having a magnitude frequency responses shown in Fig. 18.24 is produced.

Computer Study. The following sequence of Siglab commands was used to produce the data shown in Fig. 18.24.

```
> showinc on
> include "chap18\dec6.s"        # Execute script.
h=rf("chap18\halfband.imp")
mask=mkcos(1,1/2,0,len(h))       # +/-1 modulation signal
hmirror=mask.*h
pad=zeros(1024-len(h))
ograph(mag(pfft(h&pad)),mag(pfft(hmirror&pad)))
```
❖

In the study of FIR filters, the virtues of linear phase performance were established on numerous occasions. Suppose it is desired to de-

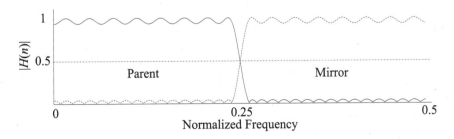

Figure 18.24 Original and mirror magnitude frequency responses.

sign an Nth-order linear phase QMF FIR filter. If N is odd, then it can be shown that a null will be placed in the output spectrum at the normalized frequency $\omega = \pi/2$. While an odd-order linear phase FIR filter can remove the aliasing errors from the output, it does not have a flat magnitude frequency response and therefore is distorting. If N is even, the only flat linear phase realization is trivial two-coefficient FIR filter having the form

$$H_0(z) = c_0 z^{-2n_0} + c_1 z^{-2(n_1 + 1)}$$

$$H_1(z) = c_0 z^{-2n_0} - c_1 z^{-2(n_1 + 1)}$$

$$(18.38)$$

for some n_0 and n_1. The frequency response of an even-order linear phase QMF, except for the trivial two-coefficient FIR filter case, is likewise non-flat. Since one of the objectives of a QMF is to split the input spectrum into subbands that can be more readily communicated across bandlimited channels, trivial QMFs have a very limited practical value.

Example 18.11: Linear Phase QMF

Problem Statement. Analyze the QMF filtering of $x[k] = \{0,1,2,3,0,0,...\}$ using a two-coefficient QMF linear phase filter based on the Haar basis functions $h_0[k] = [1,1]$ and $h_1[k] = [1,-1]$.

Analysis and Conclusions. The system shown in Fig. 18.20, called a *filter bank*, consists of decimators, approximation and detail filters, and various scale factors. The basis functions $h_0 = [1,1]$ and $h_1 = [1,-1]$ are second-order FIR filters that satisfy the condition $H_1(z) = H_0(-z)$. The filter h_0 is sometimes called the *approximation filter* since is smooths (averages) the two successive sample values. The filter h_1 is called the *detail filter* and responds to small detailed changes in $x[k]$ (i.e., differentiator). Refer to Fig.18.22, where a multirate system is partitioned into two sections called the *analysis section* and *synthesis section*. The synthesis filters are given by $G_0(z) = H_1(-z) = H_0(z)$ and $G_1(z) = -H_0(-z) = -H_1(z)$. From Eq. (18.34), it follows that $T(z) = [H_0^2(z) - H_0^2(-z) = 4z^{-1}$, which suggests that the output is scaled and delayed by one sample delay.

Computer Study. Using the sequence of Siglab commands shown below, the output of a linear phase QMF is computed, and then the original signal is reconstructed from the two subbands.

```
> include "chap18\interp.s"      # Get interpolation function.
> x=[0,1,2,3,0,0]; h0=[1,1]; h1=[1,-1]   # Test signal & Haar basis.
> xlow=x$h0; xlow                # Analysis filters.
     0   1   3   5   3   0   0
> xhi=x$h1; xhi
```

```
     0    1    1    1   -3    0    0
> xlowdec=dec(xlow,2); xlowdec    # Decimate data by two.
     0    3    3    0
> xhidec=dec(xhi,2); xhidec
     0    1   -3    0
> ylo=interp(xlowdec,2)            # Interpolate data by two.
> yhi=interp(xhidec,2)
> g0=[1,1]; g1=[-1,1]              # Define inverse Haar basis.
> zlo=g0$ylo; zlo                  # Synthesis filters.
     0    0    3    3    3    3    0
> zhi=g1$yhi; zhi
     0    0   -1    1    3   -3    0    0
> (zlo+zhi)/2                      # Synthesized signal.
     0    0    1    2    3    0    0    0
```

The above Siglab commands demonstrate the analysis of $x[k] = \{0,1,2,3,0,0,...\}$ into two subbands using a QMF, and the subsequent reconstruction operations, producing $z[k] = \{0, 0,1,2,3,0,0,...\}$, as predicted. The reconstruction powers of a QMF filter can also be studied acoustically using a soundboard. The sequence of Siglab commands below creates a long sinusoid that is passed through the QMF filter, performs all the requisite QMF operations, reconstructs the signal, and plays it through a sixteen-bit soundboard sampled at a 11.025 kHz rate. It can be observed that reconstructed signal is a good facsimile of the original.

```
> include "chap18\interp.s"       # Get interpolation function.
> x=mkcos(2^14,1/16,0,10000)       # Create test signal.
> wavout(x)                        # Play original test signal.
> h0=[1,1]; h1=[1,-1]              # Define analysis basis.
> xlow=x$h0; xhi=x$h1              # Perform QMF.
> xlowdec=dec(xlow,2); xhidec=dec(xhi,2)    # Decimate by two.
> ylo=interp(xlowdec,2)            # Interpolate by two.
> hyi=interp(xhidec,2)
> g0=[1,1]; g1=[-1,1]              # Define synthesis basis.
> zlo=g0$ylo; zhi=g1$yhi          # Perform QMF.
> qmf=(zlo+zhi)/2                  # Complete reconstruction.
> wavout(qmf)                      # Play reconstructed signal.    ❖
```

Designing a high-order QMF, unfortunately, is a challenging process. It is known that there does not exist any non-trivial, or physically meaningful, flat-response linear phase QMF filter. Most QMF designs represent some compromise. If the linear phase *or* perfect mirror condition [i.e., $H_1(z) = H_0(-z)$] is relaxed, then a magnitude or phase distortionless QMF system can be realized. A popular design paradigm is called the *perfect reconstruction QMF* (PRQMF). The output of a PRQMF system is equal to the input with a known delay. The PRQMF design procedure is given by the following recipe:

1. Define a linear phase FIR filter $F(z)$ to be a $2N - 1$-order halfband FIR filter having a ripple deviation δ.

2. Classify the zeros of the filter as being interior or exterior to the unit circle. Unfortunately many of the zeros of $F(z)$ lie on the unit circle and cannot be readily classified as being interior or exterior with respect to the unit circle. Therefore, add δ to the center tap weight of $F(z)$ to form $F_+(z) = F(z) + q\delta$ where $q > 1.0$ but close to unity. This action makes the minimum passband gain of $F_+(z)$ bounded from below by unity. $F_+(z)$ is also halfband and the factors of $F_+(z)$ (i.e., zeros) are moved slightly off the unit circle. The biasing of $F(z)$ in this manner lifts the zeros off the unit circle and forces them to be either interior of exterior.

3. Define two $(N - 1)$th order FIR filters, $H_0(z)$ and $H_1(z)$, satisfying $H_0(z) = H(z)$ and $H_1(z) = (-1)^{N-1}z^{-(N-1)}H(-z^{-1})$ with $F_+(z) = H(z)H(z^{-1})$, where the interior zeros belong to $H(z)$.

4. Let $G_0(z) = H_1(-z)$ and $G_1(z) = -H(-z)$

The result is an all-pass PFQMF system having an input-output transfer function $T(z) = Kz^{-(N-1)}$ where K is a constant.

Example: 18.12 PFQMF Filter

Problem Statement. Design a PRQMF system based on a 15th-order linear phase halfband filter satisfying

$$F(z) = -0.02648z^7 + 0.0441z^5 - 0.0934z^3 - 0.3139z^1 + 0.5$$
$$- 0.3139z^{-1} - 0.0934z^{-3} + 0.0441z^{-5} - 0.02648z^{-7}$$

Analysis and Conclusions. Creation of the PRQMF follows the given step-by-step processes.

Step 1. $F(z)$ is given, and $\delta = 0.0238$.

Step 2. Let $q = 1.01$ and produce $F_+(z)$ such that $F_+(z) = F(z)$ except at $z = 0$, where $F_+(0) = F(0) + q\delta$. The magnitude frequency response and zero distribution of $F_+(z)$ is shown in Fig. 18.25.

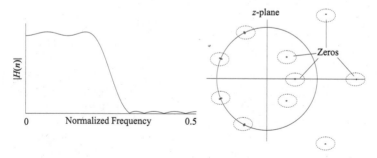

Figure 18.25 Halfband parent FIR filter and zero distribution of $F_+(z)$.

Step 3. The factors of $F_+(z)$ (up to the precision of the computing routine) are as shown in the following table:

z_i	$|z_i|$	Interior/exterior
$-0.939 \pm j0.398$	1.02	Exterior
$-0.903 \pm j0.382$	0.98	Interior
$-0.451 \pm j0.907$	1.013	Exterior
$-0.439 \pm j0.884$	0.987	Interior
$-0.394 \pm j0.427$	0.581	Interior
0.561	0.561	Interior
$-0.167 \pm j1.264$	1.72	Exterior
1.782	1.782	Exterior

Collecting the zeros residing within the unit circle and multiplying them together, one obtains (up to the level of precision of the computing routine), filters having the following approximate values:

$$H(z)=1.0+1.34z^{-1}+0.68z^{-2}-0.24z^{-3}-0.34z^{-4}+0.099z^{-5}+0.239z^{-6}-0.17z^{-7}=H_0(z)$$

$$H_1(z)=-0.17-0.24z^{-1}+0.099z^{-2}+0.31z^{-3}-0.24z^{-4}-0.68z^{-5}+1.34z^{-6}-1.0z^{-7}$$

$$G_0(z)=-0.17+0.24z^{-1}+0.099z^{-2}-0.31z^{-3}-0.24z^{-4}+0.68z^{-5}+1.34z^{-6}+1.0z^{-7}$$

$$G_1(z)=-1.0+1.34z^{-1}-0.68z^{-2}-0.24z^{-3}+0.34z^{-4}+0.099z^{-5}-0.24z^{-6}-0.17z^{-7}$$

The spectral shape and location of these FIR filters in the system are graphically reported in Fig. 18.26.

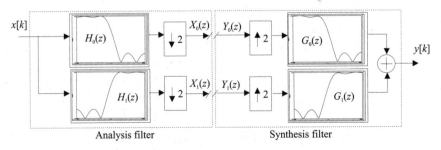

Analysis filter   Synthesis filter

Figure 18.26 $H_0(z)$, $H_1(z)$, $G_0(z)$, and $G_1(z)$ of a QMF filter.

Computer Study. The s-file "dec7.s" contains a script that analyzes the zero distribution of the target QMF FIR filter. The script is executed using the Siglab command shown below, with the annotated results following.

```
> include "chap18\dec7.s"
# Complex locations of the zeros of F + (z)
Columns 0 - 0
0: -0.939-0.398j  -0.939+0.398j  -0.903+0.382j  -0.903-0.382j
   -0.451+0.907j
Columns 5 - 0
0: -0.451-0.907j  -0.439+0.884j  -0.439-0.884j  0.394-0.427j 0.394+0.427j
Columns 10 - 0
0: 0.561+4e-16j  1.167-1.264j  1.167+1.264j  1.782+9e-09j
# Distance from the origin of the complex zeros of F + (z)
Columns 0 - 8
  0:  1.02  1.02  0.981  0.981  1.013  1.013  0.987  0.987  0.581
Columns 9 - 13
  0:  0.581  0.561  1.72  1.72  1.782
# Code shown for clarity.
poly1=re(poly([r[2],r[3]]))
poly2=re(poly([r[6],r[7]]))
poly3=re(poly([r[8],r[9]]))
poly4=re(poly([r[10]]))
polyfin=poly1$poly2$poly3$poly4 # polynomial multiplication
polyfin # H0(z)
  1  1.335  0.679  -0.245  -0.338  0.1  0.237  -0.178
```
❖

18.9 Wavelets

In Sec. 3.9, the STFT was introduced. At that time, it was noted that the STFT has value in the time-frequency studies of nonstationary signal processes. In practice, the STFT would be implemented using a DFT, or FFT, which would result in the uniform spectral coverage of the time-frequency plane. That is, every new N sample signal record is mapped by a DFT or FFT into a spectral image consisting of N harmonics distributed uniformly on $\Delta f = f_s/N$ Hz centers over $f \in [-f_s/2, f_s/2]$. This can be undesirable, since the energy found in many physically important signal processes is distributed in a $1/f$ manner. That is, the energy is proportionally concentrated at the lower frequencies. Thus, an individual high-frequency harmonic, by itself, may have little significance. Another problem with the DFT/FFT format is a general inability to localize signal behavior. A small temporal event, which may occupy only a few sample instances of an N-sample data record, would be averaged into insignificance by an N-sample DFT. Similarly, the inability of the STFT to support *local analysis* in the time-frequency plane is one its the shortcomings. A more logical framework for analyzing time-frequency processes is called the *wavelet transform*.

Before the wavelet transform is formally defined, it will be functionally compared to a standard Fourier transform, which is given by

$$X(\omega) = \int\limits_{-\infty}^{\infty} x(t)e^{-j\omega t}dt \qquad (18.39)$$

The Fourier transform is defined in terms of an infinite set of complex exponential basis functions $e^{j\omega t}$ that persist for all time. The reconstruction of $x(t)$ is accomplished by integrating the Fourier components as shown in Fig. 18.27, where c_i denotes a Fourier coefficient.

For a wavelet transform, frequency is replaced by concept of *scale size*. Scale size can be related to a period of oscillation or the geometric size of an object. The scaling of a signal, say $\lambda(t) = \cos(2\pi a\omega_0 t)$, by some scale factor a will stretch or dilate the time axis. As the scaling parameter increases, the frequency at which $\lambda(t)$ oscillates will decrease and increase. This concept is extensively used in the study of Fourier transforms where $\lambda(t)$ would be considered a basis function for a Fourier transformable signal $x(t)$. It should be apparent that, due to the infinite duration of $\lambda(t)$, such basis functions are geared to evaluate global rather than local phenomena.

A wavelet transform is also defined in terms of a set of basis functions. Wavelet basis functions, unlike their Fourier counterparts, are of finite duration (called *compact support*). As such, wavelet basis functions can be moved up and down the time axis to locally analyze a

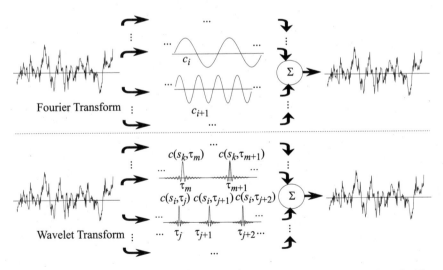

Figure 18.27 Comparison of a Fourier and wavelet transform with $c(\tau,s)$ given by Eq. (18.40).

signal. This cannot be done with the Fourier basis functions since $\lambda(t)$ and $\lambda(t - \tau)$ cover the same data record. In addition, wavelet basis functions can be scaled in a manner that alters the frequency response of the basis functions. For example, suppose $\phi(\tau,s,t)$ is a basis function of compact support for an integral expansion of some $x(t)$ given by

$$c(\tau, s) = \int_{-\infty}^{\infty} x(t)\phi(\tau, s, t)dt \qquad (18.40)$$

By shifting $\phi(\tau,s,t)$ by an amount τ, the analysis is localized to values of $x(t)$ near $t = \tau$. Scaling $\phi(\tau,s,t)$ by s will localize the analysis $x(t)$ to those scale sizes concentrated about a given scale size. Thus, the analysis of $x(t)$ can be tuned to meet a variety of needs through the choice of ϕ and s and τ, as shown in Fig. 18.27.

As previously noted, in practice, an STFT is computed using an N sample DFT or FFT. The frequency cover provided by an FFT is uniform with harmonics spaced on $\Delta f = f_s/N$ Hz centers. To an analog filter engineer, this would be called a constant bandwidth cover and would be the result of processing a signal $x(t)$ through a bank of N constant bandwidth band-selectable filters. Analog engineers would consider this to be an unnatural method to display time-frequency information. Audio and vibration engineers have, for decades, analyzed signal spectra in a log-frequency domain. Sometimes called *octave analysis*, each distinct spectral band is defined by a constant Q filter, where Q is defined by

$$Q = \frac{\omega_{HI} - \omega_{LO}}{\sqrt{\omega_{HI} - \omega_{LO}}} \qquad (18.41)$$

and ω_{HI} and ω_{LO} are the upper and lower -3 dB filter frequencies, respectively.

A family of analog constant Q filters, having constant energy passbands, can be defined in terms of a prototype filter having an impulse response $h(t)$. The constant Q filters can be defined in terms of $h(t)$ using scaling relationships. If the scaled filters are to have constant energy, then

$$\int_{-\infty}^{\infty} h_m^2(t)dt = \int_{-\infty}^{\infty} h^2(t)dt \qquad (18.42)$$

and scaled filters satisfy

$$h_m(t) = a^{-m/2} h(a^{-m}t) \qquad (18.43)$$

or, in the frequency domain

$$H_m(j\omega) = a^{-m/2} H(ja^m\omega) \qquad (18.44)$$

as shown in Fig. 18.28. Notice also that, as m increased, the filter's impulse response $h_m(t)$ spreads (elongates) and vice versa. Furthermore, the filter's center frequencies and bandwidth proportionally contract.

A discrete-time version of the analog model gives rise to the *discrete wavelet transform* (DWT). Assume that the output of the filter, $H(j\omega)$, is sampled every T seconds and $H_m(j\omega)$ every $T_m = a^m T$ seconds. The DWT of a signal $x(t)$ is given by

$$X_{DWT}(m, k) = \int_{-\infty}^{\infty} x(t)h_m(a^m kT - t)dt = a^{-m/2} \int_{-\infty}^{\infty} x(t)h_m(kT - a^{-m}t)dt$$

$$(18.45)$$

The popular DWT is seen to be a two-parameter transform that represents scaling and translation. Their effect can be analyzed in the time-frequency plane in the manner introduced for the STFT. Refer again to Fig. 18.28, which displays the DWT along logarithmically partitioned

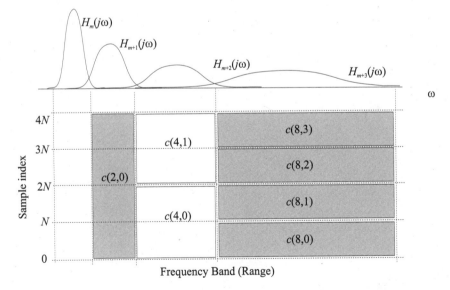

Figure 18.28   Bank of constant energy, constant Q filters derived from a prototype for $a = 2$.

time and frequency axes. It can be seen that short signal records give
rise to spectral components having a small scale size (i.e., high frequency). Alternatively, large scale size spectral components can come
only from long signal records. This property makes the wavelet transform sensitive to small local signal changes in signal behavior.

The continuous wavelet transform (CWT) is based on a special interpretation of Eq. (18.45) for real values of p and q, where $p = a^m$ and
$q = a^m kT$ with

$$X_{CWT}(p, q) = \frac{1}{\sqrt{p}} \int_{-\infty}^{\infty} x(t)w[p^{-1}(t-q)]dt \qquad (18.46)$$

where $w(t) = h(-t)$.

The DWT *analysis filters* $h_m(t)$ are interrelated to $h(t)$ through Eq.
(18.43). The output of the mth filter is the mth element of the wavelet
transform and is denoted $X_{DWT}(m,k)$. Implementation is generally considered to be a filter bank consisting of a collection of frequency-selective highpass and lowpass filters. A lowpass filter can quantify the
coarse behavior of a signal, whereas the high-frequency filters provides the details. The analysis section of a DWT is generally configured in what is called a *wavelet decomposition tree,* shown in Fig.
18.29. The output of each analysis filter, in this example, is decimated
(downsampled) by two. The decimation-by-two of the lowpass filter
can be accomplished without (appreciable) aliasing errors. The decimation-by-two of the highpass filter, however, introduces major aliasing errors. Recall that aliasing errors are, in fact, equivalent to a
signal's spectrum wrapping around the Nyquist frequency. A similar
situation was encountered in the study of quadrature mirror filters

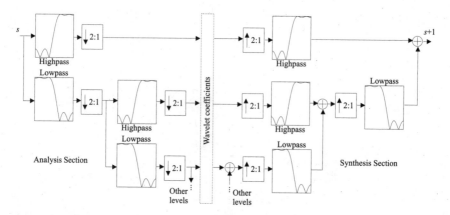

Figure 18.29 Wavelet filter bank decomposition tree.

(QMFs) where restoration filters unwrapped the aliasing distortion introduced by the filters. As was found in the study of QMFs, the analysis and synthesis (reconstruction) filters must satisfy a number of restrictive equations to control aliasing.

The reconstruction is performed using a *synthesis filter*, another filter bank defined in terms of filters having a frequency response $G_m(j\omega)$ and an impulse response $g_m(t)$. The output of the mth synthesis filter is given by

$$\hat{x}(m, k, t) = \sum_{k=-\infty}^{\infty} X_{DWT}(m, k)g_m(t - a^m kT) \qquad (18.47)$$

and the reconstructed output is

$$x(t) = \sum_{m=0}^{M} \hat{x}(m, k, t) \qquad (18.48)$$

where it is assumed that

$$g_m(t) = a^{-m/2}g(a^{-m}t) \qquad (15.49)$$

It turns out that there is a natural connection between $h(t)$ and $g(t)$. Define $\varphi(t) = h(t)$ and $\varphi_{mk}(t)$ as

$$\varphi_{mk}(t) = a^{-m/2}\varphi(t - a^m kT) \qquad (15.50)$$

where $\varphi(t)$ is called the *mother wavelet*. It can be seen that $\varphi_{mk}(t)$ can be obtained from $\varphi(t)$ through dilation and translation operations, namely dilatation by $t = a^{-m/2}t$ and translation by $t = t - a^m kT$. The shape of the mother wavelet can be derived by expanding a sparse time series defined by low order filter model using the simple routine developed below.

Assume that an N sample lowpass filter is given by $\varphi_1[k]$ in the time-domain and $\Phi_1(z)$ in the z-domain. That is,

$$\varphi_1[k] = \{\varphi_1[0], \varphi_1[1], ..., \varphi_1[N-1]\}$$

$$\Phi_1(z) = \sum_{m=0}^{N-1} \varphi_1[m]z^{-m}$$

$$\qquad (15.51)$$

A highpass (mirror) version of $\varphi_1[k]$, say $\theta_1[k]$, can be created by changing the sign of every other (odd) coefficient of $\varphi_1[k]$ (i.e., modulation by $(-1)^k$). This results in

$$\theta_1[k] = \{\varphi_1[N-1], -\varphi_1[N-2], \ldots, \varphi_1[0]\}$$

$$\Theta_1(z) = \sum_{m=0}^{N-1} (-1)^m \varphi_1[N-m]z^{-m}$$

(18.52)

Interpolating $\theta_1[k]$ by two results in $\theta_2[k]$, which satisfies

$$\theta_2[k] = \{\varphi_1([N-1], 0, -\varphi_1[N-2], 0, \ldots, \varphi_1[0])\}$$

$$\Theta_2(z) = \sum_{m=0}^{N-1} (-1)^m \varphi_1[N-m]z^{-2m}$$

(18.53)

This, in turn, defines the filter sections of a DWT.

Example 18.13: Mother Wavelet

Problem Statement. An example of a Daubechies orthonormal compactly supported wavelet is the D_4 wavelet which is defined in terms of the D_4 scaling function

$$D_4 = \frac{1}{4\sqrt{2}}\{1 + \sqrt{3}, 3 + \sqrt{3}, 1 - \sqrt{3}\}$$

Compute the corresponding D_4 scaling function and wavelet equations.

Analysis and Conclusions. Upon recursively computing the D_4 scaling function and wavelet using the computational procedure, the time series shown in Fig. 18.30 result.

Computer Study. The s-file "mother.s" defines a function that iterates the computing algorithm for producing the D_4 scaling function using a format "wavelet(h,n)" where h is the scaling function and n is the number of iterations to be computed. The following sequence of Siglab commands were used to produce the data shown in Fig. 18.30.

```
> include "chap18\mother.s"
> h0=[1+sqrt(3),3+sqrt(3),3-sqrt(3),1-sqrt(3)]/(4*sqrt(2)) #D4 function.
> x=wavelet(h0/sum(h0),7)          # Compute seven iterations.
> graph(x/max(x))                  # Graph results.            ❖
```

From a practical standpoint, the discrete-time wavelet transform (DTWT) is the implementation of choice. There are some fundamental differences between the DWT and the DTWT. One, for example, is a requirement that time dilation steps $t = a^{-m/2}t$ be integer multiples of

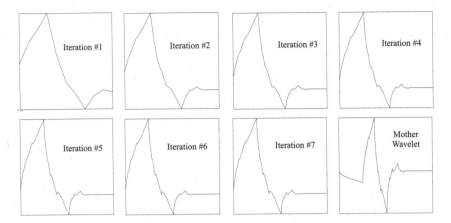

Figure 18.30 Iteration on the Daubechies D_4 scaling function and mother wavelet.

the sample period. The same is true for translations. The DTWT can be modeled as a collection of half-band filters as shown in Fig. 18.31. Fig. 18.31 diagrams a popular *dyadic wavelet transform*, which is based on setting $a = 2$. It consists of a collection of highpass and low-pass filters whose outputs are decimated by two (down-sampled) in the analysis section and interpolated by two (up-sampled) in the synthesis section. The components denoted $y_m[k]$ define the wavelet transform. The filters denoted $H(z)$ correspond to highpass filters relative to their Nyquist frequency. The filters labeled $G(z)$ are complementary lowpass filters. Notice that the filters $G(z^r)$ and $H(z^r)$, for $r = 2^n$, carry a polyphase representation. The filters identified as $H_i(z)$ are bandpass filters satisfying

$$H_3(z) = G(z)G(z^2)G(z^4)$$

$$H_2(z) = G(z)G(z^2)H(z^4)$$

$$H_1(z) = G(z)H(z^2)$$

$$H_0(z) = H(z) \tag{18.54}$$

Notice also that the filters $H_i(z)$ provide the DTWT with a logarithmic cover of the baseband frequencies.

Example 18.14: Perfect Reconstruction

Problem Statement. Show that $x[k] = \{0,1,2,3\}$ can be perfectly reconstructed using the unit energy Haar bases as analysis and synthesis functions. A two-

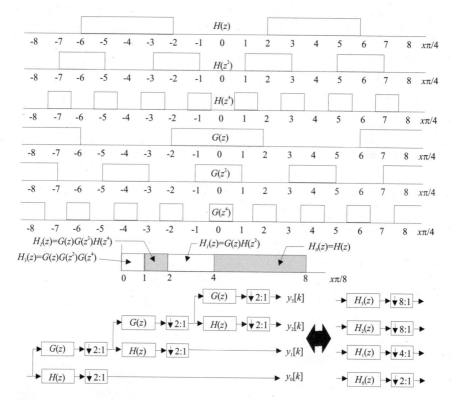

Figure 18.31   Half-band filter model of a dyadic DTWT.

sample set of Haar basis functions is given by $\varphi_1[k] = \{1,1\}$ and $\varphi_2[k] = \{-1,1\}$, which for the purpose of discussion will be considered to be an averager (Σ) and differentiator (Δ).

Analysis and Conclusions. The basis functions $\varphi_1 = \{1,1\}$ and $\varphi_2 = \{-1,1\}$ can be thought of as second-order FIR filters. The filter φ_1 is called the *approximation* filter since it smooths (averages) the two successive sample values. The filter φ_2 is called the *detail filter* and responds to small detail changes in $x[k]$ (i.e., differentiator). Refer to Fig. 18.32, where the multirate system is partitioned into two sections called the *analysis section* and *synthesis section*. Notice that the filter bank consists of decimators, approximation filters, and detail filters for various scale factors. The output of the analysis section can be transmitted to the synthesis section for reconstruction.

Computer Study. The transform and reconstruction of time series $x[k] = \{0,1,2,3,0,0\}$ is performed using the sequence of Siglab commands shown below:

```
> include "chap18\interp.s"          # Load interpolator function.
> x=[0,1,2,3,0,0]; h0=[1,1]; h1=[1,-1]   # Define signal & Haar basis.
> xlow=x$h0; xlow                    # Perform analysis filtering.
```

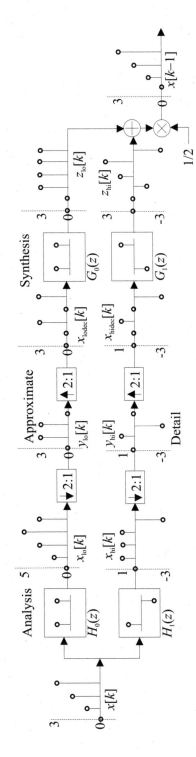

Figure 18.32 Multirate perfect reconstruction example.

```
          0   1   3   5   3   0   0
     > hi=x$h1; xhi
          0   1   1   1  -3   0   0
     > xlowdec=dec(xlow,2); xlowdec      # Decimate by two.
          0   3   3   0
     > xhidec=dec(xhi,2); xhidec
          0   1  -3   0
     > ylo=interp(xlowdec,2)             # Interpolate by two.
     > yhi=interp(xhidec,2)
     > g0=[1,1]; g1=[-1,1]               # Define reconstruction filters.
     > zlo=g0$ylo                        # Perform reconstruction filtering.
          0   0   3  -3   3  -3   0
     > zhi=yhi$g1;zhi
          0   0  -1   1   3  -3   0   0
     > (zlo + zhi)/2                      # Perform final reconstruction step.
          0   0   1   2   3   0   0   0                                ❖
```

18.10 Frequency-Masking FIR Filter Method

The frequency-masking method employs equiripple FIR filter and complementary FIR filter designs methods to create an FIR filter which can have a very steep skirt with a relatively low order filter. Recall that the order of an equiripple FIR filter was estimated by Eq. (11.29) as a function of the design parameters (δ_p, δ_a, δ_p, δ_a), where

$$N_{FIR} \approx \frac{-10\log_{10}(\delta_a\delta_p) - 15}{14\Delta\omega} + 1 \qquad (18.55)$$

and $\Delta\omega = (\omega_a-\omega_p)/\omega_s$. Here $\Delta\omega$ corresponds to the steepness of the filter's skirt (i.e., transition bandwidth). It can be immediately seen that for steep skirt filter, $\Delta\omega \approx 0$, the resulting filter order is going to be excessively high. The frequency-masking method achieves the effect of a high-order FIR filter by integrating systems of low-order FIR filters.

Define an FIR filter of order L (L odd) with a z-transform given by $H_1(z)$ and a magnitude frequency response shown in Fig. 18.33a. Assume, for the sake of discussion, that the transition bandwidth of $H_1(z)$ is $\Delta\omega$. The complement of $H_1(z)$, denoted $H_2(z)$, is easily constructed using an adder and shift-register delay of length $(L-1)/2$ as shown in Fig. 18.33b. The magnitude frequency response of $H_2(z) = z^{-(L-1)/2} - H_1(z)$ satisfies $H_1(z) + H_2(z) \equiv 1$ as shown in Fig. 18.33a. The transition bandwidth of $H_2(z)$ is again $\Delta\omega$. Suppose that each unit delay in both systems is replaced by an N sample delay. The resulting filters are called the *compressed* and *compressed complement* filters and are given by

$$\hat{H}_1(z) = H_1(z^N) \text{ (compressed)} \qquad (18.56)$$

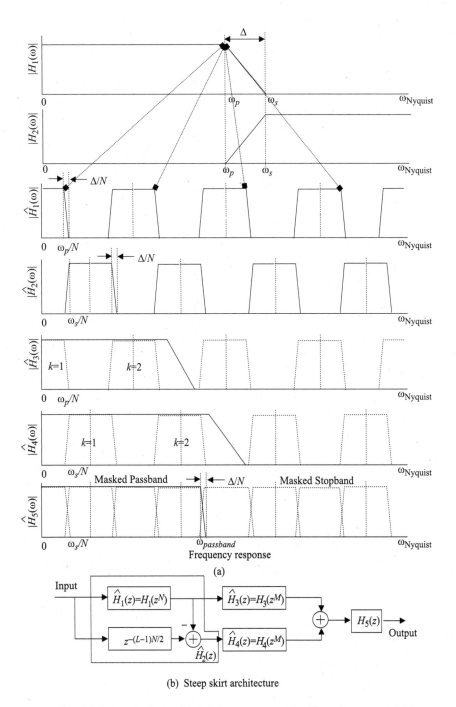

Frequency response

(a)

(b) Steep skirt architecture

Figure 18.33 (a) Spectral relationships of frequency-masking filter elements and (b) system architecture for $N = 8$, $M = 1$, and $k = 2$.

and

$$\hat{H}_2(z) = H_2(z^N) \text{ (compressed complement)} \tag{18.57}$$

The compressed filters are interpreted for $N = 8$ in Fig. 18.33. Observe that the critical frequencies of both $H_1(z)$ and $H_2(z)$ are scaled by a factor of N. The most important observation, however, is the realization that the transition bandwidth of each filter is also scaled from $\Delta\omega$ to $\Delta\omega' = \Delta\omega/N$. That is, there is an N-fold increase in the filter's transition band slope. Thus, an FIR filter can technically be designed to meet a steep skirt specification simply by choosing the appropriate value of N. The problem, of course, is to also meet the posted passband specifications and remove the unwanted copies of $\hat{H}_1(z)$ and $\hat{H}_2(z)$ shown in Fig. 18.33a. The removal of the unwanted spectral versions of $\hat{H}_1(z)$ and $\hat{H}_2(z)$ is the responsibility of *frequency masking filters* $H_3(z)$ and $H_4(z)$. These filters can also be compressed by a factor M, if desired, to form

$$\hat{H}_3(z) = H_3(z^M) \text{ (compressed)} \tag{9.58}$$

and

$$\hat{H}_3(z) = H_3(z^M) \text{ (compressed complement} \tag{9.59}$$

where the delays found in $H_3(z)$ and $H_4(z)$ are replaced by $z = z^M$. The mission of these filters is to leave k replicas of $\hat{H}_1(z)$ and k or $k-1$ replicas of $\hat{H}_2(z)$ in the desired passband and to eliminate the unwanted copies found in the stopband. For the case presented in Fig. 18.33, $k = 2$ replicas of $\hat{H}_1(z)$ and $\hat{H}_2(z)$ are reported for $N = 8$. They define the region of the passband support of the desired FIR filter. The frequency masking filters $\hat{H}_3(z)$ and $\hat{H}_4(z)$ must have their transition bands fall in the stopbands of $\hat{H}_1(z)$ and $\hat{H}_2(z)$, respectively, in a manner that supports the passband and provides high attenuation to the stopband. It should be appreciated that these frequency ranges are relatively large compared to the desired transition bandwidth $\Delta\omega'$. As a result, the design of the frequency masking filters $\hat{H}_3(z)$ and $\hat{H}_4(z)$ is considered to be an easily managed task. The responses obtained from the cascaded filter sections $\hat{H}_1(z)\hat{H}_3(z)$ and $\hat{H}_2(z)\hat{H}_4(z)$ are finally combined and passed through an optional filter $H_5(z)$, which provides final housekeeping and clean-up services. The filter $H_5(z)$ is definitely necessary if $M > 1$ to remove the unwanted multiple copies of the compressed versions $\hat{H}_3(z)$ and $\hat{H}_4(z)$.

The ultimate design objective is to create a linear-phase FIR filter that has prespecified passband and stopband critical frequencies, plus a narrow transition bandwidth $\Delta\omega'$. It is assumed that $\Delta\omega'$ is sufficiently small so as to make a single equiripple design impractical. The design procedure must account for several realities. At issue are the critical frequency locations and the resulting passband deviation of a multi-equiripple FIR filter design. It should be remembered that the passband deviations of a cascaded filter are additive, in decibels and multiplicative in real units. For example, cascading two –3 dB filters will result in a –6 dB passband. This means that the individual filter passband deviations must be less that the final target passband deviation value. In addition, it is important to also realize that cascading FIR filter also affects the resulting transition bandwidth (filter skirt steepness). In general, the resulting transition band of two FIR filters having transition bandwidths Δ_1 and Δ_2, respectively, is given by the equation (analogous to an electrical conductance relationship)

$$\frac{1}{\Delta_{\text{FIR}}} \cong \frac{1}{\Delta_1} + \frac{1}{\Delta_2} \qquad (18.60)$$

Thus, choosing the transition bandwidth of a collection of cascaded FIR filters so that the final bandwidth achieves a specific value is challenging in general and often requires that an initial design be tested and then manipulated to meet the final design objectives.

Example 18.15: Cascade FIR Filter

Problem Statement. Evaluate the effect of cascading two FIR filters on the resulting passband and stopband deviations along with the transition bandwidth.

Analysis and Results. The study of cascaded system will be conducted experimentally using a basic equiripple lowpass filter having a normalized passband range [0,0.2] and transition band of $\Delta = 0.05$. Suppose that the FIR filter's impulse response is denoted $h[k]$, then compute and evaluate $h_1[k] = h[k]$, $h_2[k] = h[k]*h[k]$, and $h_4[k] = h[k]*h[k]*h[k]*h[k]$.

Computer Study. The following sequence of Siglab commands creates the comparative database using a saved FIR filter function "h1.imp."

```
> h=rf("chap\h1.imp")            # Load impulse response.
> h1=mag(pfft(h&zeros(512-len(h))))   # Zero pad to increase resolution.
> h=h$h; h2=mag(pfft(h&zeros(512-len(h))))  # Linear convolution - 2.
> h=h$h; h4=mag(pfft(h&zeros(512-len(h))))  # linear convolution - 4.
> ograph(h1,h2,h4); ograph(log(h1), log(h2), log(h4)) # Plot results.
```

Refer to Fig. 18.34 and note that the passband deviation increases as the function of levels of cascading. The width of the transition band is seen to decrease

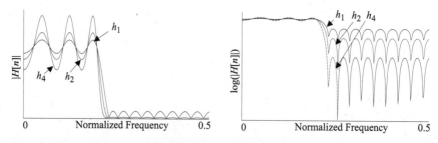

Figure 18.34 Linear and logarithmic degradation in passband and stopband deviation and transition bandwidth due to convolution.

by a factor of two when two FIR filters are cascaded, a factor of four when four FIR filters are cascaded, and so forth.   ❖

The design of a frequency-masked filter requires that a number of design parameters be simultaneously manipulated. Design guidelines exist that can assist in developing a workable solution. They are:

- The component FIR filters, namely $H_1(z)$, $H_3(z)$, $H_4(z)$, and $H_5(z)$, should be designed to have their transition bands in the middle of the baseband range $f(\in 0, f_s/2)$. This ensures that no unusual passband or stopband widths are imposed on the design.

- The filter $H_1(z)$ must have a designed passband critical frequency of ω_p which is a rational factor of the desired passband critical frequency ω_p', where $\omega_p' = (k\omega_s + \omega_p)/N$ if working from a passband critical frequency of $\hat{H}_1(z)$ and $\omega_p' = (k\omega_s - \omega_p)/N$ if working from a passband critical frequency of $\hat{H}_2(z)$ (the case for $N = 8$ and $k = 2$ is shown in Fig. 18.33).

- The individual filter transition bands should minimize the metric

$$\gamma = \frac{1}{\Delta_{H_1(z)}} + \frac{1}{\Delta_{H_2(z)}} + \frac{1}{\Delta_{H_3(z)}} + \frac{1}{\Delta_{H_4(z)}} + \frac{1}{\Delta_{H_5(z)}} \qquad (18.61)$$

The transition band objective function γ is satisfied experimentally. If all the filters are essentially the same order, then γ can be minimized by designing all the filters to have essentially the same transition bandwidth. As a design guideline, for the case where the component filters are of differing order, the value of {order[$H_i(z)$] × $\Delta_{H_i(z)}$} should be similar for $i \in \{1,...,5\}$ since, according to Eq. (18.54), FIR filter order and transition bandwidth are inversely related. As a rule, the highest-order FIR filter section in a frequency-masked system is generally $H_1(z)$ followed by $H_3(z)$, $H_4(z)$, and $H_5(z)$ [$H_2(z)$ is a multiplierless FIR filter]. This suggests that width of their transition bands would be in the reverse order. Finally, as a general rule, the passband deviation of each filter can be chosen to be 25 to 33 percent of the target deviation to account for the effects of passband ripple deviation degradation due to cascading.

Example 18.16: Frequency Masking FIR Filter

Problem Statement. Design a steep skirt FIR filter having a normalized passband gain on the order of 0.1 dB and a transition bandwidth of $\Delta f = 0.0025 \ll 0.04$. The passband is assumed to range from $f \in [0.0, 0.1]$. The stopband ranges over $f \in [0.1025, 0.5]$, which defines the transition band to have the narrow normalized value 0.0025. Based on the order estimator found in Eq. (18.54), an equiripple filter of order in excess of 700 would be required.

Analysis and Results. The frequency masking filter elements are summarized below. They are all equiripple FIR filters with bandweights w_p and w_s for the passband and stopband, respectively. The filters are generally designed to have a passband deviation on the order of -40 dB and an $N\Delta$ product of approximately 3.

Subfilter $H_1(z)$		Subfilter $H_4(z)$	
Passband edge	0.257200	Passband edge	0.300000
Stopband edge	0.300000	Stopband edge	0.398382
Weights w_p and w_s	[1,5]	Weights w_p and w_s	[1,5]
Passband ripple	-43 dB	Passband ripple	-58 dB
Stopband ripple	-56 dB	Stopband ripple	-71 dB
Filter order N	63	Filter order N	37
Δ	0.0428	Δ	0.0984
$N\Delta$	2.8676	$N\Delta$	3.64

Subfilter $H_3(z)$		Subfilter $H_5(z)$	
Passband edge	0.229412	Passband edge	0.100000
Stopband edge	0.307500	Stopband edge	0.200599
Weights w_p and w_s	[1,5]	Weights w_p and w_s	[1,1]
Passband ripple	-45 dB	Passband ripple	-42 dB
Stopband ripple	-58 dB	Stopband ripple	-42 dB
Filter order N	37	Filter order N	23
Δ	0.0781	Δ	0.1006
$N\Delta$	2.8897	$N\Delta$	2.438

Total filter order = 164

There is considerable design latitude in the creation of a frequency-masked filter, and no unique solution exists. The elemental filters $H_i(z)$ were chosen to correspond to values of $N = 17$ and $M = 3$. Based in the given data, the choice of N maps the critical frequencies of $\hat{H}_2(z)$ to the specified passband critical frequency of 0.1 as shown in Fig. 18.35. It can be seen that the filters $\hat{H}_3(z)$ and $\hat{H}_4(z)$ pass the first $k = 2$ copies of $\hat{H}_1(z)$ and $\hat{H}_2(z)$, respectively. Notice that the target normalized critical passband frequency, namely 0.1, is located on the trailing edge of the second copy of $\hat{H}_2(z)$, whose critical frequency was chosen to be 0.3. The estimated transition bandwidth, based on a contiguous path through the filter system, is

$$\frac{1}{\Delta_{\text{steepskirt}}} = \frac{1}{\Delta_{\hat{H}_1}} + \frac{1}{\Delta_{\hat{H}_2}} + \frac{1}{\Delta_{\hat{H}_3}}$$

which results in a value of $\Delta_{\text{steepskirt}} = 0.023 < 0.0025$.

Computer Study. The frequency masked filter is implemented using the s-file script "`steep.s`", which is executed using the Siglab command shown below. The results are shown in Fig. 18.36.

```
> include "chap18\steep.s"
```

Figure 18.36*a* reports the magnitude frequency responses of $H_1(z)$, $H_3(z)$, $H_4(z)$, and $H_5(z)$. Fig. 18.36*b* reports the magnitude frequency response of the

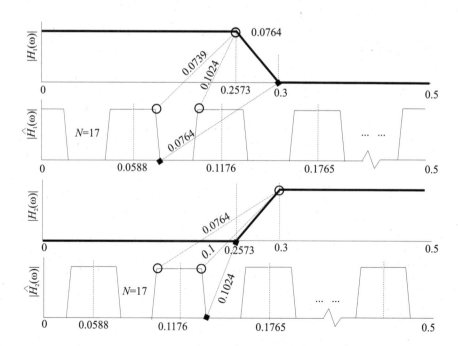

Figure 18.35 Mapping of critical frequencies of $H_1(z)$, and $H_2(z)$ for $N = 17$.

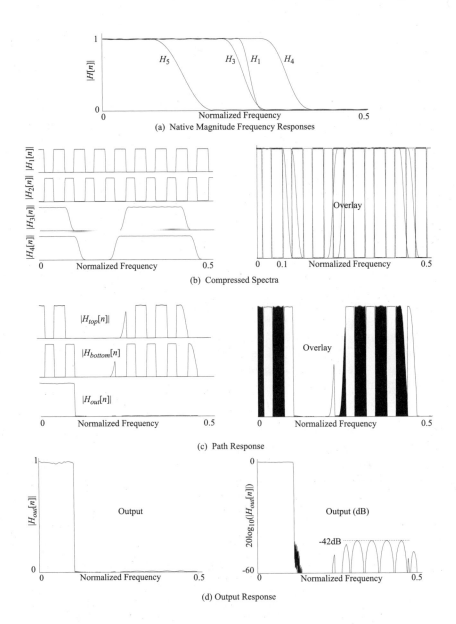

Figure 18.36 (a) Component FIR filter responses, (b) compressed spectra, (c) filter path responses, and (d) output response.

compressed filters for $N = 17$ and $M = 3$. Fig. 18.36c explores their convolved responses before and after passing them through the filter $H_5(z)$. Finally, the final frequency response is shown in Fig. 18.36d. The measured performance of the resulting filter, taken from Fig. 18.36d, are

Passband edge	$0.1f_s$
Stopband edge	$0.1025f_s$
Passband ripple	< 0.1 dB
Stopband ripple	> -42 dB

❖

18.11 Summary

In Chap. 18, the concept of multirate filtering was presented. It was shown that, when the desired frequency response of a filter occupies only a portion of the input sample rate, decimation can be used to reduce both the information bandwidth and the complexity of the resulting filter. When decimation is combined with interpolation, systems can be constructed that have the ability to reconstruct a signal or image from data found on channels of reduced bandwidth. A mathematical framework was presented, called *polyphase*, that expedited the analysis of these systems. Special cases, called *DFT filter banks, quadrature mirror filters*, and *frequency masking systems* were singled out for special attention. Wavelets, another special multirate case, were also discussed in the context of the STFT and general multirate system theory.

18.12 Self-Study Problems

1. Multirate Design In Example 18.8, an audio CD-ROM design was studied. The audio signal is assumed to be bandlimited to 20 kHz. The class of signal can be digitized above the Nyquist frequency using an industry standard 44.1 kHz and 100 kHz ADC. Compare the analog anti-aliasing filter requirements, assuming a third -order Butterworth model, for each design strategy.

2. Polyphase Representation Experimentally verify, in both the time and frequency domains, that the two filters shown in Fig. 18.15 are identical.

3. DFT Filter Bank Let $H_0(z)$ be an 11th-order equiripple lowpass FIR filter with $M = 4$. Construct a DFT filter bank having a maximum gain that approximately linearly decreases over $f \in [0, f_s/2]$. Conduct experiments to verify its performance.

4. QMF Filter Experimentally perform a frequency response test on the QMF filter studied in Example 18.11. Measure the reconstruction error variance if the input is a pure tone at $f = f_s/8$ and $f = f_s/4$ and compare. Repeat using a uniformly distributed 256 sample random time series.

5. Wavelets Analyze the frequency response of each filter found in Fig. 18.32. Measure the reconstruction error variance if the input is a pure tone at $f = f_s/8$ and $f = f_s/4$ and compare. Repeat using a uniformly distributed random 256 sample time series.

6. Frequency-Masked Filter Redesign the frequency masked filter found in Example 18.15 by placing the leading edge of $H_1(z)$ at the normalized frequency $f = 0.1$. Achieve at least a −40dB stopband attenuation.

Monarch

"What we have to learn to do we learn by doing." ARISTOTLE
Ethnica Nicomachea II

A.1 Introduction

The McGraw-Hill version of MONARCH for Windows© is derived from the standard version of MONARCH for Windows developed and marketed by The Athena Group, Inc.* The McGraw-Hill version differs from the standard version in that it

- Does not save or print (all filters and s-files used in the text have been pre-computed and saved)
- Does not support code generation for DSP microprocessors

The McGraw-Hill version contains three basic integrated modules called

- MONFIR: An FIR filter design system,
- MONIIR: An IIR filter design system
- SIGLAB: A command line interpretive language used in the analysis and study of DSP signals and systems.

A.2 Digital Filter Design

A.2.1 Introduction

The FIR and IIR filter design modules accept data in context sensitive menus to define a filter's function and parameters. The use of these menus is self-evident.

---

* To upgrade to the standard version of Monarch for Windows, contact The Athena Group, Inc., (352) 371-2567 (voice), (352) 373-5182 (fax), http://www.athena-group.com, support@athena-group.com.

A.2.2   FIR Filter Design

The FIR design options are shown in Table A.1. Upon generating an FIR filter, the standard version produces a set of output files or events that include:

1. Graphical display of the realized impulse and magnitude frequency response.

2. Saved filter definition file under "`filename.fir`".

3. Saved coefficient file under "`filename.imp`" of FIR filter tap weight coefficients that can be imported into SIGLAB using the read file command, `rf` [e.g., `h=rf("filename.imp")`], which will import the FIR filter's impulse response into SIGLAB.

4. Optional report file saved under "`filename.rpt`".

Due to the limitations of the McGraw-Hill version, only the first action is taken. Since the McGraw-Hill version does not save results, the requisite files referenced in the text are pre-computed and saved. The format of the .FIR, .IMP, and .RPT files can be viewed with a text editor.

A.2.3   IIR Filter Design

The IIR filter design options are summarized in Table A.2. Upon generating an IIR filter, the standard version of Monarch for Windows produces a set of output files or events which include:

1. Graphical display of the realized impulse and magnitude frequency response.

Table A.1   FIR Filter Design Options

FIR type	Reference	Sample rate	Maximum bands	Order
Equiripple	§11.8	Arbitrary	10	Arbitrary[*]
Direct	§11.6	Arbitrary	10	Arbitrary
Differentiator	not covered	Arbitrary	10	Arbitrary
Hilbert	§11.9	Arbitrary	10	Arbitrary
Max Flat[†]	not covered	Arbitrary	2	Internally computed

[*] If the order is specified, the predicted ripple deviation for a uniformly weighted FIR is estimated. If the ripple deviation is specified, the filter order is estimated.
[†] A Max Flat FIR filter is a baseband filter characterized by a very flat passband, but a very slow falloff in the transition and stopbands.

Table A.2   IIR Filter Design Options

IIR Type	Reference	Sample Rate	Bands	Order
Butterworth	§13.5	Arbitrary	2 or 3	Internally computed
Chebyshev I	§13.6	Arbitrary	2 or 3	Internally computed
Chebyshev II	§13.6	Arbitrary	2 or 3	Internally computed
Elliptic	§13.7	Arbitrary	2 or 3	Internally computed
State variable	§9.4, §9.8	Arbitrary	Arbitrary	User defined in SIGLAB

2. Filter definition file saved under "`filename.iir`".

3. Saved coefficient file under "`filename.arc`" of IIR coefficients, which can be imported into SIGLAB using the read architecture command rarc [e.g., `h=rarc("filename.imp",x)`, where $x \in \{0,1,2,3\}$ (0 = transfer function, 1 = state variable, 2 = pole zero, and 3 = gain constant], as described below for an Nth-order IIR filter:

- $x = 0$, h is a $2 \times N$ array containing the coefficients of the transfer function $H(z)$ [see Eq. (13.2)]:

$$H(z) = K \frac{\displaystyle\sum_{i=0}^{N} a_{N-i} z^{i}}{\displaystyle\sum_{i=0}^{N} b_{N-i} z^{i}}$$

- $x = 1$, h is an $N \times N + 4$ state variable matrix (default architecture is a direct II) having the following format (see Secs. 9.4 and 9.8):

$$\begin{bmatrix} \mathbf{A} \\ \mathbf{b} \\ \mathbf{c}^{T} \\ d \\ k \end{bmatrix} = \begin{bmatrix} a_{11} & a_{12} & \cdots & a_{1N} \\ \vdots & \vdots & \ddots & \vdots \\ a_{n1} & a_{N2} & \cdots & a_{NN} \\ b_{1} & b_{2} & \cdots & b_{N} \\ c_{1} & c_{2} & \cdots & c_{N} \\ k & 0 & \cdots & 0 \end{bmatrix}$$

- $x = 2$, h is a $2 \times N$ array containing the zeros and poles of the transfer function $H(z)$.

$$H(z) = K\frac{\displaystyle\prod_{i=1}^{N}(z_1 - z)}{\displaystyle\prod_{i=1}^{N}(P_i - z)}$$

- $x = 3$, h is the scalar value of K found in the state variable matrix.

4. Optional report file saved under "`filename.rpt`". The report file is annotated.

Because the McGraw-Hill version does not save results, the requisite files referenced in the text are precomputed and saved. The format of the .IIR and .RPT files can be viewed with a text editor. The format of the .ARC file is binary. Each IIR filter can be mapped into one of the IIR filter architectures listed in Table A.3.

Table A.3   Available IIR Filter Architectures

Architecture	Reference
Direct II	§14.3
Cascade	§14.5
Parallel	§14.6
Normal-cascade	§14.7
Normal-parallel	§14.7
Wave*	not covered
Ladder	§14.8

* A wave filter is based on a maximum power transfer analog filter.

A.3 SIGLAB Programming Guide

A.3.1 Introduction

The SIGLAB environment, although designed to be user friendly, is sophisticated and requires some time to master. The objective of the material found in this section is not to provide mastery but, rather, supply the prerequisite understanding of SIGLAB for use with the textbook.

SIGLAB interacts with the user through a menu bar, button bar of icons, an option cell (default value = Siglab), and SIGLAB programming text box. SIGLAB programs are displayed in the large program listing display window. The first-time user is strongly encouraged to review the material found under Help, found in the Menu Bar. In particular, the reader should examine the material found under the Help/Menus option.

In common use with the text, the user will be instructed either to enter a SIGLAB program, line by line, into the SIGLAB programming text box or install (include) a predefined SIGLAB program (called an s-file) as instructed in Sec. A.3.10.

A.3.2 Statement Types

There are five types of SIGLAB statements.

1. Expressions

2. Commands

3. Procedures

4. Control flow

5. Definition of new functions and procedures

Note that the "#" symbol in a SIGLAB statement indicates the beginning of a comment. Any characters following the "#" up to a new line are ignored.

Note also that, in functions and procedures, the arguments must be separated by commas. In commands, the arguments may be separated by spaces *or* commas. Text strings are set off by " " (an assigned variable is not a string). Multiple statements may be placed on a line separated by semicolons.

A.3.2.1 Expressions.

An expression is used to compute a value. Mathematical expressions and SIGLAB expressions are very similar. Expressions generate, alter, or store matrices and are either

$$name = expression$$

or

$$expression$$

In both forms, the expression is interpreted and evaluated. In the first form, the results are *assigned* to a given name. In the second form, the results are *listed* to the screen. For example, $x = 1 + 1$ creates a real scalar named x whose value is 2, and 1 + 1 simply reports the value 2. When assigning an expression, any name that is not a constant or a reserved word may be used (see sections on constants and reserved words). A name can be any letter followed by one or more digits or letters. The SIGLAB grammar is case sensitive, so that x and X are distinct.

Expressions are any combination of arithmetic operators, functions, and other defined expressions. Expressions cannot include any other statement type (i.e., commands, procedures, control flow, or definition of new functions or procedures). Examples are

```
x=mksin(1,1/16,0,128)+0.05*gn(128) # Sinewave + noise.
y=mag(fft(x)) # Magnitude frequency response of x.
```

A.3.2.2 Commands. Commands usually perform some type of environment control, such as deleting variables or typing a file. Commands can take either variable names or text strings as arguments. For commands that take arguments, parentheses are not used. Examples are

```
delvar x y z    # deletes variables x,y, and z
show            # shows active variable
```

Commands that cannot be embedded within other statements are

```
include  time  exit
```

A.3.2.3 Functions. A function behaves like a subroutine and returns a value. Functions are similar to commands and take expressions as arguments. Examples are

```
gn(3)      # lists three Gaussian random numbers
x=gn(3)    # saves three Gaussian random numbers as an array x
x=fft(x)   # saves FFT of an array x as a variable y
```

A.3.2.4 Procedures. A procedure is a function that does not return a value and thus must stand alone. Procedures are similar to commands but take expressions as arguments. Examples are

```
graph(x,y,z)                    # graphs three arrays
wf("myfile2.imp",mag(fft(x)))   # writes the FFT of x to a
                                  file
wf("myfile3.imp",y)             # writes the array to a file
```

A.3.2.5   Control Flow.   The control flow statements are while, for, and if/else. These allow conditional statements and looping. They are similar to virtually all other programming languages, such as C. Control flow statements can contain any statement types (except the unembeddable commands and definition of new functions and procedures). Examples are given in Sec. A.3.8.

A.3.2.6   Definition of New Functions and Procedures.   You can define a new function or procedure. The definition of a new function or procedure can contain any statement types (except the four unembeddable commands and definition of new functions and procedures). Once created, they are treated as any built-in function or procedure. Examples are given in Secs. A.3.9 and A.3.10.

A.3.3   Constants and Reserved Words

Constants and reserved words cannot be overwritten, and expressions cannot be assigned to constants or reserved words. Constants can be displayed using the const (show constants) command or viewed from the Show menu. The trigonometric and exponential constants $\pi = 3.14159...$ and $e = 2.71828...$ are both defined to machine precision (about 17 decimal places). SIGLAB constants are summarized in Table A.4, and SIGLAB reserved words are summarized in Table A.5 and displayed under the Constants option found under the Show menu.

Table A.4   Siglab Predefined Constant Definitions

e	pi	STVAR	TRFUNC	PZ	SCALE	XAXIS
YAXIS	ZAXIS	TITLE	NAME	SOLID	DOTTED	CENTER
DASHED	BAR	POINTS	CIRCLES	EXES	TRIANGLES	SQUARES

Table A.5   Siglab Reserved Words

for	if	else	while	exit
include	function	procedure	beep	cd
cls	const	deg	delfun	delvar
echo	pause	pwd	rad	save
show	showinc	vector	ver	end

A.3.4   Generating Matrices

All matrices are dynamically allocated. If the matrix contains real data, only the real portion is allocated. Matrices may be generated in the SIGLAB environment by

1. Using row and column construction operators,

2. Using matrix generation functions, or

3. Loading a filter file (see `rarc`, `rf`, and `rfb` commands).

A.3.4.1   Matrix Generation Operators [ ] and { }.   The row construction operator uses square brackets [...] to enter a matrix as an explicit list of elements. In the case of scalar arguments, the statement `x = [1,2,3,4]` produces the vector

$$\mathbf{x} = [1\ 2\ 3\ 4]$$

The column operator uses curly braces {...} to enter a matrix as an explicit list of elements. Typing x = {1,2,3,4}, produces the vector

$$\mathbf{x} = \begin{bmatrix} 1 \\ 2 \\ 3 \\ 4 \end{bmatrix}$$

The row and column operators may be combined. For example, the statement `x=[{1,2,3},{4,5,6},{7,8,9}]`, produces the matrix

$$\mathbf{x} = \begin{bmatrix} 1 & 4 & 7 \\ 2 & 5 & 8 \\ 3 & 6 & 9 \end{bmatrix}$$

The same matrix could be constructed as a column of rows using

`x = {[1,4,7],[2,5,8],[3,6,9]}`

The row and column operators accept arguments of any dimension, provided that their dimensions are compatible. Given the 3×3 matrix x from above,

$$[x, x] \Rightarrow \begin{bmatrix} 1 & 4 & 7 & 1 & 4 & 7 \\ 2 & 5 & 8 & 2 & 5 & 8 \\ 3 & 6 & 9 & 3 & 6 & 9 \end{bmatrix}$$

A.3.4.2 Matrix Generation Functions.

SIGLAB matrix generation functions are summarized in Table A.6.

Table A.6 Siglab Matrix Generation Functions

Usage	Action
black(k)	Creates a Blackman window of length k
eye(n)	Creates an $n \times n$ identity matrix
gn(k)	Creates a zero mean unit variance Gaussian noise signal of length k
gn(m,k)	Creates a zero mean unit variance Gaussian noise matrix of dimensions $m \times k$
ham(k)	Creates a Hamming window of length k
han(k)	Creates a Hann window of length k
impulse(k)	Creates an impulse time-series of length k
kai(b,k)	Creates a Kaiser window of a given beta value b and length k
mkcos(a,f,ϕ,k)	for (i=0:k-1); x[i]=cos(2*pi*f*i+ϕ); end
mksin(a,f,ϕ,k)	for (i=0:k-1); x[i]=sin(2*pi*f*i+ϕ); end
ones(k)	Creates a signal of ones of length k
ones(m,k)	Creates a matrix of ones of dimensions $m \times k$
rand(k)	Creates a zero mean uniform noise signal of length k
rand(m,k)	Creates a zero mean uniform noise matrix of dimensions $m \times k$
rmp(T,s,k)	Creates a ramp function with period T, slope s, and length k
sq(T,d,k)	Creates a square wave with period T, duty cycle d ($0 \leq d \leq 1$), and length k
tri(T,k)	Creates a triangle wave with period T and length k
zeros(k)	Creates a signal of zeros of length k
zeros(m,k)	Creates a matrix of zeros of dimensions $m \times k$

A.3.5   Operators

SIGLAB has three basic types of operators: point-wise, row-wise, and matrix operators.

A.3.5.1   Point-Wise Operators.

Point-wise operators operate on matrices element by element. If "φ" is a point-wise operator, and a and b are existing matrices, then the expression $a\phi b$ is valid if a and b have the same dimensions, or if either a or b is a scalar. If either a or b is a scalar, then it is promoted to a matrix of the same dimensions as the other matrix, with all elements set to the value of the original scalar. The point-wise operators are summarized in Table A.7.

Table A.7   Siglab Point-Wise Arithmetic Operators

Usage	Action
a+b	$a[i,j] + b[i,j]$
a–b	$a[i,j] - b[i,j]$
a.*b	$a[i,j] \times b[i,j]$
a./b	$a[i,j]/b[i,j]$
a%b	$a[i,j] \bmod (b[i,j])$
a^b	Raise $a[i,j]$ to the $b[i,j]$th power

A.3.5.2   Row-Wise Operators.

Row-wise operators operate on matrices row by row. If "φ" is a row-wise operator, and **a** and **b** are existing matrices, then the expression **a**φ**b** is valid if **a** and **b** have the same number of rows. The row-wise operators are summarized in Table A.8.

Table A.8   Siglab Row-Wise Arithmetic Operators

Usage	Action
a$b	Convolves matrices **a** and **b**
a@b	Correlates matrices **a** and **b**
a&b	Concatenates matrices **a** and **b**

A.3.5.3   Matrix Operators.

SIGLAB matrix operators are summarized in Table A.9. Matrix multiply requires either that the matrices are conformable or that one of the variables is a scalar. Solve requires **A** to be an $n \times n$ square matrix and **b** to be of dimensions $n \times q$. If $q > 1$, then for each column of **b**, **Ax** = **b** is solved.

Table A.9   Siglab Matrix Operators

Usage	Action
a*b	Matrix multiply **a** and **b**
a\b	Solve system **ax** = **b** for **x**
a/b	$a[i,j]/b$ where b is a scalar
a'	Transpose **a**
<a,b>	Same as cmplx(a,b), creates a complex matrix with real part **a** and imaginary part **b**, where the dimensions of **a** and **b** must agree and **a** and **b** are real

A.3.5.4   Indexing.   Once a matrix is created, any element of the matrix may be accessed by indexing. Matrix elements have a row/column index convention given by x[*row,column*]. Indexing in SIGLAB starts from zero. For example, given a=rand(m,n), the (i,j)th element of a is referenced by a[i,j], where $0 \le i < m$ and $0 \le j < n$. Omitting a row or a column index refers to all rows or all columns. For example, given x=rand(5,5), the ith column is referenced as x[,i] for $0 \le i \le 4$. The ith row is referenced as x[i,] or, in shorthand notation, x[i] for $0 \le i \le 4$.

SIGLAB can use colon notation in indexing. The expression 3:5 indexes from 3 to 5. Using colon notation, a[0:5,4] gives the first six elements of column 4 of "a", while a[4,0:5] gives the first six elements of row 4. For example, given an $n \times n$ matrix x, the principal leading 5×5 submatrix of x is x[0:4,0:4]. In general, a submatrix of x can be expressed as x[a:b,c:d] for $0 \le (a,b,c,d) < n$, and $a \le b; c \le d$. If the row argument is left blank, as in x[,c:d], then columns c through d are accessed over all rows. Similarly, if the column argument is left blank, as in x[a:b,], then rows a through b are accessed over all columns.

Indexing cannot be used to extend the dimensions of a matrix. However, indexing can be used to alter the elements of a submatrix. For example, if x is created as a zero 5×5 matrix (i.e., x=zeros(5,5)), the following operations

$$x[0:2,0:2] = eye(3)$$

$$x[3:4,3:4] = \text{ones}(2,2)$$

$$x[3:4,0:2] = 2*\text{ones}(2,3)$$

$$x[0:2,3:4] = 3*\text{ones}(3,2)$$

result in

$$x = \begin{bmatrix} 1 & 0 & 0 & 3 & 3 \\ 0 & 1 & 0 & 3 & 3 \\ 0 & 0 & 1 & 3 & 3 \\ 2 & 2 & 2 & 1 & 1 \\ 2 & 2 & 2 & 1 & 1 \end{bmatrix}$$

The SIGLAB vector (vector addressing) command allows a signal vector to be addressed in one of two modes. With vector set to on (the default value), and x=rand(5), the ith element of x is referenced as either x[,i] or x[i] for $0 \leq i \leq 4$. When set to vector off, the vector x is treated as a $1 \times n$ matrix and is referenced as the matrix entry x[,i]. Note, in this case, that x[i] results in an indexing error if $i \geq 1$.

A.3.6   Functions

A.3.6.1   Point-Wise Functions. SIGLAB's point-wise functions are, in fact, scalar functions that operate on each element of a matrix. For point-wise trigonometric operations, the angular argument can be defined to be in degrees, using **deg**, or radians, using **rad**. SIGLAB's point-wise functions are summarized in Table A.10.

Table A.10   Siglab Point-Wise Functions

Usage	Action
acos(x)	Returns the arccosine of x
acosh(x)	Returns the inverse hyperbolic cosine of x
asin(x)	Returns the arcsine of x
asinh(x)	Returns the inverse hyperbolic sine of x
atan(x)	Returns the arctangent of x
atanh(x)	Returns the inverse hyperbolic tangent of x

Table A.10 *(continued)*

Usage	Action
clrnan(x[,v])	Returns x with all values that are NAN set to v, if specified, otherwise 0. As a side effect, all values in x equal to $\pm$INF are set to $\pm$HUGE (where HUGE $\approx 1.798 \times 10^{308}$)
conj(x)	complex conjugates x
cos(x)	Returns the cosine of x
cosh(x)	Returns the hyperbolic cosine of x
exp(x)	Returns the exponential function of x
fxpt(x)	Returns the quantized value of x using the format defined by "wordlen" (see also the fixed-point operations section)
im(x)	Returns imaginary part of x
int(x)	Returns the integer part of x
ln(x)	Returns the natural logarithm (ln) of x
log(x)	Returns the common logarithm ($\log_{10}$) of x
mag(x)	Same as abs(x), returns magnitude of x
re(x)	Returns the real part of x
round(x)	Returns x rounded to the nearest integer
sign(x)	Returns the sign of x $(-1,0,+1)$
sin(x)	Returns the sine of x
sinh(x)	Returns the hyperbolic sine of x
sqrt(x)	Returns the square root of x
tan(x)	Returns the tangent of x
tanh(x)	Returns the hyperbolic tangent of x

A.3.6.2 Row-Wise Functions. SIGLAB's row-wise functions operate on each row of a matrix. These include many traditional signal processing functions, since they operate on rows (i.e., signals). These functions are summarized in Table A.11.

Table A.11   Siglab's Row-Wise Functions

Usage	Action
cshift(x,m)	x is circularly shifted right or left by m places (right where $m > 0$, left where $m < 0$)
dec(x,r)	x is decimated by r
derv(x)	Returns the backward difference derivative of x
dft(x)	Returns the DFT of x
fft(x)	Returns the two-sided FFT of x
filt(f,x)	Filters x by an IIR f (floating point)
fxpfilt(f,x,m,n)	Filters x by an IIR f using an m-bit fixed-point word with n fractional bits
grp(x)	Returns the group delay of x
idft(x)	Returns the inverse DFT of x
ifft(x)	Returns the inverse FFT of x
intg(x)	Returns the integrated values x
laguer(x,e,p,m)	Returns the roots of polynomial z using Laguerre's method to accuracy e, with or without polishing p, in m iterations or less
pfft(x)	Returns the one-sided positive FFT of x
phs(x)	Returns the phase of x
poly(x)	Returns $\Pi(x[i] - z) = \Sigma a[i]z^i$
polyval(a,x)	Returns $\Sigma a[i]z^i$ evaluated at $z = x$
rev(x)	Reverses the order of x
root(x)	Returns the roots of the polynomial whose coefficients are in x
shift(x,m)	Linearly shifts x right or left by m places (right where $m > 0$, left where $m < 0$)

It is often desirable to use row-wise functions on the columns of a matrix. To do this, simply transpose the matrix, then apply the row-wise function, and then transpose the result. For example, a 2-D FFT is given by taking the FFT of the rows, then taking the FFT of the resulting columns, and then transposing the result. Thus,

```
y = fft(fft(x)')'
```

To apply the `derv( )` function to the seventh column of **x**, type

```
dx7=derv(x[,7]')'
```

The inner transpose turns the `derv` argument into a row vector, and the outer transpose turns the row vector result of the `derv` function back into a column vector.

A.3.6.3 Matrix Functions.

SIGLAB functions that return parameters of matrices are summarized in Table A.12.

A.12   Siglab's Matrix Functions

Usage	Action
avg(x)	Returns the average value of **x**
diag(x)	Returns the diagonal elements of **x**
dim(x)	Returns the dimensions of **x**
eig(x)	Returns the eigenvalues of **x**
inv(x)	Returns the inverse of **x**
len(x)	Returns the number of elements in **x**
max(x)	Returns the maximum element of **x**
maxloc(x)	Returns the location of the first maximal element of **x**
min(x)	Returns the minimum element of **x**
minloc(x)	Returns the location of the first minimum element of **x**
ninf(x)	Returns the number of elements equal to ±INF
nnan(x)	Returns the number of elements equal to ±NAN
prod(x)	Returns the product of all elements of **x**
sum(x)	Returns the sum of all elements of **x**
var(x)	Returns the variance of **x**

A.3.7   Fixed-Point Operations

SIGLAB performs all floating-point operations in 64-bit IEEE format. Fixed-point arithmetic can also be emulated in SIGLAB. The `round(x)` function rounds a real or complex value x to the nearest real or complex integer. The `int` (integer) function returns the integer part of a real number.

A more sophisticated method of emulating fixed-point signal processing uses the `fxpt` (fixed-point) function, and the `wordlen` (wordlength), and `getword` procedures. In this scheme, an N-bit dataword is represented as one sign bit, I integer bits, and F fractional bits, where $N = I + F + 1$. The data format is defined by `wordlen(N,F,overflow)` where `overflow`$\in \{0,1\}$ enables or disables the reporting of overflow locations. The state of `wordleng` can be viewed using the `getword` function. The `fxpt(x)` function quantizes x with respect to the defined data format. For example, suppose $h(n)$ is an FIR filter's impulse response sequence to be used with a 16-bit DSP microprocessor. If 12 fractional bits of coefficient precision are required, then

```
wordlen(16,12,0); x=fxpt(h)
```

quantizes $h(n)$ to a 16-bit signed word, with 12 fractional bits of precision, and reports only the number of overflows.

Note that both the `signif` *(significant digits) and* `width` *procedures affect only the screen display, not how values are computed or saved.*

A.3.8   Control Flow

SIGLAB supports looping and conditional statements. As in any interpretive language, loops are highly discouraged in favor of embedded functions.

A.3.8.1   The `for` Statement.
The SIGLAB `for` statement operates like those in most computer languages. The `for` statement takes the multiline form

```
for(index=start:stop:step)
>statement-1
   .       .
   .       .
   .       .
>statement-n
>end
```

The SIGLAB prompt indents within the loop body. No operations are carried out until "end" is typed. The `step` parameter defaults to 1 and may be omitted.

Loops may be written as single-line statement using the form

```
for(index=start:stop:step) statement-1;...;statement-n;end
```

Sometimes, the one line form is more convenient, since it is stored as one line in the keyboard buffer.

A.3.8.2 Conditional Statements. There are two types of SIGLAB statements that use conditional (boolean) expressions: *while* and *if-then-else*. A conditional expression returns the value "true" or "false" and is used to control the while and if-then-else statements. A basic conditional expression is made up of two scalar expressions combined with any of the relational operators summarized in Table A.13.

Table A.13 Siglab's Relational Operators

e1=e2	True if e1 equals e2
e1<>e2	True if e1 does not equal e2
e1!=e2	True if e1 does not equal e2 (same as above)
e1<=e2	True if e1 is less than or equal to e2
e1<e2	True if e1 is less than e2
e1>=e2	True if e1 is greater than or equals e2
e1>e2	True if e1 is greater than e2

These relational expressions can in turn be combined with "and", "or", and "not" to form more complex expressions. Parentheses can be used for grouping. These operations are described in Table A.14.

Table A.14 Siglab's Logical Operators

(c1)	True if c1 is true
not c1	True if c1 is false
c1 and c2	True if c1 and c2 are both true
c1 or c2	True if either or both of c1 and c2 are true

Note that "not" has a higher precedence than "and", and "and" has a higher precedence than "or". All these operators have lower precedence than the relational operators listed in Table A.13. Parentheses can be used to alter precedence.

In the following examples, w, x, y, and z are all scalar expressions that evaluate as follows: w=1, x=1, y=2, z=4.

```
w<y and y<z is true.
y=z and w=x or x<y is true, but
y=z and (w=x or x<y) is false.
```

```
not y=z and not y>z is the same as
y! = z and y <= z which is true.
not (x>y and y<z) is the same as
x <= y or y >= x which evaluates to true.
(x=w or y=z) and (x<y or z=w) or not y <> z evaluates to true
```

A.3.8.3 The while Statement.

The SIGLAB while statement behaves like common while statements. Note that while statements are dangerous because they can cause inadvertent infinite looping. The while statement syntax is

```
while(condition)
>statement-1
  .     .
  .     .
  .     .
>statement-n
>end
or:
while(condition) statement-1;..;statement-n;end
```

A.3.8.4 The if-else Statement.

The if-else statement uses the same tests as the while statement to allow for the conditional execution of a statement or block of statements. The if statement takes the form

```
if(condition)
>statement-1
  .     .
  .     .
  .     .
>statement-n
>end
```

An optional else clause is added as

```
if(condition)
>statement-1
  .     .
  .     .
  .     .
>statement-n
>else
>statement-1
  .     .
```

```
. .
. .
>statement-n
>end
```

The `if` statement may be typed as a single line using

```
if(condition) statement;end
```

or

```
if(condition) statement;else;statement;end
```

The `if` statement may also be nested. Nested `if-else` statements take the following form:

```
if(condition1)
>block1
else if(condition2)
>>block2
>>else if(condition3)
>>>block3
>>>end
>>end
>end
```

A.3.9   Function and Procedure Definitions

SIGLAB has more than 140 built-in functions. When you need a function that is not built into SIGLAB but can be performed by a series of SIGLAB statements, define your own function or procedure. A function converts a list of arguments into *a value that is assigned to the function's name*. To repeat a constant set of SIGLAB statements and produce a result, use function. A procedure is similar to a function except that *it does not return a value*. When you need to repeat a constant set of SIGLAB statements that perform a service rather than produce a result, use a procedure.

A.3.9.1   The **function** Statement.   The form of a function is

```
function name(arguments)
>statement list
>end
```

Arguments, if specified, are separated by commas. One of the statements in the statement list must assign a value to the function name.

This value becomes the return value of the function. Variables assigned in a function are local to that function.

For example, to define a function that applies a Hamming window to a time-series, computes its FFT, and finds the normalized frequency location of the maximal magnitude of the FFT, one could use the following:

```
function maxi(x)
>n=len(x) # read length of x
>x=x.*ham(n) # window x
>xf=mag(pfft(x))
>maxi=maxloc(xf)/(2*len(xf)) #assignment
>end # end function definition
```

Then, maxi can be used as any other SIGLAB function.

A.3.9.2 The procedure Statement. The form of a procedure is

```
procedure name(arguments)
>statements
>end
```

Arguments, if specified, are separated by commas. Unlike the function statement, a value may or may not be assigned to the procedure name; however, no value is returned to the procedure name following execution. Variables assigned in a procedure are local to that procedure.

For example, to define a procedure that displays the positive half of the magnitude frequency response of a time-series x, one could use the following:

```
procedure grfft(x)
>graph(mag(pfft(x))) # compute mag. spectrum
>end # end procedure definition
```

Then, grfft can be used just like any other SIGLAB function.

A.3.10 s-files

To repeat a constant set of SIGLAB statements in multiple SIGLAB sessions, use the include command to import a file containing SIGLAB statements. SIGLAB statements that are imported into SIGLAB using the include command are called *s-files*. An s-file must follow SIGLAB's syntax and can be written using any ASCII text editor or by editing the keyboard buffer. The maximum number of nested includes (i.e., an include statement within an include) is nine. Do not use include within any other statement.

Import an s-file into SIGLAB using

```
include "filename"
```

The s-file can also be imported into the SIGLAB environment using the Get Includes option found under File in the Menu Bar. The presence of installed s-files can be verified using the Functions option found under Show in the Menu Bar. If the showinc (show include) command is active, all include files are displayed when they are read. It is recommended that the showinc option be enabled, as shown below, when running the s-files supplied with the text.

For example, procedure grfft shown below can be created with an ASCII editor and saved as an s-file called "myprog.s" as follows:

```
# graph magnitude spectrum of signal x
procedure grfft(x)
graph(mag(pfft(x))) # magnitude spectrum
end # end procedure definition
pause
```

To load the procedure grfft into SIGLAB and display it as it is read, use

```
showinc on # to display include files
include "myprog.s"
```

A.3.11 Environment Functions

Some SIGLAB operations, such as linear convolution, may be relatively lengthy. Use

```
z=x$y;beep
```

to issue an audible tone at the end of the convolution.

By placing time in front of a SIGLAB statement, the execution time of a statement can be measured. For example, to display the computational latency of an FFT, use

```
time y=fft(x)
```

Messages can be displayed from a SIGLAB program using the echo command. For example, if a SIGLAB loop is being indexed by i, then

```
echo "value at i=\t";i
```

prints the message followed by a tab, then the value i on the same line every time it is encountered. Other supported "backslash escape" characters are \n (newline), \a (bell), \b (backspace), \\ (backslash), and \r (return).

The pause command suspends SIGLAB until you respond to the prompt dialog box. You may echo a message to the screen within a pause command (see the pause and echo commands).

A.3.12 2-D Graphics

Arrays of data can be graphically interpreted by SIGLAB. A 2-D graph can be displayed in two forms, called *multigraph* and *overlay* graphs. A multigraph graphically displays a collection of arrays, each in its own frame. A multigraph of three arrays (say, x, y, and z) would be initiated as follows:

```
graph(x,y,z) # x=graph 0, y=graph 1, z=graph 2
```

Labeling, zoom expansion, and restoration operations are described under the Button Bar menu found under Help in the Menu Bar. Specific screens of interest are as follows:

- Comment block (T icon)
- Zoom in (+ magnifying glass icon)
- Zoom out or restore graph (– magnifying glass icon)
- Pointer enable (\icon)
- Measure graph or crosshairs (+ icon)
- Select graph (dashed box icon)

These operations can be applied only to an active or selected graph, which is denoted with a bright blue border. Some subtle operations, such as graph selection, color, style, and precision selection, are found under Graph in the Menu Bar. The range of the x- and y-axes can be controlled by invoking the MiniMax Values option, also found under Graph in the Menu Bar. The graph attributes also can be defined and saved in a SIGLAB program using the gattrib command, which is not developed in this supplement.

Multiple graphs also can be displayed in an overlay from using the command

```
ograph (x,y,z) # x = graph 0, y = graph 1, and z = graph 2
```

The graph annotation and presentation schemes applicable to the 2-D multigraph apply to overlay graphs.

A.3.13 2-D Graphics

The graphical display of a 2-D array is a 2-D graph produced using the command

```
graph3d(x)   # x = 2D data array
```

The configuration of a 2-D graph is defined by the elements found under the 2-D graphs option found in Graph of the Menu Bar.

A.3.14   File I/O

A.3.14.1   Data Files. SIGLAB can read matrix (signal) or architecture
(.ARC) files in a standard MONARCH file format or in a binary file
format. Matrix files are read into SIGLAB using the `rf` (read file) or
`rfb` (read file binary) function. In the standard version, matrix files
are written out of SIGLAB using the `wf` (write file) or `wfb` (write file
binary) procedure. The `wf` and `wfb` procedures are not available in the
McGraw-Hill version.

A.3.14.2   WAV Files. Four SIGLAB procedures (`wavin`, `wavout`,
`wavread`, and `wavwrite`) support WAV devices and files. These procedures allow SIGLAB to capture signals and to play signals with any
sound card that is supported by Windows. In addition, signals can be
read to or written from a `.WAV` file. Signals can be either mono (one
channel) or stereo (two channel), 16 or 8 bits, and can be sampled at
11.025, 22.05, or 44.1 kHz. A signal should be scaled to use the full 8-
or 16-bit soundboard dynamic range. For stereo signals, the data is
stored in an array with two rows, where row 0 contains one channel
and row 1 contains the second channel. Mono signals are stored in a
vector. Data that is read from a file must have been created in one of
these formats if it is to be played using the sound card. At this time,
only the Microsoft ADPCM file format is supported. The `wavout` and
`wavwrite` procedures are not available in the McGraw-Hill version.

Glossary

Architecture Interconnection of fundamental building block elements to define a DSP solution.

Arithmetic error An error that occurs when discarding least significant bits of a fixed-point arithmetic operation.

Aliasing Two or more distinctly different signals having identical (impersonating) sampled time-series.

Analog prototype Classic analog Butterworth, Chebyshev, or elliptic low-pass filter model having a 1 radian per second passband.

Auto-regressive moving average (ARMA) model Approximate model of an LSI system based on measured frequency response.

Bilinear z-transform Type of z-transform well suited for use with frequency-selective IIR designs.

Canonic sided digit (CSD) Special arithmetic code dense in zero digits.

Causal signal Signal produced by a causal (nonanticipatory) system.

Chirp z-transform Method of computing z-transforms using convolution methods.

Circular convolution Convolution result obtained when signals and systems are periodic.

Classic IIR IIR designed using a Butterworth, Chebyshev, or elliptic filter model.

Coefficient roundoff error An error that occurs when real coefficients are quantized into finite precision data words.

Complement filter A filter that, when additively combined with its parent, defines an all-pass filter.

Continuous-time (signal or system) Continuously resolved in time and amplitude.

Convolution A mathematical process, appearing in linear or circular form, that models the input-output filtering process.

Decimation The act of discarding samples from a time-series to produce a new sparse time-series.

Digital (signal or system) Discretely resolved in time and amplitude.

Digital filter A digital device that is capable of altering the magnitude, frequency, or phase response of a digitally encoded input signal.

Digital signal processing (DSP) A science and technology relating to the creation, modification, and manipulation of signals using digital elements.

Digital signal processing microprocessor An electronic chip or chip set designed to perform the digital signal processing operations of transformation and filtering.

Discrete cosine transform (DCT) Time-frequency transform of a finite length time-series based on a cosine expansion.

Discrete Fourier transform (DFT) Fourier transform of a periodic finite length time-series.

Discrete-time (signal or system) Discretely resolved in time, continuously resolved in amplitude.

Discrete-time Fourier series (DTFS) Fourier transform of an infinitely long periodic discrete time-series.

Discrete-time Fourier transform (DTFT) Fourier transform of an infinitely long discrete time-series.

Distributed arithmetic Special fixed-point multiply-accumulator that uses memory table look-ups.

Equiripple FIR Type of FIR satisfying an mini-max ripple error criterion.

Fast Fourier transform (FFT) A popular algorithm used to implement discrete Fourier transforms.

Filter bank A collection of individual filters designed to collectively achieve a desired effect.

Finite impulse response filter (FIR) A digital filter having a finite-duration response to an impulse.

Finite word length effects A filter or transform behavior that can be attributed to the rounding or truncation of data found in finite-length register machines.

Fixed-point A numbering and arithmetic system of finite precision.

Frequency masking filter A multirate FIR filter designed from a combination of small FIRs and their complements.

Frequency sampling filter A FIR filter implemented from a bank of frequency-selective filters.

Group delay Signal propagation delay introduced by a digital filter.

Hilbert filter An all-pass filter having quadrature phase behavior in the frequency domain.

Homogeneous solution Unforced output by a discrete-time system.

Impulse response   The response of a filter to a single input sample. If of finite duration, it is said to be produced by a finite impulse response filter; otherwise an infinite impulse response filter.

Infinite impulse response filter (IIR)   A digital filter having an infinite-duration response to an impulse.

Inhomogeneous solution   Forced output of a discrete-time system.

Interpolation   The act of synthesizing a signal from a sparse time-series.

Limit cycling   Filter behavior that results in unwanted oscillations due to data rounding, truncation, or saturation.

Linear phase FIR   A FIR whose phase shift is linearly proportional to frequency.

Linear shift invariant system (LSI)   Discrete-time system modeled by a linear difference equation with constant coefficients.

Mirror FIR   A FIR having a magnitude frequency response that is a mirrored reflection of another.

Magnitude frequency response   Absolute magnitude of a system's complex frequency response.

Multiply-accumulate (MAC)   Fundamental DSP operation of the form $S \leftarrow XY + S$.

Multirate system   A system containing two or more differing sample rates.

Noise gain   A power multiplier on the amount of uncertainty in the loss of a bit of precision measured at the system output.

Nyquist frequency   One-half the sample rate.

Nyquist sample rate   Sample rate above which a band-limited signal can be reconstructed from its sample values.

Overflow saturation   An error that occurs when a variable exceeds the dynamic range limitation of a fixed-point numbering system.

Periodogram   Power spectral estimation method.

Phase response   Phase argument of a system's complex frequency response.

Polyphase representation   A mathematical procedure of representing decimated signals.

Quadrature mirror filter (QMF)   A filter bank capable of perfectly reconstructing a signal from its component parts.

Quantization   Process by which a real variable is converted into a digital word of finite precision.

Region of convergence (ROC)   Region of the z-plane in which a z-transform is guaranteed to converge.

Sample frequency (rate)   The rate at which a signal is sampled, measured in samples per second.

Sample value   A number or value associated with a single distinct sample; it is real for a discrete systems and of finite precision in a digital signal processing system.

Short-time Fourier transform (STFT)   DFT of an aperiodic time-series over short temporal intervals that assume signal stationarity.

Standard z-transform   A z-transform used to maintain linear shift invariance in IIRs.

State variables   Information-bearing variables found at the shift register locations of a digital filter.

Subband filters   A technique of decomposing wideband signal processes into a set of narrowband channels.

Superposition   Property of linear discrete-time systems.

Time-series   Continuous string of sample values.

Transfer function   Ratio of the z-transform of a system's output and input time-series.

Wavelet   A special representation of a signal in time and scale (frequency) that obeys a number of mathematical properties.

Window   The process of isolating, localizing, and modifying data over a finite interval to the exclusion of all others.

z-transform   A mathematical method of analyzing and representing discrete signals.

Index